Communications and Control Engineering

Springer-Verlag
London Ltd.

Tokyo
Published titles include:

Fundamental Limitations in Filtering and Control
María M. Seron, Julio Braslavsky and Graham C. Goodwin

Constructive Nonlinear Control
R. Sepulchre, M. Jankovic and P.V. Kokotovic

A Theory of Learning and Generalization
M. Vidyasagar

Adaptive Control
I.D. Landau, R. Lozano and M.M'Saad

Stabilization of Nonlinear Uncertain Systems
Miroslav Krstic and Hua Deng

Passivity-based Control of Euler-Lagrange Systems
Romeo Ortega, Antonio Loría, Per Johan Nicklasson and Hebertt Sira-Ramírez

Stability and Stabilization of Infinite Dimensional Systems with Applications
Zheng-Hua Luo, Bao-Zhu Guo and Omer Morgul

Nonsmooth Mechanics (2nd edition)
Bernard Brogliato

Nonlinear Control Systems II
Alberto Isidori

L_2-Gain and Passivity Techniques in Nonlinear Control
Arjan van der Schaft

Control of Linear Systems with Regulation and Input Constraints
Ali Saberi, Anton A. Stoorvogel and Peddapullaiah Sannuti

Robust and H Control
Ben M. Chen

Computer Controlled Systems
Efim N. Rosenwasser and Bernhard P. Lampe

Dissipative Systems Analysis and Control
Rogelio Lozano, Bernard Brogliato, Olav Egeland and Bernhard Maschke

Control of Complex and Uncertain Systems
Stanislav V. Emelyanov and Sergey K. Korovin

Robust Control Design Using H Methods
Ian R.Petersen, Valery A. Ugrinovski and Andrey V.Savkin

Model Reduction for Control System Design
Goro Obinata and Brian D. O. Anderson

Harry L. Trentelman, Anton A. Stoorvogel
and Malo Hautus

Control Theory for Linear Systems

With 47 Figures

Springer

Harry L. Trentelman, PhD
Department of Mathematics, University of Groningen, PO Box 800, 9700 AV Groningen, The Netherlands

Anton A. Stoorvogel, PhD
Department of Mathematics and Computing Science, Eindhoven University of Technology, PO Box 513, 5600 MB Eindhoven, The Netherlands

Malo Hautus, PhD
Department of Mathematics and Computing Science, Eindhoven University of Technology, PO Box 513, 5600 MB Eindhoven, The Netherlands

Series Editors
E.D. Sontag M. Thoma

ISSN 0178-5354
ISBN 978-1-4471-1073-6

British Library Cataloguing in Publication Data
Trentelman, H.L. (Harry L.)
Control theory for linear systems. - (Communications and control engineering)
1.Control theory 2.Linear systems
I.Title II.Stoorvogel, Anton III.Hautus, Malo L.J.
629.8'312
ISBN 978-1-4471-1073-6

Library of Congress Cataloging-in-Publication Data
Trentelman, H.L.
Control Theory for linear systems / Harry L. Trentelman, Anton A. Stoorvogel, and Malo L.J. Hautus.
p. cm. -- (Communications and control engineering)
Includes bibliographical references and index.
ISBN 978-1-4471-1073-6 ISBN 978-1-4471-0339-4 (eBook)
DOI 10.1007/978-1-4471-0339-4
1. Control theory. 2. Linear systems. I. Stoorvogel, Anton. II. Hautus, Malo L.J., 1940- III. Title. IV. Series.
QA402.3.T69 2001
629.8'312--dc21 00-063766

Originally published by Springer-Verlag London Limited in 2001
Softcover reprint of the hardcover 1st edition 2001

Preface

This book originates from several editions of lecture notes that were used as teaching material for the course 'Control Theory for Linear Systems', given within the framework of the national Dutch graduate school of systems and control, in the period from 1987 to 1999. The aim of this course is to provide an extensive treatment of the theory of feedback control design for linear, finite-dimensional, time-invariant state space systems with inputs and outputs.

One of the important themes of control is the design of controllers that, while achieving an internally stable closed system, make the influence of certain exogenous disturbance inputs on given to-be-controlled output variables as small as possible. Indeed, in the appropriate sense this theme is covered by the classical linear quadratic regulator problem and the linear quadratic Gaussian problem, as well as, more recently, by the H_2 and H_∞ control problems. Most of the research efforts on the linear quadratic regulator problem and the linear quadratic Gaussian problem took place in the period up to 1975, whereas in particular H_∞ control has been the important issue in the most recent period, starting around 1985.

In, roughly, the intermediate period, from 1970 to 1985, much attention was attracted by control design problems that require to make the influence of the exogenous disturbances on the to-be-controlled outputs equal to zero. The static state feedback versions of these control design problems, often called disturbance decoupling, or disturbance localization, problems were treated in the classical textbook 'Linear Multivariable Control: A Geometric Approach', by W.M. Wonham. Around 1980, a complete theory on the disturbance decoupling problem by dynamic measurement feedback became available. A central role in this theory is played by the geometric (i.e., linear algebraic) properties of the coefficient matrices appearing in the system equations. In particular, the notions of (A, B)-invariant subspace and (C, A)-invariant subspace play an important role. These notions, and their generalizations, also turned out to be central in understanding and classifying the 'fine structure' of the system under consideration. For example, important dynamic properties such as system invertibility, strong observability, strong detectability, the minimum phase property, output stabilizability, etc., can be characterized in terms of these geometric concepts. The notions of (A, B)-invariance and (C, A)-invariance also turned out to be instrumental in other synthesis problems, like observer design, problems of tracking and regulation, etc.

In this book, we will treat both the ‘pre-1975’ approach represented by the linear quadratic regulator problem and the H_2 control problem, as well as the ‘post-1985’ approach represented by the H_∞ control problem and its applications to robust control. However, we feel that a textbook dedicated to control theory for linear state space systems should also contain the central issues of the ‘geometric approach’, namely a treatment of the disturbance decoupling problem by dynamic measurement feedback, and the geometric concepts around this synthesis problem. Our motivation for this is three-fold.

Firstly, in a context of making the influence of the exogenous disturbances on the to-be-controlled outputs as small as possible, it is natural to ask first under what conditions on the plant this influence can actually be made to vanish, i.e., under what conditions the closed loop transfer matrix can made zero by choosing an appropriate controller.

Secondly, as also mentioned above, the notions of controlled invariance and conditioned invariance, and their generalizations of weakly unobservable subspace and strongly reachable subspace, play a very important role in studying the dynamic properties of the system. As an example, the system property of strong observability holds if and only if the system coefficient matrices have the geometric property that the associated weakly unobservable subspace is equal to zero. As another example, the system property of left-invertibility holds if and only if the intersection of the weakly unobservable subspace and the strongly reachable subspace is equal to zero. Also, the important notions of system transmission polynomials and system zeros can be given an interpretation in terms of the weakly unobservable subspace, etc. In other words, a good understanding of the fine, structural, dynamic properties of the system goes hand in hand with an understanding of the basic geometric properties associated with the system parameter matrices.

Thirdly, also in the linear quadratic regulator problem, in the H_2 control problem, and in the H_∞ control problem, the idea of disturbance decoupling and its associated geometric concepts play an important role. For example, the notion of output stabilizability, and the associated output stabilizable subspace of the system, turn out to be relevant in establishing necessary and sufficient conditions for the existence of a positive semi-definite solution of the LQ algebraic Riccati equation. Also, by an appropriate transformation of the system parameter matrices, the H_2 control problem can be transformed into a disturbance decoupling problem. In fact, any controller that achieves disturbance decoupling for the transformed system turns out to be an optimal controller for the original H_2 problem. The same holds for the H_∞ control problem: by an appropriate transformation of the system parameter matrices, the original problem of making the H_∞ norm of the closed loop transfer matrix strictly less than a given tolerance, is transformed into a disturbance decoupling problem. Any controller that achieves disturbance decoupling for the transformed system turns out to achieve the required strict upper bound on H_∞-norm of the closed loop transfer matrix.

The outline of this book is as follows. After a general introduction in chapter 1, and a summary of the mathematical prerequisites in chapter 2, chapter 3 of this book deals with the basic material on linear state space systems. We review controllability and observability, the notions of controllable eigenvalues and observable eigenvalues, and basis transformations in state space. Then we treat the problem of stabilization by dynamic measurement feedback. As intermediate steps in this synthesis problem, we discuss state observers, detectability, the problem of pole placement by static state feedback, and the notion of stabilizability.

The central issue of chapters 4 to 6 is the problem of disturbance decoupling by dynamic measurement feedback. First, in chapter 4, we introduce the notion of controlled invariance, or (A, B)-invariance. As an immediate application, we treat the problem of disturbance decoupling by static state feedback. Next, we introduce controllability subspaces, and stabilizability subspaces. These are used to treat the static state feedback versions of the disturbance decoupling problem with internal stability, and the problem of external stabilization. In chapter 5, we introduce the central notion of conditioned invariance, or (C, A)-invariance. Next, we discuss detectability subspaces, and their application to the problem of designing estimators in the presence of external disturbances. In chapter 6, we combine the notions of controlled invariance and conditioned invariance into the notion of (C, A, B)-pair of subspaces. As an immediate, straightforward, application we treat the dynamic measurement feedback version of the disturbance decoupling problem. Next, we take stability issues into consideration, and consider (C, A, B)-pairs of subspaces consisting of a detectability subspace and a stabilizability subspace. This structure is applied to resolve the dynamic measurement feedback version of the problem of disturbance decoupling with internal stability. The final subject of chapter 6 is the application of the idea of pairs of (C, A, B)-pairs to the problem of external stabilization by dynamic measurement feedback.

Chapters 7 and 8 of this book deal with system structure. In chapter 7, we first give a review of some basic material on polynomial matrices, elementary operations, Smith form, and left- and right-unimodularity. Then we introduce the notions of transmission polynomials and zeros, in terms of the system matrix associated with the system. We then discuss the weakly unobservable subspace, and the related notion of strong observability, and finally give a characterization of the transmission polynomials and zeros in terms of a linear map associated with the weakly unobservable subspace. In chapter 8 we discuss the idea of distributions as inputs. Allowing distributions (instead of just functions) as inputs gives rise to some new concepts in state space, such as the strongly reachable subspace and the distributionally weakly unobservable subspace. The notions of system left- and right-invertibility are introduced, and characterized in terms of these new subspaces. The basic material on distributions that is used in chapter 8 is treated in appendix A of this book.

In chapter 9 we treat the problem of tracking and regulation. In this problem, certain variables of the plant are required to track an a priori given signal, regardless of the disturbance input and the initial state of the plant. Both the signal to be tracked as well as the disturbance input are modeled as being generated by an additional finite-dimensional linear system, called the exosystem. Conditions for the existence of a regulator are given in terms of the transmission polynomials of certain system matrices associated with the interconnection of the plant and the exosystem. We also address the issue of well-posedness of the regulator problem, and characterize this property in terms of right-invertibility of the plant, and the relation between the zeros of the plant and the poles of the exosystem.

In chapter 10 we give a detailed treatment of the linear quadratic regulator problem. First, we explain how to transform the general problem to a so-called standard problem. Then we treat the finite-horizon problem in terms of the solution of the Riccati differential equation. Next, we discuss the infinite-horizon problem, both the free-endpoint as well as the zero-endpoint problem, and characterize the optimal cost and optimal control laws for these problems in terms of certain solutions of the algebraic Riccati equation. Finally, the results are reformulated for the general, non-standard case.

Chapter 11 is about the H_2 control problem. First, we explain how the original stochastic linear quadratic Gaussian problem can be reformulated as the deterministic problem of minimizing the L_2 norm of the closed loop impulse response matrix, equivalently, the H_2-norm of the closed loop transfer matrix. Then we discuss the problems of minimizing this H_2-norm over the class of all internally stabilizing static state feedback controllers, and over the class of all internally stabilizing dynamic measurement feedback controllers. In both cases, the original problem is reduced to a disturbance decoupling problem by means of transformations involving real symmetric solutions of the relevant algebraic Riccati equations.

Chapters 12, 13, 14, and 15 deal with the H_∞ control problem, and its application to problems of robust stabilization. In chapter 12, the H_∞ control problem is introduced, and it is explained how it can be applied, via the celebrated small gain theorem, to problems of robust stabilization. Next, chapter 13 gives a complete treatment of the static state feedback version of the H_∞ control problem, both for the finite-horizon as well as the infinite horizon case. Then, in chapter 14, the general dynamic measurement feedback version of the H_∞ control problem is treated. Again, both the finite, as well as the infinite horizon problem are discussed. In particular, the celebrated result on the existence of H_∞ suboptimal controllers in terms of the existence of solutions of two Riccati equations, together with a coupling condition, is treated. Finally, in chapter 15, the results of chapter 14 are applied to the problem of robust stabilization introduced in chapter 12. The chapter closes with some remarks on the singular H_∞ control problem, and with a discussion on the minimum entropy H_∞ control problem.

The book closes with an appendix that reviews the basic material on distribution theory, as needed in chapter 8.

As mentioned in the first paragraph of this preface, the lecture notes that led to this book were used as teaching material for the course 'Control Theory for Linear Systems' of the Dutch graduate school of systems and control over a period of many years. During this period, many former and present Ph.D. students taking courses with the Dutch graduate school contributed to the contents of this book through their critical remarks and suggestions. Also, most of the problems and exercises in this book were used as problems in the take-home exams that were part of the course, so were tried out on 'real' students. We want to take the opportunity to thank all former and present Ph.D. students that followed our course between 1987 and 1999 for their constructive remarks on the contents of this book. Finally, we want to thank those of our colleagues that encouraged us to complete the project of converting the original set of lecture notes to this book.

Harry L. Trentelman
University of Groningen, Groningen, The Netherlands

Anton A. Stoorvogel
Eindhoven University of Technology, Eindhoven, The Netherlands

Malo Hautus
Eindhoven University of Technology, Eindhoven, The Netherlands

Contents

Chapter 1

Introduction

1.1 Control system design and mathematical control theory

Very roughly speaking, control system design deals with the problem of making a concrete physical system behave according to certain desired specifications. The ultimate product of a control system design problem is a physical device that, if connected to the to be controlled physical system, makes it behave according to the specifications. This device is called a controller.

To get from a concrete to be controlled physical system to a concrete physical device to control the system, the following intermediate steps are often taken. First, a mathematical model of the physical system is made. Such a mathematical model can take many forms. For example, the model could be in the form of a system of ordinary and/or partial differential equations, together with a number of algebraic equations, relating the relevant variables of the system. The model could also involve difference equations, some of the variables could be related by transfer functions, etc. The usual way to get a model of an actual system is to apply the basic laws that the system satisfies. Often, this method is called *first principles modeling*. For example, if one deals with an electro-mechanical system, the set of basic physical laws (Newton's laws, Kirchoff's laws, etc.) that variables in the system satisfy form a mathematical model. A second way to get a model is called *system identification*. Here, the idea is to do experiments on the physical system: certain variables in the physical system are set to particular values from the outside, and at the same time other variables are measured. In this way, one tries to estimate ('identify') the laws that the variables in the system satisfy, thus obtaining a model. Very often, a combination of first principles modeling and system identification is used to obtain a model.

The second step in a control system design problem is to decide which desirable properties we want the physical system to satisfy. Very often, these properties can be formulated mathematically by requiring the mathematical model to have certain qualitative or quantitative mathematical properties. Together, these properties form the *design specifications*.

The third, very crucial, step is to design, on the basis of the mathematical model of the physical system, and the list of design specifications, a mathematical model of the physical controller device. *It is this step in the control design problem that we deal with in this book*: it deals with mathematical control theory, in other words, with the mathematical theory of design of models of controllers. The problem of getting from a model, and a list of design specifications to a model of a controller is called a *control synthesis problem*. Of course, for a given model, each particular list of design specifications will give rise to a particular control synthesis problem. In this book we will study for a great variety of design specifications the corresponding control synthesis problems.

We restrict ourselves in this book to a particular class of mathematical models: we assume that our models (both of the physical, to be controlled systems, as well as the controllers) are linear, time-invariant, finite-dimensional state-space systems with inputs and outputs. This class of models is rich enough to treat the fundamental issues in control system design, and the resulting design techniques work remarkably well for a large class of concrete control system design problems encountered in engineering applications.

A final step in the control system design problem is, of course, to realize the mathematical model of the controller by an actual physical device, often in the form of suitable hardware and software, and to interconnect this device with the to be controlled physical system.

As an illustration of a control design problem for a concrete physical system, we consider the motion of a communications satellite. In order for a satellite to have a fixed position to an observer on the earth's surface, while moving with its jet engines switched off, it has to describe a circular orbit, say in the equator plane, at a fixed altitude of 35620 km, with the same velocity of rotation as the earth (this orbit of a satellite is called a *geostationary orbit*). We wish to be able to influence the motion of the satellite such that it remains in this geostationary orbit. In order to do this, we want to build a device that exerts forces on the satellite when needed, by means of the satellite's jet-engines.

In the actual control design problem, this physical system should first be described by a mathematical model. In this example, a first mathematical model of the satellite's motion (based on the assumption that the satellite is represented by a point mass) will consist of a set of non-linear differential equations that can be deduced using elementary laws from physics. From this model, we can obtain a simplified one in the form of a linear, time-invariant, finite-dimensional state-space system with inputs and outputs.

In this simplified model, the geostationary orbit corresponds to the zero equilibrium solution of this finite-dimensional state-space system. Of course, if initially the satellite is placed in this equilibrium position, then it will remain there for ever, as desired. However, if, for some reason, at some moment in time the position of the satellite is perturbed slightly, then it will from that moment on follow a trajectory corresponding to an undesired periodic motion in the equator plane, away from the equilibrium solution. Our desire is to design a controller such that the equilibrium solution corresponding to the geostationary orbit becomes asymptotically stable. This would guarantee that trajectories starting in small perturbations away from the equilibrium solution converge back to that equilibrium solution as time runs off to infinity. In other words, the design specification is: asymptotic stability of the equilibrium solution of the controlled system.

Based on the linear, time invariant, finite-dimensional state-space model of the satellite's motion around the geostationary orbit and on the design specification of asymptotic stability, the next step is to find a model of a controller that achieves the design specification. The controller should also be chosen from the class of linear, time-invariant, finite-dimensional state-space systems with inputs and outputs. In mathematical control theory, the mathematical model describing the physical system that we want to behave according to the specifications is called *the control system*, *the system to be controlled*, or *the plant*, and the mathematical model of the controller device that is aimed at achieving these specifications is called *the controller*. The mathematical description of the system to be controlled, together with the controller is called the *controlled system*. In our example, we want the controlled system to be asymptotically stable.

An important paradigm in control systems design, and in mathematical control theory, is *feedback*. The idea of feedback is to let the action of the physical controlling device at any moment in time depend on the actual behavior of the physical system that is being controlled. This idea imposes a certain 'smart' structure on the controlling device: it 'looks' at the system that it is influencing, and decides on the basis of what it 'sees' how it will influence the system the next moment. In the example of the communications satellite, the controlling device, at a fixed position on the earth's surface, takes a measurement of the position of the satellite. Depending on the deviation from the desired fixed position, the controlling device then exerts certain forces on the satellite by switching on or off its jet-engines using radio signals.

Any physical controller device that has this *feedback structure* is called a *feedback controller*. In terms of its mathematical model, the feedback structure of a controller is often represented by certain variables (representing what the controller 'sees') being mapped to other variables (representing the influence that the controller exerts on the system). The first kind of variables are called *measured outputs* of the system, the second kind of variables are called *control inputs* to the system. Typically, the input variables are considered to be *caused* by the measured outputs. Mathematically, the relation between the measured outputs and the control inputs can be described by a *map*. Often, designing a controller for a given system can be formulated as the problem of finding a suitable map between measured outputs and control inputs. The controlled system corresponding to the combination of a control system and a feedback controller is often called *the closed-loop system*, and one often speaks about *the interconnection* of the control system and the feedback controller. The principle of feedback is illustrated pictorially in the diagram in Figure 1.1 on the next page.

The control synthesis problems treated in this book are all concerned with the

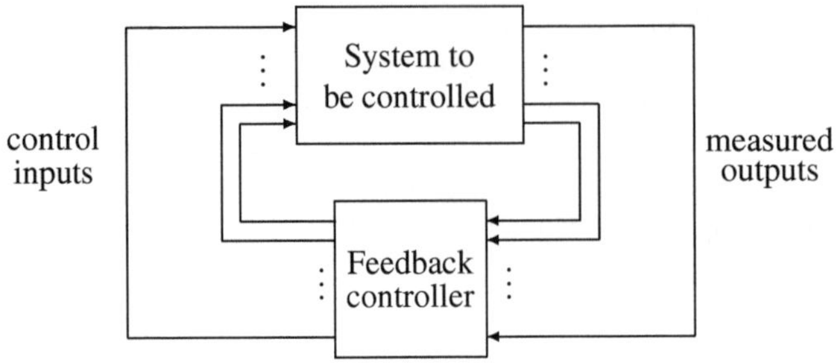

Figure 1.1: The principle of feedback

design of feedback controllers: given a linear time-invariant state-space system, and certain design specifications, find a feedback controller such that the design specifications are fulfilled by the closed-loop system, or determine that such feedback controller does not exist.

1.2 An example: instability of the geostationary orbit

In this example we will take a more detailed look at the system describing the motion of a communications satellite. The principle of such a satellite is that it serves as a mirror for electromagnetic signals. In order not to be forced to continuously aim the transmitters and receiving antennas at the satellite, it is desired that the satellite has a fixed position with respect to these devices. This also has the advantage that the satellite does not go down and rise, so that it can be used 24 hours a day.

In order to simplify the example, we will consider the motion of the satellite in the equator plane. By taking the origin at the center of the earth, the position of the satellite is given by its polar coordinates (r, θ). Introduce the following constants:

$$
\begin{aligned}
M_E &:= \text{mass of the earth,}\\
G &:= \text{earth's gravitational constant,}\\
\Omega &:= \text{earth's angular velocity,}\\
M_S &:= \text{mass of the satellite.}
\end{aligned}
$$

We assume that the satellite has on-board jets which make it possible to exert forces $F_r(t)$ and $F_\theta(t)$ to the satellite in the direction of r and θ, respectively. Using Newton's law it can be verified that the equations of motion of the satellite are given by

$$
\begin{aligned}
\ddot{r}(t) &= r(t)\dot{\theta}(t)^2 - \tfrac{GM_E}{r(t)^2} + \tfrac{F_r(t)}{M_S},\\
\ddot{\theta}(t) &= -2\tfrac{\dot{r}(t)\dot{\theta}(t)}{r(t)} + \tfrac{F_\theta(t)}{M_S r(t)}.
\end{aligned}
$$

The desired geostationary orbit is given by

$$\begin{aligned} \theta(t) &= \theta_0 + \Omega t, \\ r(t) &= R_0, \\ F_r(t) &= 0, \\ F_\theta(t) &= 0, \end{aligned}$$

where R_0 still has to be determined. We first check that this indeed yields a solution to the equations of motion for suitable R_0. By substitution in the differential equations we obtain

$$0 = R_0\Omega^2 - \frac{GM_E}{R_0^2},$$

which yields

$$R_0 = \sqrt[3]{\frac{GM_E}{\Omega^2}}.$$

By taking the appropriate values for the physical constants in this formula, we find that R_0 is approximately equal to 42000 km. Thus the geostationary orbit is a circular orbit in the equator plane, at an altitude of approximately 35620 km over the equator (the radius of the earth being approximately 6380 km). It is convenient to replace the equations of motions by an equivalent system of four first order differential equations by putting

$$\begin{aligned} x_1 &:= r(t) - R_0, \\ x_2 &:= \dot{r}(t), \\ x_3 &:= \theta(t) - (\theta_0 + \Omega t), \\ x_4 &:= \dot{\theta} - \Omega. \end{aligned}$$

Note that x_3 is the deviation from the desired angle. This value can at any time instant be measured by an observer on the equator by comparing the actual position of the satellite to the desired position. In terms of these new variables, the system is described by

$$\begin{pmatrix} \dot{x}_1(t) \\ \dot{x}_2(t) \\ \dot{x}_3(t) \\ \dot{x}_4(t) \end{pmatrix} = \begin{pmatrix} x_2(t) \\ (x_1(t) + R_0)(x_4(t) + \Omega)^2 - \frac{GM_E}{(x_1(t)+R_0)^2} + \frac{F_r(t)}{M_S} \\ x_4(t) \\ -\frac{2x_2(t)(x_4(t)+\Omega)}{x_1(t)+R_0} + \frac{F_\theta(t)}{M_S(x_1(t)+R_0)} \end{pmatrix} \tag{1.1}$$

The equations (1.1) constitute a nonlinear state-space model with inputs and outputs. The control input is $u = (F_r, F_\theta)^{\mathrm{T}}$, for the measured output one could take x_3. The geostationary orbit corresponds to the equilibrium solution given by

$$(x_1, x_2, x_3, x_4) = (0, 0, 0, 0), \qquad F_r = 0, \qquad F_\theta = 0.$$

By Kepler's law it is clear that if at time t_0 the equilibrium solution is perturbed to, say,

$$(x_1(t_0), x_2(t_0), x_3(t_0), x_4(t_0)) = (\xi_1, \xi_2, \xi_3, \xi_4),$$

then the resulting orbit will be an ellipsoid in the equator plane with the earth in one of its focuses. The angular velocity of the satellite with respect to the center of earth will then no longer be constant, so to an observer on the equator the satellite will not be in a fixed position, but will actually go down and rise periodically. Mathematically this can be expressed by saying that the equilibrium solution is not locally asymptotically stable. *What can we do about this?* We still have the possibility to exert forces to the satellite in the r-direction and in the θ-direction. Also, the variable x_3 can be measured. The control synthesis problem can now be formulated as: find a feedback controller that generates a control input $u = (F_r, F_\theta)^{\mathrm{T}}$ on the basis of the measured output x_3, in such a way that the equilibrium solution corresponding to the geostationary orbit becomes locally asymptotically stable. Of course, it is not clear a priori whether such controller exists.

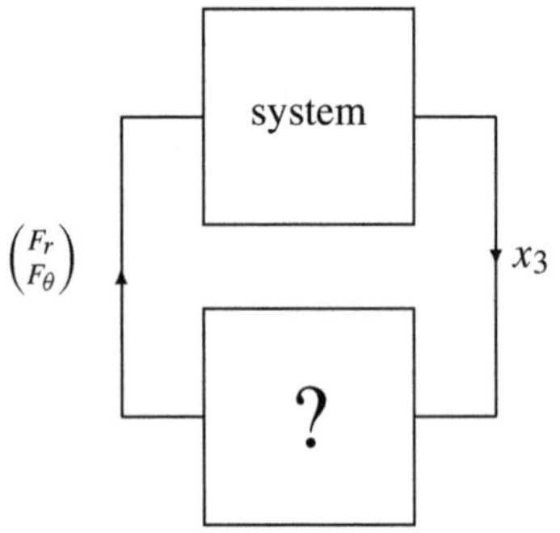

1.3 Linear control systems

The above is an example in which the system to be controlled is a nonlinear system. The vector $(x_1, x_2, x_3, x_4)^{\mathrm{T}}$ is called the *state variable* of the system. Solutions to the differential equations (1.1) take their values in the *state space* $\mathbb{R}^4$. The control input $u = (F_r, F_\theta)^{\mathrm{T}}$ takes its values in the *input space* $\mathbb{R}^2$, the measured output x_3 takes its values in the *output space* $\mathbb{R}$. More generally, a control system with state variable x, state space $\mathbb{R}^n$, input variable u, input space $\mathbb{R}^m$, output variable y, and output space $\mathbb{R}^p$ is given by the following equations:

$$\dot{x}(t) = f(x(t), u(t)),$$
$$y(t) = g(x(t), u(t)).$$

Here, f is a function from $\mathbb{R}^n \times \mathbb{R}^m$ to $\mathbb{R}^n$ and g is a function from $\mathbb{R}^n \times \mathbb{R}^m$ to $\mathbb{R}^p$. If f and g are linear, then we obtain a *linear control system*. If f is linear then there exist linear maps $A : \mathbb{R}^n \to \mathbb{R}^n$ and $B : \mathbb{R}^m \to \mathbb{R}^n$ such that $f(x, u) = Ax + Bu$. If

g is linear then there exist linear maps $C : \mathbb{R}^n \to \mathbb{R}^p$ and $D : \mathbb{R}^m \to \mathbb{R}^p$ such that $g(x, u) = Cx + Du$. The equations of the corresponding control system are then

$$\begin{aligned}\dot{x}(t) &= Ax(t) + Bu(t),\\ y(t) &= Cx(t) + Du(t).\end{aligned}$$

This is called a *linear, time-invariant, finite-dimensional state-space system.* In this book we will exclusively deal with the latter kind of control system models. Many real life systems can be modeled very well by this kind of system models. Often, the behavior of a nonlinear system can, at least in the neighborhood of an equilibrium solution, be approximately modeled by such a linear system.

1.4 Example: linearization around the geostationary orbit

Again consider the motion of the satellite. We arrived at a nonlinear control system described by the equations

$$\begin{aligned}\dot{x}(t) &= f(x(t), u(t)),\\ y(t) &= g(x(t)),\end{aligned} \tag{1.2}$$

where $u = (F_r, F_\theta)^{\mathrm{T}}$ is the control input, $x = (x_1, x_2, x_3, x_4)^{\mathrm{T}}$ is the state variable, and where y denotes the measured output x_3. The function $f : \mathbb{R}^4 \times \mathbb{R}^2 \to \mathbb{R}^4$ is given by

$$f((x_1, x_2, x_3, x_4)^{\mathrm{T}}, (F_r, F_\theta)^{\mathrm{T}}) = \begin{pmatrix} x_2 \\ (x_1 + R_0)(x_4 + \Omega)^2 - \frac{GM_E}{(x_1+R_0)^2} + \frac{F_r}{M_S} \\ x_4 \\ -\frac{2x_2(x_4+\Omega)}{x_1+R_0} + \frac{F_\theta}{M_S(x_1+R_0)} \end{pmatrix},$$

and the function $g : \mathbb{R}^4 \to \mathbb{R}$ is simply given by $g(x_1, x_2, x_3, x_4) = x_3$. Using a Taylor expansion in a neighborhood of 0, we know that for x and F small

$$f(x, u) \approx f(0, 0) + D_x f(0, 0)x + D_u f(0, 0)u.$$

Here, $D_x f$ and $D_u f$ are the derivatives of f with respect to $x = (x_1, x_2, x_3, x_4)^{\mathrm{T}}$ and $u = (F_r, F_\theta)^{\mathrm{T}}$. In our example we have $f(0, 0) = 0$ and

$$D_x f(0, 0) = \begin{pmatrix} 0 & 1 & 0 & 0 \\ 3\Omega^2 & 0 & 0 & 2\Omega R_0 \\ 0 & 0 & 0 & 1 \\ 0 & -\frac{2\Omega}{R_0} & 0 & 0 \end{pmatrix},$$

$$D_u f(0, 0) = \begin{pmatrix} 0 & 0 \\ \frac{1}{M_S} & 0 \\ 0 & 0 \\ 0 & \frac{1}{M_S R_0} \end{pmatrix}.$$

Here, we have used the fact that $\frac{GM_E}{R_0^3} = \Omega^2$. Thus, for small $x = (x_1, x_2, x_3, x_4)^{\mathrm{T}}$ and $u = (F_r, F_\theta)^{\mathrm{T}}$, the original nonlinear control system can be approximated by the linear control system

$$\begin{aligned} \dot{x}_1(t) &= x_2(t), \\ \dot{x}_2(t) &= 3\Omega^2 x_1(t) + 2\Omega R_0 x_4(t) + \tfrac{F_r}{M_S}, \\ \dot{x}_3(t) &= x_4(t), \\ \dot{x}_4(t) &= -\tfrac{2\Omega}{R_0} x_2(t) + \tfrac{F_\theta}{M_S R_0}. \end{aligned} \tag{1.3}$$

Of course this can be written as

$$\begin{aligned} \dot{x}(t) &= Ax(t) + Bu(t), \\ y(t) &= Cx(t), \end{aligned} \tag{1.4}$$

with $A := D_x f(0,0)$, $B := D_u f(0,0)$, and $C := (0\ 0\ 1\ 0)$. This linear control system is called the *linearization* of the original system around the equilibrium solution $(u, x, y) = (0, 0, 0)$.

1.5 Linear controllers

As explained, a feedback controller for a given control system is a mathematical model that generates control input signals for the system to be controlled on the basis of measured outputs of this system. If we are dealing with a system in state space form given by the equations

$$\begin{aligned} \dot{x}(t) &= f(x(t), u(t)), \\ y(t) &= g(x(t), u(t)), \end{aligned} \tag{1.5}$$

then a possible choice for the form of such a mathematical model is to mimic the form of the control system, and to consider pairs of equations of the form

$$\begin{aligned} \dot{w}(t) &= h(w(t), y(t)), \\ u(t) &= k(w(t), y(t)). \end{aligned} \tag{1.6}$$

Any such pair of equations will be called a feedback controller for the system (1.5). The variable w is called the state variable of the controller, it takes its values in $\mathbb{R}^\ell$ for some ℓ. The controller is completely determined by the integer ℓ, together with the functions h and k. The measured output y is taken as an input for the controller. On the basis of y the controller determines the control input u.

If we are dealing with a linear control system given by the equations

$$\begin{aligned} \dot{x}(t) &= Ax(t) + Bu(t), \\ y(t) &= Cx(t) + Du(t), \end{aligned}$$

then it is reasonable to consider feedback controllers of a form that is compatible with the linearity of these equations. This means that we will consider controllers of the form (1.6) in which the functions h and k are linear. Such controllers are also represented by linear, time-invariant, finite-dimensional systems in state space form, given by

$$\begin{aligned} \dot{w}(t) &= Kw(t) + Ly(t), \\ u(t) &= Mw(t) + Ny(t), \end{aligned} \tag{1.7}$$

where K, L, M and N are linear maps. The state variable of the controller is w. Any pair of equations (1.7) is called a *linear feedback controller* .

1.6 Example: stabilizing the geostationary orbit

Our design specification is local asymptotic stability of the geostationary orbit. Without going into the details, we mention the following important result on local asymptotic stability of a given stationary solution of a system of first order nonlinear differential equations: *if the linearization around the stationary solution is asymptotically stable, then the stationary solution itself is locally asymptotically stable.* This means that if we succeed in finding a linear controller for the linearization (1.4), then the same linear controller applied to the original nonlinear control system will make the geostationary orbit locally asymptotically stable!

Consider the linearization (1.4) around this stationary solution. If we interconnect this linear system with a linear controller of the form,

$$\begin{aligned} \dot{w}(t) &= Kw(t) + Ly(t), \\ u(t) &= Mw(t) + Ny(t), \end{aligned} \tag{1.8}$$

then the resulting closed-loop system is obtained by substituting $u = Mw + Ny$ into (1.4) and $y = Cx$ into (1.8). This yields

$$\begin{pmatrix} \dot{x}(t) \\ \dot{w}(t) \end{pmatrix} = \begin{pmatrix} A + BNC & BM \\ LC & K \end{pmatrix} \begin{pmatrix} x(t) \\ w(t) \end{pmatrix}. \tag{1.9}$$

This system of first order linear differential equations is asymptotically stable if and only if all eigenvalues λ of the matrix

$$A_e := \begin{pmatrix} A + BNC & BM \\ LC & K \end{pmatrix}$$

satisfy $\Re\mathrm{e}\ \lambda < 0$, i.e., have negative real parts. A matrix with this property is referred to as a *stability matrix*. The problem is to find matrices K, L, M and N such that this property holds. Once we have found these matrices, the corresponding linear controller applied to the original nonlinear system will make the geostationary orbit locally asymptotically stable. Indeed, the interconnection of (1.2) with (1.8) is given by

$$\begin{aligned} \dot{x}(t) &= f(x(t), Mw(t) + Ng(x(t))), \\ \dot{w}(t) &= Kw(t) + Lg(x(t)), \end{aligned}$$

and its linearization around the stationary solution (0, 0) is exactly given by (1.9).

1.7 Example: regulation of the satellite's position

In the previous section we discussed the problem of making the geostationary orbit asymptotically stable. This property guarantees that the state $x = (x_1, x_2, x_3, x_4)^T$, after an initial perturbation $x(0) = (\xi_1, \xi_2, \xi_3, \xi_4)^T$ away from the zero equilibrium, converges back to zero as time runs off to infinity. In the satellite example, we are very much interested in two particular variables, namely $x_1(t) = r(t) - R_0$ and $x_3(t) = \theta(t) - (\theta_0 + \Omega t)$. The values of these variables express the deviation from the satellite's required position. It is very important that, after a possible initial perturbation away from the zero solution, these values return to zero as quickly as possible. The design specification of asymptotic stability alone is too weak to achieve this quick return to zero.

This motivates the following approach. Given the linear model (1.3) of the satellite's motion around the geostationary equilibrium and a possible perturbation $x(0) = \xi = (\xi_1, \xi_2, \xi_3, \xi_4)^T$ away from the equilibrium, express the *performance* of the system by the following functional of the control input $u = (F_r, F_\theta)^T$:

$$J(\xi, F_r, F_\theta) = \int_0^\infty \alpha x_1^2(t) + \beta x_2^2(t) + \gamma F_r^2(t) + \delta F_\theta^2(t) \mathrm{d}t \tag{1.10}$$

Here, α, β, γ, and δ are non-negative constants, called *weighting coefficients*. A reasonable control synthesis problem is now to design a feedback controller that generates input functions $u = (F_r, F_\theta)^T$, for example on the basis of measurements of the state variables $x = (x_1, x_2, x_2, x_4)^T$, such that the performance functional $J(\xi, F_r, F_\theta)$ is as small as possible, while at the same time the closed-loop system is asymptotically stable. More concretely, one could try to find a feedback controller of the form

$$u = Fx,$$

with F a linear map from $\mathbb{R}^4$ to $\mathbb{R}^2$ that has to be determined, such that (1.10) is minimal, and such that the closed-loop system is asymptotically stable. Such feedback controller where u is a static linear function of the state variable x is called a *static state feedback control law*. By a suitable choice of the weighting coefficients, it is expected that in the closed-loop system both $\int_0^\infty x_1^2(t)\mathrm{d}t$ and $\int_0^\infty x_3^2(t)\mathrm{d}t$ are small, so that $x_1(t)$ and $x_3(t)$ will return to a small neighborhood of zero quickly, as desired. A feedback controller that minimizes the quadratic performance functional (1.10) is called a *linear quadratic regulator*. This terminology comes from the fact that the underlying control system is linear, and the performance functional depends quadratically on the state and input variables.

1.8 Exogenous inputs and outputs to be controlled

Often, if we make a mathematical model of a real life physical system, we do not only want to specify the control inputs, but also a second kind of inputs, the *exogenous inputs*. These can be used, for example, to model unknown disturbances that act on the system. Also, they can be used to 'inject' into the system the description of given time functions that certain variables in the system are required to track. In this case the exogenous inputs are called *reference signals*. Often, apart from the measured output, we want to include in our mathematical model a second kind of outputs, the *outputs to be controlled*, also called the *exogenous outputs*. Typically, the outputs to be controlled include those variables in the control system that we are particularly interested in, and that, for example, we want to keep close to certain, a priori given, values.

A general control system in state space form with exogenous inputs and outputs to be controlled is described by the following equations:

$$\begin{aligned} \dot{x}(t) &= f(x(t), u(t), d(t)), \\ y(t) &= g(x(t), u(t), d(t)), \\ z(t) &= h(x(t), u(t), d(t)). \end{aligned} \tag{1.11}$$

Here, d represents the exogenous inputs. The functions d are assumed to take their values in some fixed finite-dimensional linear space, say, $\mathbb{R}^r$. The variable z represents the outputs to be controlled, which are assumed to take their values in, say, $\mathbb{R}^q$. The variables x, u and y are as before, and the functions f, g and h are smooth functions mapping between the appropriately dimensioned linear spaces. Typically, the function h is chosen in such a way that z represents those variables in the system that we want to keep close to, or at, some prespecified value z^*, regardless of the disturbance inputs d that happen to act on the system.

Again, if f, g and h are linear functions, then the equations take the following form

$$\begin{aligned} \dot{x}(t) &= Ax(t) + Bu(t) + Ed(t), \\ z(t) &= C_1x(t) + D_{11}u(t) + D_{12}d(t), \\ y(t) &= C_2x(t) + D_{21}u(t) + D_{22}d(t), \end{aligned}$$

for given linear maps $A, B, E, C_1, D_{11}, D_{12}, C_2, D_{21}$ and D_{22}. These equations are said to constitute a *linear control system in state space form with exogenous inputs and outputs*. Many real life systems can be modelled quite satisfactorily in this way. Moreover, the behavior of nonlinear systems around equilibrium solutions is often modelled by such linear systems.

1.9 Example: including the moon's gravitational field

In the equations of motion of our satellite we did not include gravitational forces acting on the satellite caused by other bodies than the earth. Now suppose that in our satellite model we want to include the forces caused by the gravitational field of the moon. We can do this by including into our system the forces exerted by the moon on the satellite as disturbance inputs, whose values are unknown. Let $F_{M,r}$ and $F_{M,\theta}$ be the forces applied by the moon in the r and θ direction, respectively. Including these into the model, we obtain

$$\begin{pmatrix} \dot{x}_1(t) \\ \dot{x}_2(t) \\ \dot{x}_3(t) \\ \dot{x}_4(t) \end{pmatrix} = \begin{pmatrix} x_2(t) \\ (x_1(t)+R_0)(x_4(t)+\Omega)^2 - \frac{GM_E}{(x_1(t)+R_0)^2} + \frac{F_r(t)}{M_S} + \frac{F_{M,r}(t)}{M_S} \\ x_4(t) \\ -\frac{2x_2(t)(x_4(t)+\Omega)}{x_1(t)+R_0} + \frac{F_\theta(t)}{M_S(x_1(t)+R_0)} + \frac{F_{M,\theta}(t)}{M_S(x_1(t)+R_0)} \end{pmatrix} \tag{1.12}$$

We are particularly interested in the variables $x_1(t)$ and $x_3(t)$, describing the deviation from the desired geostationary orbit $(R_0, \theta_0+\Omega t)$. Thus, as output to be controlled we can take the vector (x_1, x_3). In this way we exactly obtain a model of the form (1.11), with control input $u = (F_r, F_\theta)^{\mathrm{T}}$, exogenous input $d = (F_{M,r}, F_{M,\theta})^{\mathrm{T}}$, measured output $y = x_3$, and output to be controlled $z = (x_1, x_3)^{\mathrm{T}}$.

Of course, an equilibrium solution is given by

$$\begin{aligned} &(x_1, x_2, x_3, x_4) = (0, 0, 0, 0), \\ &(F_r, F_\theta) = (0, 0), \\ &(F_{M,r}, F_{M,\theta}) = (0, 0). \end{aligned}$$

By linearization around this stationary solution, we find that for small (x_1, x_2, x_3, x_4), small (F_r, F_θ) and small $(F_{M,r}, F_{M,\theta})$, our original control system is approximated by

$$\begin{aligned} \dot{x}_1(t) &= x_2(t), \\ \dot{x}_2(t) &= 3\Omega^2 x_1(t) + 2\Omega R_0 x_4(t) + \frac{F_r(t)}{M_S} + \frac{F_{M,r}(t)}{M_S}, \\ \dot{x}_3(t) &= x_4(t), \\ \dot{x}_4(t) &= -\frac{2\Omega}{R_0} x_2(t) + \frac{F_\theta(t)}{M_S R_0} + \frac{F_{M,\theta}(t)}{M_S R_0} . y(t) = x_3(t), \\ z(t) &= \begin{pmatrix} x_1(t) \\ x_3(t) \end{pmatrix}. \end{aligned}$$

Of course, these equations constitute a linear control system in state space form with exogenous inputs and outputs:

$$\begin{aligned} \dot{x}(t) &= Ax(t) + Bu(t) + Ed(t), \\ y(t) &= C_1 x(t), \\ z(t) &= C_2 x(t), \end{aligned} \tag{1.13}$$

with A and B as before, $E := B$, $C_1 := \begin{pmatrix} 0 & 0 & 1 & 0 \end{pmatrix}$ and

$$C_2 := \begin{pmatrix} 1 & 0 & 0 & 0 \\ 0 & 0 & 1 & 0 \end{pmatrix}.$$

If we interconnect this system with a linear controller of the form (1.8), then the closed-loop system will be given by

$$\begin{aligned} \begin{pmatrix} \dot{x}(t) \\ \dot{w}(t) \end{pmatrix} &= \begin{pmatrix} A + BNC_1 & BM \\ LC_1 & K \end{pmatrix} \begin{pmatrix} x(t) \\ w(t) \end{pmatrix} + \begin{pmatrix} E \\ 0 \end{pmatrix} d(t), \\ z(t) &= \begin{pmatrix} C_2 & 0 \end{pmatrix} \begin{pmatrix} x(t) \\ w(t) \end{pmatrix}. \end{aligned} \tag{1.14}$$

Our control synthesis problem might now be to invent a linear controller such that, in the closed-loop system, the disturbance input d $(= (F_{M,r}, F_{M,\theta}))$ does not influence the output z. If a controller achieves this design specification, then it is said to achieve *disturbance decoupling*. If this design specification is fulfilled, then, at least according to the linear model, the satellite will remain in its geostationary orbit once it has been put there, regardless of the gravitational forces of the moon. Of course, part of the design problem would be to answer the question whether such controller actually exists. If it does not exist, one could weaken the design specification, and require that the influence of the disturbances on the outputs to be controlled be as small as possible, in some appropriate sense. One could, of course, also ask for combinations of design specifications to be satisfied, for example both disturbance decoupling and asymptotic stability of the closed-loop system, in the sense of section 1.6. Alternatively, one could try to design a controller that makes the influence of the disturbances on the output to be controlled as small as possible, while making the closed-loop system asymptotically stable, again in the sense of section 1.6.

1.10 Robust stabilization

In general, a mathematical model of a real life physical system is based on many idealizing assumptions. Thus, in general, the control system that models a certain real life phenomenon will not be a precise description of that phenomenon. Thus it might happen that a controller that asymptotically stabilizes the control system that we are working with, does not make the real life system behave in a stable way at all, simply because the control system we are working with is not a good description of this real life system. Sometimes, it is not unreasonable to assume that the correct description rather lies *in a neighborhood* (in some appropriate sense) of the control system that we are working with (this control system is often called *the nominal system*). In order to assure that a controller also stabilizes our real life system, we could formulate the following design specification: given the nominal control system, together with a fixed neighborhood of this system, find a controller that stabilizes all systems in that neighborhood. If a controller achieves this design objective, we say that it *robustly stabilizes* the nominal system.

As an example, consider the linear control system that models the motion of the satellite around its stationary solution. This model was obtained under several idealizing assumptions. For example, we have neglected the dynamics of the satellite that are caused by the fact that, in reality, it is not a point mass. If these additional dynamics were taken into account in the nonlinear control system, then we would obtain a different linearization, lying in a neighborhood (in an appropriate sense) of the original (nominal) linearization, described in section 1.4. One could then try to design a robustly stabilizing controller for the nominal linearization. Such controller will not only stabilize the nominal control system, but also all systems in a neighborhood of the nominal one.

1.11 Notes and references

Many textbooks on control systems design, and the mathematical theory of systems and control are available. Among the more recent engineering oriented textbooks we mention the books by Van de Vegte [202], Phillips and Harbor [145], Franklin, Power and Emami-Naeini [49], and Kuo [101]. Among the textbooks that concentrate more on the mathematical aspects of systems and control we mention the classical textbooks by Kwakernaak and Sivan [105] and Brockett [25]. The satellite example that was discussed in this chapter is a standard example in several textbooks; see for instance the book by Brockett [25]. We also mention the seminal book by Wonham [223], which was the main source of inspiration for the geometric ideas and methods used in this book. Other relevant textbooks are the books by Kailath [90], Sontag [181], Maciejowski [118], and Doyle, Francis and Tannenbaum, [40]. As more more recent textbooks on control theory for linear systems we mention the books by Green and Limebeer [66], Zhou, Doyle and Glover [232], and Dullerud and Paganini [42]. For textbooks on system identification and modelling we would like to refer to the books by Ljung [112] and Ljung and Glad [113]. For textbooks on systems and control theory for nonlinear systems, we refer to the books by Isidori [86], Nijmeijer and Van der Schaft [134], Khalil [99], and Vidyasagar [208].

Chapter 2

Mathematical preliminaries

In this chapter we start from the assumption that the reader is familiar with the concept of vector spaces (or linear spaces) and with linear maps. The objective of this chapter is to give a short summary of the standard linear-algebra tools to be used in this book with special emphasis on the geometric (as opposed to matrix) properties.

2.1 Linear spaces and subspaces

Linear spaces are typically denoted by script symbols like $\mathcal{V}$, $\mathcal{X}$, ... We will only be dealing with finite-dimensional spaces. Let $\mathcal{X}$ be a linear space and $\mathcal{V}$, $\mathcal{W}$ subspaces of $\mathcal{X}$. Then $\mathcal{V} \cap \mathcal{W}$ and $\mathcal{V} + \mathcal{W} := \{x + y \mid x \in \mathcal{V}, y \in \mathcal{W}\}$ are also subspaces. The diagram in Figure 2.1 symbolizes the various inclusion relations between these spaces. It is easily seen that $\mathcal{V} + \mathcal{W}$ is the smallest subspace containing both $\mathcal{V}$ and

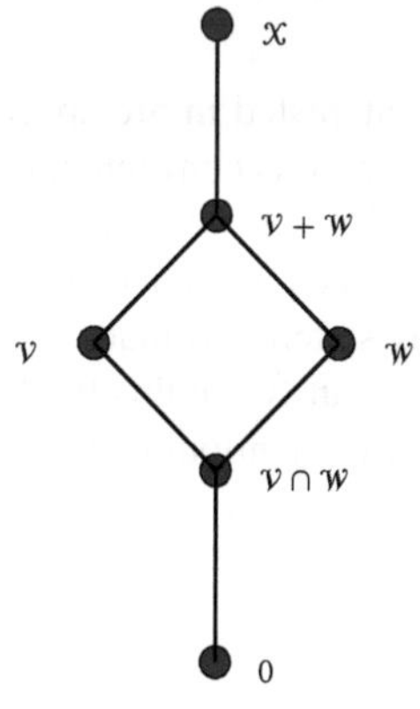

Figure 2.1: Lattice diagram

$\mathcal{W}$, i.e., if a subspace $\mathcal{L}$ satisfies $\mathcal{V} \subset \mathcal{L}$ and $\mathcal{W} \subset \mathcal{L}$ then $\mathcal{V} + \mathcal{W} \subset \mathcal{L}$. Similarly, $\mathcal{V} \cap \mathcal{W}$ is the largest subspace contained in both $\mathcal{V}$ and $\mathcal{W}$. The fact that for every pair of subspaces there exists a smallest subspace containing both subspaces and a largest subspace contained in both spaces, is expressed by saying that the set of subspaces of $\mathcal{X}$ forms a *lattice*, and diagram in Figure 2.1 is called a *lattice diagram.*

If $\mathcal{V}$, $\mathcal{W}$ and $\mathcal{R}$ are subspaces and $\mathcal{V} \subset \mathcal{R}$ then

$$\mathcal{R} \cap (\mathcal{V} + \mathcal{W}) = \mathcal{V} + (\mathcal{R} \cap \mathcal{W}). \tag{2.1}$$

This formula, which can be proved by direct verification (see exercise 2.1), is called the *modular rule*. Let $\mathcal{V}_1, \mathcal{V}_2, \dots, \mathcal{V}_k$ be subspaces. Then they are called *(linearly) independent* if every $x \in \mathcal{V}_1 + \mathcal{V}_2 + \cdots + \mathcal{V}_k$ has a *unique* representation of the form $x = x_1 + x_2 + \cdots + x_k$ with $x_i \in \mathcal{V}_i$ $(i = 1, \dots, k)$, equivalently, if $x_i \in \mathcal{V}_i$ $(i = 1, \dots, k)$ and $x_1 + \cdots + x_k = 0$ imply $x_1 = \cdots = x_k = 0$. Still another characterization is

$$\mathcal{V}_i \cap \sum_{j \neq i} \mathcal{V}_j = 0 \quad (i = 1, \dots, k). \tag{2.2}$$

Here, the symbol 0 is used to denote the null subspace of a vector space, i.e. the subspace consisting only of the element 0. If $\mathcal{V}_1, \dots, \mathcal{V}_k$ are independent subspaces, their sum $\mathcal{V}$ is called the *direct sum* of $\mathcal{V}_1, \dots, \mathcal{V}_k$ and it is written

$$\mathcal{V} = \mathcal{V}_1 \oplus \cdots \oplus \mathcal{V}_k = \bigoplus_{i=1}^{k} \mathcal{V}_k.$$

If $\mathcal{V}$ is a subspace then there exists a subspace $\mathcal{W}$ such that $\mathcal{V} \oplus \mathcal{W} = \mathcal{X}$. Such a subspace is called a *(linear) complement* of $\mathcal{V}$. It can be constructed by first choosing a basis $q_1, \dots, q_k$ of $\mathcal{V}$ and then extending it to a basis $q_1, \dots, q_n$ of $\mathcal{X}$. Then the span of $q_{k+1}, \dots, q_n$ is a complement of $\mathcal{V}$, as can easily be verified. Obviously, a complement is not unique.

The linear spaces we are interested in are spaces over the field $\mathbb{R}$ of real numbers. For some purposes, however, it is convenient to allow also complex vectors and coefficients. We denote by $\mathbb{C}$ the field of complex numbers. We use the complex extension $\mathcal{X}_{\mathbb{C}}$ of a given linear space $\mathcal{X}$, consisting of all vectors of the form $v + iw$, where v and w are in $\mathcal{X}$. Many statements made in terms of $\mathcal{X}_{\mathbb{C}}$ can easily be translated to corresponding results about $\mathcal{X}$. In this book, we will freely use the complex extension, often without explicitly mentioning it.

2.2 Linear maps

For a linear map $A : \mathcal{X} \to \mathcal{Y}$, we define

$$\begin{aligned} \ker A &:= \{x \in \mathcal{X} \mid Ax = 0\}, \\ \operatorname{im} A &:= \{Ax \mid x \in \mathcal{X}\}, \end{aligned} \tag{2.3}$$

called the *kernel* and *image* of A, respectively. These are linear spaces. We say that A is *surjective* if $\operatorname{im} A = \mathcal{Y}$ and *injective* if $\ker A = 0$. Also, A is called *bijective* (or an *isomorphism*) if A is injective and surjective. In this case, A has an inverse map, usually denoted by A^{-1}. It is also known that a bijection $A : \mathcal{X} \to \mathcal{Y}$ exists if and only if $\dim \mathcal{X} = \dim \mathcal{Y}$ (supposing, as we always do, that the linear spaces are finite dimensional).

In general, if $A : \mathcal{X} \to \mathcal{Y}$ is a, not necessarily invertible, linear map and if $\mathcal{V}$ is a subspace of $\mathcal{Y}$, then the *inverse image of* $\mathcal{V}$ is the subspace of $\mathcal{X}$ defined by

$$A^{-1}\mathcal{V} := \{x \in \mathcal{X} \mid Ax \in \mathcal{V}\}$$

Given $A : \mathcal{X} \to \mathcal{X}$ and a subspace $\mathcal{V}$ of $\mathcal{X}$, we say that $\mathcal{V}$ is A-*invariant* (or, if the map A is supposed to be obvious, simply *invariant*) if for all $x \in \mathcal{V}$ we have $Ax \in \mathcal{V}$, which can be written as $A\mathcal{V} \subset \mathcal{V}$. The concept of invariance will play a crucial role in this book.

One can consider various types of restriction of a linear map:

- If $B : \mathcal{U} \to \mathcal{X}$ is a linear map satisfying $\operatorname{im} B \subset \mathcal{V}$ where $\mathcal{V}$ is a subspace of $\mathcal{X}$ then the *codomain restriction* of B is the map $\underline{B} : \mathcal{U} \to \mathcal{V}$ satisfying $\underline{B}u := Bu$ for all $u \in \mathcal{U}$. We will not use a special notation for this type of restriction.
- If $C : \mathcal{X} \to \mathcal{Y}$ and $\mathcal{V} \subset \mathcal{X}$ then the map $\underline{C} : \mathcal{V} \to \mathcal{Y}$ defined by $\underline{C}x := Cx$ for $x \in \mathcal{V}$, is called the *(domain) restriction* of C to $\mathcal{V}$, and it is denoted $C \mid \mathcal{V}$.
- If $A : \mathcal{X} \to \mathcal{X}$ and $\mathcal{V} \subset \mathcal{X}$ is an A-invariant subspace then the map $\underline{A} : \mathcal{V} \to \mathcal{V}$ defined by $\underline{A}x := Ax$ for $x \in \mathcal{V}$ is called the *restriction* of A to $\mathcal{V}$, notation $A \mid \mathcal{V}$.

These somewhat abstract definitions will be clarified later on in terms of matrix representations. If $\mathcal{X}$ is an n-dimensional space and $q_1, \ldots, q_n$ is a basis then every vector $x \in \mathcal{X}$ has a unique representation of the form

$$x = x_1 q_1 + \cdots + x_n q_n.$$

The coefficients of this representation, written as a column vector in $\mathbb{R}^n$, form *the column of x with respect to $q_1, \ldots, q_n$*:

$$\underline{x} = \begin{pmatrix} x_1 \\ \vdots \\ x_n \end{pmatrix}.$$

For typographical reasons, we often write this column as $(x_1, \ldots, x_n)^{\mathrm{T}}$, where the 'T' denotes transposition. If $q_1, \ldots, q_n$ is a basis of $\mathcal{X}$ and $r_1, \ldots, r_p$ is a basis of $\mathcal{Y}$ and $C : \mathcal{X} \to \mathcal{Y}$ is a linear map, the *matrix of* C with respect to the bases $q_1, \ldots, q_n$ and

$r_1, \dots, r_p$ is formed by writing next to each other the columns of $Cq_1, \dots, Cq_n$ with respect to the basis $r_1, \dots, r_p$. The result will look like

$$(C) := \begin{pmatrix} c_{11} & \cdots & c_{1n} \\ \vdots & & \vdots \\ c_{p1} & \cdots & c_{pn} \end{pmatrix},$$

where $(c_{1i}, \dots, c_{pi})^{\mathrm{T}}$ is the column of Cq_i with respect to $r_1, \dots, r_p$. We use brackets around C here to emphasize that we are talking about the matrix of the map. We will use the notation $\mathbb{R}^{p \times n}$ for the set of $p \times n$ matrices. Hence $(C) \in \mathbb{R}^{p \times n}$. Once the bases are fixed, the map C and its matrix determine each other uniquely. Also, the operations of matrix addition, scalar multiplication, product of matrices correspond to addition of maps, scalar multiplication of a map, composition of maps, respectively. For this reason it is customary to identify maps with matrices, once the bases of the given spaces are given and fixed. The advantage of the matrix formulation over the more abstract linear-map formulation is that it allows for much more explicit calculations. On the other hand, if one works with linear maps, one does not have to specify bases, which sometimes makes the treatment much more elegant and transparent.

If $\mathcal{V} \subset \mathcal{X}$ then a basis $q_1, \dots, q_n$ of $\mathcal{X}$ for which $q_1, \dots, q_k$ is a basis of $\mathcal{V}$ (where $k = \dim \mathcal{V}$) is called a basis of $\mathcal{X}$ *adapted* to $\mathcal{V}$. More generally, if $\mathcal{V}_1, \mathcal{V}_2, \dots, \mathcal{V}_r$ is a chain of subspaces (i.e., $\mathcal{V}_1 \subset \mathcal{V}_2 \subset \cdots \subset \mathcal{V}_r$), then a basis $q_1, \dots, q_n$ of $\mathcal{X}$ is said to be *adapted to this chain* if there exist numbers $k_1, \dots, k_r$ such that $q_1, \dots, q_{k_i}$ is a basis of $\mathcal{V}_i$ for $i = 1, \dots, r$. Finally, if $\mathcal{V}_1, \dots, \mathcal{V}_r$ are subspaces of $\mathcal{X}$ such that $\mathcal{X} = \mathcal{V}_1 \oplus \mathcal{V}_2 \oplus \cdots \oplus \mathcal{V}_r$, we say that a basis $q_1, \dots, q_n$ is *adapted* to $\mathcal{V}_1, \dots, \mathcal{V}_r$ if there exist numbers $k_1, \dots, k_{r+1}$ such that $k_1 = 1$, $k_{r+1} = n+1$ and $q_{k_i}, \dots, q_{k_{i+1}-1}$ is a basis of $\mathcal{V}_i$ for $i = 1, \dots, r$.

We illustrate the use of matrix representations for the restriction operations introduced earlier in this section.

- If $B : \mathcal{U} \to \mathcal{X}$ is a linear map satisfying $\operatorname{im} B \subset \mathcal{V}$, we choose a basis $q_1, \dots, q_n$ of $\mathcal{X}$ adapted to $\mathcal{V}$. Let $p_1, \dots, p_m$ be a basis of $\mathcal{U}$. Then the matrix representation of B with respect to the chosen bases is of the form

$$(B) = \begin{pmatrix} B_1 \\ 0 \end{pmatrix}, \tag{2.4}$$

  where $B_1 \in \mathbb{R}^{k \times m}$ ($k := \dim \mathcal{V}$). This particular form is a consequence of the condition $\operatorname{im} B \subset \mathcal{V}$. The matrix of the codomain restriction $\underline{B} : \mathcal{U} \to \mathcal{V}$ with respect to the bases $p_1, \dots, p_m$ and $q_1, \dots, q_k$ is B_1.

- Let $C : \mathcal{X} \to \mathcal{Y}$ and $\mathcal{V} \subset \mathcal{X}$. Let $q_1, \dots, q_n$ be a basis of $\mathcal{X}$ adapted to $\mathcal{V}$. Furthermore, let $r_1, \dots, r_p$ be a basis of $\mathcal{Y}$. The matrix of C with respect to $q_1, \dots, q_n$ and $r_1, \dots, r_p$ is $(C) = (C_1 \;\; C_2)$, where $C_1 \in \mathbb{R}^{p \times k}$ and $C_2 \in \mathbb{R}^{p \times (n-k)}$. The matrix of $C \mid \mathcal{V}$ with respect to these bases is C_1.

- Let $A : \mathcal{X} \to \mathcal{X}$, $\mathcal{V} \subset \mathcal{X}$ such that $A\mathcal{V} \subset \mathcal{V}$. Let $q_1, \dots, q_n$ be a basis of $\mathcal{X}$

adapted to $\mathcal{V}$. The matrix of A with respect to this basis is

$$(A) = \begin{pmatrix} A_{11} & A_{12} \\ 0 & A_{22} \end{pmatrix}. \tag{2.5}$$

The property $A_{21} = 0$ is a consequence of the A-invariance of $\mathcal{V}$. The matrix of $A \mid \mathcal{V}$ is A_{11}.

2.3 Inner product spaces

We assume that the reader is familiar with the concept of inner product. A linear space over the field $\mathbb{R}$ with a real inner product is called a *real inner product space.* A linear space over the field $\mathbb{C}$ with a complex inner product is called a *complex inner product space.* The most commonly used real inner product space is the linear space $\mathbb{R}^n$ with the inner product $(x, y) := x^{\mathrm{T}}y$. The most commonly used complex inner product space is the linear space $\mathbb{C}^n$ with the inner product $(x, y) := x^*y$ (here, '$*$' denotes the conjugate transposition, i.e. $x^* = \bar{x}^{\mathrm{T}}$).

If $\mathcal{X}$ is a (real or complex) inner product space and if $\mathcal{V}$ is a subspace of $\mathcal{X}$, then $\mathcal{V}^{\perp}$ will denote the orthogonal complement of $\mathcal{V}$. It is easy to see that for any pair of subspaces $\mathcal{V}$, $\mathcal{W}$ of $\mathcal{X}$ the following equality holds:

$$(\mathcal{V} \cap \mathcal{W})^{\perp} = \mathcal{V}^{\perp} + \mathcal{W}^{\perp}$$

Let $\mathcal{X}$ and $\mathcal{Y}$ be (real or complex) inner product spaces with inner products $(\ ,\ )_{\mathcal{X}}$ and $(\ ,\ )_{\mathcal{Y}}$, respectively. If $C : \mathcal{X} \to \mathcal{Y}$ is a map, then the *adjoint* $C^* : \mathcal{Y} \to \mathcal{X}$ of C is the map defined by

$$(x, C^*y)_{\mathcal{X}} = (Cx, y)_{\mathcal{Y}} \tag{2.6}$$

for all x in $\mathcal{X}$ and y in $\mathcal{Y}$. It can easily be seen that there exists a unique map satisfying these properties. If $\mathcal{X}$ is a (real or complex) inner product space and if $A : \mathcal{X} \to \mathcal{X}$ is a map, then it can be shown by direct verification that the following holds:

$$(A^{-1}\mathcal{V})^{\perp} = A^*\mathcal{V}^{\perp}.$$

It is also easy to verify that if $\mathcal{V}$ is A-invariant, then $\mathcal{V}^{\perp}$ is A^*-invariant.

If $\mathcal{X}$ and $\mathcal{Y}$ are *real* inner product spaces, if $q_1, \ldots, q_n$ and $r_1, \ldots, r_p$ are orthonormal bases of $\mathcal{X}$ and $\mathcal{Y}$, respectively, and if (C) is the matrix of C with respect to these bases, then the matrix of the adjoint map C^* is equal to the transposed $(C)^{\mathrm{T}}$ of (C). Indeed, if $\underline{x}$ and $\underline{y}$ denote the columns of x and y, respectively, with respect to the given bases, then (2.6) is equivalent to

$$\underline{x}^{\mathrm{T}}(C^*)\underline{y} = \underline{x}^{\mathrm{T}}(C)^{\mathrm{T}}\underline{y}$$

for all $\underline{x}$ in $\mathbb{R}^n$ and $\underline{y}$ in $\mathbb{R}^p$.

In the same way, one can show that if $\mathcal{X}$ and $\mathcal{Y}$ are *complex* inner product spaces, if $q_1, \ldots, q_n$ and $r_1, \ldots, r_p$ are orthonormal bases of $\mathcal{X}$ and $\mathcal{Y}$, respectively and if

(C) is the matrix of C with respect to these bases, then the matrix of the adjoint map C^* is equal to the conjugate transposed $(C)^*$ of (C).

As noted in section 2.2, once the bases are fixed, we will identify maps with matrices. In this book we will usually work with the real inner product space $\mathbb{R}^n$ with inner product $(x, y) := x^{\top}y$. Therefore, instead of using the terminology 'adjoint' of a map and using the notation C^*, we will in this book often use the terminology 'transposed' and use the notation $C^{\top}$.

2.4 Quotient spaces

A linear subspace $\mathcal{V}$ gives rise to the equivalence relation:

$$x \sim y \quad :\Longleftrightarrow \quad x - y \in \mathcal{V}.$$

on $\mathcal{X}$. The set of equivalence classes is called the *quotient space* or *factor space* of $\mathcal{X}$ modulo $\mathcal{V}$, and it is denoted $\mathcal{X}/\mathcal{V}$. For any $x \in \mathcal{X}$, the equivalence class of which x is an element is often denoted by $\bar{x}$. There is a natural mapping $\Pi : \mathcal{X} \to \mathcal{X}/\mathcal{V}$, called the *canonical projection* of $\mathcal{X}$ onto $\mathcal{X}/\mathcal{V}$ and defined by $\Pi x := \bar{x}$. The set $\mathcal{X}/\mathcal{V} =: \bar{\mathcal{X}}$ is made into a linear space by

$$\bar{x} + \bar{y} := \overline{x + y}, \quad \lambda\bar{x} := \overline{\lambda x}.$$

It can be shown that addition and scalar multiplication are well defined by these formulas.

Also, these formulas state that Π is a linear map. Obviously, Π is surjective and $\ker \Pi = \mathcal{V}$. The following result is of importance:

$$\dim \mathcal{V} + \dim \mathcal{X}/\mathcal{V} = \dim \mathcal{X}.$$

In fact, let $q_1, \ldots, q_n$ be a basis of $\mathcal{X}$ adapted to $\mathcal{V}$ and let $\dim \mathcal{V} = k$. Then $\bar{q}_1 = \cdots = \bar{q}_k = 0$, where the bar denotes the projection. Hence, if $\bar{x}$ is an arbitrary element of $\mathcal{X}/\mathcal{V}$ and x is a representative, we write $x = \lambda_1 q_1 + \cdots + \lambda_n q_n$. Taking the image in $\mathcal{X}/\mathcal{V}$, we get

$$\bar{x} = \lambda_1\bar{q}_1 + \cdots + \lambda_n\bar{q}_n = \lambda_{k+1}\bar{q}_{k+1} + \cdots + \lambda_n\bar{q}_n.$$

It follows that every element of $\mathcal{X}/\mathcal{V}$ can be written as a linear combination of $\bar{q}_{k+1}, \ldots, \bar{q}_n$. On the other hand $\bar{q}_{k+1}, \ldots, \bar{q}_n$ are independent. In fact, if $\lambda_{k+1}\bar{q}_{k+1} + \cdots + \lambda_n\bar{q}_n = 0$, then $q := \lambda_{k+1}q_{k+1} + \cdots + \lambda_n q_n \in \mathcal{V}$, hence q can be written as a linear combination of $q_1, \ldots, q_k$. The result is

$$\lambda_{k+1}q_{k+1} + \cdots + \lambda_n q_n = \lambda_1 q_1 + \cdots + \lambda_k q_k.$$

Now it follows from the independence of $q_1, \ldots, q_n$ that $\lambda_{k+1} = \cdots = \lambda_n = 0$.

Let $A : \mathcal{X} \to \mathcal{X}$, let $\mathcal{V}$ denote an A-invariant subspace and let $\bar{\mathcal{X}} := \mathcal{X}/\mathcal{V}$. Then we define the *quotient map* $\bar{A} : \bar{\mathcal{X}} \to \bar{\mathcal{X}}$ by $\bar{A}\bar{x} := \overline{Ax}$. It is easily verified that the map $\bar{A}$ is well defined, i.e., that $\bar{A}\bar{x}$ is independent of the particular choice of the representative of $\bar{x}$. Also, this map is linear. Note that the defining formula of $\bar{A}$ can be rewritten as $\bar{A}\Pi = \Pi A$. The following commutative diagram illustrates the above definition:

$$\begin{array}{ccc} \mathcal{X} & \xrightarrow{A} & \mathcal{X} \\ \Pi \big\downarrow & & \big\downarrow \Pi \\ \mathcal{X}/\mathcal{V} & \xrightarrow{\bar{A}} & \mathcal{X}/\mathcal{V} \end{array}$$

That this diagram commutes (i.e. that if we go from top-left to the bottom-right, it yields the same solution whatever route one takes) is obviously equivalent to $\bar{A}\Pi = \Pi A$.

We will use the notation $A \mid \mathcal{X}/\mathcal{V}$ for $\bar{A}$. More generally, if $\mathcal{V}$ and $\mathcal{W}$ are A-invariant subspaces satisfying $\mathcal{V} \subset \mathcal{W}$, then we define $A \mid \mathcal{W}/\mathcal{V}$ to be the map obtained by first restricting A to $\mathcal{W}$ and then taking the quotient over $\mathcal{V}$ as described a moment ago.

Let $\mathcal{V} \subset \mathcal{X}$ be k-dimensional and A-invariant. Let $q_1, \ldots, q_n$ be an adapted basis. The matrix of A with respect to $q_1, \ldots, q_n$ has the form

$$(A) = \begin{pmatrix} A_{11} & A_{12} \\ 0 & A_{22} \end{pmatrix}.$$

We have seen before that A_{11} is the matrix of $A \mid \mathcal{V}$ with respect to the basis $q_1, \ldots, q_k$. It can easily be verified that A_{22} is the matrix of $A \mid \mathcal{X}/\mathcal{V}$ with respect to the basis $\bar{q}_{k+1}, \ldots, \bar{q}_n$ of $\mathcal{X}/\mathcal{V}$.

Let $\mathcal{V} \subset \mathcal{X}$ and $C : \mathcal{X} \to \mathcal{Y}$ be such that $\mathcal{V} \subset \ker C$ (equivalently, $C\mathcal{V} = 0$). Then we define the quotient map $\bar{C} : \mathcal{X}/\mathcal{V} \to \mathcal{Y}$ by $\bar{C}\bar{x} := Cx$. Again, this is easily seen to be well defined. The defining formula for $\bar{C}$ can be written as $C = \bar{C}\Pi$.

2.5 Eigenvalues

If $A : \mathcal{X} \to \mathcal{X}$ is a linear map then $\lambda \in \mathbb{C}$ is called an *eigenvalue* of A if there exists a nonzero vector $v \in \mathcal{X}$ such that $Av = \lambda v$. The set of eigenvalues, which contains at most n elements, is called the *spectrum* of A and denoted $\sigma(A)$. Necessary and sufficient for λ to be an eigenvalue of A is $\det(\lambda I - A) = 0$. The *spectral radius* $\rho(A)$ is defined as follows:

$$\rho(A) = \max\{\, |\lambda| \mid \lambda \in \sigma A \,\}$$

The function

$$\chi_A(s) := \det(sI - A) \tag{2.7}$$

is a polynomial of degree n. This polynomial is *monic*, i.e., the leading coefficient is 1. Hence the polynomial is of the form $\chi_A(s) = s^n + a_1 s^{n-1} + \cdots + a_n$. It is called the *characteristic polynomial* of A. Its zeros are exactly the eigenvalues of A. Any nonzero vector v such that $Av = \lambda v$ is called an *eigenvector* of A corresponding to the eigenvalue λ. The set of eigenvectors corresponding to λ is a linear space equal to $\ker(\lambda I - A)$ with the zero element removed.

Let $\mathcal{V}$ be an A-invariant subspace and let $B := A \mid \mathcal{V}$ (the restriction, see section 2.2). Then, obviously, every eigenvalue of B is an eigenvalue of A. We can make a stronger statement: χ_B is a divisor of χ_A. In order to see this, we choose a basis of $\mathcal{X}$ adapted to $\mathcal{V}$, and we obtain the matrix representation

$$(A) = \begin{pmatrix} A_{11} & A_{12} \\ 0 & A_{22} \end{pmatrix}. \tag{2.8}$$

Hence $\chi_A(s) = \det(sI - (A)) = \det(sI - A_{11})\det(sI - A_{22})$. On the other hand, A_{11} is the matrix of B with respect to the chosen basis. Hence

$$\det(sI - A_{11}) = \chi_B(s),$$

which proves the statement. We also see that the quotient map $\bar{A} : \mathcal{X}/\mathcal{V} \to \mathcal{X}/\mathcal{V}$ has a divisor of χ_A as characteristic polynomial, because A_{22} is the matrix of $\bar{A}$ with respect to a suitable basis (see section 2.4).

From section 2.3, recall that if $\mathcal{X}$ is an inner product space and if $\mathcal{V}$ is A-invariant, then $\mathcal{V}^{\perp}$ is A^{T}-invariant. Let $C := A^{\mathrm{T}} \mid \mathcal{V}^{\perp}$. We claim that the characteristic polynomial $\chi_{\bar{A}}$ of the quotient map $\bar{A}$ is equal to the characteristic polynomial χ_C of the restricted map C. To see this, choose an orthonormal basis of $\mathcal{X}$ adapted to $\mathcal{V}, \mathcal{V}^{\perp}$ to obtain the matrix representation (2.8). Obviously,

$$(A)^{\mathrm{T}} = \begin{pmatrix} A_{11}^{\mathrm{T}} & 0 \\ A_{12}^{\mathrm{T}} & A_{22}^{\mathrm{T}} \end{pmatrix}$$

is a matrix representation of the adjoint map A^{T}. Thus, A_{22} is a matrix representation of $\bar{A}$, while A_{22}^{T} is a matrix representation of C. Hence,

$$\chi_{\bar{A}} = \det(sI - A_{22}) = \det(sI - A_{22}^{\mathrm{T}}) = \chi_C.$$

In particular, we have now proven the following useful formula:

$$\sigma(A \mid \mathcal{X}/\mathcal{V}) = \sigma(A^{\mathrm{T}} \mid \mathcal{V}^{\perp}). \tag{2.9}$$

Let $A : \mathcal{X} \to \mathcal{X}$ and let $p(s) = a_0 s^m + a_1 s^{m-1} + \cdots + a_m$ be a polynomial. We define $p(A) := a_0 A^m + a_1 A^{m-1} + \cdots + a_m I$. This substitution has the following properties (here p and q are polynomials):

$$p(A)q(A) = (pq)(A), \quad p(A) + q(A) = (p+q)(A). \tag{2.10}$$

In particular, $p(A)$ and $q(A)$ commute. Furthermore, we have the *spectral mapping theorem* (see exercise 2.3):

$$\sigma(p(A)) = \{p(\lambda) \mid \lambda \in \sigma(A)\}, \tag{2.11}$$

which is sometimes abbreviated to $\sigma(p(A)) = p(\sigma(A))$. Another very famous and important result is the *Cayley-Hamilton theorem*:

$$\chi_A(A) = 0. \tag{2.12}$$

Theorem 2.1 *Let $A : \mathcal{X} \to \mathcal{X}$ and suppose that χ_A is factorized as $\chi_A = pq$, where p and q are monic coprime polynomials. Define $\mathcal{V} := \ker p(A)$ and $\mathcal{W} := \ker q(A)$. Then we have*

(i) $\mathcal{V} = \operatorname{im} q(A)$, $\mathcal{W} = \operatorname{im} p(A)$,

(ii) $\mathcal{V} \oplus \mathcal{W} = \mathcal{X}$,

(iii) $\mathcal{V}$ *and* $\mathcal{W}$ *are* A*-invariant,*

(iv) $\chi_{A|\mathcal{V}} = p, \quad \chi_{A|\mathcal{W}} = q.$

We say that two polynomials are *coprime* (or *relatively prime*) if they do not have a common factor, or equivalently, a common (complex) zero. An equivalent condition is: there exist polynomials r and s such that $rp + sq = 1$. Furthermore, $p \mid q$ means that p is a divisor of q.

Proof of theorem 2.1 : Let r and s be such that $rp + sq = 1$. Substituting A into this equation, we find that $I = r(A)p(A) + s(A)q(A)$ (see (2.10)). Hence, for every $x \in \mathcal{X}$, the following equations hold:

$$x = r(A)p(A)x + s(A)q(A)x = p(A)r(A)x + q(A)s(A)x. \tag{2.13}$$

(i) Note that if $x \in \operatorname{im} q(A)$ then $x = q(A)y$ for some y. Hence

$$p(A)x = \chi_A(A)y = 0,$$

so that $x \in \mathcal{V}$. Conversely, if $x \in \mathcal{V}$ then $p(A)x = 0$ and hence, by (2.13), $x = s(A)q(A)x = q(A)s(A)x \in \operatorname{im} q(A)$. The proof of $\mathcal{W} = \operatorname{im} p(A)$ is similar.

(ii) If $x \in \mathcal{V} \cap \mathcal{W}$ then $p(A)x = 0$ and $q(A)x = 0$. Hence, by (2.13), $x = 0$. That is, $\mathcal{V}$ and $\mathcal{W}$ are independent. It also follows from (2.13) that every $x \in \mathcal{X}$ is an element of $\operatorname{im} p(A) + \operatorname{im} q(A) = \mathcal{W} + \mathcal{V}$.

(iii) If $p(A)x = 0$ then $p(A)Ax = Ap(A)x = 0$.

(iv) We use lemma 2.2 which is stated directly after the current proof. Now let $\alpha := \chi_{A|\mathcal{V}}$. Then α and q are coprime. In fact, if $\alpha(\lambda) = 0$ there exists $x \in \mathcal{V}, x \neq 0$ such that $Ax = \lambda x$. Then $p(A)x = p(\lambda)x = 0$ and hence $p(\lambda) = 0$. It follows that $q(\lambda) \neq 0$ (since p and q are coprime). As $\alpha \mid \chi_A, \;\; q \mid \chi_A$ and α and q are coprime, the lemma implies that $\alpha q \mid \chi_A$. As a consequence, $\alpha \mid p$. Similarly, we have that $\beta := \chi_{A|\mathcal{W}}$ divides q. Since $\deg(\alpha\beta) = \deg(pq) = n$, we must have $\alpha = p$ and $\beta = q$. This completes the proof of theorem 2.1 ■

Lemma 2.2 *If p, q and r are polynomials satisfying $p \mid r$ and $q \mid r$, and p and q are coprime then $pq \mid r$.*

Proof : Let u and v be such that $up + vq = 1$. Then $pq \mid vqr$ and $pq \mid upr$ and hence $pq \mid r$ since $r = upr + vqr$. ■

Note that the matrix of A with respect to a basis $q_1, \ldots, q_n$ adapted to $\mathcal{V}$, $\mathcal{W}$ is of the form

$$\begin{pmatrix} A_{11} & 0 \\ 0 & A_{22} \end{pmatrix}.$$

where $\chi_{A_{11}} = p$ and $\chi_{A_{22}} = q$.

2.6 Differential equations

In this section, we consider the linear homogeneous time-invariant differential equation

$$\dot{x}(t) = Ax(t), \tag{2.14}$$

where $A : \mathcal{X} \to \mathcal{X}$ is a linear map and $\mathcal{X}$ an n-dimensional space. The equation is called *linear* because the right-hand side is a linear function of $x(t)$. It is called *homogeneous* because the right-hand side is zero for $x(t) = 0$. (An equation of the form $\dot{x} = Ax(t) + f(t)$ with $f(t) \neq 0$ is called *inhomogeneous*). The equation is called *time invariant* because A is independent of t. (One also says that the equation has *constant coefficients*). If A would be allowed to be time dependent, the equation would be called *time varying*.

In order to specify the solution of (2.14), one has to provide an initial value. The solution of the *initial value problem*: find a function x satisfying

$$\dot{x} = Ax, \quad x(0) = x_0, \tag{2.15}$$

is denoted by $x(t, x_0)$. In the scalar case ($n = 1$, $A = a$), it is well known that $x(t, x_0) = e^{at} x_0$.

In order to have a similar result for the multivariable case, we introduce the matrix exponential function

$$e^{At} := \exp(At) := \sum_{k=0}^{\infty} \frac{t^k A^k}{k!}. \tag{2.16}$$

Since $\|A^k\| \leqslant \|A\|^k$ for all k, the sum is dominated by $\sum(|t|\|A\|)^k/k!$. This series converges for every t and hence so does the series (2.16). Also, (2.16) is a power series, so that term by term differentiation is allowed. Hence, $\frac{\mathrm{d}}{\mathrm{d}t} e^{At} = Ae^{At} = e^{At}A$. It follows that $x(t, x_0) = e^{At}x_0$.

The concept of invariant subspace has an important significance for systems of differential equations:

Definition 2.3 *A subspace $\mathcal{V}$ is called an* invariant subspace *of (2.15) if every solution of (2.15) that starts in $\mathcal{V}$, remains in $\mathcal{V}$ for all $t \geqslant 0$. More explicitly, if for every $x_0 \in \mathcal{V}$, we have that $x(t, x_0) \in \mathcal{V}$ for all $t > 0$.*

Theorem 2.4 *A subspace $\mathcal{V}$ is an invariant subspace of (2.15) if and only if $\mathcal{V}$ is A-invariant.*

Proof : If $x_0 \in \mathcal{V}$, then $Ax_0 \in \mathcal{V}$, $A^2x_0 \in \mathcal{V}$, and, by induction, $A^k x_0 \in \mathcal{V}$ for all k. It follows that for all m, the vector

$$\sum_{k=0}^{m} \frac{t^k A^k}{k!} x_0 \in \mathcal{V}.$$

Since $\mathcal{V}$ is a subspace of the finite dimensional linear space $\mathcal{X}$, it is closed in the Euclidean topology. Thus we infer that $e^{At}x_0 \in \mathcal{V}$.

Conversely, if $x(t, x_0) \in \mathcal{V}$ for all $t > 0$ and all $x_0 \in \mathcal{V}$, then

$$Ax_0 = \lim_{t \downarrow 0} t^{-1}(e^{tA} - I)x_0 = \lim_{t \downarrow 0} t^{-1}(x(t, x_0) - x_0) \in \mathcal{V},$$

again because $\mathcal{V}$ is closed. ■

The definition of e^{At} is not very suitable for the investigation of the behavior of $x(t)$ for large values of t. This behavior is more adequately given by the spectrum of A. We will not give a complete picture of the behavior in the most general case, because it is rather complicated and also not necessary. A convenient formula for the exponential function is given by Cauchy's integral in the complex plane:

$$e^{At} = \frac{1}{2\pi i} \int_C e^{zt}(zI - A)^{-1}\,\mathrm{d}z, \tag{2.17}$$

where the contour of integration C encloses all eigenvalues of A. This formula can be proven by using the Neumann expansion

$$(zI - A)^{-1} = z^{-1}I + z^{-2}A + z^{-3}A^2 + \cdots$$

for $|z| \geqslant \|A\|$. Because this is a power series, we can perform integration term wise. Hence:

$$\frac{1}{2\pi i} \int_C e^{zt}(zI - A)^{-1}\,\mathrm{d}z = \sum_{k=0}^{\infty} f_k(t),$$

where

$$f_k(t) := \frac{1}{2\pi i} \int_C e^{zt} z^{-k-1}\,\mathrm{d}z\, A^k = \frac{t^k A^k}{k!}.$$

It follows from Cramer's rule that $(zI - A)^{-1}$ is of the form

$$B(z)/\det(zI - A),$$

where $B(z)$ is a matrix of which the entries are polynomials of z. Hence, each entry of $(zI - A)^{-1}$ is a rational function and the poles of this function are eigenvalues of A. If we perform partial-fraction decomposition, we find that the entries of $(zI - A)^{-1}$ can be written as a linear combination of functions of the form $(z - \lambda)^{-k}$. As a consequence, e^{At} is a linear combination of functions of the form

$$\frac{1}{2\pi i}\int_C e^{zt}(z-\lambda)^{-k}\,dz = \frac{t^{k-1}e^{\lambda t}}{k!},$$

where the λ's are eigenvalues of A.

Definition 2.5 *A function that is a linear combination of functions of the form $t^k e^{\lambda t}$, where the k's are nonnegative integers and $\lambda \in \mathbb{C}$, is called a* Bohl function. *The numbers λ that appear in this linear combination (and cannot be canceled) are called the* characteristic exponents *of the Bohl function. The set of characteristic exponents of a Bohl function p is called the* spectrum *of the function and denoted $\sigma(p)$.*

The previous discussion implies:

Theorem 2.6 *The entries of e^{At} are Bohl functions. Their characteristic exponents are eigenvalues of A. Each eigenvalue of A appears as a characteristic exponent of some entry of e^{At}.*

The last statement of the theorem follows from the fact that for every $\lambda \in \sigma(A)$ with corresponding eigenvector v, the function $x(t) := e^{\lambda t}v$ is a solution of (2.14) and hence equal to $x(t, v)$. On the other hand, according to (2.16) we have $x(t, v) = e^{At}v$. Hence there must be at least one entry of e^{At} with exponent λ. Note that, if one insists on having real functions and vectors, one has to define Bohl functions as linear combinations of the functions of the form $t^k e^{\alpha t}\cos\omega t$ and $t^k e^{\alpha t}\sin\omega t$, where k is an integer and α and ω are real numbers.

The following result is easily verified:

Theorem 2.7 *If p and q are Bohl functions then $p + q$, pq and $\dot{p}$ are also Bohl functions. Furthermore, $\sigma(pq) \subset \sigma(p) + \sigma(q), \sigma(p + q) \subset \sigma(p) \cup \sigma(q)$ and $\sigma(\dot{p}) \subset \sigma(p)$. Also, if $r(t) := \int_0^t p(\tau)\,d\tau$ then r is a Bohl function with $\sigma(r) \subset \sigma(p) \cup \{0\}$.*

The concept of Bohl function can be extended to vector-valued and matrix-valued functions in the obvious way. Every vector or matrix whose entries are Bohl functions will also be called a Bohl function.

Next we consider the inhomogeneous initial value problem

$$\dot{x} = Ax + f(t), \quad x(0) = x_0. \tag{2.18}$$

The solution to this problem can also be expressed in terms of the matrix exponential function. For this, we introduce the function $z(t) := e^{-At}x(t)$. We note that this function satisfies

$$\dot{z}(t) = -Az(t) + e^{-At}(Ax(t) + f(t)) = e^{-At} f(t),$$

a differential equation which can be solved by quadrature. The solution is

$$z(t) = z(0) + \int_0^t e^{-A\tau} f(\tau)\,\mathrm{d}\tau.$$

Since $x(t) = e^{At}z(t)$, it follows that

$$x(t) = e^{At}x_0 + \int_0^t e^{A(t-\tau)} f(\tau)\,\mathrm{d}\tau, \tag{2.19}$$

where we have used that $z(0) = x_0$. Formula (2.19) is called the *variations of constants formula.* It is valid for any continuous function f. If f is an arbitrary integrable function, the expression in the right-hand size of (2.19) is still defined. We will view the function defined by the right-hand side of (2.19) as the solution of (2.18) in a generalized sense. As a corollary of theorem 2.7 and this formula we have

Theorem 2.8 *If $f(t)$ is a Bohl function then the solution of (2.18) is also a Bohl function.*

2.7 Stability

The concept of stability plays an important role in the theory of differential equations. In the general theory, where time varying and nonlinear systems are considered, the definitions are rather involved and the distinctions are subtle. For the very simple situation we are considering here, a drastic simplification of the treatment is possible. In particular, we will not need the concept that is called (Liapunov) stability, although in section 3.8 we will briefly introduce the Liapunov equation and its relation to stability will be presented.

What is important in control theory is a concept that is usually called *asymptotic stability* in the theory of ordinary differential equations. We will refer to this concept as stability, omitting the adjective 'asymptotic'. For linear time-invariant systems, this concept can be defined very simply:

Definition 2.9 *The system $\dot{x} = Ax$ is called* stable *if every solution tends to zero for $t \to \infty$.*

A condition for stability can be derived from the following result

Theorem 2.10 *Let p be a Bohl function and let the* spectral abscis *of p be defined by*

$$\Lambda(p) := \max\{\, \Re\mathrm{e}\,\lambda \mid \lambda \in \sigma(p)\,\}. \tag{2.20}$$

Then the following statements hold:

(i) *If $\Lambda(p) < 0$ or, equivalently, every exponent λ of p satisfies $\Re\mathrm{e}\,\lambda < 0$ then $p(t) \to 0$ $(t \to \infty)$. More specifically, for every $\gamma > 0$ there exists $M > 0$ such that for all $t \geqslant 0$ we have $|p(t)| \leqslant Me^{-\gamma t}$.*

(ii) *For all $\alpha > \Lambda(p)$ there exists $M > 0$ such that for all $t \geqslant 0$ we have $|p(t)| \leqslant Me^{\alpha t}$.*

Proof : (i) It is well known that $t^k e^{\lambda t} \to 0$ $(t \to \infty)$ if $\Re\mathrm{e}\,\lambda < 0$. Hence a Bohl function converges to zero if all its exponents have a negative real part. In particular, such a function is bounded. For γ, take any positive real number such that $\Lambda(p) < -\gamma < 0$. Then the function $q(t) = e^{\gamma t}p(t)$ still is a Bohl function satisfying that $\Lambda(q) < 0$. There exists $M > 0$ such that $|q(t)| \leqslant M$, which implies that $|p(t)| \leqslant Me^{-\gamma t}$.

(ii) If p is a Bohl function and $\beta = \Lambda(p)$ then for every $\alpha > \beta$, the function $q(t) = e^{-\alpha t}p(t)$ is a Bohl function satisfying $\Lambda(q) < 0$. Hence q is bounded and there exists a number M such that $|p(t)| \leqslant Me^{\alpha t}$. ■

We use the notation

$$\Lambda(A) := \max\{\Re\mathrm{e}\,\lambda \mid \lambda \in \sigma(A)\} \tag{2.21}$$

for the *spectral abscis of a linear map*. Then we have

Corollary 2.11 *System (2.14) is stable if and only if $\Lambda(A) < 0$. In this case, there exist positive numbers M and γ such that the inequality $\|e^{At}\| \leqslant Me^{-\gamma t}$ holds for all $t > 0$.*

Proof : In view of the above, we only need to prove necessity. Suppose $\Lambda(A) \geqslant 0$ and let $\lambda \in \sigma(A)$, $Av = \lambda v$, $v \neq 0$ and $\Re\mathrm{e}\,\lambda \geqslant 0$. Then $e^{\lambda t}v$ is a solution of (2.14) that does not tend to zero. ■

For some purposes, it is desirable to have different notions of stability. For instance, one might be interested in having a special rate of convergence, i.e., one might want the solutions to satisfy conditions like $e^{\alpha t}x(t) \to 0$ $(t \to \infty)$, for a prescribed α. Or, one wants to avoid high frequencies in the functions, i.e., one imposes on the exponents a condition like $|\Im\mathrm{m}\,\lambda| \leqslant \beta$ for a prescribed β. For discrete-time systems, one can prove that the stability condition is $|\lambda| < 1$ for all eigenvalues. These examples suggest a generalization of the stability condition giving rise to conditions on the eigenvalues and exponents of the form $\lambda \in \mathbb{C}_g$, where $\mathbb{C}_g$ is a prescribed part of the complex plane, thought of as the 'good' part. The complement of $\mathbb{C}_g$ is often denoted $\mathbb{C}_b$. For the original stability condition, we have $\mathbb{C}_g = \mathbb{C}^- := \{\lambda \in \mathbb{C} \mid \Re\mathrm{e}\,\lambda < 0\}$. Similarly we will use:

$$\mathbb{C}^0 := \{\lambda \in \mathbb{C} \mid \Re\mathrm{e}\,\lambda = 0\}$$
$$\mathbb{C}^+ := \{\lambda \in \mathbb{C} \mid \Re\mathrm{e}\,\lambda > 0\}$$

We introduce a term for $\mathbb{C}_g$:

Definition 2.12 *A subset* $\mathbb{C}_g$ *of* $\mathbb{C}$ *is called a* stability domain *if* $\mathbb{C}_g \cap \mathbb{R}$ *is nonempty and* $\mathbb{C}_g$ *is symmetric with respect to the real axis (i.e.,* $\lambda \in \mathbb{C}_g \Longrightarrow \bar{\lambda} \in \mathbb{C}_g$*). If* $\mathbb{C}_g$ *is a stability domain, a Bohl function p is called* $\mathbb{C}_g$-stable *if* $\sigma(p) \subset \mathbb{C}_g$*. The system* $\dot{x} = Ax$ *is called* $\mathbb{C}_g$-stable *if every solution is a* $\mathbb{C}_g$*-stable Bohl function. Finally, a linear map A is called* $\mathbb{C}_g$-stable *if* $\sigma(A) \subset \mathbb{C}_g$*. A linear map is called* $\mathbb{C}_g$-antistable *if* $\sigma(A) \cap \mathbb{C}_g$ *is empty.*

If the stability domain is obvious from the context, we omit the specification $\mathbb{C}_g$. Some important properties are not valid without this condition. The condition imposed on a stability domain that there exists an element in $\mathbb{R}$ is for technical reasons. We will study the spectrum of real matrices which should be contained in the stability domain. Since the spectrum is always symmetric, it is natural to also require the stability domain to be symmetric. Note also that the condition of $\mathbb{C}_g$-stability is only defined for Bohl functions. There seems to be no sensible way to extend this condition to general functions.

Definition 2.13 *Let* $\mathbb{C}_g$ *be a stability domain and let* $A : \mathcal{X} \to \mathcal{X}$ *be a linear map. The* $\mathbb{C}_g$-stable subspace $\mathcal{X}_g(A)$ *is the set of all* x_0 *such that the solution* $x(t, x_0)$ *of the differential equation* $\dot{x} = Ax$ *with initial value* x_0 *is* $\mathbb{C}_g$*-stable. The* $\mathbb{C}_g$-unstable subspace $\mathcal{X}_b(A)$ *of A is the set of all* x_0 *such that* $x(t, x_0)$ *has only exponents in* $\mathbb{C}_b$*.*

We can obtain explicit expressions for $\mathcal{X}_g(A)$ and $\mathcal{X}_b(A)$ using the characteristic polynomial of A. To this extent, we factorize χ_A as $\chi_A = \chi_A^g \cdot \chi_A^b$, where χ_A^g and χ_A^b are monic polynomials such that χ_A^g contains the factors of χ_A with zeros in $\mathbb{C}_g$ and χ_A^b the factors with zeros in $\mathbb{C}_b$. Then we have

Theorem 2.14 $\mathcal{X}_g(A) = \ker \chi_A^g(A)$, $\mathcal{X}_b(A) = \ker \chi_A^b(A)$.

Proof : We use the notation $\bar{\mathcal{X}}_g(A) := \ker \chi_A^g(A)$, $\bar{\mathcal{X}}_b(A) := \ker \chi_A^b(A)$. Since χ_A^g and χ_A^b are coprime, it follows from theorem 2.1 that $\bar{\mathcal{X}}_g(A)$ and $\bar{\mathcal{X}}_b(A)$ are A-invariant, that $\bar{\mathcal{X}}_g(A) \oplus \bar{\mathcal{X}}_b(A) = \mathcal{X}$ and that $\chi_{A|\bar{\mathcal{X}}_g(A)} = \chi_A^g$, $\chi_{A|\bar{\mathcal{X}}_b(A)} = \chi_A^b$. I.e., $A \mid \bar{\mathcal{X}}_g(A)$ is $\mathbb{C}_g$-stable and $A \mid \bar{\mathcal{X}}_b(A)$ is $\mathbb{C}_g$-antistable. Hence, if $x_0 \in \bar{\mathcal{X}}_g(A)$ then $x(t, x_0)$ is stable, since it satisfies the restricted differential equation $\dot{x} = (A \mid \bar{\mathcal{X}}_g(A))x$. The argument is similar if $x_0 \in \bar{\mathcal{X}}_b(A)$. On the other hand, if $x_0 \in \mathcal{X}_g(A)$ then we can write $x_0 = x_g + x_b$ with $x_g \in \bar{\mathcal{X}}_g(A)$ and $x_b \in \bar{\mathcal{X}}_b(A)$. Then $x(t, x_b) = x(t, x_0) - x(t, x_g)$. Both terms in the right-hand side have only exponents in $\mathbb{C}_g$ and hence so has $x(t, x_b)$. But since $x_b \in \bar{\mathcal{X}}_b(A)$ it follows that $x_b = 0$ and hence $x_0 \in \bar{\mathcal{X}}_g(A)$. Again one argues similarly when $x_0 \in \mathcal{X}_b(A)$. ■

An intuitive way of thinking of $\mathcal{X}_g(A)$ is as of the space generated by the eigenvectors of A corresponding to stable eigenvalues. This characterization is generally not correct if A has multiple eigenvalues. In that case one has to use generalized eigenvectors, which are a lot less intuitive. This is why we have avoided using this as a formal definition.

Corollary 2.15 *Let $\mathbb{C}_g$ be a stability domain. The system $\dot{x} = Ax$ is $\mathbb{C}_g$-stable if and only if $\sigma(A) \subset \mathbb{C}_g$, i.e., the map A is $\mathbb{C}_g$-stable.*

Proof : ($\Rightarrow$) If $\dot{x} = Ax$ is $\mathbb{C}_g$-stable then $\mathcal{X}_g(A) = \mathcal{X}$. Thus, according to theorem 2.14, $\chi_A^g(A) = 0$. Let $\lambda \in \sigma(A)$. By applying the spectral mapping theorem (see section 2.5, with $p(s) = \chi_A^g(s)$, we get $\chi_A^g(\lambda) \in \sigma(\chi_A^g(A)) = \{0\}$, so $\lambda \in \mathbb{C}_g$.

($\Leftarrow$) Let $\sigma(A) \subset \mathbb{C}_g$. Then $\chi_A(s) = \chi_A^g(s)$. Thus, by the Cayley-Hamilton theorem, $\mathcal{X}_g(A) = \ker \chi_A^g(A) = \ker \chi_A(A) = \mathcal{X}$, so $\dot{x} = Ax$ is $\mathbb{C}_g$-stable. ∎

Definition 2.16 *Given a linear map $A : \mathcal{X} \to \mathcal{X}$, an A-invariant subspace $\mathcal{V}$ of $\mathcal{X}$ is called* inner $\mathbb{C}_g$-stable *if the map $A \mid \mathcal{V}$ is $\mathbb{C}_g$-stable and* outer $\mathbb{C}_g$-stable *if $A \mid \mathcal{X}/\mathcal{V}$ is $\mathbb{C}_g$-stable.*

If the stability domain is fixed we often simply refer to inner stable and outer stable respectively.

In the case $\mathbb{C}_g = \mathbb{C}^-$, these concepts can be interpreted as follows: $\mathcal{V}$ is inner stable if and only if solutions starting in $\mathcal{V}$ (which will remain in $\mathcal{V}$ because $\mathcal{V}$ is invariant) converge to zero. In fact, when we restrict the differential equation to $\mathcal{V}$, the differential equation has a stable coefficient map, viz. $A \mid \mathcal{V}$. On the other hand, $\mathcal{V}$ is outer stable if and only if the distance of arbitrary solutions to $\mathcal{V}$ converges to zero as $t \to \infty$. It is easy to give conditions for a subspace to be inner stable.

Theorem 2.17 *Let $\mathcal{V}$ be an A-invariant subspace. Then the following statements are equivalent:*

(i) *$\mathcal{V}$ is inner stable,*

(ii) *$\mathcal{V} \subset \mathcal{X}_g(A)$,*

(iii) $\forall \lambda \in \mathbb{C}_b \;\; (\lambda I - A)\mathcal{V} = \mathcal{V}$.

Proof : (i) $\Leftrightarrow$ (ii). This equivalence is immediate from the definitions.

(i) $\Leftrightarrow$ (iii). Because of the invariance of $\mathcal{V}$ it follows that $(\lambda I - A)\mathcal{V} \subset \mathcal{V}$ is always true. Therefore we concentrate on the converse inclusion. If $\mathcal{V}$ is inner stable, we know that no $\lambda \in \mathbb{C}_b$ is an eigenvalue of $A \mid \mathcal{V}$. Hence for every $\lambda \in \mathbb{C}_b$, the map $(\lambda I - A) \mid \mathcal{V}$ is invertible. Hence, we must have $(\lambda I - A)\mathcal{V} = \mathcal{V}$. Conversely, if $\mathcal{V}$ is not inner stable, there is an eigenvalue λ of $A \mid \mathcal{V}$ in $\mathbb{C}_b$. For this λ, the map $(\lambda I - A) \mid \mathcal{V}$ is not invertible and hence $(\lambda I - A)\mathcal{V} \neq \mathcal{V}$. ∎

Similar characterizations for outer stability are not so easy to obtain. One way of dealing with outer stability is via orthogonal complements. It is easily verified that $\mathcal{V}$ is outer stable with respect to A if and only if $\mathcal{V}^{\perp}$ is inner stable with respect to A^{T}.

2.8 Rational matrices

If $p(s)$ and $q(s)$ are polynomials with real coefficients, then their quotient $p(s)/q(s)$ is called a *rational function*. This rational function is called *proper* if $\deg p \leqslant \deg q$ and *strictly proper* if $\deg p < \deg q$. Here $\deg p$ denotes the degree of p. Obviously, a proper rational function has a limit as $|s| \to \infty$. For a strictly proper rational function this limit is zero.

If $g(s)$ is the rational function obtained as the quotient of the polynomial $p(s)$ and $q(s)$, then $p(s)$ and $q(s)$ can have common factors. After cancellation of these common factors, we obtain $g(s) = p_1(s)/q_1(s)$, with $p_1(s)$ and $q_1(s)$ coprime polynomials. The *poles* of $g(s)$ are the (complex) zeros of the polynomial $q_1(s)$. The set of all rational functions (with pointwise addition and multiplication) forms a field, denoted by $\mathbb{R}(s)$.

A *rational matrix* is a matrix whose entries are rational functions. A rational matrix is called *proper* if all its entries are proper, and *strictly proper* if all its entries are strictly proper. A complex number λ is called a *pole* of the rational matrix if it is a pole of at least one of its entries.

A rational matrix $G(s)$ is called *left-invertible*, if there exists a rational matrix $G_L(s)$ such that $G_L(s)G(s) = I$, the identity matrix. The rational matrix $G_L(s)$ is called a left-inverse of $G(s)$. $G(s)$ is left-invertible if and only if for every column vector of rational functions $q(s)$ we have: $G(s)q(s) = 0 \Leftrightarrow q(s) = 0$. In other words, $G(s)$ is left-invertible if and only if its columns (interpreted as elements of the linear space of column vectors with rational functions as components) are linearly independent. The rational matrix $G(s)$ is called *right-invertible* if there exists a rational matrix $G_R(s)$ such that $G(s)G_R(s) = I$. Any such rational matrix $G_R(s)$ is called a right-inverse of $G(s)$. $G(s)$ is right-invertible if and only if for every row-vector of rational functions $p(s)$ we have: $p(s)G(s) = 0 \Leftrightarrow p(s) = 0$. Thus $G(s)$ is right-invertible if and only if its rows (interpreted as elements of the linear space of row vectors with rational functions as components) are linearly independent. A square rational matrix $G(s)$ is called *invertible* if there exists a rational matrix, $G^{-1}(s)$, such that $G^{-1}(s)G(s) = G(s)G^{-1}(s) = I$, the identity matrix. If such $G^{-1}(s)$ exists, then it is unique, and it is called the inverse of $G(s)$. $G(s)$ is invertible if and only if it is both right and left-invertible. Also, $G(s)$ is invertible if and only if the rational function $\det G(s)$ is non-zero.

The *normal rank* of a, not necessarily square, rational matrix is defined as

$$\operatorname{normrank} G := \max\{\operatorname{rank} G(\lambda) \mid \lambda \in \mathbb{C}\}.$$

Except for a finite number of points $\lambda \in \mathbb{C}$, one has $\operatorname{normrank} G = \operatorname{rank} G(\lambda)$. A $p \times m$ rational matrix $G(s)$ is left-invertible if and only if $\operatorname{normrank} G = m$, and right-invertible if and only if $\operatorname{normrank} G = p$.

2.9 Laplace transformation

A function $u(t)$ defined for $t \geqslant 0$ is called *exponentially bounded* if there exist numbers M and α such that $|u(t)| \leqslant Me^{\alpha t}$ for all $t \geqslant 0$. The number α is called a *bounding exponent* of u. It is of course not uniquely defined. Obviously, Bohl functions are exponentionally bounded. Examples of functions that are not exponentionally bounded are t^t and e^{t^2}. For exponentionally bounded functions we can define the Laplace transform:

Definition 2.18 *If $u(t)$ is exponentionally bounded with bounding exponent α then*

$$L(u)(s) := \int_0^\infty e^{-st} u(t)\, dt.$$

for $\Re\mathrm{e}\, s > \alpha$ is called the Laplace transform *of u. The operation of forming the Laplace transform is called the* Laplace transformation.

Typically, one denotes the Laplace transform of a function u by $\hat{u}$. It is easily shown that the integral defining the Laplace transform converges uniformly in any domain of the form $\Re\mathrm{e}\, s \geqslant \beta$, if $\beta > \alpha$. The Laplace transform is an analytic function in that area. Furthermore, the following fundamental properties hold:

Theorem 2.19

(i) *The Laplace transformation is linear, i.e.,*

$$L(u + v) = L(u) + L(v), \quad L(\lambda u) = \lambda L(u),$$

for every $\lambda \in \mathbb{C}$, and exponentionally bounded functions u and v. (The sum and scalar multiple of exponentionally bounded functions is of course also exponentionally bounded)

(ii) *If $u(t)$ is continuously differentiable and $\dot{u}$ is exponentionally bounded then u is exponentionally bounded and*

$$L(\dot{u}) = sL(u) - u(0).$$

(iii) *If u and v are exponentionally bounded then the* convolution

$$w(t) := \int_0^t u(\tau)v(t - \tau)\, d\tau$$

is exponentionally bounded and $L(w) = L(u)L(v)$.

(iv) *The Laplace transformation is injective, hence if $L(u) = L(v)$ then $u = v$.*

(v) *Let u be exponentionally bounded and $\hat{u} := L(u)$. Then $\hat{u}(s) \to 0$ for $\Re\mathrm{e}\, s \to \infty$.*

As an example we take the function $u(t) := e^{at}$. A direct computation of the integral yields $\hat{u} = 1/(s-a)$. Alternatively, using the previous theorem, we have $\dot{u} = au(t)$ and hence, $L(\dot{u}) = s\hat{u}(s) - u(0) = a\hat{u}(s)$. Solving for $\hat{u}$ gives the same result. More generally, if $u_k(t) := t^k e^{at}$ and $k > 0$ then $\dot{u}_k(t) = ku_{k-1}(t) + au_k(t)$ and hence $s\hat{u}_k = k\hat{u}_{k-1} + a\hat{u}_k$ (since $u_k(0) = 0$). This yields $\hat{u}_k = k(s-a)^{-1}\hat{u}_{k-1}$ for $k = 1, 2, \ldots$. By induction, it follows that $\hat{u}_k = k!/(s-a)^{k+1}$. Similarly to scalar functions, the Laplace transformation can be applied to vector and matrix valued functions. Therefore by the same reasoning as above, we can prove that $L(e^{At}) = (sI - A)^{-1}$.

The following property is fundamental for linear system theory:

Theorem 2.20 *An exponentionally bounded function u is a Bohl function if and only if $L(u)$ is rational, i.e., is the quotient of two polynomials.*

Proof : As we saw a moment ago, $L(t^k e^{at}) = k!/(s-a)^{k+1}$, which is rational. Hence, by linearity, every Bohl function has a rational Laplace transform. Conversely, let u be exponentionally bounded and its Laplace transform $\hat{u}$ rational. By theorem 2.19 (v), the degree of the numerator is less than the degree of the denominator. Using partial fraction decomposition, we can write $\hat{u}$ as a linear combination of functions of the form $c(s-a)^{-k}$ with k a positive integer and c a complex number. Each of these functions is the Laplace transform of a Bohl function (viz. $ct^{k-1}e^{at}/(k-1)!$) and hence so is their linear combination. ■

It follows from the proof of theorem 2.20 that the Laplace transform of a Bohl function is a strictly proper rational function.

Obviously, the Laplace transform of a Bohl matrix function (i.e. a matrix whose coefficients are Bohl functions) is a strictly proper rational matrix.

2.10 Exercises

2.1 Let $\mathcal{X}$ be a linear space and $\mathcal{V}, \mathcal{W}, \mathcal{S}$ subspaces of $\mathcal{X}$. Prove the following statements:

a. $(\mathcal{V} + \mathcal{W}) \cap \mathcal{S} \supset \mathcal{V} \cap \mathcal{S} + \mathcal{W} \cap \mathcal{S}$ and give an example where one does not have equality

b. If $\mathcal{V} \subset \mathcal{S}$ then equality holds in (a). (Note: this is the modular rule.)

c. $\dim(\mathcal{V} + \mathcal{W}) = \dim \mathcal{V} + \dim \mathcal{W} - \dim(\mathcal{V} \cap \mathcal{W})$

d. $\dim((\mathcal{V} + \mathcal{W})/\mathcal{W}) = \dim(\mathcal{V}/(\mathcal{V} \cap \mathcal{W}))$.

2.2 Let $B : \mathcal{X} \to \mathcal{Y}$ be a surjective linear map, and let $v : [0, \infty) \to \mathcal{Y}$ be continuous. Show that there exists a continuous function $w : [0, \infty) \to \mathcal{X}$ such that $v(t) = Bw(t)$ for all $t \geqslant 0$.

2.3 Let p be a polynomial.

a. Show that $Av = \lambda v$ implies $p(A)v = p(\lambda)v$.

b. Give an example where $p(A)v = p(\lambda)v$ but $Av \neq \lambda v$.

c. Show that if μ is an eigenvalue of $p(A)$ then there exists an eigenvalue λ of A such that $p(\lambda) = \mu$ (spectral mapping theorem).

d. Show that for an eigenvalue λ of A, $p(A)v = p(\lambda)v$ does imply $Av = \lambda v$ if p satisfies the conditions:

1. $p'(\lambda) \neq 0$ for $\lambda \in \sigma(A)$. (Here $'$ denotes the derivative)
2. $p(\lambda) \neq p(\mu)$ if $\lambda, \mu \in \sigma(A)$.

2.4 Let $A : \mathcal{X} \to \mathcal{X}$ and assume that all eigenvalues of A are simple. Prove that for every q that divides χ_A, the space $\ker q(A)$ is spanned by the eigenvectors of A corresponding to the eigenvalues that are roots of q.

2.5 Show that $\mathcal{X}_b(A^{\mathrm{T}}) = \mathcal{X}_g(A)^{\perp}$.

2.6 Consider the discrete-time equation

$$x(t+1) = Ax(t)$$

for $t = 0, 1, 2, \ldots$. Develop a treatment of the system similar to the treatment of systems of differential equations given in section 2.6. In particular, give the analogs of e^{At}, the variation-of-constants formula, Bohl functions, stability.

2.7 The z-transform of a sequence $u(t), t = 0, 1, 2, \ldots$ is given by

$$\hat{u}(z) := \sum_{t=0}^{\infty} u(t) z^{-t}.$$

Investigate the convergence domain and prove that $\hat{u}(z)$ is rational if and only if $u(t)$ is a discrete Bohl function.

2.8 Let $A : \mathcal{X} \to \mathcal{X}$ be a linear map and let $\mathcal{V}$ and $\mathcal{W}$ be A-invariant subspaces. Show that the characteristic polynomials of $A \mid (\mathcal{V} + \mathcal{W})/\mathcal{V}$ and $A \mid \mathcal{W}/(\mathcal{V} \cap \mathcal{W})$ are equal. Note that this implies:

$$\sigma(A \mid (\mathcal{V} + \mathcal{W})/\mathcal{V}) = \sigma(A \mid \mathcal{W}/(\mathcal{V} \cap \mathcal{W})).$$

2.9 Let f be a Bohl function which is not identically zero.

a. Show that $\sigma(f) \neq \varnothing$.

Define $\Lambda(f) := \max\{\Re\mathrm{e}\,\lambda \mid \lambda \in \sigma(f)\}$.

b. Show that there exists a $\lambda \in \sigma(f)$ satisfying $\Re\mathrm{e}\,\lambda = \Lambda(f)$ and a $k \in \mathbb{N}$ such that

$$\lim_{T \to \infty} T^{-1} \int_0^T t^{-k} e^{-\lambda t} f(t) \mathrm{d}t$$

exists and is nonzero.

c. Show that $f \to 0$ $(t \to \infty)$ if and only if $\Lambda(f) < 0$.

2.10 Let f be a Bohl function and $\mathbb{C}_g \subset \mathbb{C}$ a stability domain.

a. Show that f can be uniquely decomposed as $f = f_1 + f_2$, where f_1 is stable (i.e. $\sigma(f_1) \subset \mathbb{C}_g$) and f_2 is antistable (i.e., $\sigma(f_2) \cap \mathbb{C}_g = \varnothing$).

b. Let f be stable and antistable. Show that $f = 0$.

Hint: Use the Laplace transform.

2.11 Let $\mathcal{V}$ and $\mathcal{W}$ be as in theorem 2.1. Let $\mathcal{U}$ be A-invariant. Prove

$$\mathcal{V} \subset \mathcal{U} \quad \Leftrightarrow \quad \mathcal{U} + \mathcal{W} = \mathcal{X}.$$

2.12 **a.** Let $A : \mathcal{X} \to \mathcal{X}$ be invertible and $\mathcal{V} \subset \mathcal{X}$ be A-invariant. Show that $\mathcal{V}$ is also A^{-1}-invariant.

b. Let $A : \mathcal{X} \to \mathcal{X}$ and $\mathcal{V} \subset \mathcal{X}$ be A-invariant. Show that $\mathcal{V}$ is $(\lambda I - A)^{-1}$-invariant for all $\lambda \notin \sigma(A)$.

2.11 Notes and references

Linear algebra became a major tool in linear system theory ever since the state space description of linear systems became the standard tool for developing the theory. In the earlier literature, the results were expressed in rather explicit matrix terms, and in a very basis-dependent way. Wonham in [223] stressed the basis-independent geometric approach using subspaces and quotient spaces rather than matrix decompositions. This approach is largely followed in this book, although in some situations, matrix operations have been used to simplify the treatment.

Bohl functions as introduced in this chapter are named after the Latvian mathematician Piers Bohl and his early work as presented in [24].

There are numerous books available on the topics introduced in this chapter. For the first five sections a good reference is Halmos [67]. Differential equations and stability are also treated in many textbooks, see for instance Perko [143], Arnold [8] and Goldberg and Potter [64]. Finally the Laplace transform is treated in for instance Kwakernaak and Sivan [107] or Kamen and Heck [97]. In particular properties as listed in theorem 2.19 can be found in these references.

Chapter 3

Systems with inputs and outputs

3.1 Introduction

The following equations will be the main object of study of this chapter:

$$\begin{aligned} \dot{x}(t) &= Ax(t) + Bu(t), \\ y(t) &= Cx(t) + Du(t). \end{aligned} \tag{3.1}$$

These equations are said to describe a *system*, which we usually denote by Σ. Here $A,\ B,\ C$ and D are maps between suitable spaces (or matrices of suitable dimensions) and the functions $x,\ u$ and y are usually considered to be defined on the real axis $\mathbb{R}$ or on any subinterval of it. In particular, one often assumes the domain of definition to be the nonnegative part of $\mathbb{R}$. The function u is called the *input*, and its values are assumed to be given from outside the system. The class of admissible input functions will be denoted **U**. Often, **U** will be the class of piecewise continuous or locally integrable functions, but for most purposes, the exact class from which the input functions are chosen is not important. However, it is important that **U** has the *slicing property*, i.e., if $u_1 \in \mathbf{U}$ and $u_2 \in \mathbf{U}$, then for any $\theta > 0$, the function u_3 defined by $u_3(t) := u_1(t)\ (0 \leqslant t < \theta)$ and $u_3(t) := u_2(t)\ (t \geqslant \theta)$, is in **U**. We will assume that input functions take values in an m-dimensional space $\mathcal{U}$, which we often identify with $\mathbb{R}^m$. The first equation of Σ is a differential equation for the variable x. For a given initial value of x and input function u, the function is completely determined by this equation. The variable x is called the *state variable* and it is assumed to have values in an n-dimensional space $\mathcal{X}$. The space $\mathcal{X}$ will be called the *state space*. It will usually be identified with $\mathbb{R}^n$. Finally, y is called the *output* of the system, and has values in a p-dimensional space $\mathcal{Y}$, which we identify with $\mathbb{R}^p$. Since the system Σ is completely determined by the maps (or matrices) $A,\ B,\ C$ and D, we identify Σ with the quadruple (A, B, C, D).

In many cases, the map D is irrelevant in the given consideration. Therefore, it is often assumed to be absent or zero. This gives rise to a slightly simpler theory. In this case the system is often denoted (C, A, B), the order of the matrices reflecting the order in which the matrices mostly appear in products. For some applications and problems, however, the presence of the term Du, called the *feedthrough term*, is essential.

The solution of the differential equation of Σ with initial value $x(0) = x_0$ will be denoted as $x_u(t, x_0)$. It can be given explicitly using the variation-of-constants formula (2.19). The result is

$$x_u(t, x_0) = e^{At}x_0 + \int_0^t e^{A(t-\tau)}Bu(\tau)\,\mathrm{d}\tau. \tag{3.2}$$

The corresponding value of y is denoted by $y_u(t, x_0)$. As a consequence of (3.2) we have

$$y_u(t, x_0) = Ce^{At}x_0 + \int_0^t K(t-\tau)u(\tau)\,\mathrm{d}\tau + Du(t). \tag{3.3}$$

where $K(t) := Ce^{At}B$. In the case $D = 0$, it is customary to call $K(t)$ the *impulse-response matrix*. In the general case one would call the distribution $K(t) + D\delta(t)$ (see appendix A) the impulse response, but we will not use distributions at this point. The integral in the right-hand side is a convolution integral (see theorem 2.19). This suggests the use of the Laplace transformation. Using the properties of the Laplace transformation as described in section 1.7, we find that the equations of Σ reduce to the algebraic equations:

$$s\hat{x} = A\hat{x} + B\hat{u} + x_0, \quad \hat{y} = C\hat{x} + D\hat{u},$$

from which we can eliminate $\hat{x}$. In the case $x_0 = 0$, we obtain:

$$\hat{y}(s) = T(s)\hat{u}(s),$$

where

$$T(s) := C(sI - A)^{-1}B + D. \tag{3.4}$$

The function $T(s)$ is called the *transfer matrix* of the system. If $D = 0$, it is easily seen to be the Laplace transform of the impulse response. This is in correspondence with (3.3) because of the convolution theorem. Note that $T(s)$ is a proper rational matrix and strictly proper if $D = 0$. We call the system Σ proper or strictly proper if the transfer matrix is proper or strictly proper respectively. If λ is a *pole* of the transfer matrix, then the matrix $\lambda I - A$ must be singular, i.e., λ must be an eigenvalue of A. Therefore, the eigenvalues of A are often called the poles of the system. Note however, that an eigenvalue of A is not necessarily a pole of $T(s)$ because there may be pole-zero cancellations. Conditions for every eigenvalue of A to be pole of Σ can be expressed in terms of controllability and observability, to be introduced in the next sections.

3.2 Controllability

In this section, we concentrate on the differential equation of (3.1), hence on the relation between u and x. We investigate to what extent one can influence the state by a suitable choice of the control. For this purpose, we introduce the (at time T) *reachable space* $\mathcal{W}_T$, defined as the space of points x_1 for which there exists a control u such that $x_u(T, 0) = x_1$, i.e. the set of points that can be reached from the origin at time T. It follows from the linearity of the differential equation that $\mathcal{W}_T$ is a linear subspace of $\mathcal{X}$. In fact, (3.2) implies:

$$\mathcal{W}_T = \left\{ \int_0^T e^{A(T-\tau)} Bu(\tau)\, \mathrm{d}\tau \;\middle|\; u \in \mathbf{U} \right\}. \tag{3.5}$$

We call system Σ *reachable at time* T if every point can be reached from the origin, i.e., if $\mathcal{W}_T = \mathcal{X}$. It follows from (3.2) that if the system is reachable, every point can be reached from every point, because the condition for the point x_1 to be reachable from x_0 is

$$x_1 - e^{AT} x_0 \in \mathcal{W}_T.$$

The property that every point is reachable from any point in a given time interval $[0, T]$ is called *controllability (at T)*. Finally we have the concept of *null controllability*, i.e., the possibility to reach the origin from an arbitrary initial point. According to (3.2), for a point x_0 to be null controllable at T we must have

$$e^{AT} x_0 + \int_0^T e^{A(T-\tau)} Bu(\tau)\, \mathrm{d}\tau = 0$$

for some $u \in \mathbf{U}$. We observe that x_0 is null controllable at T (by the control u) if and only if $-e^{AT} x_0$ is reachable at T (by the control u). Since e^{AT} is invertible we see that Σ is null controllable if and only if Σ is reachable at T. Henceforth, we refer to the equivalent properties reachability, controllability, null controllability simply as controllability (at T). It should be remarked that the equivalence of these concepts breaks down in other situations, e.g. for discrete-time systems. Now we intend to obtain an explicit expression for the space $\mathcal{W}_T$ and based on this an explicit condition for controllability. This is provided by the following result.

Theorem 3.1 *Let η be an n-dimensional row vector and $T > 0$. Then the following statements are equivalent:*

(i) $\eta \perp \mathcal{W}_T$ *(i.e., $\eta x = 0$ for all $x \in \mathcal{W}_T$),*

(ii) $\eta e^{tA} B = 0$ *for* $0 \leqslant t \leqslant T$,

(iii) $\eta A^k B = 0$ *for* $k = 0, 1, 2, \ldots$

(iv) $\eta \begin{pmatrix} B & AB & \cdots & A^{n-1}B \end{pmatrix} = 0.$

Here, for given $n \times m$ matrices $B_1, B_2, \ldots, B_n$, the expression

$$\begin{pmatrix} B_1 & B_2 & \cdots & B_n \end{pmatrix}$$

denotes the $n \times nm$ block matrix we obtain by putting the matrices next to each other.

Proof : (i) ⇔ (ii) If $\eta \perp \mathcal{W}_T$, then (see (3.5)):

$$\int_0^T \eta e^{A(T-\tau)} Bu(\tau)\,\mathrm{d}\tau = 0 \tag{3.6}$$

for every $u \in \mathbf{U}$. Choosing $u(t) = B^{\mathrm{T}} e^{A^{\mathrm{T}}(T-t)} \eta^{\mathrm{T}}$ for $0 \leqslant t \leqslant T$, we obtain

$$\int_0^T \|\eta e^{A(T-\tau)} B\|^2 \mathrm{d}\tau = 0,$$

from which (ii) follows. Conversely, assume that (ii) holds. Then (3.6) holds and hence (i) follows.

(ii) ⇔ (iii) This is obtained by power series expansion of e^{At}.

(iii) ⇒ (iv) This follows immediately from the evaluation of the vector-matrix product.

(iv) ⇒ (iii) This implication is based on the Cayley-Hamilton Theorem (see 2.12). According to this theorem, A^n is a linear combination of $I, A, \ldots, A^{n-1}$. By induction, it follows that A^k $(k > n)$ is a linear combination of $I, A, \ldots, A^{n-1}$ as well. Therefore, $\eta A^k B = 0$ for $k = 0, 1, \ldots, n-1$ implies that $\eta A^k B = 0$ for all $k \in \mathbb{N}$. ■

Corollary 3.2 *$\mathcal{W}_T$ is independent of T for $T > 0$. Specifically:*

$$\mathcal{W}_T = \operatorname{im}\begin{pmatrix} B & AB & \cdots & A^{n-1}B \end{pmatrix}.$$

Because of the above corollary we will often refer to $\mathcal{W}$ instead of $\mathcal{W}_T$.

Corollary 3.3 *$\mathcal{W}$ is the* A-invariant subspace generated *by $\mathcal{B} := \operatorname{im} B$, i.e., $\mathcal{W}$ is the smallest A-invariant subspace containing $\mathcal{B}$. Explicitly, $\mathcal{W}$ is A-invariant, $\mathcal{B} \subset \mathcal{W}$, and any A-invariant subspace $\mathcal{L}$ satisfying $\mathcal{B} \subset \mathcal{L}$ also satisfies $\mathcal{W} \subset \mathcal{L}$. We will denote the A-invariant subspace generated by $\mathcal{B}$ by $\langle A \mid \mathcal{B}\rangle$, so that we can write $\mathcal{W} = \langle A \mid \mathcal{B}\rangle$. For the space $\langle A \mid \mathcal{B}\rangle$ we have the following explicit formula*

$$\langle A \mid \mathcal{B}\rangle = \mathcal{B} + A\mathcal{B} + \cdots + A^{n-1}\mathcal{B},$$

Corollary 3.4 *The following statements are equivalent:*

(i) There exists $T > 0$ such that system Σ is controllable at T,

(*ii*) $\langle A \mid \mathcal{B} \rangle = \mathcal{X}$,

(*iii*) $\operatorname{rank}\begin{pmatrix} B & AB & \cdots & A^{n-1}B \end{pmatrix} = n$,

(*iv*) *the system* Σ *is controllable at* T *for all* $T > 0$.

We will say that the matrix pair (A, B) is *controllable* if one of these equivalent conditions is satisfied.

Example 3.5 Let A and B be defined by

$$A := \begin{pmatrix} -2 & -6 \\ 2 & 5 \end{pmatrix}, \quad B := \begin{pmatrix} -3 \\ 2 \end{pmatrix}.$$

Then $(B \quad AB) = \begin{pmatrix} -3 & -6 \\ 2 & 4 \end{pmatrix}$, $\operatorname{rank}(B \quad AB) = 1$, and consequently, (A, B) is not controllable. The reachable space is the span of $(B \quad AB)$, i.e., the line with parameter representation $x = \alpha(-3, 2)^{\mathrm{T}}$, equivalently, the line given by the equation $2x_1 + 3x_2 = 0$. This can also be seen as follows: When introducing $z := 2x_1 + 3x_2$, we see immediately that $\dot{z} = z$. Hence, if $z(0) = 0$, which is the case if $x(0) = 0$, we must have $z(t) = 0$ for all $t \geqslant 0$.

3.3 Observability

In this section, we include the second equation of (3.1), $y = Cx + Du$, in our considerations. Specifically, we want to investigate to what extent it is possible to reconstruct the state x when the input u and the output y are known. The motivation is that we often can measure the output and prescribe (and hence know) the input, whereas the state variable is hidden.

Definition 3.6 *Two states* x_0 *and* x_1 *in* $\mathcal{X}$ *are called* indistinguishable *on the interval* $[0, T]$ *if for any input* u *we have* $y_u(t, x_0) = y_u(t, x_1)$ *for* $0 \leqslant t \leqslant T$.

Hence x_0 and x_1 are indistinguishable if they give rise to the same output values for every input u. According to (3.3), for x_0 and x_1 to be indistinguishable on $[0, T]$ we must have that

$$Ce^{At}x_0 + \int_0^t K(t-\tau)u(\tau)\,\mathrm{d}\tau + Du(t) = Ce^{At}x_1 + \int_0^t K(t-\tau)u(\tau)\,\mathrm{d}\tau + Du(t)$$

for $0 \leqslant t \leqslant T$ and for any input function u. We notice that the input function is of no relevance to distinguishability, i.e. if one u is able to distinguish between two states, then any input is. In fact, x_0 and x_1 are indistinguishable if and only if $Ce^{At}x_0 = Ce^{At}x_1 \quad (0 \leqslant t \leqslant T)$. Obviously, x_0 and x_1 are indistinguishable if and only if $v := x_0 - x_1$ and 0 are indistinguishable. We apply theorem 3.1 with $\eta = v^{\mathrm{T}}$ and we transpose the equations. Then it follows that $Ce^{At}x_0 = Ce^{At}x_1 \quad (0 \leqslant t \leqslant T)$ if and only if $Ce^{At}v = 0 \quad (0 \leqslant t \leqslant T)$ and hence if and only if $CA^k v = 0 \quad (k =$

$0, 1, 2, \ldots$). Using the Cayley-Hamilton theorem, we can show that we only need to consider the first n terms, i.e.

$$\begin{pmatrix} C \\ CA \\ CA^2 \\ \vdots \\ CA^{n-1} \end{pmatrix} v = 0. \tag{3.7}$$

As a consequence, the distinguishability of two vectors does not depend on T. The space of vectors v for which (3.7) holds is denoted $\langle \ker C \mid A \rangle$, and called the *unobservable subspace*. It is equivalently characterized as the intersection of the spaces $\ker CA^k$ for $k = 0, \ldots, n-1$, i.e.

$$\langle \ker C \mid A \rangle = \bigcap_{k=0}^{n-1} \ker CA^k.$$

One can also say: $\langle \ker C \mid A \rangle$ is the largest A-invariant subspace contained in $\ker C$. Still another characterization is: '$v \in \langle \ker C \mid A \rangle$ if and only if $y_0(t, v)$ is identically zero', where the subscript refers to the zero input.

Definition 3.7 *System Σ is called* observable *if any two distinct states are distinguishable.*

The previous considerations immediately lead to the following result:

Theorem 3.8 *The following statements are equivalent:*

(*i*) *system Σ is observable,*

(*ii*) *every state is distinguishable from the origin,*

(*iii*) $\langle \ker C \mid A \rangle = 0$,

(*iv*) $Ce^{At}v = 0 \;\; (0 \leqslant t \leqslant T) \Longrightarrow v = 0$,

(*v*) $\operatorname{rank} \begin{pmatrix} C \\ CA \\ CA^2 \\ \vdots \\ CA^{n-1} \end{pmatrix} = n.$

Since observability is completely determined by the matrix pair (C, A), we will often say '(C, A) is observable' instead of 'system Σ is observable'.

There is a remarkable relation between controllability and observability properties, which is referred to as *duality*. This property is most conspicuous from the conditions in corollary 3.4 (iii) and theorem 3.8 (v), respectively. Specifically, (C, A) is observable if and only if $(A^{\mathrm{T}}, C^{\mathrm{T}})$ is controllable. As a consequence of duality, many theorems on controllability can be translated to theorems on observability, and vice versa by mere transposition of matrices. Duality will play an important role in the remainder of this book.

Example 3.9 Let

$$A := \begin{pmatrix} -11 & 3 \\ -3 & -5 \end{pmatrix}, \quad B := \begin{pmatrix} 1 \\ 1 \end{pmatrix}, \quad C := \begin{pmatrix} 1 & -1 \end{pmatrix},$$

Then,

$$\operatorname{rank} \begin{pmatrix} C \\ CA \end{pmatrix} = \operatorname{rank} \begin{pmatrix} 1 & -1 \\ -8 & 8 \end{pmatrix} = 1,$$

hence (C, A) is not observable. Notice that if $v \in \langle \ker C \mid A \rangle$ and $u = 0$ identically then $y = 0$ identically. In this example $\langle \ker C \mid A \rangle$ is the span of $(1, 1)^{\mathrm{T}}$.

3.4 Basis transformations

A choice of a basis in $\mathcal{X}$ induces matrix representations for A, B, C and D. Very often, one chooses the standard basis for $\mathcal{X}$ and identifies the triple of linear maps with their matrices. A change of basis results in a modification of the matrix representations. Here we want to investigate this in more detail.

Let S be a basis-transformation matrix. Then the relation between the 'old' and the 'new' state vector is: $x = S\bar{x}$. When we substitute this into the equations of Σ, we obtain:

$$\begin{aligned} \dot{\bar{x}}(t) &= \bar{A}\bar{x}(t) + \bar{B}u(t), \\ y(t) &= \bar{C}\bar{x}(t) + Du(t), \end{aligned} \tag{3.8}$$

where

$$\bar{A} := S^{-1}AS, \ \bar{B} := S^{-1}B, \ \bar{C} := CS. \tag{3.9}$$

Obviously, D is not modified by such a transformation. As a result, the matrix D is irrelevant in the considerations of this section and we will assume D to be absent.

We will say that the systems (C, A, B) and $(\bar{C}, \bar{A}, \bar{B})$ are *isomorphic* if there exists a nonsingular matrix S for which (3.9) holds. The following result shows that isomorphic systems have similar properties.

Theorem 3.10 *Let (C, A, B) and $(\bar{C}, \bar{A}, \bar{B})$ be isomorphic. Then we have:*

(i) *$(\bar{A}, \bar{B})$ is controllable if and only if (A, B) is controllable,*

(ii) *$(\bar{C}, \bar{A})$ is observable if and only if (C, A) is observable,*

(*iii*) $\bar{C}e^{\bar{A}t}\bar{B} = Ce^{At}B$ *for all* t,

(*iv*) $\bar{C}(sI - \bar{A})^{-1}\bar{B} = C(sI - A)^{-1}B$ *as rational functions.*

(*v*) $\sigma(\bar{A}) = \sigma(A)$.

We see that isomorphic systems have the same controllability and observability properties, the same impulse response and transfer matrix (i.e., the same i/o behavior) and the same eigenvalues.

Proof : (i) Follows from

$$\begin{pmatrix}\bar{B} & \bar{A}\bar{B} & \cdots & \bar{A}^{n-1}\bar{B}\end{pmatrix} = S^{-1}\begin{pmatrix}B & AB & \cdots & A^{n-1}B\end{pmatrix}$$

so that

$$\text{rank}\begin{pmatrix}\bar{B} & \bar{A}\bar{B} & \cdots & \bar{A}^{n-1}\bar{B}\end{pmatrix} = \text{rank}\begin{pmatrix}B & AB & \cdots & A^{n-1}B\end{pmatrix}.$$

(ii) is dual to (i).

(iii) and (iv) follow by straightforward substitution of (3.9) into the formulas.

(v) is a well-known result from linear algebra. ■

Basis transformations can be used to simplify the system in such a way that various problems can be solved more easily. For instance, one might want to choose S such that A has a diagonal form (if possible). Such transformations are well known in linear algebra. For system theory however, different transformations have shown to be useful. These will not be discussed here in detail. We restrict ourselves to the derivation of a transformation that displays the controllability properties of Σ very clearly. Assume that Σ is not controllable and that $B \neq 0$. Then the reachable space $\mathcal{W} = \langle A \mid \text{im}\, B\rangle$ is not zero and not equal to $\mathcal{X}$. Hence, it is a proper subspace of dimension, say k. Let $q_1, \ldots, q_n$ be a basis of $\mathcal{X}$ adapted to $\mathcal{W}$. The column of the state variable x with respect to this basis will be denoted $\bar{x}$. We can decompose this column into

$$\bar{x} = \begin{pmatrix}x_1\\ x_2\end{pmatrix},$$

where x_1 has k and x_2 has $n-k$ components. Hence $x \in \mathcal{W}$ if and only if $x_2 = 0$. We decompose the new matrices $\bar{A}$ and $\bar{B}$ accordingly ($\bar{C}$ is irrelevant in this consideration):

$$\bar{A} = \begin{pmatrix}A_{11} & A_{12}\\ 0 & A_{22}\end{pmatrix}, \quad \bar{B} = \begin{pmatrix}B_1\\ 0\end{pmatrix}.$$

Here $B_2 = 0$ because $\text{im}\, B \subset \mathcal{W}$ and $A_{21} = 0$ because $\mathcal{W}$ is A-invariant (see (2.4) and (2.5)). It follows that x_1 and x_2 satisfy the following system of differential equations:

$$\begin{aligned}\dot{x}_1 &= A_{11}x_1 + A_{12}x_2 + B_1u,\\ \dot{x}_2 &= A_{22}x_2.\end{aligned} \tag{3.10}$$

We observe that the variable x_2 is completely uncontrollable. The input does not have any effect on it. On the other hand, the equation for x_1 is completely controllable, since $\mathcal{W}$ is the reachable subspace. Thus we have the following result:

Theorem 3.11 *Let (A, B) be not controllable and B not zero. Then there exists an invertible matrix S such that the pair $(\bar{A}, \bar{B})$ given by $\bar{A} := S^{-1}AS$, $\bar{B} := S^{-1}B$, has the form*

$$\bar{A} = \begin{pmatrix} A_{11} & A_{12} \\ 0 & A_{22} \end{pmatrix}, \quad \bar{B} = \begin{pmatrix} B_1 \\ 0 \end{pmatrix},$$

where (A_{11}, B_1) is controllable.

The above theorem basically splits the system described by the differential equation $\dot{x} = Ax + Bu$ into two subsystems as illustrated in figure 3.1.

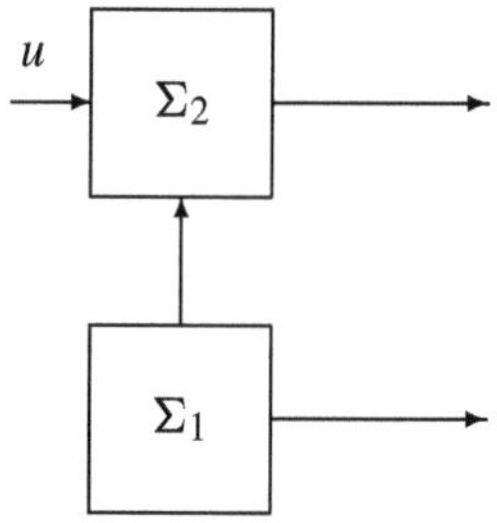

Figure 3.1

The result can also be formulated in a more geometric fashion, using subspaces and quotient spaces as introduced in chapter 2. Since $\mathcal{W}$ is A-invariant, it is possible to consider the restriction $\underline{A}$ of A to $\mathcal{W}$. We also introduce $\underline{B}$, the codomain restriction of B to $\mathcal{W}$ and $\underline{C}$, the (domain) restriction of C to $\mathcal{W}$. If we assume that $x_0 \in \mathcal{W}$, we have that $x(t) \in \mathcal{W}$ for all $t \geqslant 0$. Then we obtain the equations

$$\begin{aligned} \dot{x}(t) &= \underline{A}x(t) + \underline{B}u(t), \\ y(t) &= \underline{C}x(t), \end{aligned}$$

which represents a controllable system with state space $\mathcal{W}$. On the other hand, we can introduce the quotient space $\bar{\mathcal{X}} := \mathcal{X}/\mathcal{W}$ and the corresponding state vector $\bar{x} = \Pi x$ (see section 2.4). They give rise to the quotient maps

$$\bar{A} : \bar{\mathcal{X}} \to \bar{\mathcal{X}}, \quad \bar{B} : \mathcal{U} \to \bar{\mathcal{X}},$$

defined by the relations $\bar{A}\Pi = \Pi A$, $\bar{B} = \Pi B$. As $\bar{B} = 0$, the equation for this quotient system is:

$$\dot{\bar{x}} = \bar{A}\bar{x}, \tag{3.11}$$

which is obviously completely uncontrollable. When comparing with the matrix decomposition formulation, we note that A_{11} corresponds to $\underline{A}$, and A_{22} to $\bar{A}$. In particular, $\sigma(A_{11}) = \sigma(\underline{A})$ and $\sigma(A_{22}) = \sigma(\bar{A})$. It is customary to call the system for x_1

given by the equations of (3.10) (under the assumption that $x_2 = 0$) the *controllable subsystem* of Σ. Similarly one would like to call the part describing the equation for x_2 the uncontrollable subsystem. Unfortunately, this is not really possible, since this part of the system is not unique, but depends on a particular choice of a basis of the state space. It is possible, however, to talk about the *uncontrollable quotient system*, i.e., the system given by (3.11).

Dual results can be given for observability (see exercise 3.7).

3.5 Controllable and observable eigenvalues

We consider again the system (3.1).

Definition 3.12 *An eigenvalue* λ *of* A *is called* (A, B)-controllable *if*

$$\operatorname{rank}\begin{pmatrix} A - \lambda I & B \end{pmatrix} = n.$$

The eigenvalue λ *is called* (C, A)-observable *if*

$$\operatorname{rank}\begin{pmatrix} A - \lambda I \\ C \end{pmatrix} = n.$$

When there is no danger of confusion we omit the prefix (A, B), and we write controllable instead of (A, B)-controllable, and similarly for observability. Note that instead of the rank condition for controllability we can write:

$$\text{‘for every row vector } \eta \text{, we have: } \eta A = \lambda \eta, \ \ \eta B = 0 \implies \eta = 0.\text{’}$$

I.e., there does not exist a left eigenvector of A corresponding to the eigenvalue λ which is orthogonal to $\operatorname{im} B$. Still another interpretation can be given in terms of subspaces:

$$(A - \lambda I)\mathcal{X} + B\mathcal{U} = \mathcal{X}, \tag{3.12}$$

or, equivalently,

$$\operatorname{im}(A - \lambda I) + \operatorname{im} B = \mathcal{X}.$$

Dual interpretations are possible for the observability of an eigenvalue. E.g., a condition for observability of an eigenvalue is

$$\text{‘for every vector } p \text{ we have: } Ap = \lambda p, \ \ Cp = 0 \implies p = 0\text{’},$$

equivalently,

$$\ker(A - \lambda I) \cap \ker C = 0.$$

The definitions will be motivated in theorem 3.15, but first we give a controllability and an observability condition in terms of the new concepts:

Theorem 3.13

(i) (A, B) is controllable if and only if every eigenvalue of A is (A, B)-controllable, i.e. if and only if

$$\forall \lambda \in \sigma(A) \quad \operatorname{rank}\begin{pmatrix} A - \lambda I & B \end{pmatrix} = n. \tag{3.13}$$

(ii) (C, A) is observable if and only if every eigenvalue of A is (C, A)-observable, i.e. if and only if

$$\forall \lambda \in \sigma(A) \quad \operatorname{rank}\begin{pmatrix} A - \lambda I \\ C \end{pmatrix} = n. \tag{3.14}$$

Proof : We prove the second statement. The first result is then obtained by dualization. Suppose that (C, A) is not observable. Then $\mathcal{N} := \langle \ker C \mid A \rangle$ is a nontrivial A-invariant subspace. Hence, the map A restricted to $\mathcal{N}$ has an eigenvalue λ and a corresponding eigenvector $v \in \mathcal{N}$. Since $\mathcal{N} \subset \ker C$, we have, in addition to $Av = \lambda v$ that $Cv = 0$, so that (3.14) is violated. Conversely, if condition (3.14) is not satisfied, there exists an eigenvalue λ of A and a corresponding eigenvector v such that $Av = \lambda v$ and $Cv = 0$. But then we have $A^k v = \lambda^k v$ and hence $CA^k v = 0$ for all $k \geqslant 0$, contradicting the observability of (C, A). ■

In order to give a justification of the terms controllable and observable eigenvalues, we are going to perform a basis transformation in the state space. First we note that controllability and observability of eigenvalues are invariant under such transformations.

Lemma 3.14 *If $(\bar{C}, \bar{A}, \bar{B})$ and (C, A, B) are isomorphic, then λ is $(\bar{A}, \bar{B})$-controllable if and only if λ is (A, B)-controllable. Dually, λ is $(\bar{C}, \bar{A})$-observable if and only if λ is (C, A)-observable.*

Proof : This is a consequence of

$$\begin{pmatrix} \bar{A} - \lambda I & \bar{B} \end{pmatrix} = S^{-1} \begin{pmatrix} A - \lambda I & B \end{pmatrix} \begin{pmatrix} S & 0 \\ 0 & I \end{pmatrix}$$

and its dual. ■

Theorem 3.15 *Let (A, B) be given and let $(\bar{A}, \bar{B})$ be isomorphic to (A, B) and of the form described in theorem 3.11. Then $\lambda \in \sigma(A)$ is a controllable eigenvalue of (A, B) if and only if $\lambda \notin \sigma(A_{22})$ $(= \sigma(A \mid \mathcal{X}/\langle A \mid \operatorname{im} B \rangle))$.*

Proof : Because of the invariance of the controllability of eigenvalues under isomorphism, we may assume that $(C, A, B) = (\bar{C}, \bar{A}, \bar{B})$. Assume that $\lambda \in \sigma(A_{22})$. Let $\eta_2 \neq 0$ be a row vector such that $\eta_2 A_{22} = \lambda \eta_2$. Then the vector $\eta := (0 \;\; \eta_2)$ satisfies $\eta A = \lambda \eta$, $\eta B = 0$ and $\eta \neq 0$. Hence, λ is not (A, B)-controllable. Conversely, if there exists a row vector $\eta \neq 0$ such that $\eta A = \lambda \eta$ and $\eta B = 0$, we decompose η into $(\eta_1 \;\; \eta_2)$, according to the decomposition of A and B. Then

$$\begin{aligned}
&\eta_1 A_{11} = \lambda \eta_1, \\
&\eta_1 A_{12} + \eta_2 A_{22} = \lambda \eta_2, \\
&\eta_1 B_1 = 0.
\end{aligned}$$

Since (A_{11}, B_1) is controllable, this implies that $\eta_1 = 0$. Hence $\eta_2 \neq 0$ and $\eta_2 A_{22} = \lambda \eta_2$, so that $\lambda \in \sigma(A_{22})$. ■

A dual result can be given for observable eigenvalues.

3.6 Single-variable systems

In this section, we consider the situation where $m = p = 1$, i.e., where there is one input and one output. Systems for which this is the case are called *single-variable* (or SISO, or monovariable) systems. When one wants to emphasize that the system under consideration is not single variable, one speaks of *multivariable* (or MIMO) systems. Single-variable systems arise when one has a higher order differential equation describing the relation between scalar inputs and outputs. For instance,

$$y^{(n)} + a_1 y^{(n-1)} + \cdots + a_n y = u. \tag{3.15}$$

Introducing the variables $x_1 := y$, $x_2 := \dot{y}$, $x_3 := y^{(2)}, \ldots, x_n := y^{(n-1)}$, one can rewrite (3.15) in matrix-vector form:

$$\begin{aligned}
\dot{x} &= Ax + bu, \\
y &= cx,
\end{aligned} \tag{3.16}$$

where

$$A := \begin{pmatrix} 0 & 1 & 0 & \cdots & 0 \\ 0 & 0 & 1 & \ddots & \vdots \\ \vdots & \vdots & \ddots & \ddots & 0 \\ 0 & 0 & \cdots & 0 & 1 \\ -a_n & -a_{n-1} & \cdots & -a_2 & -a_1 \end{pmatrix}, \quad b := \begin{pmatrix} 0 \\ \vdots \\ 0 \\ 0 \\ 1 \end{pmatrix}, \tag{3.17}$$

$$c := \begin{pmatrix} 1 & 0 & \cdots & 0 & 0 \end{pmatrix}.$$

A system of this form is called a system in *control canonical form*. The matrix A is said to be a *companion matrix*. We have written lower case letters for the B and C

matrices to emphasize the fact that they are vectors. Somewhat more general is the following equation.

$$y^{(n)} + a_1 y^{(n-1)} + \cdots + a_n y = c_n u^{(n-1)} + \cdots + c_1 u. \tag{3.18}$$

Such a system will not fit in our framework, because the input function must satisfy differentiability conditions that conflict with the slicing property (see section 3.1). However, if we introduce a variable x_1 satisfying

$$x_1^{(n)} + a_1 x_1^{(n-1)} + \cdots + a_n x_1 = u, \tag{3.19}$$

the function $y := c_n x_1^{(n-1)} + \cdots + c_1 x_1$ satisfies (3.18). Introducing the rest of the state variables $x_2, \ldots, x_n$ as before, we obtain the system (3.16) with

$$A = \begin{pmatrix} 0 & 1 & 0 & \cdots & 0 \\ 0 & 0 & 1 & \ddots & \vdots \\ \vdots & \vdots & \ddots & \ddots & 0 \\ 0 & 0 & \cdots & 0 & 1 \\ -a_n & -a_{n-1} & \cdots & -a_2 & -a_1 \end{pmatrix}, \qquad b := \begin{pmatrix} 0 \\ \vdots \\ 0 \\ 0 \\ 1 \end{pmatrix}, \tag{3.20}$$

$$c = \begin{pmatrix} c_1 & c_2 & \cdots & c_{n-1} & c_n \end{pmatrix}.$$

Remark 3.16 If y satisfies an equation of the form

$$y^{(n)} + a_1 y^{(n-1)} + \cdots + a_n y = d_0 u^{(n)} + \cdots + d_n u, \tag{3.21}$$

we introduce the variable $z := y - d_0 u$ and notice that z satisfies (3.18) with

$$c_n := d_1 - d_0 a_1, \ldots, c_1 := d_n - d_0 a_n.$$

Hence, equation (3.21) corresponds to the system $\Sigma_s := (A, b, c, d)$ where (c, A, b) is given by (3.20) and $d := d_0$. In the sequel of this section, we will assume that we deal with equation (3.18).

We are now going to investigate the controllability and observability properties of the system Σ_s given by (3.20). First of all, it is easily seen that (A, b) is controllable. In fact,

$$\begin{pmatrix} A - \lambda I & b \end{pmatrix} = \begin{pmatrix} -\lambda & 1 & 0 & \cdots & 0 & 0 \\ 0 & -\lambda & 1 & \ddots & \vdots & \vdots \\ \vdots & \ddots & \ddots & \ddots & 0 & 0 \\ 0 & \cdots & 0 & -\lambda & 1 & 0 \\ -a_n & \cdots & -a_3 & -a_2 & -a_1 - \lambda & 1 \end{pmatrix}.$$

Omitting the first column, we get a square matrix which is obviously nonsingular for all $\lambda \in \sigma(A)$. The question of observability is somewhat more subtle. Assume that

$\lambda \in \sigma(A)$ and that v is a corresponding eigenvector. Writing $v = (v_1, \ldots, v_n)^{\mathrm{T}}$ and substituting (3.20) into the equation $Av = \lambda v$, we find

$$\begin{aligned} \lambda v_1 &= v_2 \\ &\vdots \\ \lambda v_{n-1} &= v_n \\ \lambda v_n &= -a_n v_1 - a_{n-1} v_2 - \cdots - a_1 v_n. \end{aligned}$$

Hence $v_k = \lambda^{k-1} v_1$ for $k = 1, \ldots, n$, and $q(\lambda) v_1 = 0$, where

$$q(s) := s^n + a_1 s^{n-1} + \cdots + a_n. \tag{3.22}$$

Obviously, $v_1 \neq 0$, since otherwise the vector v would be zero. By homogeneity, we may assume $v_1 = 1$, so that $v_k = \lambda^{k-1}$ and $q(\lambda) = 0$. Conversely, if for some $\lambda \in \mathbb{C}$ we have $q(\lambda) = 0$, then $\lambda \in \sigma(A)$ and $v := (1, \lambda, \ldots, \lambda^{n-1})^{\mathrm{T}}$ is a corresponding eigenvector. (Note that this eigenvector is unique apart from scalar multiplication). Assume now that λ is not observable. Then there exists a nonzero vector v such that $Av = \lambda v$, $Cv = 0$. According to the previous calculations, this means that $q(\lambda) = 0$ and, in addition, $p(\lambda) = 0$, where

$$p(s) := c_n s^{n-1} + \cdots + c_1. \tag{3.23}$$

We conclude that the unobservable eigenvalues are the common zeros of $p(s)$ and $q(s)$. Consequently, (c, A) is observable if and only if $p(s)$ and $q(s)$ are coprime. The polynomials p and q are fundamental for the system. In fact, when we apply the Laplace transformation to equation (3.18), and we assume that we have homogeneous initial values, we get

$$q(s)\hat{y}(s) = p(s)\hat{u}(s).$$

This implies that $T(s) := p(s)/q(s)$ is the transfer function of Σ_s, i.e., $c(sI - A)^{-1}b = p(s)/q(s)$. This equality implies that $q(s)$ is the characteristic polynomial of A. Indeed, it is known that $(sI - A)^{-1} = N(s)/\psi(s)$, where the numerator polynomial $N(s)$ is the adjoint matrix of $sI - A$ and $\psi(s) := \det(sI - A)$ (see section 2.6). It follows that $T(s) = \phi(s)/\psi(s)$ for some polynomial ϕ. If we take the case of the system given by (3.17), we have also $T(s) = 1/q(s)$. Hence $q(s)\phi(s) = \psi(s)$. Since both $q(s)$ and $\psi(s)$ are monic polynomials of degree n, this implies that $q(s) = \psi(s)$. We formulate the results about Σ_s in the following:

Theorem 3.17 *Consider the system Σ_s given by (3.20) and define the polynomials p and q by (3.22) and (3.23). Then*

(i) Σ_s is controllable,

(ii) Σ_s is observable if and only if $p(s)$ and $q(s)$ are coprime,

(iii) $q(s)$ is the characteristic polynomial of A,

(iv) *$p(s)/q(s)$ is the transfer function of Σ_s.*

Note that Σ_s is observable if and only if $p(s)/q(s)$ is an irreducible representation of the transfer function, equivalently, if and only if each eigenvalue of A is a pole of the transfer function with the same multiplicity. In case of unobservability, we have pole-zero cancellation. The poles that cancel are exactly the unobservable eigenvalues.

If $\Sigma = (c, A, b)$ is a single-variable system isomorphic to Σ_s given by (3.20) then Σ is obviously controllable. The following result states the converse of this observation. (We omit c, because it is not relevant.)

Theorem 3.18 *Every controllable single-variable system (A, b) is isomorphic to a system of the form (3.20).*

Proof : Let $q(s) := s^n + a_1 s^{n-1} + \cdots + a_n$ be the characteristic polynomial of A. Because (A, b) is controllable, the vectors $b, Ab, \ldots, A^{n-1}b$ are independent. Therefore, they form a basis of $\mathbb{R}^n$. The matrix of the map A and the column of the vector b with respect to this basis have the following form:

$$\hat{A} := \begin{pmatrix} 0 & 0 & 0 & \cdots & -a_n \\ 1 & \ddots & \ddots & \vdots & \vdots \\ 0 & \ddots & \ddots & \ddots & \vdots \\ \vdots & \cdots & \ddots & 0 & -a_2 \\ 0 & \cdots & \cdots & 1 & -a_1 \end{pmatrix}, \quad \hat{b} := \begin{pmatrix} 1 \\ 0 \\ \vdots \\ \vdots \\ 0 \end{pmatrix}. \tag{3.24}$$

We conclude that the matrix pair (A, b) is transformed into $(\hat{A}, \hat{b})$ by the transformation $S := (b \; Ab \; \cdots A^{n-1}b)$. Note that $(\hat{A}, \hat{b})$ is uniquely determined by the characteristic polynomial of A. We denote the matrix pair in (3.20) by $(\bar{A}, \bar{b})$ and note that it is also transformed to (3.24) by the corresponding transformation. Hence (A, b) and $(\bar{A}, \bar{b})$ are isomorphic. ∎

3.7 Poles, eigenvalues and stability

In the present section, the D matrix is assumed to be zero, because it is irrelevant for the discussion. As noticed in section 2.6, $(sI - A)^{-1}$ is of the form

$$B(s)/\det(sI - A),$$

where $B(s)$ is a matrix of which the entries are polynomials of s. This implies that every pole of $T(s) = C(sI - A)^{-1}B + D$ is an eigenvalue of A, but that the converse is not always true. In the previous section, we saw that for controllable and observable single-variable systems, every eigenvalue is necessarily a pole. Here we show that this statement is also valid for multivariable systems. For this purpose we need the following auxiliary result.

Lemma 3.19 *Given $\Sigma = (C, A, B)$, there exist polynomial matrices $P(s)$, $Q(s)$ and $R(s)$ such that*

$$\begin{pmatrix} C \\ CA \\ \vdots \\ CA^{n-1} \end{pmatrix} (sI - A)^{-1} \begin{pmatrix} B & AB & \cdots & A^{n-1}B \end{pmatrix}$$

$$= P(s)T(s)Q(s) + R(s). \tag{3.25}$$

Proof : For every k, the following holds:

$$A^k(sI - A)^{-1} = s^k(sI - A)^{-1} - P_k(s),$$

for some polynomial matrix $P_k(s)$. In fact,

$$P_k(s) = s^{k-1}I + s^{k-2}A + \cdots + A^{k-1}.$$

Hence

$$CA^k(sI - A)^{-1}B = s^kT(s) + Q_k(s),$$

for some $Q_k(s)$. Since the (i, j) block entry of the left-hand side of (3.25) equals $CA^{i+j-2}(sI - A)^{-1}B$, the desired result follows. ■

If Σ is controllable and observable, there exists a left inverse P^+ of the matrix

$$\begin{pmatrix} C \\ CA \\ \vdots \\ CA^{n-1} \end{pmatrix},$$

and a right inverse Q^+ of

$$\begin{pmatrix} B & AB & \cdots & A^{n-1}B \end{pmatrix}.$$

Hence,

$$(sI - A)^{-1} = U(s)T(s)W(s) + V(s),$$

for certain polynomial matrices $U(s)$, $V(s)$ and $W(s)$. Since the poles of $(sI - A)^{-1}$ are exactly the eigenvalues of A, we have

Theorem 3.20 *If the system $\Sigma = (C, A, B)$ is controllable and observable, every eigenvalue of A is a pole of the transfer function $T(s)$.*

In chapter 2, we introduced the concept of stability of an autonomous system $\dot{x} = Ax$. It was seen that stability is equivalent to the property that all eigenvalues of A have a negative real part. Here we will call the control system $\Sigma = (C, A, B)$ *internally stable* if the corresponding autonomous system is stable, i.e., if $\sigma(A) \subset \mathbb{C}^- := \{s \in \mathbb{C} \mid \Re e\, s < 0\}$. On the other hand, we are also interested in the external asymptotic behavior of the system. We will say that Σ is *externally (or BIBO) stable* if for every positive number M, there exists a positive number N such that for every input u, satisfying $\|u(t)\| \leqslant M$ for all $t \geqslant 0$, we have for the corresponding output with zero initial state that $\|y(t)\| \leqslant N$ for all $t \geqslant 0$ (BIBO stands for *Bounded Input, Bounded Output*). The following result gives a condition for external stability in terms of the impulse response and the transfer function.

Theorem 3.21 *Let the system Σ have impulse response matrix $K(t)$ and transfer function $T(s)$. Then the following statements are equivalent:*

(*i*) *Σ is externally stable,*

(*ii*) *$\int_0^\infty \|K(t)\|\, dt < \infty$,*

(*iii*) *all poles of $T(s)$ are in $\mathbb{C}^-$.*

Proof : (ii) $\Rightarrow$ (i) Let $\|u(t)\| \leqslant M$ for all $t \geqslant 0$. Then

$$\|y(t)\| \leqslant \int_0^t \|K(t-\tau)\| M \,\mathrm{d}\tau = M \int_0^t \|K(\tau)\| \,\mathrm{d}\tau$$

$$\leqslant N := M \int_0^\infty \|K(t)\| \,\mathrm{d}t,$$

for all $t \geqslant 0$.

(i) $\Rightarrow$ (ii) If Σ is externally stable there exists N such that $\|u(t)\| \leqslant 1$ $(t \geqslant 0)$ implies $\|y(t)\| \leqslant N$ for $t \geqslant 0$. Fix $T > 0$ and a pair (i, j) of indices satisfying $1 \leqslant i \leqslant p$, $1 \leqslant j \leqslant m$. If

$$\begin{array}{lll} u_j(t) & := & \operatorname{sgn} K_{ij}(T-t) \quad (0 \leqslant t \leqslant T), \\ u_j(t) & := & 0 \quad t > T, \\ u_k(t) & := & 0 \quad (k \neq j,\ t \geqslant 0), \end{array}$$

then $\|u(t)\| \leqslant 1$. Therefore, the corresponding output $y(t)$ with initial state 0 satisfies $\|y(t)\| \leqslant N$. For this output we have

$$y_i(T) = \int_0^T K_{ij}(T-t) u_j(t) \,\mathrm{d}t = \int_0^T |K_{ij}(t)| \,\mathrm{d}t.$$

Hence, $\int_0^T |K_{ij}(t)| \,\mathrm{d}t \leqslant N$ for all $T \geqslant 0$ and all pairs of indices (i, j). This implies (ii).

(ii) $\Rightarrow$ (iii) Obviously, (ii) implies that the integral $\int_0^\infty K(t)e^{-st}\,dt$ converges for all s with $\Re\mathrm{e}\, s \geqslant 0$. Hence, $T(s)$ has all its poles in $\mathbb{C}^-$.

(iii) $\Rightarrow$ (ii) One can reconstruct $K(t)$ using partial-fraction decomposition. Since $T(s)$ is a strictly proper rational matrix and $T(s)$ has only poles in $\mathbb{C}^-$, the entries of $K(t)$ are Bohl functions with exponents in $\mathbb{C}^-$. Consequently, (see theorem 2.10) the integral of the absolute value of each term converges. Hence, we have (ii). ∎

Corollary 3.22 *If system Σ is externally stable and $x(0) = 0$, we have that $u(t) \to 0$ $(t \to \infty)$ implies $y(t) \to 0$ $(t \to \infty)$.*

Proof : Suppose that $u(t) \to 0$ $(t \to \infty)$. Define

$$\mu(t) := \sup\{\|u(\tau)\| \mid \tau \geqslant t\}.$$

Then μ is decreasing and $\mu(t) \to 0$ $(t \to \infty)$. We have

$$\begin{aligned}
\|y(t)\| &\leqslant \int_0^t \|K(t-\tau)\|\|u(\tau)\|\,d\tau \\
&\leqslant \mu(0)\int_0^{t/2} \|K(t-\tau)\|\,d\tau + \mu(t/2)\int_{t/2}^t \|K(t-\tau)\|\,d\tau \\
&\leqslant \mu(0)\int_{t/2}^{\infty} \|K(\tau)\|\,d\tau + \mu(t/2)\int_0^\infty \|K(\tau)\|\,d\tau \to 0 \quad (t \to \infty).
\end{aligned}$$

∎

Because of the estimate of the matrix exponential $\|e^{tA}\| \leqslant Le^{-\gamma t}$, for some positive γ and L (see corollary 2.11), and hence $\|K(t)\| \leqslant Me^{-\gamma t}$, it follows from theorem 3.21 that internal stability implies external stability. If the system is controllable and observable, the opposite statement is also true. In fact, in this case, external stability implies that there are no poles of $T(s)$ outside $\mathbb{C}^-$, and because of theorem 3.20 this implies that all eigenvalues of A are in $\mathbb{C}^-$. Hence:

Theorem 3.23 *If Σ is internally stable then Σ is externally stable. Conversely, if Σ is controllable, observable and externally stable then Σ is internally stable.*

These concepts can easily be generalized to the more general stability concept introduced in definition 2.12. The only restriction is that we only can allow Bohl functions as input functions. In this case, it follows from theorem 2.8 that the state and output variable are also Bohl functions. Hence, it is possible to replace internal stability by *internal $\mathbb{C}_g$-stability* and external stability by *external $\mathbb{C}_g$-stability*, the latter concept meaning: 'if the input is a $\mathbb{C}_g$-stable Bohl function then, with initial state $x(0) = 0$, the output is a $\mathbb{C}_g$-stable Bohl function'. Then the analog of theorem 3.23 can easily be shown.

3.8 Liapunov functions

The stability of a system can be investigated with the aid of a Liapunov function. For linear systems, we can restrict ourselves to quadratic Liapunov functions. First we recall some properties of positive symmetric matrices.

Definition 3.24 *A symmetric matrix P is called* positive semidefinite *if $x^T Px \geqslant 0$ for all $x \in \mathbb{R}^n$ and* positive definite *if $x^T Px > 0$ for all nonzero $x \in \mathbb{R}^n$. We use the notation $P \geqslant 0$ and $P > 0$, respectively. The notations $P > Q$ and $\geqslant Q$, where P and Q are symmetric matrices, stand for $P - Q > 0$ and $P - Q \geqslant 0$, respectively.*

The following results are well known from linear algebra:

Lemma 3.25 *Let P be a symmetric matrix. Then*

(i) *$P \geqslant 0 \Leftrightarrow$ There exists a matrix D such that $P = D^T D$.*

(ii) *$P > 0 \Leftrightarrow$ There exists a nonsingular matrix D such that $P = D^T D$.*

Although we only require the positivity property for real vectors, the quadratic form is also positive for complex vectors. Specifically,

Lemma 3.26 *Let P be a symmetric matrix. Then*

(i) *$P \geqslant 0 \Leftrightarrow z^* Pz \geqslant 0$ for all complex vectors z.*

(ii) *$P > 0 \Leftrightarrow z^* Pz > 0$ for all complex vectors $z \neq 0$.*

This result follows by direct computation, or by using lemma 3.25. Also the following result is an easy consequence of lemma 3.25:

Lemma 3.27 *Let P be a positive semidefinite matrix and z a complex vector satisfying $z^* Pz = 0$. Then $z = 0$.*

Now suppose that we are given the differential equation

$$\dot{x} = Ax. \tag{3.26}$$

If P is any symmetric matrix, we consider the function

$$V(x) := x^{\mathrm{T}} Px. \tag{3.27}$$

Then, a simple computation yields

$$\frac{\mathrm{d}}{\mathrm{d}t} V(x) = -x^{\mathrm{T}}(t) Q x(t), \tag{3.28}$$

where

$$Q := -(A^{\mathrm{T}} P + PA). \tag{3.29}$$

If $P > 0$, the function V can serve as a measure for the length of the vector x. If, in addition, $Q > 0$, we see that the value of this function is a decreasing function of t along any trajectory $x(t)$. This suggests that the quantity $V(x(t))$ and hence the vector $x(t)$ tends to zero as $x \to \infty$, and hence that (3.26) is stable.

This turns out to be a correct suggestion. We can even relax the condition $Q > 0$ to $Q = C^{\mathrm{T}}C$, where (C, A) is observable. And we can give a converse of this result.

Theorem 3.28 *Let the system (3.26) be given, and assume that C is an $p \times n$ matrix. Consider the following statements:*

(i) A is a stability matrix.

(ii) (C, A) is observable.

(iii) Equation

$$A^{\mathrm{T}}P + PA = -C^{\mathrm{T}}C \tag{3.30}$$

has a positive definite solution P.

Then any two of these statements imply the third.

Proof :

(i) $\wedge$ (ii) $\to$ (iii): because $\Lambda(A) < 0$, we have $\|e^{At}\| \leqslant Me^{-\gamma t}$ for some positive numbers M and γ (see corollary 2.11). Consequently, the integral

$$P := \int_0^\infty e^{A^{\mathrm{T}}t}C^{\mathrm{T}}Ce^{At}\,\mathrm{d}t \tag{3.31}$$

converges. It is a solution of (3.30), because

$$\begin{aligned} A^{\mathrm{T}}P + PA &= \int_0^\infty \left\{A^{\mathrm{T}}e^{A^{\mathrm{T}}t}C^{\mathrm{T}}Ce^{At} + e^{A^{\mathrm{T}}t}C^{\mathrm{T}}Ce^{At}A\right\}\,\mathrm{d}t = \\ &= \int_0^\infty \frac{\mathrm{d}}{\mathrm{d}t}\left\{e^{A^{\mathrm{T}}t}C^{\mathrm{T}}Ce^{At}\right\}\,\mathrm{d}t = \left[e^{A^{\mathrm{T}}t}C^{\mathrm{T}}Ce^{At}\right]_0^\infty = -C^{\mathrm{T}}C. \end{aligned}$$

Furthermore, $x^{\mathrm{T}}Px = \int_0^\infty x^{\mathrm{T}}e^{A^{\mathrm{T}}t}C^{\mathrm{T}}Ce^{At}x\,\mathrm{d}t = \int_0^\infty |Ce^{At}x|^2\,dt \geqslant 0$ for all x, so that $P \geqslant 0$. In addition, if $x_0^{\mathrm{T}}Px_0 = 0$ for some x_0, then the previous calculation implies that $Ce^{At}x_0 = 0$ for all $t \geqslant 0$. That is, the output of the system $\dot{x} = Ax$, $y = Cx$, $x(0) = x_0$ is identically zero. Because the system is supposed to be observable, it follows that $x_0 = 0$. Hence, $P > 0$. Thus we have shown (iii).

(ii) $\wedge$ (iii) $\to$ (i): Let $\lambda \in \sigma(A)$, and let v be a corresponding (possibly complex) eigenvector. Multiplying (3.30) from the right with v and from the left with v^*, we obtain

$$(2\,\Re\mathrm{e}\,\lambda)v^*Pv = (\lambda + \bar{\lambda})v^*Pv = -|Cv|^2. \tag{3.32}$$

Because $P > 0$ and (C, A) is observable, and therefore $Cv \neq 0$, it follows that $\Re e\, \lambda < 0$. Hence (i).

(iii) $\wedge$ (i) $\to$ (ii): Let $\lambda \in \sigma(A)$, and let $v \neq 0$ be a corresponding eigenvector. We have to show that λ is observable, hence that $Cv \neq 0$. Again we have (3.32). Now we know that $\Re e\, \lambda < 0$ and $P > 0$. This implies $Cv \neq 0$. ■

Equation (3.30) is sometimes referred to as the *Liapunov equation*. Note that the solution of the equation (3.30) is sometimes called the *observability gramian*. Dually, the solution of the equation

$$AP + PA^{\mathrm{T}} = -BB^{\mathrm{T}} \tag{3.33}$$

is often referred to as the *controllability gramian*.

It can be shown that the solution of equation (3.30) is unique if A is stable. As a matter of fact, this is a consequence of Sylvester's theorem (see section 9.3)

In order to investigate stability without computing the eigenvalues of A, we can proceed as follows: Choose an arbitrary positive definite matrix Q, e.g. $Q = I$, investigate whether the linear equation $A^{\mathrm{T}}P + PA = -Q$ has a solution. If no solution exists, or if the solution is not positive definite, the matrix A is not stable.

3.9 The stabilization problem

A large number of problems studied in control theory arise from our desire to modify the characteristic properties of a system by means of feedback controllers. The simplest version of this is the $\mathbb{C}_g$-stabilization problem, which can be described as follows:
'given a stability domain $\mathbb{C}_g$, and a system Σ, described by the equations

$$\begin{aligned} \dot{x}(t) &= Ax(t) + Bu(t), \\ y(t) &= Cx(t), \end{aligned} \tag{3.34}$$

determine a system Γ which takes the output y of Σ as input and produces an output that can be used as input for Σ, in such a way that the interconnection Σ_c of Σ and Γ is internally $\mathbb{C}_g$-stable.'

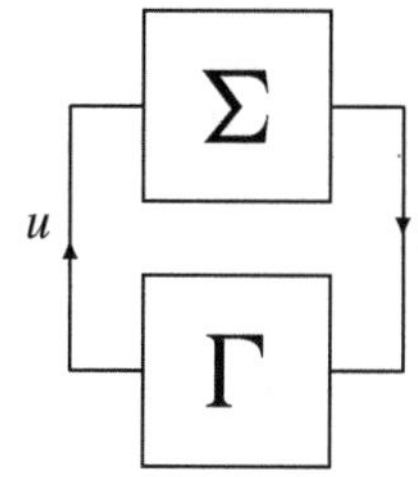

In order to write down the equations of the interconnection Σ_c, we combine the equations for Σ with the equations of the *feedback controller* Γ:

$$\begin{aligned} \dot{w}(t) &= Kw(t) + Ly(t), \\ u(t) &= Mw(t) + Ny(t), \end{aligned} \tag{3.35}$$

Note that we allow a feedthrough term in the output equation. The combination of these equations with (3.34) yields the equations of Σ_c:

$$\begin{pmatrix} \dot{x}(t) \\ \dot{w}(t) \end{pmatrix} = \begin{pmatrix} A + BNC & BM \\ LC & K \end{pmatrix} \begin{pmatrix} x(t) \\ w(t) \end{pmatrix}$$

The interconnection Σ_c is called the *closed-loop system* of Σ and Γ. In this particular case, Σ_c is a homogeneous linear system with coefficient matrix

$$A_e := \begin{pmatrix} A + BNC & BM \\ LC & K \end{pmatrix}. \tag{3.36}$$

(The subscript 'e' stands for 'extended'.) So the $\mathbb{C}_g$-stabilization problem can be reformulated as:
'given the matrix triple (C, A, B), construct a quadruple (K, L, M, N), such that the matrix A_e defined by (3.36) has all its eigenvalues in $\mathbb{C}_g$.'
If the matrix quadruple (K, L, M, N) meets this requirement then the controller Γ given by (3.35) is called *an internally stabilizing controller for* Σ. In the remainder of this chapter we will study under what conditions an internally stabilizing controller exists.

3.10 Stabilization by state feedback

A direct solution of the stabilization problem formulated in the previous section is not so obvious. First we restrict ourselves to the case where the state is available for measurement, i.e, where $C = I$. Also, we only allow a *static feedback* as controller. This means that our controller will have the form: $u = Fx$, for some linear map $F : \mathcal{X} \to \mathcal{U}$. When we apply this feedback, i.e., substitute it into the state equation $\dot{x} = Ax + Bu$, we obtain $\dot{x} = (A + BF)x$. Therefore, the problem of stabilization by state feedback reads:
'given a stability domain $\mathbb{C}_g$, and maps $A : \mathcal{X} \to \mathcal{X}$, $B : \mathcal{U} \to \mathcal{X}$, determine a map $F : \mathcal{X} \to \mathcal{U}$ such that $\sigma(A + BF) \subset \mathbb{C}_g$.'

The solution to this problem is provided by the following celebrated result.

Theorem 3.29 (pole-placement theorem) *Let $A : \mathcal{X} \to \mathcal{X}$, $B : \mathcal{U} \to \mathcal{X}$. Then there exists for every monic polynomial $p(s)$ of degree n a map $F : \mathcal{X} \to \mathcal{U}$ such that $\chi_{A+BF}(s) = p(s)$, if and only if (A, B) is controllable.*

For the definition of χ_A and the concept of 'monic', we refer to section 2.5. Most of the remainder of this section will be devoted to a proof of this result. Then we will apply it to the stabilization problem.

Proof of the necessity in theorem 3.29: : If (A, B) is not controllable, there is an uncontrollable eigenvalue λ. For this eigenvalue, there exists a nonzero row vector η such that $\eta A = \lambda\eta$, $\eta B = 0$. Then we have $\eta(A + BF) = \lambda\eta$ for all F, hence $\lambda \in \sigma(A + BF)$ for all F. So, if $p(s)$ is any polynomial such that $p(\lambda) \neq 0$, there does not exist a feedback F such that $p(s) = \chi_{A+BF}(s)$. ■

We formulate the proof of this result separately:

Corollary 3.30 *If λ is an uncontrollable eigenvalue of (A, B) then λ is an eigenvalue of $A + BF$ for all F.*

Before we start with the proof of the sufficiency part we note that we may perform a basis transformation $(A, B) \mapsto (\bar{A}, \bar{B})$, by $\bar{A} = S^{-1}AS$, $\bar{B} = S^{-1}B$. If we can find the desired feedback $\bar{F}$ for the transformed system, the matrix $F := \bar{F}S^{-1}$ solves the original problem, since $A + BF = S(\bar{A} + \bar{B}\bar{F})S^{-1}$ and hence $\chi_{A+BF} = \chi_{\bar{A}+\bar{B}\bar{F}}$.

Proof of the sufficiency in theorem 3.29 for the case $m = 1$: :

According to theorem 3.18, (A, B) is isomorphic to a system of the form

$$\bar{A} := A := \begin{pmatrix} 0 & 1 & 0 & \cdots & 0 \\ 0 & 0 & 1 & \ddots & \vdots \\ \vdots & \vdots & \ddots & \ddots & 0 \\ 0 & 0 & \cdots & 0 & 1 \\ -a_n & -a_{n-1} & \cdots & -a_2 & -a_1 \end{pmatrix}, \quad \bar{b} := \begin{pmatrix} 0 \\ \vdots \\ 0 \\ 0 \\ 1 \end{pmatrix}. \tag{3.37}$$

(Here $\bar{c}$ is irrelevant.) We try to find a feedback matrix for the pair $(\bar{A}, \bar{b})$. The sought feedback matrix $\bar{F}$ is a $1 \times n$ matrix, i.e., a row vector. We write $\bar{f}$ instead of $\bar{F}$:

$$\bar{f} = \begin{pmatrix} f_n & \cdots & f_1 \end{pmatrix}.$$

It follows that

$$\bar{A} + \bar{b}\bar{f} = \begin{pmatrix} 0 & 1 & 0 & \cdots & 0 \\ 0 & 0 & 1 & \ddots & \vdots \\ \vdots & \vdots & \ddots & \ddots & 0 \\ 0 & 0 & \cdots & 0 & 1 \\ f_n - a_n & f_{n-1} - a_{n-1} & \cdots & f_2 - a_2 & f_1 - a_1 \end{pmatrix}$$

which is again a companion matrix with characteristic polynomial

$$\chi_{\bar{A}+\bar{b}\bar{f}}(s) := s^n + (a_1 - f_1)s^{n-1} + \cdots + (a_n - f_n).$$

If the desired characteristic polynomial is

$$p(s) = s^n + p_1 s^{n-1} + \cdots + p_n,$$

we can choose $f_k := a_k - p_k$ for $k = 1, \dots, n$. ■

In order to prove the multi-input case, we need the following auxiliary result:

Lemma 3.31 *If (A, B) is controllable there exist vectors $u_0, \dots, u_{n-1}$ such that the vectors defined by*

$$x_0 := 0, \; x_{k+1} := Ax_k + Bu_k \;\; (k = 0, \dots, n-1)$$

are independent.

Proof : Since (A, B) is controllable, we must have $B \neq 0$. Hence, u_0 exists such that $Bu_0 \neq 0$. Therefore, x_1 is independent (as a single vector). Now suppose we have that $x_1, \dots, x_k$ are independent. Let $\mathcal{L}$ be the space generated by $x_1, \dots, x_k$. The next step consists of finding a vector u_k such that $x_{k+1} := Ax_k + Bu_k \notin \mathcal{L}$. If this is not possible, we must have that

$$Ax_k + Bu \in \mathcal{L} \tag{3.38}$$

for all $u \in \mathcal{U}$. This implies in particular that

$$Ax_k \in \mathcal{L}. \tag{3.39}$$

Combining (3.38) and (3.39), we find $Bu \in \mathcal{L}$ for all $u \in \mathcal{U}$. Hence $\operatorname{im} B \subset \mathcal{L}$. In addition, $\mathcal{L}$ is A-invariant. This follows from (3.39) and the fact that for $i < k$, we have

$$Ax_i = x_{i+1} - Bu_i \in \mathcal{L}.$$

Since $\mathcal{L}$ is an A-invariant space containing $\operatorname{im} B$, it follows that the inclusion $\langle A \mid \operatorname{im} B\rangle \subset \mathcal{L}$ holds. Since (A, B) is controllable, this means $\mathcal{L} = \mathcal{X}$ and consequently $k = n$. ■

Proof of the sufficiency in theorem 3.29: : We first construct $u_0, \dots, u_{n-1}$ and $x_1, \dots, x_n$ according to the above lemma and we choose u_n arbitrary. There exists a (unique) map $F_0 : \mathcal{X} \to \mathcal{U}$ such that $F_0 x_k = u_k$ for $k = 1, \dots, n$. Furthermore we define $b := Bu_0$. Then

$$x_{k+1} = Ax_k + BF_0 x_k = (A + BF_0)x_k$$

and hence $x_k = (A + BF_0)^{k-1} b$ for $k = 1, \dots, n$. Since $x_1, \dots, x_n$ are independent this implies that $(A + BF_0, b)$ is controllable. The matrix b consists of one column, so that we may apply the theorem for the case $m = 1$. This yields a row vector f such that $\chi_{A+BF_0+bf}(s) = p(s)$. Obviously, $F := F_0 + u_0 f$ satisfies the desired property $\chi_{A+BF} = p$. ■

The pole-placement theorem suggests that the controllability of (A, B) is much stronger than we actually need for stabilization. The following result expresses a necessary and sufficient condition for stabilization.

Theorem 3.32 *Given the matrix pair* (A, B) *and a stability domain* $\mathbb{C}_g$*, there exists a matrix* F *such that* $\sigma(A + BF) \subset \mathbb{C}_g$ *if and only if every eigenvalue* $\lambda \notin \mathbb{C}_g$ *is controllable.*

We note that for this result, the condition $\mathbb{C}_g \cap \mathbb{R} \neq \varnothing$ is needed (see def. 2.12). For real A, B and F we need some condition that guarantees the existence of real polynomials of arbitrary degree with zeros in $\mathbb{C}_g$. If one allows complex matrices F, this condition can be omitted.

Proof : The 'only if' part is an immediate consequence of corollary 3.30. Now suppose that the condition is satisfied. If $B = 0$, all eigenvalues are uncontrollable and the condition implies that $\sigma(A) \subset \mathbb{C}_g$, so that we can take $F = 0$. If (A, B) is controllable we choose any $p(s)$ with zeros in $\mathbb{C}_g$ and apply the pole-placement theorem. Otherwise we can apply theorem 3.11 to transform (A, B) to a system of the form

$$\bar{A} = \begin{pmatrix} A_{11} & A_{12} \\ 0 & A_{22} \end{pmatrix}, \quad \bar{B} = \begin{pmatrix} B_1 \\ 0 \end{pmatrix}.$$

Eigenvalues of A_{22} are uncontrollable and hence $\sigma(A_{22}) \subset \mathbb{C}_g$. We apply the pole-placement theorem to the controllable pair (A_{11}, B_1). Thus we can find a matrix F_1 such that $\sigma(A_{11} + B_1F_1) \subset \mathbb{C}_g$. Then the matrix $\bar{F} := (F_1 \;\; 0)$ satisfies

$$\sigma(\bar{A} + \bar{B}\bar{F}) = \sigma(A_{11} + B_1F_1) \cup \sigma(A_{22}) \subset \mathbb{C}_g.$$ ■

The pair (A, B) is called *stabilizable* (or more explicitly, $\mathbb{C}_g$-stabilizable) if there exists a map F such that $A + BF$ is $\mathbb{C}_g$-stable, equivalently, if all uncontrollable eigenvalues are in $\mathbb{C}_g$. If $\mathbb{C}_g$ has not been specified, our $\mathbb{C}_g$ is understood to be $\mathbb{C}^-$.

We have assumed up to now that the stabilizing control has the form of a constant, linear, static state feedback. Conceivably, if the conditions of theorem 3.32 are not satisfied, stabilization might be possible using more general types of stabilizing controllers, such as nonconstant ($u = F(t)x$), nonlinear ($u = g(x, t)$) or dynamic state feedback of the type described in section 3.9. However, the following result shows that if stabilization is possible at all, it can be achieved by means of state feedback as described in theorem 3.32.

Theorem 3.33 *If there exists for every* x_0 *an input* $u \in \boldsymbol{U}$ *such that the resulting state* x *is a* $\mathbb{C}_g$*-stable Bohl function, then* (A, B) *is* $\mathbb{C}_g$*-stabilizable.*

Proof : Let $\lambda \in \sigma(A)$ be uncontrollable and let $\eta \neq 0$ satisfy $\eta A = \lambda\eta$, $\eta B = 0$. If we choose x_0 such that $\mu := \eta x_0 \neq 0$, the equation $\dot{x} = Ax + Bu$ implies $\frac{\mathrm{d}}{\mathrm{d}t}\eta x = \lambda\eta x$, and hence $\eta x(t) = e^{\lambda t}\mu$. There exists an input u for which x is a $\mathbb{C}_g$-stable Bohl function. Then also ηx must be a $\mathbb{C}_g$-stable Bohl function so that $\lambda \in \mathbb{C}_g$. ■

3.11 State observers

When the state is not available for measurement, one often tries to reconstruct the state using a system, called observer, that takes the input and the output of the original system as inputs and yields an output that is an estimate of the state of the original system. The following diagram illustrates the situation:

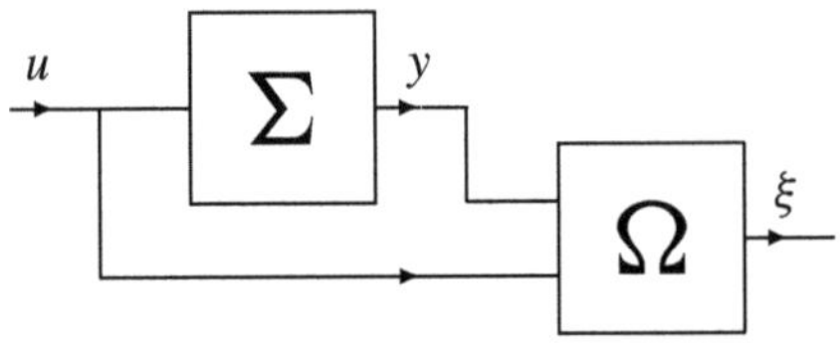

The quantity ξ is supposed to be an estimate in some sense of the state and w is the state variable of the observer. The observer, denoted by Ω has equations of the following form:

$$\begin{aligned} \dot{w}(t) &= Pw(t) + Qu(t) + Ry(t), \\ \xi(t) &= Sw(t). \end{aligned} \tag{3.40}$$

If into these equations, we substitute the equations (3.34) of the system Σ, we obtain:

$$\begin{aligned} \dot{x}(t) &= Ax(t) + Bu(t), \\ \dot{w}(t) &= Pw(t) + Qu(t) + RCx(t), \\ \xi(t) &= Sw(t). \end{aligned}$$

Hence, if we introduce the error of the estimate by $e := \xi - x$, we obtain the following equation for e

$$\dot{e}(t) = SPw(t) + SQu(t) + SRCx(t) - Ax(t) - Bu(t).$$

We substitute $x = Sw - e$ into the right-hand side of the equation in order to eliminate x:

$$\dot{e}(t) = (SP + SRCS - AS)w(t) - (SRC - A)e(t) + (SQ - B)u(t) \tag{3.41}$$

Definition 3.34 *System Ω given by (3.40) is called a* state observer *for Σ if for any pair of initial values x_0, w_0 satisfying $e(0) = 0$, and for arbitrary input function u, we have $e(t) = 0$ for all $t > 0$.*

Hence, once the observer produces an exact estimate for the state at a certain time instant, it will produce an exact estimate for all larger times, irrespective of what the input is.

Definition 3.35 *A state observer Ω is called* stable *if for each pair of initial values x_0, w_0 of Σ and Ω we have $e(t) \to 0$ $(t \to \infty)$.*

Now let Ω be a state observer of the form (3.40). The equation for e, (3.41), can not contain the input u, since otherwise e cannot remain zero if $e(0) = 0$. Consequently, $B = SQ$. Also the coefficient of w has to be zero, because of the same argument. Hence $SP = AS - SRCS$. The equation for e then simplifies to

$$\dot{e}(t) = (A - SRC)e(t).$$

It follows immediately from this equation that $e(0) = 0$ implies $e(t) = 0$ for all $t > 0$. Hence a necessary and sufficient condition for Ω to be a state observer is: $B = SQ$ and $SP = AS - SRCS$. If this condition is satisfied, we can eliminate w from our consideration and use ξ as state variable for the observer. In fact

$$\begin{aligned}\dot{\xi}(t) &= S\dot{w}(t)\\ &= SPw(t) + SQu(t) + SRy(t)\\ &= (A - SRC)\xi(t) + Bu(t) + SRy(t).\end{aligned}$$

Hence, ξ satisfies a linear differential equation. If S is a nonsingular matrix, the equations for w and ξ are isomorphic, i.e., obtained from each other by a basis transformation in the state space of Ω. Otherwise, w has a larger dimension than ξ, hence the dimension of the observer is unnecessarily high. Finally, we note that in the new formulation of the observer, R only appears in the combination SR. Hence we introduce $G = SR$. The following theorem formulates our statements about observers:

Theorem 3.36 *The general form of a state observer for Σ is*

$$\dot{\xi}(t) = (A - GC)\xi(t) + Bu(t) + Gy(t). \tag{3.42}$$

The equation for the error function $e := \xi - x$ is

$$\dot{e}(t) = (A - GC)e(t). \tag{3.43}$$

Hence the state observer is stable if and only if $A - GC$ is a stability matrix.

The equation for ξ can be rewritten using an artificial output $\eta = C\xi$, viz., $\dot{\xi} = A\xi + Bu + G(y - \eta)$. The interpretation of this is as follows: If ξ is the exact state then $\eta = y$, and hence ξ obeys exactly the same differential equation as x. Otherwise, the equation for ξ has to be corrected by a term determined by the *output error* $y - \eta$. Consequently, the state observer consists of an exact replica Σ_{dup} of the original system with an extra input channel for incorporating the output error and an extra output, the state of the observer, which serves as the desired estimate for the state of the original system. The following diagram depicts the situation.

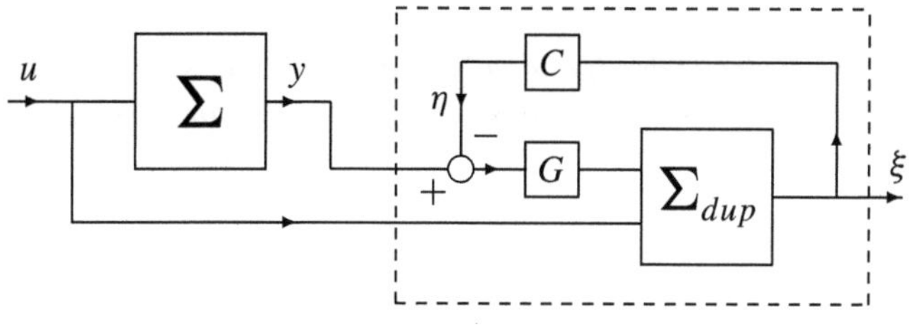

Obviously, state observers always exist. They are parameterized by the matrix G. A stable state observer exists if and only if G can be found such that $\sigma(A - GC) \subset \mathbb{C}^-$. More generally, one replaces $\mathbb{C}^-$ by $\mathbb{C}_g$. The problem of finding such a G is dual to the problem of finding a matrix F to a pair (A, B) such that $A + BF$ is $\mathbb{C}_g$-stable. Hence we are led to the introduction of the dual concept of stabilizability:

Definition 3.37 *The pair (C, A) is called* $\mathbb{C}_g$-detectable *if there exists a matrix G such that $\sigma(A - GC) \subset \mathbb{C}_g$.*

Based on the previous considerations and on theorem 3.32 we find the following result.

Theorem 3.38 *Given system $\Sigma = (C, A, B)$, the following statements are equivalent:*

(i) Σ has a $\mathbb{C}_g$-stable observer,

(ii) (C, A) is $\mathbb{C}_g$-detectable,

(iii) every (C, A)-unobservable eigenvalue is in $\mathbb{C}_g$.

We have chosen a specific type of $\mathbb{C}^-$-stable state observer, viz., a finite-dimensional, time-invariant, linear system with no feedthrough term. An obvious question is of course, whether in the case of a nondetectable system, for which such an observer does not exist, a stable state observer can be constructed of a more general type. Let us call a *generalized state observer* any system taking y and u as the only information about the state of Σ and yielding as output an estimate ξ of the state with the property that $\xi(t) - x(t) \to 0$ $(t \to \infty)$ for all possible u. We claim that the existence of such an observer implies the detectability of Σ. In fact, suppose that λ is an unobservable eigenvalue with $p \neq 0$ such that $Ap = \lambda p$ and $Cp = 0$. Assume that $x_0 = 0$ and $u(t) = 0$ for all $t \geqslant 0$. Then $y(t) = 0$ $(t > 0)$. The observer has zero input and must yield an output $\xi(t)$ which converges to zero, since $x(t) = 0$ and $x(t) - \xi(t) \to 0$ $(t \to \infty)$. On the other hand if $x_0 = p$ and still $u(t) = 0$ for all t, we also have $y(t) = 0$ and consequently, $\xi(t) \to 0$, since nothing has changed for the observer. It follows from $x(t) - \xi(t) \to 0$ that we must have $x(t) \to 0$. Since $x(t) = e^{\lambda t} p$ and $p \neq 0$, we conclude that $\Re\mathrm{e}\,\lambda < 0$. Hence every unobservable eigenvalue is in $\mathbb{C}^-$, so that Σ is detectable.

As a consequence of our results about state observers, we obtain another condition for $\mathbb{C}^-$-detectability:

Corollary 3.39 *Σ is $\mathbb{C}^-$-detectable if and only if for any initial state and any input, we have: $u(t) \to 0$ and $y(t) \to 0$ for $t \to \infty$ imply $x(t) \to 0$ for $t \to \infty$.*

Proof : ($\Rightarrow$) Choose G such that $\sigma(A - GC) \subset \mathbb{C}^-$. The state satisfies the equation

$$\dot{x}(t) = (A - GC)x(t) + Bu(t) + Gy(t),$$

which can be viewed as an internally stable equation with inputs u and y. Since u and y both converge to zero, corollary 3.22 implies that $x(t) \to 0$ $(t \to \infty)$.

($\Leftarrow$) If λ is an eigenvalue with $p \neq 0$ such that $Ap = \lambda p$ and $Cp = 0$ and we choose p as initial state and $u(t) = 0$ for all t then $y(t) = 0$ for all t. It then follows from the given property that $x(t) \to 0$ $(t \to \infty)$. But $x(t) = e^{\lambda t} p$. Hence $\Re e\, \lambda < 0$. ■

3.12 Stabilization by dynamic measurement feedback

We combine the results of the previous two sections in order to construct a dynamic feedback controller that stabilizes the system using only the output. That is, we assume that we know how to stabilize by state feedback and how to build a state observer. If we have a plant of which we do not have the state available for measurement, we use a state observer to obtain an estimate of the state and we apply the state feedback to this estimate rather than to the true state. This is illustrated by the following picture:

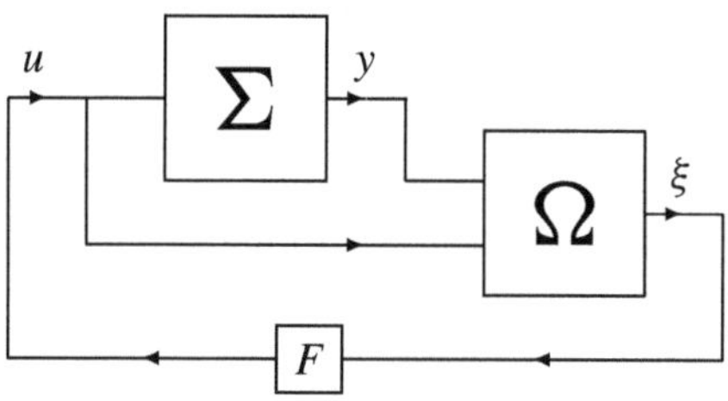

Again, consider the system Σ given by (3.34) and let the observer Ω be given by (3.42). Combining this with $u = F\xi$, we obtain

$$\begin{aligned} \dot{x}(t) &= Ax(t) + BF\xi, \\ \dot{\xi}(t) &= (A - GC + BF)\xi(t) + GCx(t). \end{aligned} \tag{3.44}$$

Introducing again $e := \xi - x$, we obtain, in accordance with the previous section, $\dot{e} = (A - GC)e$. Hence we have the following system:

$$\begin{aligned} \dot{x}(t) &= (A + BF)x(t) + BFe(t), \\ \dot{e}(t) &= (A - GC)e(t). \end{aligned}$$

That is, the equation $\dot{x}_e = A_e x_e$ with

$$x_e := \begin{pmatrix} x \\ e \end{pmatrix}, \quad A_e := \begin{pmatrix} A + BF & BF \\ 0 & A - GC \end{pmatrix}.$$

Suppose we are given a stability domain $\mathbb{C}_g$ and assume that $\Sigma = (C, A, B)$ is $\mathbb{C}_g$-stabilizable and $\mathbb{C}_g$-detectable. Then F and G can be found such that $A + BF$ and

$A - GC$ are $\mathbb{C}_g$-stable. Since $\sigma(A_e) = \sigma(A + BF) \cup \sigma(A - GC)$, it follows that $\sigma(A_e) \subset \mathbb{C}_g$. Consequently, the system $\dot{x}_e = A_e x_e$ is $\mathbb{C}_g$ stable, equivalently, every solution $x_e = (x, e)$ is a $\mathbb{C}_g$-stable Bohl function. Of course, if (x, ξ) is a solution of (3.44) then $\xi = x + e$, with $x_e = (x, e)$ a solution of $\dot{x}_e = A_e x_e$. Hence (x, ξ) is also a $\mathbb{C}_g$-stable Bohl function. Thus we have proved the 'if' part of the following theorem:

Theorem 3.40 *Let $\mathbb{C}_g$ be a stability domain. Then there exists a $\mathbb{C}_g$-stabilizing feedback controller for Σ if and only if Σ is $\mathbb{C}_g$-stabilizable and $\mathbb{C}_g$-detectable.*

Proof (of the 'only if' part) : Let Γ be a $\mathbb{C}_g$-stabilizing controller. Then, for any initial state, the output of Γ will be such that the state x is a $\mathbb{C}_g$-stable Bohl function. It follows from theorem 3.33 that Σ must be $\mathbb{C}_g$-stabilizable. Now let λ be an unobservable eigenvalue, with corresponding $p \neq 0$ such that $Ap = \lambda p$ and $Cp = 0$. We consider first the situation where the initial state of Σ is $x_0 = 0$ and the initial state of the controller arbitrary. The resulting state is $x_u(t, 0)$, where u is the input resulting from the controller. We know that $x_u(t, 0)$ is a $\mathbb{C}_g$-stable Bohl function. The corresponding output is $y_u(t, 0)$. Next we assume that $x_0 = p$. We claim that $\tilde{x}(t) := e^{\lambda t} p + x_u(t, 0)$ and the same state trajectory for Γ as before satisfy the differential equations. In fact, this is a consequence of $C\tilde{x}(t) = Cx_u(t, 0)$. Because we must also have that $\tilde{x}$ is a $\mathbb{C}_g$-stable Bohl function, it follows that $\lambda \in \mathbb{C}_g$. Thus we have shown that Σ is $\mathbb{C}_g$-detectable. ■

The proof in the previous theorem has be formulated in such a way that it remains valid if the controller is of a more general type than finite-dimensional, linear, time-invariant.

3.13 Well-posed interconnection

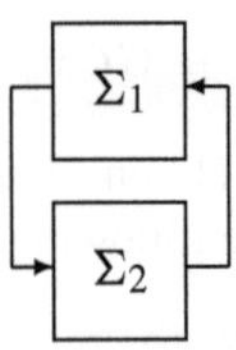

Figure 3.2

Assume we have the interconnection in figure 3.2. Assume Σ_1 and Σ_2 are given by the following two state space models:

$$\bar{\Gamma}: \begin{array}{l} \dot{x}_1 = A_1 x_1 + B_1 u_1, \\ y_1 = C_1 x_1 + D_1 u_1. \end{array} \tag{3.45}$$

and

$$\bar{\Gamma}: \begin{aligned} \dot{x}_2 &= A_2 x_2 + B_2 u_2, \\ y_2 &= C_2 x_2 + D_2 u_2. \end{aligned} \tag{3.46}$$

The interconnection implies $y_1 = u_2$ and $y_2 = u_1$. If we try to solve this set of equations we get:

$$u_2 = y_1 = C_1 x_1 + D_1 u_1 = C_1 x_1 + D_1(C_2 x_2 + D_2 u_2). \tag{3.47}$$

and we can solve u_2 from the above equation if $I - D_1 D_2$ is an invertible matrix. In that case the dynamics of the interconnection are given by:

$$\begin{pmatrix} \dot{x}_1 \\ \dot{x}_2 \end{pmatrix} = \begin{pmatrix} A_1 + B_1(I - D_2 D_1)^{-1} D_2 C_1 & B_1(I - D_2 D_1)^{-1} C_2 \\ B_2(I - D_1 D_2)^{-1} C_1 & A_2 + B_2(I - D_1 D_2)^{-1} D_1 C_2 \end{pmatrix} \begin{pmatrix} \dot{x}_1 \\ \dot{x}_2 \end{pmatrix}.$$

What happens if $I - D_1 D_2$ is not invertible? Then there either exist initial states for which the equation (3.47) is not solvable for u_2 or there exist multiple solutions for u_2. Therefore either the interconnection in Figure 3.2 on the facing page has no solution for certain initial states or the solution exists for for certain initial states but is not unique.

Therefore we call the interconnection *well posed* if for all possible initial states the interconnection has a solution which is moreover unique. Otherwise we call the interconnection *ill posed*.

We obtain the obvious result that the interconnection in figure 3.2 on the facing page is well posed if and only if $I - D_1 D_2$ is invertible. Note that $I - D_1 D_2$ is invertible if and only if $I - D_2 D_1$ is invertible.

In many chapters we will have interconnections where we do not mention the obvious requirement that we want to have a well-posed interconnection. This is due to the fact that the interconnection of a system and a controller is always well posed if either the system or the controller is strictly proper. For instance in the previous section we stabilized the system with a controller which used an observer in combination with a state feedback. However, this type of controller is always strictly proper and therefore the interconnection is always well posed.

3.14 Exercises

3.1 Consider the system $\Sigma : \dot{x} = Ax + Bu,\ y = Cx + Du$. If, with $x(0) = 0$, the output is bounded for every input function u, show that Σ has transfer function zero.

3.2 An $n \times n$ matrix A is called *cyclic* if there exists a column vector (i.e., an $n \times 1$ matrix) b such that (A, b) is controllable. Show that the following statements are equivalent:

a. A is cyclic,

b. there exists an invertible matrix S such that $S^{-1}AS$ has a companion form,

c. for all $\lambda \in \mathbb{C}$, $\operatorname{rank}(\lambda I - A) \geqslant n - 1$,

d. χ_A is the monic polynomial $p(z)$ of minimal degree for which $p(A) = 0$. (That is, χ_A is the minimal polynomial of A.)

e. Any $n \times n$ matrix B that commutes with A is a polynomial of A, i.e., if $AB = BA$, there exists a polynomial $p(z)$ such that $p(A) = B$.

3.3 Let $A \in \mathbb{R}^{n\times n}$, $B \in \mathbb{R}^{n\times m}$. Show that there exists a vector $p \in \mathbb{R}^m$ such that (A, Bp) is controllable if and only if

a. (A, B) is controllable,

b. A is cyclic.

For the definition of a cyclic matrix see exercise 3.2.

3.4 Prove that the pair (A, B) is controllable if and only if $a_{i,i+1} \neq 0$ $(i = 1, \ldots, n-1)$ where

$$A = \begin{pmatrix} a_{11} & a_{12} & 0 & \cdots & 0 \\ a_{21} & a_{22} & a_{32} & \ddots & \vdots \\ \vdots & \vdots & \vdots & \ddots & 0 \\ a_{n-1,1} & a_{n-1,2} & a_{n-1,2} & \cdots & a_{n-1,n} \\ a_{n1} & a_{n2} & a_{n2} & \cdots & a_{nn} \end{pmatrix}, \quad B = \begin{pmatrix} 0 \\ 0 \\ \vdots \\ 0 \\ 1 \end{pmatrix}$$

3.5 Let $M \in \mathbb{R}^{n\times n}$ and $N \in \mathbb{R}^{n\times m}$. Show that

$$\left[\begin{pmatrix} 0 & M \\ M & 0 \end{pmatrix}, \begin{pmatrix} 0 \\ N \end{pmatrix}\right]$$

is controllable if and only if M is nonsingular and (M^2, N) is controllable.

3.6 Which matrices A have the property that (A, B) is controllable for every non-zero B?

3.7 Consider the system Σ given by $\dot{x}(t) = Ax(t),\ y(t) = Cx(t)$ with state space $\mathcal{X}$ and output space $\mathcal{Y}$. Denote the unobservable subspace $\langle \ker C \mid A\rangle$ by $\mathcal{N}$. Let $\bar{\mathcal{X}}$ denote the quotient space $\mathcal{X}/\mathcal{N}$ and let Π be the canonical projection.

a. Let $\bar{A} : \bar{\mathcal{X}} \to \bar{\mathcal{X}}$ and $\bar{C} : \bar{\mathcal{X}} \to \mathcal{Y}$ be the quotient maps defined by $\Pi A = \bar{A}\Pi$ and $\bar{C}\Pi = C$. Show that $(\bar{C}, \bar{A})$ is observable. (This system is usually called the *observable quotient system* of Σ.)

b. Assume that (C, A) is not observable and C is not zero. Show that there exists an invertible matrix S such that $\bar{A} := S^{-1}AS$ and $\bar{C} := CS$ have the form

$$\bar{A} = \begin{pmatrix} A_{11} & A_{12} \\ 0 & A_{22} \end{pmatrix}, \quad \bar{C} = \begin{pmatrix} 0 & C_2 \end{pmatrix},$$

where (C_2, A_{22}) is observable.

c. Explain why it is reasonable to call the system $\dot{x}_1 = A_{11}x_1$ the *unobservable subsystem* of Σ.

3.8 Consider the system Σ given by $\dot{x}(t) = Ax(t) + Bu(t)$, $y(t) = Cx(t)$ with input space $\mathcal{U}$, state space $\mathcal{X}$ and output space $\mathcal{Y}$. Let $\mathcal{V} \subset \ker C$ be an A-invariant subspace. Denote the quotient space $\mathcal{X}/\mathcal{V}$ by $\bar{\mathcal{X}}$ and let Π be the canonical projection. Let $\bar{A} : \bar{\mathcal{X}} \to \bar{\mathcal{X}}$, $\bar{B} : \mathcal{U} \to \mathcal{X}$ and $\bar{C} : \bar{\mathcal{X}} \to \mathcal{Y}$ be the quotient maps defined by $\Pi A = \bar{A}\Pi$, $\bar{B} = \Pi B$ and $\bar{C}\Pi = C$. The system $\bar{\Sigma}$ given by $\dot{\bar{x}}(t) = \bar{A}\bar{x}(t) + \bar{B}u(t)$, $y(t) = \bar{C}\bar{x}(t)$ (with state space $\bar{\mathcal{X}}$) is often called the *quotient system* Σ *modulo* $\mathcal{V}$. Show that the transfer functions of Σ and $\bar{\Sigma}$, respectively, coincide

3.9 Show that if (A, B) is controllable then $(A + BF, B)$ is controllable for any map $F : \mathcal{X} \to \mathcal{U}$.

3.10 Consider the system (C, A) and let $(\bar{C}, \bar{A})$ be isomorphic to (C, A) and of the form described in exercise 3.7. Show that $\lambda \in \sigma(A)$ is an observable eigenvalue of (C, A) if and only if $\lambda \notin \sigma(A_{11})$.

3.11 Consider a cascade connection of two systems:

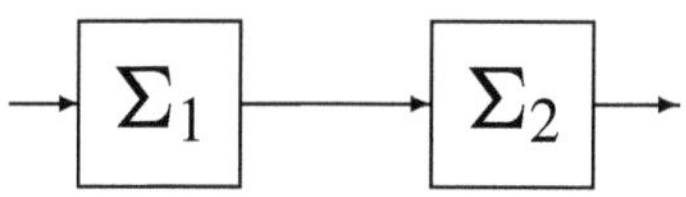

where $\Sigma_i = (A_i, B_i, C_i, D_i)$, and assume that $\sigma(A_1) \cap \sigma(A_2) = \varnothing$.

a. Determine a state space representation of the cascaded system Σ_c.

b. Prove that Σ_c is controllable if and only if (A_1, B_1) is controllable and

$$\operatorname{rank}\begin{pmatrix} A_2 - \lambda I & B_2 T_1(\lambda) \end{pmatrix} = n_2$$

for all $\lambda \in \sigma(A_2)$, where n_2 is the dimension of the state space of Σ_2 and where $T_1(\lambda)$ denotes the transfer function of Σ_1.

3.12 Let

$$A := \begin{pmatrix} 0 & 1 & 0 \\ -2 & -3 & 0 \\ 0 & 0 & \alpha \end{pmatrix}, \quad B := \begin{pmatrix} 0 \\ 1 \\ \beta \end{pmatrix}; \quad \alpha, \beta \in \mathbb{R}.$$

a. Determine all values of α and β for which (A, B) is controllable.

b. For those values of α and β for which (A, B) is not controllable, determine the uncontrollable eigenvalues.

c. Determine all values of α and β for which (A, B) is $\mathbb{C}^-$-stabilizable.

3.13 Let A_1 and A_2 be $n \times n$ matrices with no common eigenvalue, and let b and c be an $1 \times n$ and an $n \times 1$ matrix, respectively. Consider the matrix equation

$$A_1 X - X A_2 = bc.$$

a. Show that the above equation has exactly one solution X.

b. Show that, if (A_1, b) is controllable and (c, A_2) is observable, then X is invertible.

c. Use (b) to give a proof of the pole-placement theorem for the single-input case, independent of the control-canonical form.

3.14 Show that (A, B) is controllable iff the equations $XA - AX = 0$, $XB = 0$ only have the trivial solution $X = 0$.

3.15 Given the system $\Sigma : \dot{x} = Ax + Bu$, $y = Cx + Du$, what condition is necessary and sufficient in order that $y(t) \to 0$ $(t \to \infty)$ holds whenever the input is identically zero?

3.16 Let $A \in \mathbb{R}^{n\times n}$, $b \in \mathbb{R}^n$ be such that (A, b) is controllable. Let $p(z) = z^n + c_1 z^{n-1} + \cdots + c_n$ be given. Let f be an n-dimensional row vector such that $\chi_{A+bf} = p$.

a. Show that there exists exactly one n-dimensional row vector η such that $\eta A^k b = 0$ for $k = 0, \ldots, n-2$ and $\eta A^{n-1} b = 1$.

b. Show that $f = -\eta p(A)$.

c. Show that the following algorithm yields the correct f.

$$\begin{aligned} &\eta_0 := \eta \\ &\eta_{k+1} := \eta_k A + c_{k+1}\eta \ (k = 0, \ldots, n-1) \\ &f := -\eta_n . \end{aligned}$$

3.17 Let $A \in \mathbb{R}^{n\times n}$, $C \in \mathbb{R}^{p\times n}$. Show that there exists for every monic polynomial p of degree n a matrix $G \in \mathbb{R}^{n\times p}$ such that $\chi_{A+GC} = p$ if and only if (C, A) is observable.

3.18 Let $\Sigma : \dot{x} = Ax + Bu$ and let $\mathbb{C}_g$ be a stability domain. Show that (A, B) is $\mathbb{C}_g$-stabilizable if and only if for every x_0 there exists a Bohl function u such that the resulting state x is a $\mathbb{C}_g$-stable Bohl function.

3.19 Consider the discrete-time system Σ_d:

$$\begin{aligned} x(t+1) &= Ax(t) + Bu(t), \\ y(t) &= Cx(t). \end{aligned}$$

a. Determine a formula that for any initial state x_0 and any input sequence $(u(t))_{t=0}^{\infty}$ yields the corresponding state trajectory $x(t)$.

b. Define the concepts of controllability, null-controllability, reachability, and observability for the given discrete-time system.

c. Prove that the following statements are equivalent:

1. Σ_d is controllable,
2. Σ_d is reachable,

3. The pair (A, B) is controllable (in other words $\langle A \mid \operatorname{im} B\rangle = \mathcal{X}$).

d. Show that Σ_d is observable if and only if the pair (C, A) is observable (in other words $\langle \ker C \mid A\rangle = 0$).

e. Show that Σ_d is null controllable if and only if every nonzero eigenvalue of A is (A, B)-controllable.

f. Show that Σ_d is null controllable if and only if

$$\operatorname{rank}\begin{pmatrix} B & AB & \cdots & A^{n-1}B \end{pmatrix} = \operatorname{rank}\begin{pmatrix} B & AB & \cdots & A^{n-1}B & A^n \end{pmatrix}.$$

3.20 In the continuous-time system $\Sigma : \dot{x} = Ax + Bu, y = Cx$, a fixed sampling interval T is chosen. The control function u is taken constant, say u_k, on each of the intervals $[kT, (k+1)T]$, where $k \in \mathbb{N}$. Also, the output variable y is measured only at the time instances kT. The values thus measured are denoted by y_k. In this way, one obtains a discrete-time system Σ_d.

a. Give a state-space representation for Σ_d.

b. Assume that Σ is controllable and that the following condition is satisfied:

$$T(\lambda - \mu) \neq 2k\pi i \ (k \in \mathbb{Z},\ k \neq 0) \tag{3.48}$$

for every pair λ, μ of eigenvalues of A. Prove that Σ_d is controllable.

c. Show that condition (3.48) is necessary in the case that $m = 1$, i.e., if there is only one input.

d. Give an example showing that condition (3.48) is not necessary if $m > 1$.

e. Assume that condition (3.48) is satisfied and that Σ is observable. Show that Σ_d is observable.

3.21 Show that (A,B) is controllable if and only if for every nonzero matrix C, there exists $s \notin \sigma(A)$ such that $C(sI - A)^{-1}B \neq O$.

3.22 The system Σ given by the equations

$$\begin{aligned} \dot{x}(t) &= Ax(t) + Bu(t), \\ y(t) &= Cx(t) + Du(t), \end{aligned}$$

with $x(t) \in \mathbb{R}^n$ and $y(t) \in \mathbb{R}^p$, is called *output-controllable* if for any $x_0 \in \mathbb{R}^n$ and $y_1 \in \mathbb{R}^p$ there exists an input function $u \in \mathbf{U}$ and a $T > 0$ such that $y_u(T, x_0) = y_1$. Show that Σ is output-controllable if and only if

$$\operatorname{rank}\begin{pmatrix} D & CB & CAB & \cdots & CA^{n-1}B \end{pmatrix} = p.$$

3.23 Suppose that n persons, $P_1, \ldots, P_n$ are sitting at a round table each with an amount of money, say $x_1, \ldots, x_n$. Let α, β be real mumbers satisfying $0 \leqslant \alpha, \beta$ and $\alpha + \beta \leqslant 1$. Each person P_k gives αx_k to his left neighbour P_{k-1}, and βx_k to his right neighbour P_{k+1}, where we make the identifications $P_0 = P_n$, $P_{n+1} = P_1$. Suppose that these actions are repeated $n - 1$ times. Each person P_k knows the numbers α and β, and he knows the numbers x_k at each step. Is he able to find out what the values of all x_i's were, when the actions started?

3.24 Consider the system $\Sigma : \dot{x} = Ax + Bu$, with $A \in \mathbb{R}^{n\times n}$ and $B \in \mathbb{R}^{n\times m}$. Let $\mathcal{W}$ be the reachable subspace of (A, B), i.e., $\mathcal{W} = \langle A \mid \operatorname{im} B\rangle$. Define the $\mathbb{C}^-$-stabilizable subspace of (A, B) by:

$$\mathcal{X}_{\text{stab}} := \{x_0 \in \mathbb{R}^n \mid \exists u \in \mathbf{U} \;:\; \lim_{t\to\infty} x(t, x_0, u) = 0\}.$$

a. Show that $\mathcal{X}_{\text{stab}}$ is a subspace of $\mathbb{R}^n$

b. Show that $\mathcal{W} \subset \mathcal{X}_{\text{stab}}$.

c. Find a system Σ for which the inclusion $\mathcal{W} \subset \mathcal{X}_{\text{stab}}$ is a strict inclusion.

d. Assume that (A, B) is $\mathbb{C}^-$-stabilizable. Show that $\mathcal{X}_{\text{stab}} = \mathbb{R}^n$.

e. Let $\lambda \in \mathbb{C}$ and let η be a nonzero row vector such that

$$\eta A = \lambda\eta, \quad \eta B = 0.$$

Given $x_0 \in \mathbb{R}^n$, $u \in \mathbf{U}$, define $z : \mathbb{R} \to \mathbb{C}$ by $z(t) := \eta x(t, x_0, u)$. Show that z satisfies the differential equation $\dot{z} = \lambda z$.

f. Show that (A, B) is $\mathbb{C}^-$-stabilizable if and only if $\mathcal{X}_{\text{stab}} = \mathbb{R}^n$.

3.25 An autonomous discrete-time system $\Sigma_a : x(t+1) = Ax(t)$ is called a *dead-beat* system if for any initial value x_0, the state $x(t)$ equals zero for $t \geqslant n$. What is a necessary and sufficient condition for the existence of a feedback $u = Fx$ in the controlled system $\Sigma : x(t+1) = Ax(t) + Bu(t)$ such that the resulting (autonomous) system is dead-beat?

3.26 Consider the system

$$\dot{x} = Ax + Bu \tag{3.49}$$

and assume that K, M, N are matrices satisfying $M \geqslant 0$, $N^{\mathrm{T}} = -N$, $K > 0$, $A^{\mathrm{T}}K + KA < 0$. Show that the feedback $u = Fx$, where $F := (N - M)B^{\mathrm{T}}K$, stabilizes (3.49).

3.27 Let P be a positive semidefinite solution of (3.30) and let (C, A) be observable. Show that P is positive definite.

3.28 Let $n \times n$ matrices A, P, Q satisfy (3.30), and let $P > 0$, $Q \geqslant 0$. Show that

$$\Lambda(A) \leqslant \frac{\Lambda(-Q)}{2\Lambda(P)}.$$

3.29 Let A be an $n \times n$ matrix. If the equation $A^{\mathrm{T}}P + PA = -I$ has a positive definite solution, then the equation $A^{\mathrm{T}}P + PA = -Q$ has a positive solution for every $Q > 0$.

3.30 (*Stabilization algorithm by Bass*) Let (A, B) be stabilizable. Show that the following algorithm leads to a stabilizing feedback.

a. Choose α such that $\Lambda(-\alpha I - A) < 0$, e.g. by taking $\alpha > \|A\|$.

b. Solve P from the equation

$$(\alpha I + A)P + P(\alpha I + A^{\mathrm{T}}) = 2BB^{\mathrm{T}}.$$

c. Compute F from $FP = -B^{\mathrm{T}}$.

What can be said about the location of the controllable eigenvalues? Verify in particular that equations mentioned have a solution.

3.15 Notes and references

The description of linear systems in terms of a state space representation was particularly stressed by R.E. Kalman in the early sixties (see Kalman [91, 93, 94] and Kalman, Narendra and Ho [96]). See also Zadeh and Desoer [229], Gilbert [59]). In particular, Kalman introduced the concepts of controllability and observability, and gave the conditions expressed in Corollary 3.4, part (iii) and Theorem 3.6, part (v). Also, the decomposition of the state space, given in Theorem 3.11 is due to Kalman.

The concept of controllable and observable eigenvalues was introduced in Hautus [68]. The controllability and observability condition expressed in Theorem 3.13 were independently found by a number of authors. The first reference is Popov [148] (see also [149]). Other references are Belevitch [15] and Rosenbrock [155]. In Hautus [68, 69], the applicability of these conditions are demonstrated. The stabilizability and detectability condition of Theorem 3.32 and Theorem 3.38, part (iii), was proved in [69].

The pole-placement was proved in increasing generality by a number of authors: In 1960, J. Rissanen [153] proved the result for $m = 1$ (one input). The result was generalized to the case $m > 1$ by Popov [147] in 1964. This result was only valid, however, for complex systems. Hence, even if the matrices A, B are real, the resulting feedback matrix F could not be guaranteed to be real. A proof valid for real systems was given in 1967 by Wonham [220]. A drastic simplification in the proof was obtained by Heymann [77] in 1968. The proof in this chapter was given in Hautus [71].

State observers were introduced right with the concept of state-space representation (Kalman [96], Zadeh and Desoer [229]). They were viewed as a deterministic, nonoptimal version of the Kalman filter. The construction of the observer given in this chapter follows the pattern of the Kalman filters. In the literature, much attention is spent on reduced observers, i.e., observer with a lower dimensional state space (see Luenberger [115]).

The results of exercises 3.13 and 3.16 were given in Luenberger [116] and Ackermann [1], respectively.

Chapter 4

Controlled invariant subspaces

In this chapter we introduce controlled invariant subspaces (which are also called (A, B)-invariant subspaces) and the concepts of controllability subspace and stabilizability subspace. The notion of controlled invariance is of fundamental importance in many of the feedback design problems that the reader will encounter in this book. We will apply these concepts to a number of basic feedback design problems, among which the problem of disturbance decoupling by state feedback. The design problems treated in this chapter have in common that the entire state vector of the control system is assumed to be available for control purposes, and we confine ourselves to the design of static state feedback control laws. Dynamic feedback will be discussed in chapter 6.

4.1 Controlled invariance

In this section we will introduce the concept of controlled invariant subspace and prove the most important properties of these subspaces. Again consider the system

$$\dot{x}(t) = Ax(t) + Bu(t). \tag{4.1}$$

The input functions u are understood to be elements of the class $\mathbf{U}$ of admissible input functions (see section 3.1). A subspace of the state space will be called controlled invariant if it has the following property: for every initial condition in the subspace there exists an input function such that the resulting state trajectory remains in the subspace for all times. More explicitly:

Definition 4.1 *A subspace $\mathcal{V} \subset \mathcal{X}$ is called* controlled invariant *if for any $x_0 \in \mathcal{V}$ there exists an input function u such that $x_u(t, x_0) \in \mathcal{V}$ for all $t \geqslant 0$.*

It follows immediately from the definition that the sum of any number of controlled invariant subspaces is a controlled invariant subspace. In order to stress the dependence on the underlying system, we will often use the terminology (A, B)-*invariant subspace* instead of controlled invariant subspace. It is easily seen that if $F : \mathcal{X} \to \mathcal{U}$ is a linear map and $G : \mathcal{U} \to \mathcal{U}$ is an isomorphism then a given subspace $\mathcal{V}$ is (A, B)-invariant if and only if it is $(A + BF, BG)$-invariant. Stated differently: the classes of controlled invariant subspaces associated with the systems (A, B) and $(A + BF, BG)$, respectively, coincide. Sometimes this is expressed by saying that the property of controlled invariance is invariant under state feedback and isomorphism of the input space. The following theorem gives several equivalent characterizations of controlled invariance:

Theorem 4.2 *Consider the system (4.1). Let $\mathcal{V}$ be a subspace of $\mathcal{X}$. The following statements are equivalent:*

(*i*) *$\mathcal{V}$ is controlled invariant,*

(*ii*) $A\mathcal{V} \subset \mathcal{V} + \operatorname{im} B$,

(*iii*) *there exists a linear map $F : \mathcal{X} \to \mathcal{U}$ such that $(A + BF)\mathcal{V} \subset \mathcal{V}$.*

Proof : (i) $\Rightarrow$ (ii). Let $x_0 \in \mathcal{V}$ and let u be an input function such that $x_u(t, x_0) \in \mathcal{V}$ for all $t \geqslant 0$. Since $\mathcal{V}$ is a linear subspace, for all $t > 0$ we have $\frac{1}{t}(x_u(t, x_0) - x_0) \in \mathcal{V}$. Being a subspace of $\mathcal{X}$, $\mathcal{V}$ is closed in the Euclidean topology. Thus $\dot{x}(0^+) := \lim_{t \downarrow 0} \frac{1}{t}(x_u(t, x_0) - x_0) \in \mathcal{V}$. Since $Ax_0 = \dot{x}(0^+) - Bu(0^+)$ it follows that $Ax_0 \in \mathcal{V} + \operatorname{im} B$.

(ii) $\Rightarrow$ (iii). Choose a basis $q_1, \ldots, q_n$ of $\mathcal{X}$ adapted to $\mathcal{V}$. For all $1 \leqslant i \leqslant n$ there exist vectors $\bar{q}_i \in \mathcal{V}$ and $u_i \in \mathcal{U}$ such that $Aq_i = \bar{q}_i + Bu_i$. Define $F : \mathcal{X} \to \mathcal{U}$ as follows: for $1 \leqslant i \leqslant k$ define $Fq_i := -u_i$ and for $k + 1 \leqslant i \leqslant n$ let Fq_i be arbitrary vectors in $\mathcal{X}$. Then for $i = 1, \ldots, k$ we have $(A + BF)q_i = \bar{q}_i \in \mathcal{V}$ and hence $(A + BF)\mathcal{V} \subset \mathcal{V}$.

(iii) $\Rightarrow$ (i). Let $x_0 \in \mathcal{V}$. We claim that if the system is controlled by the feedback control law $u = Fx$, then the resulting trajectory remains in $\mathcal{V}$. Indeed, using this control law the trajectory $x_u(t, x_0)$ is equal to the solution of $\dot{x} = (A+BF)x,\ x(0) = x_0$. The claim then follows immediately from theorem 2.4. ■

In the above, the characterization (i) is typically an *open loop* characterization: the input functions are allowed to depend on the initial condition arbitrarily. In this vein, the characterization (iii) is called a *closed-loop* characterization: it turns out to be possible to remain within a controlled invariant subspace using a state feedback control law. As an intermediate between these two we stated (ii), a *geometric* characterization of controlled invariance.

If $\mathcal{V}$ is controlled invariant, then we will denote by $\underline{F}(\mathcal{V})$ the set of all linear maps F such that $(A + BF)\mathcal{V} \subset \mathcal{V}$. In the sequel we will often use the notation $A_F := A + BF$.

Let $\mathcal{V}$ be a controlled invariant subspace and let $F \in \underline{F}(\mathcal{V})$. Consider the equation (4.1). If we represent the control u as $u = Fx + v$, we obtain the equation

$$\dot{x}(t) = A_F x(t) + Bv(t).$$

Let $x_0 \in \mathcal{V}$. We know that if we choose $v = 0$, then the state trajectory starting in x_0 remains in $\mathcal{V}$. We now ask ourselves the question: what other control inputs v have the property that the resulting state trajectory remains in $\mathcal{V}$? We claim that the trajectory $x(t)$ starting in x_0 remains in $\mathcal{V}$ if and only if $Bv(t) \in \mathcal{V}$ for all $t \geqslant 0$. Indeed, if $x(t) \in \mathcal{V}$ for $t \geqslant 0$, then also $A_F x(t) \in \mathcal{V}$ and $\dot{x}(t) \in \mathcal{V}$ for $t \geqslant 0$. Thus $Bv(t) = \dot{x}(t) - A_F x(t) \in \mathcal{V}$ for $t \geqslant 0$. Conversely, if $Bv(t) \in \mathcal{V}$ for $t \geqslant 0$ then

$$x(t) = e^{A_F t} x_0 + \int_0^t e^{A_F (t-\tau)} Bv(\tau)\, \mathrm{d}\tau \in \mathcal{V}$$

for all $t \geqslant 0$, since $e^{A_F t} x_0 \in \mathcal{V}$ for $t \geqslant 0$. Consider the linear subspace

$$B^{-1}\mathcal{V} := \{u \in \mathcal{U} \mid Bu \in \mathcal{V}\}.$$

Then $Bv(t) \in \mathcal{V}$ is equivalent to $v(t) \in B^{-1}\mathcal{V}$. Let L be a linear map such that $\operatorname{im} L = B^{-1}\mathcal{V}$. Obviously, $v(t) \in B^{-1}\mathcal{V}$ for all $t \geqslant 0$ if and only if $v(t) = Lw(t),\ t \geqslant 0$, for some function w (compare exercise 2.2). Thus we have proven:

Theorem 4.3 *Let $\mathcal{V}$ be a controlled invariant subspace. Assume that $F \in \underline{F}(\mathcal{V})$ and let L be a linear map such that* $\operatorname{im} L = B^{-1}\mathcal{V}$. *Let $x_0 \in \mathcal{V}$ and let u be an input function. Then the state trajectory resulting from x_0 and u remains in $\mathcal{V}$ for all $t \geqslant 0$ if and only if u has the form*

$$u(t) = Fx(t) + Lw(t) \tag{4.2}$$

for some input function w.

Note that (A_F, BL) can be viewed as the restriction of the system Σ to the subspace $\mathcal{V}$. After all if we stay inside $\mathcal{V}$ then u must be of the form (4.2). Therefore, the dynamics must be of the form

$$\dot{x}(t) = A_F x(t) + BLw(t).$$

If $\mathcal{K}$ is a subspace of $\mathcal{X}$ which is not controlled invariant, then we are interested in a controlled invariant subspace contained in $\mathcal{K}$ which is as large as possible.

Definition 4.4 *Let $\mathcal{K}$ be a subspace of $\mathcal{X}$. Then we define*

$$\mathcal{V}^*(\mathcal{K}) := \{x_0 \in \mathcal{X} \mid \textit{there exists an input function } u \textit{ such that } x_u(t, x_0) \in \mathcal{K} \textit{ for all } t \geqslant 0\}.$$

Two things follow immediately from this definition. First, it is easy to see that $\mathcal{V}^*(\mathcal{K})$ is a linear subspace of $\mathcal{X}$. Indeed, if $x_0, y_0 \in \mathcal{V}^*(\mathcal{K})$ then there are inputs u and v such that $x_u(t, x_0) \in \mathcal{K}$ and $x_v(t, y_0) \in \mathcal{K}$ for all $t \geqslant 0$. Let $\lambda, \mu \in \mathbb{R}$. Define $w(t) := \lambda u(t) + \mu v(t)$. Then $x_w(t, \lambda x_0 + \mu y_0) \in \mathcal{K}$ for all $t \geqslant 0$ (see (3.2)). Secondly, it is clear that $\mathcal{V}^*(\mathcal{K}) \subset \mathcal{K}$. In fact, we have the following result:

Theorem 4.5 *Let $\mathcal{K}$ be a subspace of $\mathcal{X}$. Then $\mathcal{V}^*(\mathcal{K})$ is the largest controlled invariant subspace contained in $\mathcal{K}$, i.e.*

(i) $\mathcal{V}^*(\mathcal{K})$ is a controlled invariant subspace,

(ii) $\mathcal{V}^*(\mathcal{K}) \subset \mathcal{K}$,

(iii) if $\mathcal{V} \subset \mathcal{K}$ is a controlled invariant subspace then $\mathcal{V} \subset \mathcal{V}^*(\mathcal{K})$.

Proof : We first show that $\mathcal{V}^*(\mathcal{K})$ is controlled invariant. Assume $x_0 \in \mathcal{V}^*(\mathcal{K})$. There is an input u such that $x_u(t, x_0) \in \mathcal{K}$ for all $t \geqslant 0$. We claim that, in fact, $x_u(t, x_0) \in \mathcal{V}^*(\mathcal{K})$ for all $t \geqslant 0$. To show this, take a fixed but arbitrary $t_1 \geqslant 0$. Let $x_1 := x_u(t_1, x_0)$. It will be shown that $x_1 \in \mathcal{V}^*(\mathcal{K})$. Indeed, if we define $v(t) := u(t_1 + t)$ $(t \geqslant 0)$, then using (3.2) we have $x_v(t, x_1) = x_u(t + t_1, x_0) \in \mathcal{K}$ for all $t \geqslant 0$. This proves that $x_1 = x_u(t_1, x_0)$ lies in $\mathcal{V}^*(\mathcal{K})$. Since t_1 was arbitrary, $x_u(t, x_0) \in \mathcal{V}^*(\mathcal{K})$ for *all* $t \geqslant 0$ and hence $\mathcal{V}^*(\mathcal{K})$ is controlled invariant.

Next, we show that $\mathcal{V}^*(\mathcal{K})$ is the largest controlled invariant subspace in $\mathcal{K}$. Let $\mathcal{V} \subset \mathcal{K}$ be controlled invariant. Let $x_0 \in \mathcal{V}$. There is an input u such that $x_u(t, x_0) \in \mathcal{V}$ for all $t \geqslant 0$. Consequently, $x_u(t, x_0) \in \mathcal{K}$ for all $t \geqslant 0$ and hence $x_0 \in \mathcal{V}^*(\mathcal{K})$. This completes the proof. ■

In order to display the dependence on the underlying system we will sometimes denote $\mathcal{V}^*(\mathcal{K})$ by $\mathcal{V}^*(\mathcal{K}, A, B)$.

4.2 Disturbance decoupling

Consider the system

$$\begin{aligned} \dot{x}(t) &= Ax(t) + Ed(t), \\ z(t) &= Hx(t). \end{aligned} \tag{4.3}$$

In the differential equation above, d represents an unknown disturbance which is assumed to be an element of a given function space. For a given initial condition x_0 and disturbance d, the output of the system is given by

$$z(t) = He^{At}x_0 + \int_0^t T(t-\tau)d(\tau)\,\mathrm{d}\tau. \tag{4.4}$$

Here $T(t) := He^{At}E$ is the impulse response between the disturbance and the output. The system (4.3) will be called *disturbance decoupled* if $T = 0$ or, equivalently, if the transfer function $G(s) = H(Is - A)^{-1}E$ is equal to zero. If this is the case then for any given initial condition x_0 the output is equal to $z(t) = He^{At}x_0$ for all disturbances d. This means that in a system which is disturbance decoupled, the output does not depend on the disturbance. The following theorem will play an important role in the sequel:

Theorem 4.6 *The system (4.3) is disturbance decoupled if and only if there exists an A-invariant subspace $\mathcal{V}$ such that* $\operatorname{im} E \subset \mathcal{V} \subset \ker H$.

Proof : ($\Rightarrow$) If $T = 0$ then also all its time derivatives $T^{(k)}$ are identically equal to 0. Thus, $T^{(k)}(t) = HA^k e^{At}E = 0$ for all t. By taking $t = 0$ this yields $HA^kE = 0$, $k = 0, 1, 2, \ldots$ Define $\mathcal{V} := \operatorname{im}\begin{pmatrix}E & AE & \cdots & A^{n-1}E\end{pmatrix}$. Then $\mathcal{V} \subset \ker H$. By corollary 3.3, $\mathcal{V}$ is equal to $\langle A \mid \operatorname{im} E\rangle$, the smallest A-invariant subspace that contains $\operatorname{im} E$.

($\Leftarrow$) If $\operatorname{im} E \subset \mathcal{V}$ and $\mathcal{V}$ is A-invariant, then also $\operatorname{im} A^kE \subset \mathcal{V}$ for $k = 0, 1, 2, \ldots$. Since we know that $\mathcal{V} \subset \ker H$ this yields $HA^kE = 0$ for all k. Thus $T(t) = \sum_{k=0}^{\infty}(t^k/k!)HA^kE = 0$ for all t. It follows that the system is disturbance decoupled. ■

If the system (4.3) is not disturbance decoupled, then one can try to *make* it disturbance decoupled. In order to do this one needs the possibility to change the system's dynamics by using a control input. This possibility is modelled by adding a control term to the right hand side of the original differential equation in (4.3). Thus we consider the system

$$\begin{aligned}\dot{x}(t) &= Ax(t) + Bu(t) + Ed(t),\\ z(t) &= Hx(t).\end{aligned} \tag{4.5}$$

In this description, the variable u represents a control input (see also section 2.8). Let $F : \mathcal{X} \to \mathcal{U}$ be a linear map. If in (4.5) we substitute $u(t) = Fx(t)$, the system's equations change into

$$\begin{aligned}\dot{x}(t) &= (A + BF)x(t) + Ed(t),\\ z(t) &= Hx(t),\end{aligned} \tag{4.6}$$

the closed-loop system obtained from the state feedback control law $u = Fx$. The impulse response matrix of (4.6) is called the *closed-loop impulse response* and is equal to

$$T_F(t) := He^{(A+BF)t}E.$$

The corresponding transfer function $G_F(s) := H(Is - A - BF)^{-1}E$ is called the *closed-loop transfer function.* The problem of disturbance decoupling by state feedback is to find a linear map $F : \mathcal{X} \to \mathcal{U}$ such that the closed-loop system (4.6) is disturbance decoupled:

Definition 4.7 *Consider the system (4.3).* The problem of disturbance decoupling by state feedback, DDP, *is to find a linear map* $F : \mathcal{X} \to \mathcal{U}$ *such that* $T_F = 0$ *(or, equivalently, such that* $G_F = 0$*).*

The following result establishes the connection between the concept of controlled invariance and the problem of disturbance decoupling.

Theorem 4.8 *There exists a linear map* $F : \mathcal{X} \to \mathcal{U}$ *such that* $T_F = 0$ *if and only if there exists a controlled invariant subspace* $\mathcal{V}$ *such that* $\operatorname{im} E \subset \mathcal{V} \subset \ker H$.

Proof : ($\Rightarrow$) If $T_F = 0$ then (4.6) is disturbance decoupled. By theorem 4.6 there is an $(A + BF)$-invariant subspace $\mathcal{V}$ such that $\operatorname{im} E \subset \mathcal{V} \subset \ker H$. By theorem 4.2, $\mathcal{V}$ is controlled invariant.

($\Leftarrow$) Let $\mathcal{V}$ be a controlled invariant subspace such that $\operatorname{im} E \subset \mathcal{V} \subset \ker H$. By theorem 4.2 there exists a linear map $F : \mathcal{X} \to \mathcal{U}$ such that $\mathcal{V}$ is $(A+BF)$-invariant. It then follows from theorem 4.6 that the system (4.6) is disturbance decoupled. ■

Corollary 4.9 *There exists a linear map* $F : \mathcal{X} \to \mathcal{U}$ *such that* $T_F = 0$ *if and only if*

$$\operatorname{im} E \subset \mathcal{V}^*(\ker H). \tag{4.7}$$

Formula (4.7) provides a very compact necessary and sufficient condition for the existence of a state feedback control law that achieves disturbance decoupling. However, in order to be able to check this condition for an actual system, we would like to have an algorithm. In the next section we will describe an algorithm that, starting from a system (4.5), calculates the associated subspace $\mathcal{V}^*(\ker H)$.

4.3 The invariant subspace algorithm

In this section we give an algorithm to compute the subspace $\mathcal{V}^*(\mathcal{K})$. Consider the system (A, B) and let $\mathcal{K}$ be a subspace of the state space $\mathcal{X}$. The algorithm we give is most easily understood if one thinks in terms of the discrete-time system

$$x_{t+1} = Ax_t + Bu_t, \quad t = 0, 1, 2, \ldots. \tag{4.8}$$

Given an input sequence $u = (u_0, u_1, u_2, \ldots)$ and an initial condition x_0, the resulting discrete-time state trajectory is denoted by $x = (x_0, x_1, x_2, \ldots)$. The discrete-time analogue $\mathcal{V}_d^*(\mathcal{K})$ of the subspace $\mathcal{V}^*(\mathcal{K})$ defined by definition 4.4 is obviously the subspace of all $x_0 \in \mathcal{X}$ for which there exists an input sequence u such that all terms of the resulting state trajectory lie in $\mathcal{K}$:

$$\mathcal{V}_d^*(\mathcal{K}) := \{x_0 \in \mathcal{X} \mid \text{there is an input sequence } u \text{ such that } x_t \in \mathcal{K} \text{ for } t = 0, 1, 2, \ldots\}.$$

Define a sequence of subspaces $\mathcal{V}_0, \mathcal{V}_1, \mathcal{V}_2, \ldots$ by

$$\mathcal{V}_t := \{x_0 \in \mathcal{X} \mid \text{there is an input sequence } u \text{ such that } x_0, x_1, x_2, \ldots, x_t \in \mathcal{K}\}.$$

Thus, $\mathcal{V}_t$ consists of those points in which a state trajectory starts for which the first $t+1$ terms lie in $\mathcal{K}$. It is easily verified that $\mathcal{V}_t$ is indeed a subspace, that $\mathcal{V}_0 = \mathcal{K}$ and that $\mathcal{V}_0 \supset \mathcal{V}_1 \supset \mathcal{V}_2 \supset \cdots$. It turns out to be possible to derive a recurrence relation for $\mathcal{V}_t$. Indeed, $x_0 \in \mathcal{V}_{t+1}$ if and only if $x_0 \in \mathcal{K}$ and there exists $u_0 \in \mathcal{U}$ such that $Ax_0 + Bu_0 \in \mathcal{V}_t$. Hence, $x_0 \in \mathcal{V}_{t+1}$ if and only if $x_0 \in \mathcal{K}$ and $Ax_0 \in \mathcal{V}_t + \operatorname{im} B$ or, equivalently, $x_0 \in A^{-1}(\mathcal{V}_t + \operatorname{im} B)$. It follows that

$$\mathcal{V}_0 = \mathcal{K}, \quad \mathcal{V}_{t+1} = \mathcal{K} \cap A^{-1}(\mathcal{V}_t + \operatorname{im} B). \tag{4.9}$$

From this recurrence relation it follows immediately that if $\mathcal{V}_k = \mathcal{V}_{k+1}$ for some k, then $\mathcal{V}_k = \mathcal{V}_t$ for all $t \geqslant k$. Now, recall that $\mathcal{V}_0 \supset \mathcal{V}_1 \supset \mathcal{V}_2 \supset \cdots$. If we have strict inclusion, the dimension must decrease by at least one. Hence the inclusion chain must have the form

$$\mathcal{V}_0 \supset \mathcal{V}_1 \supset \cdots \supset \mathcal{V}_k = \mathcal{V}_{k+1} = \mathcal{V}_{k+2} = \cdots$$

for some integer $k \leqslant \dim \mathcal{K}$ $(\leqslant n-1)$. In the above formula, $\supset$ stands for *strict* inclusion. We claim that $\mathcal{V}_d^*(\mathcal{K}) = \mathcal{V}_k$. Indeed, on the one hand it follows immediately from the definition that $\mathcal{V}_d^*(\mathcal{K}) \subset \mathcal{V}_t$ for all t. Conversely, assume $x_0 \in \mathcal{V}_k$. We want to construct an input sequence $(u_0, u_1, \ldots)$ such that the corresponding state trajectory $x = (x_0, x_1, \ldots)$ lies in $\mathcal{K}$. Since $x_0 \in \mathcal{V}_k = \mathcal{V}_{k+1}$, there is u_0 such that $x_1 = Ax_0 + Bu_0 \in \mathcal{V}_k$. Thus, in particular we have $x_0, x_1 \in \mathcal{K}$. We now proceed inductively. Assume $u_0, u_1, \ldots, u_{s-1}$ have been found such that $x_0, x_1, \ldots, x_{s-1} \in \mathcal{K}$, while $x_s \in \mathcal{V}_k$. Again using $\mathcal{V}_k = \mathcal{V}_{k+1}$ we can find u_s such that $x_{s+1} = Ax_s + Bu_s \in \mathcal{V}_k$. This proves our claim.

The above is meant to provide some intuitive background for the introduction of the recurrence relation (4.9). This recurrence relation will henceforth be called *the invariant subspace algorithm*, ISA. Of course, we still have to show its relevance in continuous-time systems. In the following result we will collect the properties of the sequence $\{\mathcal{V}_t\}$ we established above and prove that it can be used to calculate the largest controlled invariant subspace contained in $\mathcal{K}$ for the continuous-time system (4.1).

Theorem 4.10 *Consider the system (4.1). Let $\mathcal{K}$ be a subspace of $\mathcal{X}$. Let $\mathcal{V}_t$, $t = 0, 1, 2, \ldots$, be defined by the algorithm (4.9). Then we have*

(i) $\mathcal{V}_0 \supset \mathcal{V}_1 \supset \mathcal{V}_2 \supset \cdots$,

(ii) *there exists $k \leqslant \dim \mathcal{K}$ such that $\mathcal{V}_k = \mathcal{V}_{k+1}$,*

(iii) *if $\mathcal{V}_k = \mathcal{V}_{k+1}$ then $\mathcal{V}_k = \mathcal{V}_t$ for all $t \geqslant k$,*

(iv) *if $\mathcal{V}_k = \mathcal{V}_{k+1}$ then $\mathcal{V}^*(\mathcal{K}) = \mathcal{V}_k$.*

Proof : The statements (i), (ii) and (iii) were proven above. To prove (iv), note that $\mathcal{V}_k \subset \mathcal{K}$. Moreover, it follows immediately from (4.9) that

$$A\mathcal{V}_k = A\mathcal{V}_{k+1} \subset \mathcal{V}_k + \operatorname{im} B.$$

Hence $\mathcal{V}_k$ is a controlled invariant subspace contained in $\mathcal{K}$ and therefore contained in $\mathcal{V}^*$. To prove the converse inclusion, we show that in fact $\mathcal{V}^* \subset \mathcal{V}_t$ for all t. Obviously this is true for $t = 0$. Assume $\mathcal{V}^* \subset \mathcal{V}_{t-1}$. Since $\mathcal{V}^*$ is controlled invariant we have $A\mathcal{V}^* \subset \mathcal{V}^* + \operatorname{im} B$. Thus $A\mathcal{V}^* \subset \mathcal{V}_{t-1} + \operatorname{im} B$ and hence $\mathcal{V}^* \subset A^{-1}(\mathcal{V}_{t-1} + \operatorname{im} B)$. Finally, $\mathcal{V}^* \subset \mathcal{K}$ so we conclude that $\mathcal{V}^* \subset \mathcal{V}_t$. ■

Of course item (iv) is crucial because it tells us that there exists a *finite* algorithm to compute $\mathcal{V}^*$. As a matter of fact it is easily seen that we need at most n steps where n is the dimension of the state space.

4.4 Controllability subspaces

Consider the system (4.1). If a subspace of the state space has the property that every point in that subspace can be steered to the origin in finite time without leaving the subspace, it is called a controllability subspace.

Definition 4.11 *A subspace $\mathcal{R} \subset \mathcal{X}$ is called a* controllability subspace *if for every $x_0 \in \mathcal{R}$ there exists $T > 0$ and an input function u such that $x_u(t, x_0) \in \mathcal{R}$ for $0 \leqslant t \leqslant T$ and $x_u(T, x_0) = 0$.*

It is immediately clear from this definition that every controllability subspace is controlled invariant. Indeed, if one chooses the control input to be equal to zero for $t > T$, the state trajectory also remains zero and hence does not leave $\mathcal{R}$. As was the case with controlled invariant subspaces, it can be shown that the sum of any (finite or infinite) number of controllability subspaces is a controllability subspace. Also, the class of all controllability subspaces associated with a given system is invariant under state feedback and isomorphisms of the input space. That is, if $\mathcal{R}$ is a controllability subspace with respect to (A, B), it is a controllability subspace with respect to $(A + BF, BG)$ for all linear maps $F : \mathcal{X} \to \mathcal{U}$ and isomorphisms G of $\mathcal{U}$.

We can give the following characterization of controllability subspaces:

Theorem 4.12 *A subspace $\mathcal{R} \subset \mathcal{X}$ is a controllability subspace if and only if there exist linear maps F and L such that*

$$\mathcal{R} = \langle A + BF \mid \operatorname{im} BL \rangle. \tag{4.10}$$

Proof : ($\Rightarrow$) Let $F \in \underline{F}(\mathcal{R})$ and L be a linear map such that $\operatorname{im} L = B^{-1}\mathcal{R}$. We claim that (4.10) holds. Let $x_0 \in \mathcal{R}$. There is $T > 0$ and an input u such that $x_u(t, x_0) \in \mathcal{R}$ for all $t \geqslant 0$ and $x_u(T, x_0) = 0$. By theorem 4.3 there exists w such that $u(t) = Fx_u(t, x_0) + Lw(t)$. Hence, $x_u(t, x_0)$ is a state trajectory of the system $\dot{x}(t) = A_F x(t) + BLw(t)$ with state space $\mathcal{R}$. Along this trajectory, x_0 is steered to 0 at time $t = T$. Since this is possible for all $x_0 \in \mathcal{R}$, it follows that the latter system is null-controllable. Consequently, it is reachable so $\mathcal{R} = \langle A_F \mid \operatorname{im} BL\rangle$ (see section 3.2).

($\Leftarrow$) Assume that (4.10) holds. Then we have $A_F\mathcal{R} \subset \mathcal{R}$ and $\operatorname{im} BL \subset \mathcal{R}$. Thus $\dot{x}(t) = A_F x(t) + BLw(t)$ defines a system with state space $\mathcal{R}$. By corollary 3.4, this system is controllable. Hence every point in $\mathcal{R}$ can be controlled to the origin in finite time while remaining in $\mathcal{R}$. ■

It follows from the proof of the above theorem that if $\mathcal{R}$ is a controllability subspace then for the maps F and L in the representation (4.10) we can take any $F \in \underline{F}(\mathcal{R})$ and any map L such that $\operatorname{im} L = B^{-1}\mathcal{R}$. Since the latter equality implies $\operatorname{im} BL = \operatorname{im} B \cap \mathcal{R}$ we obtain the following:

Corollary 4.13 *Let $\mathcal{R}$ be a controllability subspace. Then for any $F \in \underline{F}(\mathcal{R})$ we have*

$$\mathcal{R} = \langle A + BF \mid \operatorname{im} B \cap \mathcal{R}\rangle.$$

If $\mathcal{K}$ is a subspace of the state space $\mathcal{X}$ then we are interested in the largest controllability subspace that is contained in $\mathcal{K}$ (see also definition 4.4).

Definition 4.14 *Let $\mathcal{K}$ be a subspace of $\mathcal{X}$. Then we define*

$$\mathcal{R}^*(\mathcal{K}) := \{x_0 \in \mathcal{X} \mid \text{there exists an input function } u \text{ and } T > 0 \text{ such that } x_u(t, x_0) \in \mathcal{K} \text{ for all } 0 \leqslant t \leqslant T \text{ and } x_u(T, x_0) = 0\}.$$

Clearly, $\mathcal{R}^*(\mathcal{K})$ is contained in $\mathcal{K}$. In fact, we have

Theorem 4.15 *Let $\mathcal{K}$ be a subspace of $\mathcal{X}$. Then $\mathcal{R}^*(\mathcal{K})$ is the largest controllability subspace contained in $\mathcal{K}$, i.e.*

(i) *$\mathcal{R}^*(\mathcal{K})$ is a controllability subspace,*

(ii) *$\mathcal{R}^*(\mathcal{K}) \subset \mathcal{K}$,*

(iii) *if $\mathcal{R} \subset \mathcal{K}$ is a controllability subspace then $\mathcal{R} \subset \mathcal{R}^*(\mathcal{K})$.*

Proof : We first show that $\mathcal{R}^*(\mathcal{K})$ is a subspace. Let $x_0, y_0 \in \mathcal{R}^*(\mathcal{K})$ and $\lambda, \mu \in \mathbb{R}$. There exist controls u, v and numbers $T_1, T_2 > 0$ such that $x_u(T_1, x_0) = 0$, $x_v(T_2, y_0) = 0$, $x_u(t, x_0) \in \mathcal{K}$ for all $0 \leqslant t \leqslant T_1$ and $x_v(t, y_0) \in \mathcal{K}$ for all $0 \leqslant t \leqslant T_2$. Without loss of generality, assume that $T_1 \leqslant T_2$. Define a new control function $\tilde{u}$ by $\tilde{u}(t) = u(t)$ $(0 \leqslant t \leqslant T_1)$ and $\tilde{u}(t) = 0$ $(t > T_1)$. Then $x_{\tilde{u}}(T_2, x_0) = 0$ and $x_{\tilde{u}}(t, x_0)(t) \in \mathcal{K}$ for all $0 \leqslant t \leqslant T_2$. Define now $w(t) := \lambda\tilde{u}(t) + \mu v(t)$. Then $x_w(T_2, \lambda x_0 + \mu y_0) = 0$ and $x_w(t, \lambda x_0 + \mu y_0) \in \mathcal{K}$ for all $0 \leqslant t \leqslant T_2$.

Next, we prove that $\mathcal{R}^*(\mathcal{K})$ is a controllability subspace. Let $x_0 \in \mathcal{R}^*(\mathcal{K})$. There is a control input u and a number $T > 0$ such that $x_u(T, x_0) = 0$ and $x_u(t, x_0) \in \mathcal{K}$ for all $0 \leqslant t \leqslant T$. We contend that, in fact, $x_u(t, x_0) \in \mathcal{R}^*(\mathcal{K})$. To prove this, take a fixed but arbitrary $t_1 < T$. Let $x_1 := x_u(t_1, x_0)$. Define a new input function v by $v(t) := u(t + t_1)$ $(t \geqslant 0)$. Then $x_v(t, x_1) = x_u(t_1 + t, x_0) \in \mathcal{K}$ for all $0 \leqslant t \leqslant T - t_1$ and $x_v(T - t_1, x_1) = x_u(T, x_0) = 0$. Consequently, x_1 can be controlled to the origin in finite time while remaining in $\mathcal{K}$ and hence $x_1 \in \mathcal{R}^*(\mathcal{K})$. Since t_1 was arbitrary we find that $x_u(t, x_0) \in \mathcal{R}^*(\mathcal{K})$ for all $0 \leqslant t \leqslant T$. Finally, the fact that $R^*(\mathcal{K})$ is the largest controllability subspace in $\mathcal{K}$ is proven completely similarly as the corresponding part of theorem 4.5. ■

Sometimes, we will denote $\mathcal{R}^*(\mathcal{K})$ by $\mathcal{R}^*(\mathcal{K}, A, B)$. Starting with a subspace $\mathcal{K}$ of the state space, we have now defined $\mathcal{V}^*(\mathcal{K})$ (see definition 4.4) and $\mathcal{R}^*(\mathcal{K})$. Since $\mathcal{R}^*(\mathcal{K})$ is controlled invariant and contained in $\mathcal{K}$, it must be contained in the largest controlled invariant subspace in $\mathcal{K}$. Thus

$$\mathcal{R}^*(\mathcal{K}) \subset \mathcal{V}^*(\mathcal{K}) \subset \mathcal{K}. \tag{4.11}$$

More specifically, we have $\mathcal{R}^*(\mathcal{V}^*(\mathcal{K})) = \mathcal{R}^*(\mathcal{K})$. In the following, whenever this is convenient, we denote $\mathcal{R}^*(\mathcal{K})$ and $\mathcal{V}^*(\mathcal{K})$ by $\mathcal{R}^*$ and $\mathcal{V}^*$, respectively.

Lemma 4.16 *Let $\mathcal{K}$ be a subspace of $\mathcal{X}$. Then* $\operatorname{im} B \cap \mathcal{V}^*(\mathcal{K}) \subset \mathcal{R}^*(\mathcal{K})$.

Proof : Let L be a linear map such that $\operatorname{im} L = B^{-1}\mathcal{V}^*$. Then $\operatorname{im} BL = \operatorname{im} B \cap \mathcal{V}^*$. Choose $F \in \underline{F}(\mathcal{V}^*)$. Then we have

$$\operatorname{im} B \cap \mathcal{V}^* \subset \langle A_F \mid \operatorname{im} BL \rangle \subset \mathcal{V}^* \subset \mathcal{K}.$$

Since $\langle A_F \mid \operatorname{im} BL \rangle$ is a controllability subspace (see theorem 4.12) it must be contained in $\mathcal{R}^*(\mathcal{K})$. This proves the lemma. ■

The above lemma will be used to prove the following, stronger, result:

Theorem 4.17 *Let $\mathcal{K}$ be a subspace of $\mathcal{X}$. Then $\underline{F}(\mathcal{V}^*) \subset \underline{F}(\mathcal{R}^*)$ and*

$$\mathcal{R}^* = \langle A + BF \mid \operatorname{im} B \cap \mathcal{V}^* \rangle \tag{4.12}$$

for all $F \in \underline{F}(\mathcal{V}^*)$.

In the above, $\underline{F}(\mathcal{R}^*)$ denotes the set of all linear maps F with the property that $\mathcal{R}^*$ is $A + BF$ invariant. This is consistent with our earlier notation $\underline{F}(\mathcal{V})$ with respect to the controlled invariant subspace $\mathcal{V}$, since every controllability subspace is controlled invariant.

Proof : Let $F \in \underline{F}(\mathcal{V}^*)$. Since $\mathcal{R}^* \subset \mathcal{V}^*$ we have that $A_F\mathcal{R}^* \subset \mathcal{V}^*$. On the other hand, since $\mathcal{R}^*$ is controlled invariant, $A_F\mathcal{R}^* \subset \mathcal{R}^* + \operatorname{im} B$. Thus we have

$$A_F\mathcal{R}^* \subset (\operatorname{im} B + \mathcal{R}^*) \cap \mathcal{V}^* = (\operatorname{im} B \cap \mathcal{V}^*) + \mathcal{R}^* \subset \mathcal{R}^*.$$

where we used the modular rule for the equality in the middle. This shows that $F \in \underline{F}(\mathcal{R}^*)$. Next, by corollary 4.13, $\mathcal{R}^* = \langle A_F \mid \operatorname{im} B \cap \mathcal{R}^*\rangle$. Moreover, it follows from lemma 4.16 that $\operatorname{im} B \cap \mathcal{R}^* = \operatorname{im} B \cap \mathcal{V}^*$. This completes the proof of the theorem. ■

The above theorem has the following interpretation. By taking $F \in \underline{F}(\mathcal{V}^*)$ and a linear map L such that $\operatorname{im} L = B^{-1}\mathcal{V}^*$ we obtain a new system

$$\dot{x}(t) = (A + BF)x(t) + BLw(t)$$

with state space $\mathcal{V}^*$. This system can be considered as being obtained from the original system by restricting the trajectories to the subspace $\mathcal{V}^*$ and by restricting the input functions to take their values in $B^{-1}\mathcal{V}^*$. Since $\operatorname{im} BL = \operatorname{im} B \cap \mathcal{V}^*$, (4.12) expresses the fact that $\mathcal{R}^*$ is just the reachable subspace of this restricted system.

If $\mathcal{V}$ is a controlled invariant subspace then of course $\mathcal{V} = \mathcal{V}^*(\mathcal{V})$. Let $\mathcal{R} := \mathcal{R}^*(\mathcal{V})$, the largest controllability subspace contained in $\mathcal{V}$. Then theorem 4.17 says that if $F \in \underline{F}(\mathcal{V})$ and L is a linear map such that $\operatorname{im} L = B^{-1}\mathcal{V}$ then

$$\mathcal{R} = \langle A + BF \mid \operatorname{im} BL\rangle. \tag{4.13}$$

Finally we note that it follows from theorem 4.17 that $\mathcal{R}^*(\mathcal{X})$, the largest controllability subspace of the system (4.1), is equal to the reachable subspace $\langle A \mid \operatorname{im} B\rangle$. Indeed, the state space $\mathcal{X}$ itself is of course a controlled invariant subspace so $\mathcal{V}^*(\mathcal{X}) = \mathcal{X}$ and $\underline{F}(\mathcal{X}) = \{F : \mathcal{X} \to \mathcal{U} \mid F \text{ is linear}\}$. It also follows from this that every controllability subspace $\mathcal{V}$ is contained in $\langle A \mid \operatorname{im} B\rangle$.

4.5 Pole placement under invariance constraints

In section 3.10 we have discussed to what extent one can assign the spectrum of the system map using state feedback. In section 4.1 we introduced the class of controlled invariant subspaces and showed that these are characterized by the property that they can be made invariant by state feedback. In the present section we will combine these two issues and ask ourselves the question: how much freedom is left in the assignment of the spectrum of the system map if it is required that a given controlled invariant subspace should be made invariant? More concretely: given a controlled invariant subspace $\mathcal{V}$, what freedom do we have in the assignment of the spectrum of $A + BF$ if we restrict ourselves to $F \in \underline{F}(\mathcal{V})$. The following result gives a complete solution.

Theorem 4.18 *Consider the system (4.1). Let $\mathcal{V}$ be a controlled invariant subspace. Let $\mathcal{R} := \mathcal{R}^*(\mathcal{V})$ be the largest controllability subspace contained in $\mathcal{V}$. Let $\mathcal{S} := \mathcal{V} + \langle A \mid \operatorname{im} B\rangle$. Then we have* (*i*) $\underline{F}(\mathcal{V}) \subset \underline{F}(\mathcal{R}) \cap \underline{F}(\mathcal{S})$.

(*ii*) *Given any pair of real monic polynomials* (p_1, p_2) *with* $\deg p_1 = \dim \mathcal{R}$ *and* $\deg p_2 = \dim \mathcal{S}/\mathcal{V}$ *there exists* $F \in \underline{F}(\mathcal{V})$ *such that the characteristic polynomials of* $A_F \mid \mathcal{R}$ *and* $A_F \mid \mathcal{S}/\mathcal{V}$ *equal* p_1 *and* p_2, *respectively.*

(*iii*) *The map* $A_F \mid \mathcal{V}/\mathcal{R}$ *is independent of* F *for* $F \in \underline{F}(\mathcal{V})$. *The map* $A_F \mid \mathcal{X}/\mathcal{S}$ *is equal to* $A \mid \mathcal{X}/\mathcal{S}$ *for all* F.

The results concerning the freedom of spectral assignability under the constraint that a given controlled invariant subspace should be made invariant is depicted in the lattice diagram in Figure 4.1. Before we establish a proof of this theorem, let us

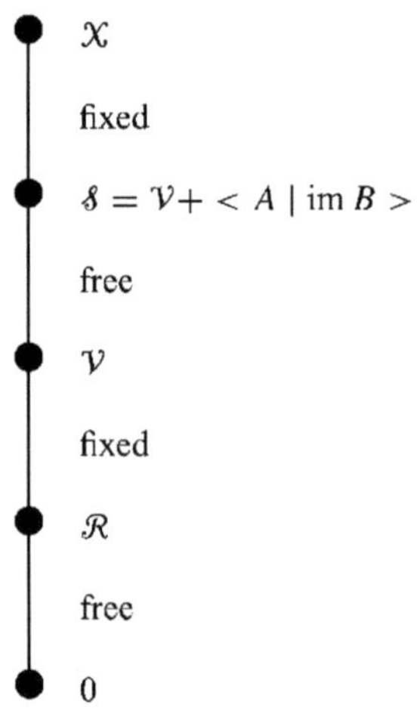

$A + BF,\ F \in \underline{F}(\mathcal{V})$

Figure 4.1

make some remarks. The theorem states that if a feedback map F makes $\mathcal{V}$ invariant under $A + BF$, then it must do the same with $\mathcal{R}$ and $\mathcal{S}$ (see also theorem 4.17). The subspaces $\mathcal{R}$, $\mathcal{V}$ and $\mathcal{S}$ form a chain, that is, they are related by the inclusion relation

$$\mathcal{R} \subset \mathcal{V} \subset \mathcal{S}. \tag{4.14}$$

In order to appreciate the content of theorem 4.18 it is useful to see what it says in terms of partitioned matrices. Choose a basis of the state space $\mathcal{X}$ adapted to the chain (4.14). Accordingly, we can split

$$A = \begin{pmatrix} A_{11} & A_{12} & A_{13} & A_{14} \\ A_{21} & A_{22} & A_{23} & A_{24} \\ A_{31} & A_{32} & A_{33} & A_{34} \\ 0 & 0 & 0 & A_{44} \end{pmatrix}, \quad B = \begin{pmatrix} B_1 \\ B_2 \\ B_3 \\ 0 \end{pmatrix}.$$

Here, the zero blocks appear due to the facts that the subspace $\mathcal{S}$ is A-invariant and contains the subspace im B. A map $F = \begin{pmatrix} F_1 & F_2 & F_3 & F_4 \end{pmatrix}$ is an element of $\underline{F}(\mathcal{V})$ if and only if

$$A_{31} + B_3 F_1 = 0 \text{ and } A_{32} + B_3 F_2 = 0. \tag{4.15}$$

If we restrict ourselves to maps F satisfying (4.15) then automatically $A_{21} + B_2 F_1 = 0$ (see theorem 4.17). Theorem 4.18 asserts that, under the restriction that F_1 should satisfy (4.15), the spectrum of $A_{11} + B_1 F_1$ is freely assignable. Also, the eigenvalues of $A_{33} + B_3 F_3$ can be placed arbitrarily by appropriate choice of F_3. Finally, the theorem states that, under the restriction (4.15) on F_2, the map $A_{22} + B_2 F_2$ is fixed. More specifically, if F_2^1 and F_2^2 satisfy $A_{32} + B_3 F_2^i = 0$ $(i = 1, 2)$ then we have $A_{22} + B_2 F_2^1 = A_{22} + B_2 F_2^2$ and, a fortiori, also $\sigma(A_{22} + B_2 F_2^1) = \sigma(A_{22} + B_2 F_2^2)$. The block A_{44} is not affected by any feedback map F and consequently also $\sigma(A_{44})$ is fixed.

As a consequence of the above theorem, given a pair of real monic polynomials (p_1, p_2) with $\deg p_1 = \dim \mathcal{R}$ and $\deg p_2 = \dim \mathcal{S}/\mathcal{V}$, we can find a linear map $(F_1\ F_2\ F_3\ F_4) = F \in \underline{F}(\mathcal{V})$ such that the characteristic polynomial of $A + BF$ becomes equal to the product $p_1 \cdot q \cdot p_2 \cdot r$. Here q is equal to the characteristic polynomial of $A_{22} + B_2 F_2$ which, as noted before, is the same for all maps F_2 such that $F \in \underline{F}(\mathcal{V})$. The polynomial r is equal to the characteristic polynomial of A_{44}.

In the proof of theorem 4.18 the following lemma will be useful.

Lemma 4.19 *Let $\mathcal{V}$ be a controlled invariant subspace and let $F_0 \in \underline{F}(\mathcal{V})$. Let $F : \mathcal{X} \to \mathcal{U}$ be a linear map. Then $F \in \underline{F}(\mathcal{V})$ if and only if $(F_0 - F)\mathcal{V} \subset B^{-1}\mathcal{V}$.*

Proof : ($\Rightarrow$) Let $x_0 \in \mathcal{V}$. Then $(A + BF_0)x_0 \in \mathcal{V}$ and $(A + BF)x_0 \in \mathcal{V}$. Hence $B(F_0 - F)x_0 \in \mathcal{V}$.

($\Leftarrow$) Let $x_0 \in \mathcal{V}$. Then $B(F_0 - F)x_0 \in \mathcal{V}$. Also $(A + BF_0)x_0 \in \mathcal{V}$. It follows that also $(A + BF)x_0 = (A + BF_0)x_0 - B(F_0 - F)x_0 \in \mathcal{V}$. ■

We will now give a proof of theorem 4.18.

Proof of theorem 4.18: : (i) The fact that $\underline{F}(\mathcal{V}) \subset \underline{F}(\mathcal{R})$ was already proven in theorem 4.17. The subspace $\mathcal{S}$ is invariant under $A + BF$ for any F (see also exercise 4.2).

(ii) Let (p_1, p_2) be a pair of polynomials as in the statement of the theorem. We choose any $F_0 \in \underline{F}(\mathcal{V})$ and $L : \mathbb{R}^k \to \mathcal{U}$ with $\text{im}\, L = B^{-1}\mathcal{V}$, where $k := \dim B^{-1}\mathcal{V}$. Then, according to (4.13), we have

$$\mathcal{R} = \langle A + BF_0 \mid \text{im}\, BL \rangle.$$

Define $A_0 := (A + BF_0) \mid \mathcal{R}$ and $B_0 := BL$. Then the system (A_0, B_0) is controllable (see corollary 3.4) and hence, by theorem 3.29, there exists a map $F_1 : \mathcal{R} \to \mathbb{R}^k$

such that $A_0 + B_0F_1$ has characteristic polynomial p_1. Extend F_1 to a linear map from $\mathcal{X} \to \mathbb{R}^k$. Define $F_2 := F_0 + LF_1$. Since $F_2 - F_0 = LF_1$ and $\operatorname{im} L = B^{-1}\mathcal{V}$, it follows from lemma 4.19 that $\mathcal{V}$ is invariant under $A + BF_2$. Also $\mathcal{S}$ is $(A+BF_2)$-invariant. Let $\Pi : \mathcal{S} \to \mathcal{S}/\mathcal{V}$ be the canonical projection (see section 2.4). Define $A_2 := (A+BF_2) \mid \mathcal{S}/\mathcal{V}$ and let $B_2 : \mathcal{U} \to \mathcal{S}/\mathcal{V}$ be defined by $B_2 := \Pi B$. We claim that the system (A_2, B_2) is controllable. We will show that $\mathcal{S}/\mathcal{V} = \langle A_2 \mid \operatorname{im} B_2\rangle$. Let $\bar{x} \in \mathcal{S}/\mathcal{V}$, say $\bar{x} = \Pi x$ with $x \in \mathcal{S}$. Then x can be written as $x = x_1 + x_2$ with $x_1 \in \langle A \mid \operatorname{im} B\rangle = \langle A + BF_2 \mid \operatorname{im} B\rangle$ and $x_2 \in \mathcal{V}$. Since $\mathcal{V} = \ker \Pi$ we have that in fact $\bar{x} = \Pi x_1$. There are $u_0, \ldots, u_{n-1} \in \mathcal{U}$ such that $x_1 = \sum_i (A+BF_2)^i Bu_i$. Thus

$$\begin{aligned}\bar{x} = \Pi x_1 &= \sum_i \Pi(A + BF_2)^i Bu_i \\ &= \sum_i A_2^i B_2 u_i \in \langle A_2 \mid \operatorname{im} B_2\rangle.\end{aligned}$$

Here, we have used the fact that $\Pi(A + BF_2) = A_2\Pi$ and $B_2 = \Pi B$. This proves our claim. Now, by theorem 3.29 there exists a map $\bar{F}_3 : \mathcal{S}/\mathcal{V} \to \mathcal{U}$ such that the characteristic polynomial of $A_2 + B_2\bar{F}_3$ equals p_2. Define $F_3 : \mathcal{S} \to \mathcal{U}$ by $F_3 := \bar{F}_3\Pi$ and extend F_3 to a map on $\mathcal{X}$. Define $F := F_2 + F_3$. Then $(A+BF)\mathcal{V} = (A + BF_2)\mathcal{V} \subset \mathcal{V}$ so $F \in \underline{F}(\mathcal{V})$. Also, $(A + BF) \mid \mathcal{S}/\mathcal{V} = A_2 + B_2\bar{F}_3$ since

$$(A_2 + B_2\bar{F}_3)\Pi = A_2\Pi + \Pi BF_3 = \Pi(A + BF_2) + \Pi BF_3 = \Pi(A + BF).$$

Thus, the characteristic polynomial of $(A + BF) \mid \mathcal{S}/\mathcal{V}$ is equal to p_2. It remains to be shown that the characteristic polynomial of $(A + BF) \mid \mathcal{R}$ equals p_1. This however follows from the fact that $(A + BF) \mid \mathcal{R} = A_0 + B_0F_1$.

(iii) Let $F_1, F_2 \in \underline{F}(\mathcal{V})$. According to lemma 4.19 we have $(F_1 - F_2)\mathcal{V} \subset B^{-1}\mathcal{V}$. Hence $B(F_1 - F_2)\mathcal{V} \subset \mathcal{V} \cap \operatorname{im} B \subset \mathcal{R}$ (see lemma 4.16). Let $\Pi_1 : \mathcal{V} \to \mathcal{V}/\mathcal{R}$ be the canonical projection. Since $\mathcal{R} = \ker \Pi_1$ we have $\Pi_1 B(F_1 - F_2)\mathcal{V} = 0$. Denote $A_1 := (A+BF_1) \mid \mathcal{V}/\mathcal{R}$ and $A_2 := (A+BF_2) \mid \mathcal{V}/\mathcal{R}$. Let $\bar{x} \in \mathcal{V}/\mathcal{R}$, say $\bar{x} = \Pi_1 x$ with $x \in \mathcal{V}$. Then $(A_1 - A_2)\bar{x} = (A_1\Pi_1 - A_2\Pi_1)x = (\Pi_1(A+BF_1) - \Pi_1(A+BF_2))x = \Pi_1 B(F_1 - F_2)x = 0$. Thus $A_1 = A_2$ and hence the map $(A + BF) \mid \mathcal{V}/\mathcal{R}$ is independent of F for $F \in \underline{F}(\mathcal{V})$.

Finally, let $\Pi_2 : \mathcal{X} \to \mathcal{X}/\mathcal{S}$ be the canonical projection. Since $\operatorname{im} B \subset \mathcal{S}$ and $\mathcal{S} = \ker \Pi_2$ we have $\Pi_2 B = 0$. Let F_1, F_2 be linear maps $\mathcal{X} \to \mathcal{U}$. Note that $\mathcal{S}$ is $(A + BF_i)$-invariant $(i = 1, 2)$ and define $A_i := (A + BF_i) \mid \mathcal{X}/\mathcal{S}$. Let $\bar{x} \in \mathcal{X}/\mathcal{S}$, say $\bar{x} = \Pi_2 x$. Then $(A_1 - A_2)\bar{x} = (A_1\Pi_2 - A_2\Pi_2)x = (\Pi_2(A + BF_1) - \Pi_2(A + BF_2))x = \Pi_2 B(F_1 - F_2)x = 0$. Thus $A_1 = A_2$ and $(A + BF) \mid \mathcal{X}/\mathcal{S} = A \mid \mathcal{X}/\mathcal{S}$ for every F. ■

We close this section by applying theorem 4.18 to obtain the following characterization of controllability subspaces.

Theorem 4.20 *Consider the system (4.1). Let $\mathcal{V}$ be a subspace of $\mathcal{X}$. The following statements are equivalent:*

(i) *$\mathcal{V}$ is a controllability subspace,*

(ii) *for all for all $\lambda \in \mathbb{C}$ we have $(\lambda I - A)\mathcal{V} + \operatorname{im} B = \mathcal{V} + \operatorname{im} B$,*

(iii) *for each real monic polynomial p with* $\deg p = \dim \mathcal{V}$, *there exists $F \in \underline{F}(\mathcal{V})$ such that the characteristic polynomial of $A_F \mid \mathcal{V}$ equals p.*

Proof : (i) $\Rightarrow$ (iii). Follows by applying theorem 4.18 to the subspace $\mathcal{V}$. Note that $\mathcal{R}^*(\mathcal{V}) = \mathcal{V}$.

(iii) $\Rightarrow$ (ii). From the fact that $\underline{F}(\mathcal{V}) \neq \varnothing$ it follows that $\mathcal{V}$ is controlled invariant. Hence $A\mathcal{V} \subset \mathcal{V} + \operatorname{im} B$ (see theorem 4.2) and consequently $(\lambda I - A)\mathcal{V} + \operatorname{im} B \subset \mathcal{V} + \operatorname{im} B$ for all $\lambda \in \mathbb{C}$. To prove the converse inclusion it suffices to show that $\mathcal{V} \subset (\lambda I - A)\mathcal{V} + \operatorname{im} B$ for all λ. Let $\lambda \in \mathbb{C}$. Pick a real monic polynomial p with $\deg p = \dim \mathcal{V}$ such that $p(\lambda) \neq 0$. There is $F \in \underline{F}(\mathcal{V})$ such that the characteristic polynomial of $A_F \mid \mathcal{V}$ equals p. It follows that $\lambda \notin \sigma(A + BF \mid \mathcal{V})$ so the map $(\lambda I - A - BF) \mid \mathcal{V}$ must be regular. Also, $(\lambda I - A - BF)\mathcal{V} \subset \mathcal{V}$ and consequently we must in fact have $(\lambda I - A - BF)\mathcal{V} = \mathcal{V}$. It follows that $\mathcal{V} \subset (\lambda I - A)\mathcal{V} + \operatorname{im} B$.

(ii) $\Rightarrow$ (i). If (ii) holds then also $A\mathcal{V} \subset \mathcal{V} + \operatorname{im} B$ so $\mathcal{V}$ is controlled invariant. For each linear map F we have

$$(\lambda I - A - BF)\mathcal{V} + \operatorname{im} B = \mathcal{V} + \operatorname{im} B \tag{4.16}$$

for all $\lambda \in \mathbb{C}$. If in (4.16) we take $F \in \underline{F}(\mathcal{V})$ and intersect both sides of the equation with $\mathcal{V}$ we obtain

$$(\lambda I - A - BF)\mathcal{V} + (\operatorname{im} B \cap \mathcal{V}) = \mathcal{V} \tag{4.17}$$

for all $\lambda \in \mathbb{C}$. Let L be a linear map such that $\operatorname{im} L = B^{-1}\mathcal{V}$, say $L : \mathbb{R}^k \to \mathcal{U}$. Then $\operatorname{im} B \cap \mathcal{V} = \operatorname{im} BL$ and (4.17) becomes

$$(\lambda I - A - BF)\mathcal{V} + BL\mathbb{R}^k = \mathcal{V} \tag{4.18}$$

for all λ. By theorem 3.13 (compare (3.12)) this implies that the system $(A+BF, BL)$ with state space $\mathcal{V}$ and input space $\mathbb{R}^k$ is controllable. Hence, by corollary 3.4, $\mathcal{V} = \langle A + BF \mid \operatorname{im} BL \rangle$. Finally, apply theorem 4.12. ■

4.6 Stabilizability subspaces

In this section we introduce the notion of stabilizability subspace. Consider the system (4.1). From section 2.6, recall that if we choose an input function that is Bohl, then for any initial state also the resulting state trajectory is Bohl. Moreover, for a Bohl trajectory x we defined its spectrum $\sigma(x)$ and we called a Bohl trajectory x stable with respect to a given stability domain $\mathbb{C}_g$ if $\sigma(x) \subset \mathbb{C}_g$.

Let $\mathbb{C}_g$ be a stability domain. A subspace $\mathcal{V}$ of the state space is called a stabilizability subspace if it has the following property: for each initial condition in the subspace there is a Bohl input such that the resulting state trajectory remains in the subspace and is stable.

Definition 4.21 *A subspace $\mathcal{V} \subset \mathcal{X}$ is called a* stabilizability subspace *if for any $x_0 \in \mathcal{V}$ there exists a Bohl function u such that $x_u(t, x_0) \in \mathcal{V}$ for all $t \geqslant 0$ and $x_u(\cdot, x_0)$ is stable.*

An important special case is obtained by taking the stability set $\mathbb{C}_g$ to be equal to $\mathbb{C}^-$, the open left half complex plane. Since a Bohl function converges to zero as t tends to infinity if and only if its spectrum is contained in $\mathbb{C}^-$, for this particular case the requirement that $x_u(\cdot, x_0)$ should be stable is equivalent to the condition $x_u(t, x_0) \to 0$ $(t \to \infty)$.

Note that every stabilizability subspace is controlled invariant. The sum of any number of stabilizability subspaces is a stabilizability subspace. It also follows from the definition that the property of being a stabilizability subspace is invariant under state feedback and isomorphisms of the input space. In the following, for a given stability domain $\mathbb{C}_g$, let $\mathbb{C}_b$ be its complement in $\mathbb{C}$. Stabilizability subspaces can be characterized as follows:

Theorem 4.22 *Consider the system (4.1) and let $\mathbb{C}_g$ be a stability domain. Let $\mathcal{V}$ be a subspace of $\mathcal{X}$. Then the following statements are equivalent:*

(i) *$\mathcal{V}$ is a stabilizability subspace,*

(ii) *for all $\lambda \in \mathbb{C}_b$ we have $(\lambda I - A)\mathcal{V} + \operatorname{im} B = \mathcal{V} + \operatorname{im} B$,*

(iii) *there exists $F \in \underline{F}(\mathcal{V})$ such that $\sigma(A_F \mid \mathcal{V}) \subset \mathbb{C}_g$.*

Proof : (i) $\Rightarrow$ (ii). For any F, condition (ii) is equivalent to

$$(\lambda I - A_F)\mathcal{V} + \operatorname{im} B = \mathcal{V} + \operatorname{im} B \quad \text{for all } \lambda \in \mathbb{C}_b. \tag{4.19}$$

Let $F \in \underline{F}(\mathcal{V})$. We claim that in this case (4.19) is equivalent to

$$(\lambda I - A_F)\mathcal{V} + (\operatorname{im} B \cap \mathcal{V}) = \mathcal{V} \quad \text{for all } \lambda \in \mathbb{C}_b. \tag{4.20}$$

Indeed, (4.20) follows from (4.19) by taking the intersection with $\mathcal{V}$ on both sides of the equation. The converse can be verified immediately. Now assume that (ii) does not hold. Then by the previous there must be a $\lambda_0 \in \mathbb{C}_b$ for which the equality in (4.20) does not hold. Since $F \in \underline{F}(\mathcal{V})$ we do have

$$(\lambda_0 I - A_F)\mathcal{V} + (\operatorname{im} B \cap \mathcal{V}) \subset \mathcal{V}. \tag{4.21}$$

Consequently, the inclusion in (4.21) must be *strict*. Hence, there exists a nonzero row vector η such that $\eta \perp (\lambda_0 I - A_F)\mathcal{V} + (\operatorname{im} B \cap \mathcal{V})$ but *not* $\eta \perp \mathcal{V}$. Let $x_0 \in \mathcal{V}$ such that $\eta x_0 \neq 0$. Since $\mathcal{V}$ is a stabilizability subspace there exists a Bohl function u such that if $x(t)$ satisfies $\dot{x} = A_F x + Bu, x(0) = x_0$, we have $x(t) \in \mathcal{V}$ for all $t \geqslant 0$ and $\sigma(x) \subset \mathbb{C}_g$. Since $x(t) \in \mathcal{V}$ for all $t \geqslant 0$, also $\dot{x}(t) \in \mathcal{V}$ and $A_F x(t) \in \mathcal{V}$ for all $t \geqslant 0$. Hence, $Bu(t) \in \mathcal{V} \cap \operatorname{im} B$ for all $t \geqslant 0$ and therefore $\eta B u(t) = 0$, $t \geqslant 0$. Also, since $\eta \perp (\lambda_0 I - A_F)\mathcal{V}$, we have $\eta(\lambda_0 I - A_F)x(t) = 0$ for all t and hence $\eta A_F x(t) = \lambda_0 \eta x(t)$ for all t. Now define $z(t) := \eta x(t)$. Then $z(0) = \eta x_0 \neq 0$ and z satisfies the differential equation $\dot{z}(t) = \lambda_0 z(t)$. It follows that $\eta^* x(t) = z(t) = e^{\lambda_0 t}\eta^* x_0$. Thus $\lambda_0 = \sigma(\eta^* x)$. Obviously, $\sigma(\eta^* x) \subset \sigma(x) \subset \mathbb{C}_g$. Since $\lambda_0 \in \mathbb{C}_b$, this yields a contradiction.

(ii) $\Rightarrow$ (iii). If (ii) holds then $A\mathcal{V} \subset \mathcal{V} + \operatorname{im} B$ so $\mathcal{V}$ is controlled invariant. Take $F_0 \in \underline{F}(\mathcal{V})$. Then (ii) is equivalent to (4.20) with F replaced by F_0. Let L be a linear map such that $\operatorname{im} L = B^{-1}\mathcal{V}$, say $L : \mathbb{R}^k \to \mathcal{U}$. Then $\operatorname{im} B \cap \mathcal{V} = \operatorname{im} BL$ so (4.20) yields $(\lambda I - A_{F_0})\mathcal{V} + BL\mathbb{R}^k = \mathcal{V}$ for all $\lambda \in \mathbb{C}_b$. It follows from theorem 3.13 that the system (A_{F_0}, BL) with state space $\mathcal{V}$ and input space $\mathbb{R}^k$ is stabilizable. Hence there is $F_1 : \mathcal{V} \to \mathbb{R}^k$ such that $\sigma((A_{F_0} + BLF_1) \mid \mathcal{V}) \subset \mathbb{C}_g$. Extend F_1 to a linear map on $\mathcal{X}$ and define $F := F_0 + LF_1$. Then $\sigma(A_F \mid \mathcal{V}) \subset \mathbb{C}_g$ and since $\operatorname{im}(F - F_0) \subset B^{-1}\mathcal{V}$ also $F \in \underline{F}(\mathcal{V})$ (see lemma 4.19).

(iii) $\Rightarrow$ (i). Let $F \in \underline{F}(\mathcal{V})$ with $\sigma(A_F \mid \mathcal{V}) \subset \mathbb{C}_g$. Denote $A_0 := A_F \mid \mathcal{V}$ and apply state feedback $u = Fx$. The resulting state trajectory is given by $x(t) = e^{A_0 t}x_0$. Obviously for $x_0 \in \mathcal{V}$ we have that $x(t) \in \mathcal{V}$ for all t. Then it follows from theorem 2.6 that the spectrum of x must be contained in $\sigma(A_0)$ which is contained in $\mathbb{C}_g$. Finally note that x is equal to the state trajectory resulting from the Bohl input $u(t) = Fe^{A_0 t}x_0$. ■

As already noted in the proof of the previous theorem, for any $F \in \underline{F}(\mathcal{V})$ and any map L such that $\operatorname{im} L = B^{-1}\mathcal{V}$ condition (ii) is equivalent to saying that the system $\dot{x} = A_F x + BLw$ with state space $\mathcal{V}$ is stabilizable. In this sense, a stabilizability subspace can be considered as a subspace to which the original system can be restricted by suitable restriction of the input functions, such that the restricted system is stabilizable. From theorem 4.20 it follows that, in the same sense, a controllability subspace is a subspace for which the restricted system is controllable. Note that, given any stability domain $\mathbb{C}_g$, every controllability subspace is a stabilizability subspace.

Using the feedback characterization in theorem 4.22 (iii) it is possible to characterize stabilizability subspaces in terms of the spectral assignability properties of controlled invariant subspaces studied in the previous section. For a given controlled invariant subspace $\mathcal{V}$, denote $\mathcal{R} := \mathcal{R}^*(\mathcal{V})$. It was shown in theorem 4.18 that the map $A_F \mid \mathcal{V}/\mathcal{R}$ is independent of F for $F \in \underline{F}(\mathcal{V})$. If $\mathcal{V}$ is a stabilizability subspace then there exists $F \in \underline{F}(\mathcal{V})$ such that $\sigma(A_F \mid \mathcal{V}) \subset \mathbb{C}_g$. Since

$$\sigma(A_F \mid \mathcal{V}) = \sigma(A_F \mid \mathcal{R}) \cup \sigma(A_F \mid \mathcal{V}/\mathcal{R})$$

this implies that the fixed spectrum $\sigma(A_F \mid \mathcal{V}/\mathcal{R})$ is contained in $\mathbb{C}_g$. Conversely, if $\mathcal{V}$ is a controlled invariant subspace such that the fixed spectrum $\sigma(A_F \mid \mathcal{V}/\mathcal{R}) \subset \mathbb{C}_g$ then obviously one can find a $F_1 \in \underline{F}(\mathcal{V})$ such that $\sigma(A_{F_1} \mid \mathcal{V}) \subset \mathbb{C}_g$ (since the characteristic polynomial of $A_{F_1} \mid \mathcal{R}$ can be chosen arbitrarily). Thus we have shown:

Corollary 4.23 *Let $\mathcal{V}$ be a controlled invariant subspace. Then $\mathcal{V}$ is a stabilizability subspace if and only if $\sigma(A + BF \mid \mathcal{V}/\mathcal{R}) \subset \mathbb{C}_g$ for any $F \in \underline{F}(\mathcal{V})$.*

If $\mathcal{K}$ is an arbitrary subspace then we want to consider the largest stabilizability subspace contained in $\mathcal{K}$.

Definition 4.24 *Let $\mathbb{C}_g$ be a stability set and let $\mathcal{K}$ be a subspace of $\mathcal{X}$. Then we define*

$$\mathcal{V}_g^*(\mathcal{K}) := \{x_0 \in \mathcal{X} \mid \text{ there is a Bohl function } u \text{ such that } x_u(t, x_0) \in \mathcal{K} \text{ for all } t \geqslant 0 \text{ and } x_u(\cdot, x_0) \text{ is } \mathbb{C}_g\text{-stable.}\}$$

Theorem 4.25 *Let $\mathcal{K}$ be a subspace of $\mathcal{X}$. Then $\mathcal{V}_g^*(\mathcal{K})$ is the largest stabilizability subspace contained in $\mathcal{K}$, i.e.,*

(i) *$\mathcal{V}_g^*(\mathcal{K})$ is a stabilizability subspace,*

(ii) *$\mathcal{V}_g^*(\mathcal{K}) \subset \mathcal{K}$,*

(iii) *if $\mathcal{V} \subset \mathcal{K}$ is a stabilizability subspace then $\mathcal{V} \subset \mathcal{V}_g^*(\mathcal{K})$.*

The proof is similar to that of theorem 4.5 and is left as an exercise to the reader.

Sometimes we denote $\mathcal{V}_g^*(\mathcal{K})$ by $\mathcal{V}_g^*(\mathcal{K}, A, B)$. It is easily verified that for a given subspace $\mathcal{K}$ the following relation holds:

$$\mathcal{R}^*(\mathcal{K}) \subset \mathcal{V}_g^*(\mathcal{K}) \subset \mathcal{V}^*(\mathcal{K}) \subset \mathcal{K}.$$

More specifically, we have $\mathcal{R}^*(\mathcal{V}_g^*(\mathcal{K})) = \mathcal{R}^*(\mathcal{K})$ and $\mathcal{V}_g^*(\mathcal{V}^*(\mathcal{K})) = \mathcal{V}_g^*(\mathcal{K})$. In particular, if we take $\mathcal{K} = \mathcal{X}$ we obtain the largest stabilizability subspace of the system (4.1), $\mathcal{V}_g^*(\mathcal{X})$. This subspace will be called the *stabilizable subspace of* (A, B) and will be denoted by $\mathcal{X}_{\text{stab}}$ or $\mathcal{X}_{\text{stab}}(A, B)$ (see also exercise 3.24). This subspace consists exactly of those points in which a stable state trajectory starts:

$$\mathcal{X}_{\text{stab}} = \{x_0 \in \mathcal{X} \mid \text{there is a Bohl function } u \text{ such that } x_u(\cdot, x_0) \text{ is stable.}\} \tag{4.22}$$

According to the following result, the stabilizable subspace is equal to the sum of the stable subspace of A (see definition 2.13) and the reachable subspace of (A, B):

Theorem 4.26 $\mathcal{X}_{\text{stab}} = \mathcal{X}_g(A) + \langle A \mid \text{im } B \rangle$.

Proof : ($\supset$). Obviously, both $\mathcal{X}_g(A)$ and $\langle A \mid \text{im } B \rangle$ are stabilizability subspaces and hence also their sum. This sum is contained in the largest stabilizability subspace.

($\subset$) In this proof, denote $\mathcal{V} := \mathcal{X}_g(A) + \langle A \mid \text{im } B \rangle$. It is easily seen that both $\mathcal{X}_{\text{stab}}$ as well as $\mathcal{V}$ are A-invariant. Let $\Pi : \mathcal{X}_{\text{stab}} \to \mathcal{X}_{\text{stab}}/\mathcal{V}$ be the canonical projection and denote $A_0 := A \mid \mathcal{X}_{\text{stab}}/\mathcal{V}$. We claim that $\sigma(A_0) \subset \mathbb{C}_b$. Indeed,

$$\sigma(A \mid \mathcal{X}_{\text{stab}}/\mathcal{V}) \subset \sigma(A \mid \mathcal{X}/\mathcal{V}) \subset \sigma(A \mid \mathcal{X}/\mathcal{X}_g(A)).$$

The latter spectrum is of course equal to $\sigma(A \mid \mathcal{X}_b(A))$, which is contained in $\mathbb{C}_b$. Now assume that $\mathcal{V} \subset \mathcal{X}_{\text{stab}}$ with *strict* inclusion. Then there is $x_0 \in \mathcal{X}_{\text{stab}}$ with $x_0 \notin \mathcal{V}$. By (4.22) there is a Bohl function u such that the resulting trajectory x is stable. Let $\bar{x}(t) := \Pi x(t)$. Then $\bar{x}$ satisfies

$$\dot{\bar{x}} = \Pi\dot{x} = \Pi A x + \Pi B u = A_0 \Pi x = A_0 \bar{x}.$$

Here we used the facts that $\operatorname{im} B \subset \mathcal{V} = \ker \Pi$ and $\Pi A = A_0 \Pi$. Since $\bar{x}(0) = \Pi x_0 \neq 0$, $\bar{x}$ is unstable. This contradicts the assumption that x and hence Πx is stable. ■

The above result can also be used to obtain an expression for the largest stabilizability subspace contained in an arbitrary subspace of the state space. Let $\mathcal{K}$ be a subspace of $\mathcal{X}$ and let $\mathcal{V}^*$ be the largest controlled invariant subspace in $\mathcal{K}$. Take an arbitrary $F \in \underline{F}(\mathcal{V}^*)$ and let L be a linear map such that $\operatorname{im} L = B^{-1}\mathcal{V}^*$, say $L : \mathbb{R}^k \to \mathcal{U}$. Consider the restricted system $\dot{x}(t) = A_F x(t) + BLw(t)$ with state space $\mathcal{V}^*$ and input space $\mathbb{R}^k$. Temporarily, denote the stabilizable subspace of the restricted system by $\tilde{\mathcal{X}}_{\text{stab}}$. By (4.22),

$$\tilde{\mathcal{X}}_{\text{stab}} = \{x_0 \in \mathcal{V}^* \mid \text{there is a Bohl function } w \text{ such that the solution } x(t) \text{ of } \dot{x} = A_F x + BLw,\ x(0) = x_0 \text{ is stable}\}.$$

We claim that the largest stabilizability subspace in $\mathcal{K}$ is equal to the stabilizable subspace of the restricted system, i.e. $\mathcal{V}_g^*(\mathcal{K}) = \tilde{\mathcal{X}}_{\text{stab}}$. To prove this, first recall that $\mathcal{V}_g^*(\mathcal{K}) = \mathcal{V}_g^*(\mathcal{V}^*)$. Let $x_0 \in \mathcal{V}_g^*(\mathcal{K})$. By definition 4.24 there exists a Bohl function u such that $x_u(t, x_0) \in \mathcal{V}^*$ for all $t \geqslant 0$ and $x_u(\cdot, x_0)$ is $\mathbb{C}_g$-stable. It follows from theorem 4.3 that the control u must be of the form

$$u(t) = F x_u(t, x_0) + Lw(t)$$

for some Bohl function w. Thus $x_u(t, x_0)$ satisfies $\dot{x} = A_F x + BLw$, $x(0) = x_0$. This shows that $x_0 \in \tilde{\mathcal{X}}_{\text{stab}}$. The converse inclusion, i.e. the inclusion $\tilde{\mathcal{X}}_{\text{stab}} \subset \mathcal{V}_g^*(\mathcal{K})$ is left as an exercise to the reader. By applying theorem 4.26 we find that

$$\mathcal{V}_g^*(\mathcal{K}) = \mathcal{X}_g(A_F \mid \mathcal{V}^*) + \langle A_F \mid \operatorname{im} BL\rangle.$$

In theorem 4.17 it was shown that $\langle A_F \mid \operatorname{im} BL\rangle$ (the reachable subspace of the restricted system) is equal to $\mathcal{R}^*(\mathcal{K})$. Moreover, since $\mathcal{V}^*$ is invariant under $A + BF$, we have

$$\mathcal{X}_g(A_F \mid \mathcal{V}^*) = \mathcal{X}_g(A_F) \cap \mathcal{V}^*.$$

Thus we obtain the following characterization of the largest stabilizability subspace contained in a given subspace:

Corollary 4.27 *Let $\mathcal{K}$ be a subspace of $\mathcal{X}$. Then for all $F \in \underline{F}(\mathcal{V}^*(\mathcal{K}))$ we have*

$$\mathcal{V}_g^*(\mathcal{K}) = \mathcal{X}_g(A + BF) \cap \mathcal{V}^*(\mathcal{K}) + \mathcal{R}^*(\mathcal{K}).$$

In theorem 4.22 we characterized stabilizability subspaces as controlled invariant subspaces $\mathcal{V}$ for which there exist $F \in \underline{F}(\mathcal{V})$ such that the restriction of $A+BF$ to $\mathcal{V}$ has all its eigenvalues in $\mathbb{C}_g$. Sometimes it will be important to know something about the spectrum of the map induced by $A+BF$ on the factor space $\mathcal{X}/\mathcal{V}$. A controlled invariant subspace $\mathcal{V}$ will be called *outer stabilizable* if there exists an $F \in \underline{F}(\mathcal{V})$ such that $\sigma(A+BF \mid \mathcal{X}/\mathcal{V}) \subset \mathbb{C}_g$. Using the terminology of section 2.7 this can be stated alternatively as: if there exists $F \in \underline{F}(\mathcal{V})$ such that the subspace $\mathcal{V}$ is outer stable with respect to the map $A+BF$. Correspondingly, stabilizability subspaces will sometimes be called *inner stabilizable* controlled invariant subspaces. In order to illustrate these concepts, let $\mathcal{V}$ be a controlled invariant subspace for the system (4.1). Choose a basis for $\mathcal{X}$ adapted to $\mathcal{V}$. Accordingly, split

$$A = \begin{pmatrix} A_{11} & A_{12} \\ A_{21} & A_{22} \end{pmatrix}, \quad B = \begin{pmatrix} B_1 \\ B_2 \end{pmatrix}, \quad x = \begin{pmatrix} x_1 \\ x_2 \end{pmatrix}.$$

Now, first assume that $\mathcal{V}$ is inner stabilizable. Then there is a map $F = (F_1 \;\; F_2)$ such that $A_{21} + B_2F_1 = 0$ and $\sigma(A_{11} + B_1F_1) \subset \mathbb{C}_g$. By taking $F_2 = 0$ we thus obtain

$$A + BF = \begin{pmatrix} A_{11} + B_1F_1 & A_{12} \\ 0 & A_{22} \end{pmatrix}.$$

Let $x_0 \in \mathcal{V}$. Then with respect to the above choice of basis for $\mathcal{X}$ we have $x_0 = (x_{10}^{\mathrm{T}}, 0)^{\mathrm{T}}$. Apply the state feedback control $u = Fx$. The trajectory resulting from x_0 and u satisfies the equations

$$\begin{aligned} \dot{x}_1(t) &= (A_{11} + B_1F_1)x_1(t) + A_{12}x_2(t), \;\; x_1(0) = x_{10}, \\ \dot{x}_2(t) &= A_{22}x_2(t), \;\; x_2(0) = 0. \end{aligned}$$

Thus $x_2(t) = 0$ $(t \geqslant 0)$ and $x_1(t) = e^{(A_{11}+B_1F_1)t}x_{10}$ $(t \geqslant 0)$. The fact that $x_2(t) = 0$ for $t \geqslant 0$ expresses the fact that the trajectory $x_u(t, x_0)$ remains in $\mathcal{V}$ for all $t \geqslant 0$. The expression for $x_1(t)$ displays the fact that the spectrum of $x_u(\cdot, x_0)$ lies in $\mathbb{C}_g$. In particular, if $\mathbb{C}_g = \mathbb{C}^-$ then $x_1(t) \to 0$ as $t \to \infty$.

Next, instead of inner stabilizable, let us assume that $\mathcal{V}$ is outer stabilizable. Then there is a map $F = (F_1 \;\; F_2)$ such that $A_{21} + B_2F_1 = 0$ and $\sigma(A_{22} + B_2F_2) \subset \mathbb{C}_g$. Thus

$$A + BF = \begin{pmatrix} A_{11} + B_1F_1 & A_{12} + B_1F_2 \\ 0 & A_{22} + B_2F_2 \end{pmatrix}.$$

Take any $x_0 = (x_{10}^{\mathrm{T}}, x_{20}^{\mathrm{T}})^{\mathrm{T}} \in \mathcal{X}$. Apply the state feedback control $u = Fx$. Then the trajectory resulting from x_0 and u is given by

$$\begin{aligned} \dot{x}_1(t) &= (A_{11} + B_1F_1)x_1(t) + (A_{12} + B_1F_2)x_2(t), \;\; x_1(0) = x_{10}, \\ \dot{x}_2(t) &= (A_{22} + B_2F_2)x_2(t), \;\; x_2(0) = x_{20}. \end{aligned}$$

Consequently, $x_2(t) = e^{(A_{22}+B_2F_2)t}x_{20}$. Thus, if we assume that $\mathbb{C}_g = \mathbb{C}^-$ then $x_2(t) \to 0$ $(t \to \infty)$. This expresses the fact that the trajectory $x_u(t, x_0)$ converges to the subspace $\mathcal{V}$ as $t \to \infty$. We see that, assuming that $\mathbb{C}_g = \mathbb{C}^-$, an outer stabilizable controlled invariant subspace has the property that there exists a state feedback such that all trajectories starting in the subspace remain in it, while all other trajectories converge to that subspace as $t \to \infty$. An inner stabilizable subspace has the property that there exists a state feedback such that all trajectories starting in the subspace remain in the subspace and converge to the origin as $t \to \infty$.

We will now establish a criterion for a controlled invariant subspace to be outer stabilizable. In the following, again let $\mathbb{C}_g$ be an arbitrary stability domain.

Lemma 4.28 *Let $\mathcal{V}$ be a controlled invariant subspace. Then $\mathcal{V}$ is outer stabilizable if and only if*

$$\sigma\,(A \mid \mathcal{X}/(\mathcal{V} + \langle A \mid \operatorname{im} B\rangle)) \subset \mathbb{C}_g.$$

Proof : Denote $\mathcal{S} := \mathcal{V} + \langle A \mid \operatorname{im} B\rangle$.

($\Rightarrow$) Let $F \in \underline{F}(\mathcal{V})$ be a map such that $\sigma(A_F \mid \mathcal{X}/\mathcal{V}) \subset \mathbb{C}_g$. Since both $\mathcal{S}$ and $\mathcal{V}$ are invariant under A_F and since, by theorem 4.18, $A_F \mid \mathcal{X}/\mathcal{S}$ is equal to $A \mid \mathcal{X}/\mathcal{S}$ we have

$$\sigma(A \mid \mathcal{X}/\mathcal{S}) = \sigma(A_F \mid \mathcal{X}/\mathcal{S}) \subset \sigma(A_F \mid \mathcal{X}/\mathcal{V}) \subset \mathbb{C}_g.$$

($\Leftarrow$) Let p be a real monic polynomial with all its zeros in $\mathbb{C}_g$, with $\deg p = \dim \mathcal{S}/\mathcal{V}$. According to theorem 4.18 there is $F \in \underline{F}(\mathcal{V})$ such that the characteristic polynomial of $A_F \mid \mathcal{S}/\mathcal{V}$ equals p. Hence $\sigma(A_F \mid \mathcal{S}/\mathcal{V}) \subset \mathbb{C}_g$. It follows that

$$\begin{aligned}\sigma(A_F \mid \mathcal{X}/\mathcal{V}) &= \sigma(A_F \mid \mathcal{X}/\mathcal{S}) \cup \sigma(A_F \mid \mathcal{S}/\mathcal{V})\\ &= \sigma(A \mid \mathcal{X}/\mathcal{S}) \cup \sigma(A_F \mid \mathcal{S}/\mathcal{V}) \subset \mathbb{C}_g.\end{aligned}$$

■

Theorem 4.29 *Let $\mathcal{V}$ be a controlled invariant subspace. Then $\mathcal{V}$ is outer stabilizable if and only if*

$$\mathcal{V} + \mathcal{X}_{\text{stab}} = \mathcal{X}. \tag{4.23}$$

Proof : ($\Rightarrow$) Assume that (4.23) does not hold. By theorem 4.26 we then have $\mathcal{S} + \mathcal{X}_g(A) \subset \mathcal{X}$ with *strict* inclusion. Thus

$$\sigma\left(A \mid \mathcal{X}/(\mathcal{S} + \mathcal{X}_g(A))\right) \neq \varnothing. \tag{4.24}$$

The spectrum (4.24) is contained in $\sigma(A \mid \mathcal{X}/\mathcal{S})$. By lemma 4.28, the latter is contained in $\mathbb{C}_g$. On the other hand, the spectrum (4.24) is contained in $\sigma(A \mid \mathcal{X}/\mathcal{X}_g(A))$, which is contained in $\mathbb{C}_b$. This yields a contradiction.

($\Leftarrow$) Assume that (4.23) holds. Then we have

$$\begin{aligned}\sigma(A \mid \mathcal{X}/\mathcal{S}) &= \sigma(A \mid (\mathcal{S} + \mathcal{X}_g(A))/\mathcal{S}) \\ &= \sigma(A \mid \mathcal{X}_g(A)/(\mathcal{S} \cap \mathcal{X}_g(A))) \\ &\subset \sigma(A \mid \mathcal{X}_g(A)) \subset \mathbb{C}_g.\end{aligned}$$

It follows from (4.28) that $\mathcal{V}$ is outer stabilizable. ■

To conclude this section we will study the connection between the concepts introduced here and stabilizability of the system (4.1). Recall from section 3.10 that (A, B) is called stabilizable if there exists an F such that $A + BF$ is stable. Obviously, (A, B) is stabilizable if and only if the state space $\mathcal{X}$ is inner stabilizable. According to theorem 4.22 this is equivalent with $(\lambda I - A)\mathcal{X} + B\mathcal{U} = \mathcal{X}$ for all $\lambda \in \mathbb{C}_b$. Thus we recover theorem 3.32. On the other hand, (A, B) is stabilizable if and only if the zero-subspace is outer stabilizable. Using this observation we obtain

Theorem 4.30 *The following statements are equivalent:*

(i) (A, B) *is stabilizable,*

(ii) $\sigma(A \mid \mathcal{X}/\langle A \mid \operatorname{im} B\rangle) \subset \mathbb{C}_g$,

(iii) $\mathcal{X}_g(A) + \langle A \mid \operatorname{im} B\rangle = \mathcal{X}$,

(iv) $\mathcal{X}_b(A) \subset \langle A \mid \operatorname{im} B\rangle$.

Proof : The equivalence of (i), (ii) and (iii) follows immediately from theorem 4.26 and lemma 4.28. The equivalence of (iii) and (iv) follows from exercise 2.11. ■

4.7 Disturbance decoupling with internal stability

In this section we again consider the disturbed control system

$$\begin{aligned}\dot{x}(t) &= Ax(t) + Bu(t) + Ed(t), \\ z(t) &= Hx(t).\end{aligned} \tag{4.25}$$

In section 4.2 we discussed the problem of disturbance decoupling by state feedback. The problem was to find a state feedback control law $u(t) = Fx(t)$ such that the impulse response matrix

$$T_F(t) := He^{(A+BF)t}E$$

of the closed-loop system (4.6) is identically equal to zero (or, equivalently, such that the closed-loop transfer function: $G_F(s)$ is equal to zero). On the other hand, in section 3.10 we discussed stabilizability of the system (4.1), that is, the existence of a state feedback such that the system controlled by means of this feedback control law becomes internally stable. In the present section we combine these two requirements into one single design problem, the problem of disturbance decoupling with internal stability by state feedback. This problem will consist of finding a state feedback control law such that the closed-loop system (4.6) is disturbance decoupled and internally stable:

Definition 4.31 *Consider the system (4.25). Let $\mathbb{C}_g$ be a stability domain.* The problem of disturbance decoupling with internal stability by state feedback, *DDPS, is to find a linear map $F : \mathcal{X} \to \mathcal{U}$ such that $T_F = 0$ and $\sigma(A + BF) \subset \mathbb{C}_g$.*

Given a feedback map F, consider the closed-loop system (4.6). Using (4.6) we see that the closed-loop system is disturbance decoupled and internally stable if and only if there exists an A_F-invariant subspace $\mathcal{V}$ between $\operatorname{im} E$ and $\ker H$ and A_F is stable. If a subspace $\mathcal{V}$ is A_F-invariant and if A_F is stable then of course $\mathcal{V}$ is a stabilizability subspace and (A, B) is stabilizable. The following result states that the converse also holds:

Theorem 4.32 *Let $\mathcal{V}$ be a subspace of $\mathcal{X}$. There exists an $F \in \underline{F}(\mathcal{V})$ such that $\sigma(A+BF) \subset \mathbb{C}_g$ if and only if $\mathcal{V}$ is a stabilizability subspace and (A, B) is stabilizable.*

Proof : ($\Rightarrow$) Of course, (A, B) is stabilizable. Also $\sigma(A_F \mid \mathcal{V}) \subset \sigma(A_F) \subset \mathbb{C}_g$ so $\mathcal{V}$ is a stabilizability subspace.

($\Leftarrow$) Denote $\mathcal{S} := \mathcal{V} + \langle A \mid \operatorname{im} B\rangle$ and $\mathcal{R} := \mathcal{R}^*(\mathcal{V})$. It follows from theorem 4.18 that there exists an $F \in \underline{F}(\mathcal{V})$ such that $\sigma(A_F \mid \mathcal{R}) \subset \mathbb{C}_g$ and $\sigma(A_F \mid \mathcal{S}/\mathcal{V}) \subset \mathbb{C}_g$. Since $\mathcal{V}$ is a stabilizability subspace, according to corollary 4.23 we have $\sigma(A_F \mid \mathcal{V}/\mathcal{R}) \subset \mathbb{C}_g$. Finally, since (A, B) is stabilizable, using theorem 4.30 we obtain

$$\sigma(A \mid \mathcal{X}/\mathcal{S}) \subset \sigma(A \mid \mathcal{X}/\langle A \mid \operatorname{im} B\rangle) \subset \mathbb{C}_g.$$ ■

Corollary 4.33 *There exists a linear map $F : \mathcal{X} \to \mathcal{U}$ such that $T_F = 0$ and $\sigma(A + BF) \subset \mathbb{C}_g$ if and only if there exists a stabilizability subspace $\mathcal{V}$ such that $\operatorname{im} E \subset \mathcal{V} \subset \ker H$ and (A, B) is stabilizable.*

Proof : ($\Rightarrow$) If $T_F = 0$ then (4.6) is disturbance decoupled. Hence there is an A_F-invariant subspace $\mathcal{V}$ with $\operatorname{im} E \subset \mathcal{V} \subset \ker H$. Since $\sigma(A_F) \subset \mathbb{C}_g$, $\mathcal{V}$ is a stabilizability subspace and (A, B) is stabilizable.

($\Leftarrow$) According to theorem 4.32 there is an F such that $\mathcal{V}$ is A_F-invariant and $\sigma(A_F) \subset \mathbb{C}_g$. Since $\operatorname{im} E \subset \mathcal{V} \subset \ker H$, the system (4.6) is disturbance decoupled. It follows that $T_F = 0$. ■

Of course, if im E is contained in a stabilizability subspace that is contained in ker H, then it is also contained in the largest stabilizability subspace contained in ker H (see theorem 4.25). Thus we obtain

Corollary 4.34 *There exists a linear map $F : \mathcal{X} \to \mathcal{U}$ such that $T_F = 0$ and $\sigma(A + BF) \subset \mathbb{C}_g$ if and only if* im $E \subset \mathcal{V}_g^*(\ker H)$ *and (A, B) is stabilizable.*

We conclude this section by noting that it is, in principle, possible to verify the subspace inclusion im $E \subset \mathcal{V}_g^*(\ker H)$ computationally. Indeed, recall from corollary 4.27 that for any $F \in \underline{F}(\mathcal{V}^*(\ker H))$ we have

$$\mathcal{V}_g^*(\ker H) = \mathcal{X}_g(A_F) \cap \mathcal{V}^*(\ker H) + \mathcal{R}^*(\ker H).$$

Thus, given the system (4.25) and a stability domain $\mathbb{C}_g$, one could first calculate $\mathcal{V}^*(\ker H)$ using the algorithm described in section 4.3. Next, one could calculate an $F \in \underline{F}(\mathcal{V}^*(\ker H))$ and compute the subspace $\mathcal{X}_g(A_F)$. Finally, the subspace $\mathcal{R}^*(\ker H)$ could be computed using theorem 4.17. Of course, the above only provides a very rough, conceptual, algorithm. If one would actually want to verify the conditions of corollary 4.34 computationally, several questions concerning numerical stability would have to be taken into account.

4.8 External stabilization

Again consider the system (4.25). In section 4.2 it was shown that the condition

$$\text{im } E \subset \mathcal{V}^*(\ker H) \tag{4.26}$$

is necessary and sufficient for the existence of a state feedback control law $u(t) = Fx(t)$ such that the transfer function of the closed-loop system becomes equal to zero. The output of the system then becomes independent of the disturbance input and, in particular, if the initial condition of the closed loop system is zero then the output will be equal to zero for all disturbances. Suppose now that condition (4.26) does not hold, so that disturbance decoupling by state feedback is not possible. In this section we will set ourselves a more modest objective and ask ourselves the question: when can we find a state feedback control law $u(t) = Fx(t)$ such that the closed-loop transfer function becomes stable. Equivalently: when can we make the closed-loop system (4.6) input/output stable by choosing F appropriately? The rationale behind this objective is of course that if the closed-loop system is stable then at least stable disturbances will result in stable outputs. If for example we take the stability set to be equal to $\mathbb{C}^-$ and if the initial condition of the closed-loop system is equal to zero, then $d(t) \to 0$ $(t \to \infty)$ will imply $z(t) \to 0$ $(t \to \infty)$ (see corollary 3.22). Also, bounded disturbances will at least result in bounded outputs.

Let us first consider the uncontrolled system

$$\begin{aligned} \dot{x}(t) &= Ax(t) + Ed(t), \\ z(t) &= Hx(t). \end{aligned} \tag{4.27}$$

Let $G(s) := H(Is - A)^{-1}E$ be the transfer function from d to z. The following lemma provides a useful sufficient condition for $G(s)$ to be stable:

Lemma 4.35 *Let $\mathbb{C}_g$ be a stability domain. Assume that there exist A-invariant subspaces $\mathcal{V}_1 \subset \mathcal{V}_2$ of $\mathcal{X}$ such that $\mathcal{V}_1 \subset \ker H$, $\operatorname{im} E \subset \mathcal{V}_2$ and $\sigma(A \mid \mathcal{V}_2/\mathcal{V}_1) \subset \mathbb{C}_g$. Then $G(s)$ is stable.*

Proof : Let $\Pi : \mathcal{V}_2 \to \mathcal{V}_2/\mathcal{V}_1$ be the canonical projection. Denote $\bar{A} := A \mid \mathcal{V}_2/\mathcal{V}_1$. Let $\bar{H} : \mathcal{V}_2/\mathcal{V}_1 \to \mathcal{Z}$ be a linear map such that $\bar{H}\Pi = H \mid \mathcal{V}_2$ (such a map $\bar{H}$ exists since $\mathcal{V}_1 \subset \ker H$, see section 2.4). Let $\bar{E} := \Pi E$. Then we have

$$G(s) = H(Is - A)^{-1}E = \bar{H}(Is - \bar{A})^{-1}\bar{E}.$$

(see exercise 3.8). Since $\sigma(\bar{A}) \subset \mathbb{C}_g$, we conclude that $G(s)$ is stable. ■

In the following, let $G_F(s)$ be the transfer function of the closed loop system (4.6).

Definition 4.36 *Consider the system (4.3). Let $\mathbb{C}_g$ be a stability domain.* The problem of external stabilization by state feedback, *ESP, is to find a linear map $F : \mathcal{X} \to \mathcal{U}$ such that $G_F(s)$ is stable.*

Assume that F is a map such that $G_F(s)$ is stable. Then for every point $x_0 \in \operatorname{im} E$, $H(Is - A_F)^{-1}x_0$ is stable. This says that if in the system $\dot{x}(t) = Ax(t) + Bu(t)$ with initial condition $x(0) = x_0$ we use the control law $u(t) = Fx(t)$ then the resulting state trajectory $x_u(\cdot, x_0)$ has the property that $Hx_u(\cdot, x_0)$ is stable. Of course, $x_u(\cdot, x_0)$ also results from the open loop control $u(t) = Fe^{A_F t}x_0$. Thus we find that if there exists an F such that $G_F(s)$ is stable then $\operatorname{im} E$ must be contained in

$$\mathcal{W}_g(\ker H) := \{x_0 \in \mathcal{X} \mid \text{there is a Bohl function } u \text{ such that } Hx_u(\cdot, x_0) \text{ is stable}\}. \tag{4.28}$$

It is easy to verify that $\mathcal{W}_g(\ker H)$ is a subspace of $\mathcal{X}$. This subspace will turn out to play a central role in the problem of external stabilization. Often, we will denote $\mathcal{W}_g(\ker H)$ by $\mathcal{W}_g$. We have the following characterization of $\mathcal{W}_g(\ker H)$ in terms of controlled invariant subspaces introduced before:

Theorem 4.37 $\mathcal{W}_g(\ker H) = \mathcal{V}^*(\ker H) + \mathcal{X}_{\text{stab}}$.

Proof : ($\supset$) It follows immediately from definition 4.4 and (4.22) that both $\mathcal{V}^*(\ker H)$ and $\mathcal{X}_{\text{stab}}$ are contained in $\mathcal{W}_g$. Hence also their sum is contained in $\mathcal{W}_g$.

($\subset$) Assume that $x_0 \in \mathcal{W}_g$. Let u be a Bohl input such that $Hx_u(\cdot, x_0)$ is stable. Denote $x := x_u(\cdot, x_0)$. Obviously, the input u and the state trajectory x can be decomposed uniquely as

$$x = x_1 + x_2 \text{ and } u = u_1 + u_2,$$

with x_1, x_2, u_1 and u_2 Bohl, the spectrum of u_1 and x_1 contained in $\mathbb{C}_g$ and the spectrum of u_2 and x_2 contained in $\mathbb{C}_b$. Denote $x_{10} := x_1(0)$ and $x_{20} := x_2(0)$. Then we have $x_0 = x_{10} + x_{20}$. Also, since $\dot{x} = Ax + Bu$, we have

$$\dot{x}_1(t) - Ax_1(t) - Bu_1(t) = -\dot{x}_2(t) + Ax_2(t) + Bu_2(t).$$

Note that in this equation the left hand side has its spectrum contained in $\mathbb{C}_g$, whereas the right hand side has its spectrum contained in $\mathbb{C}_b$ (see (2.7)). It follows that both sides of the equation must in fact be identically equal to zero. Hence we obtain

$$\dot{x}_1(t) = Ax_1(t) + Bu_1(t), \tag{4.29}$$

$$\dot{x}_2(t) = Ax_2(t) + Bu_2(t). \tag{4.30}$$

From (4.29) it follows that $x_1 = x_{u_1}(\cdot, x_{10})$. Since x_1 is stable, according to (4.22) we have $x_{10} \in \mathcal{X}_{\text{stab}}$. On the other hand,

$$Hx_2 = Hx - Hx_1.$$

Since $\sigma(x_2) \subset \mathbb{C}_b$ we have that the spectrum of Hx_2 is contained in $\mathbb{C}_b$. However, both Hx as well as Hx_1 are stable so $\sigma(Hx_2) = \sigma(Hx - Hx_1) \subset \mathbb{C}_g$. This implies that $Hx_2(t) = 0$ for all t. It follows from (4.30) that $x_2 = x_{u_2}(\cdot, x_{20})$ and hence, by definition 4.4, that $x_{20} \in \mathcal{V}^*(\ker H)$. Thus $x_0 = x_{10} + x_{20} \in \mathcal{X}_{\text{stab}} + \mathcal{V}^*(\ker H)$. ■

It follows from the above that $\mathcal{W}_g$ is a *strongly invariant subspace* (see exercise 4.2). Indeed, by combining theorem 4.37 and theorem 4.26 it is easy to see that $\mathcal{W}_g$ is A-invariant and that $\operatorname{im} B \subset \mathcal{W}_g$. Hence, $A\mathcal{W}_g + \operatorname{im} B \subset \mathcal{W}_g$. In particular this implies that $(A + BF)\mathcal{W}_g + \operatorname{im} B \subset \mathcal{W}_g$ for *any* linear map $F : \mathcal{X} \to \mathcal{U}$.

As already noted, the subspace inclusion $\operatorname{im} E \subset \mathcal{W}_g$ is a necessary condition for the existence of a map $F : \mathcal{X} \to \mathcal{U}$ such that G_F is stable . Using the representation for $\mathcal{W}_g$ obtained in theorem 4.37 we now prove that this subspace inclusion is also sufficient.

Lemma 4.38 *There exists a linear map $F \in \underline{F}(\mathcal{V}^*(\ker H))$ such that*

$$\sigma(A_F \mid \mathcal{W}_g/\mathcal{V}^*(\ker H)) \subset \mathbb{C}_g.$$

Proof : Recall that $\mathcal{W}_g$ is A-invariant and that $\operatorname{im} B \subset \mathcal{W}_g$. Let $A_0 := A \mid \mathcal{W}_g$, the restriction of A to $\mathcal{W}_g$, and consider the system $\dot{x} = A_0x + Bu$ with state space $\mathcal{W}_g$. Denote $\mathcal{V}^* := \mathcal{V}^*(\ker H)$. We have $\mathcal{V}^* \subset \mathcal{W}_g$, and its is easily seen that $\mathcal{V}^*$ is controlled invariant with respect to the restricted system (A_0, B). Also, $\mathcal{X}_{\text{stab}} \subset \mathcal{W}_g$ and it can be verified that the stabilizable subspace of (A_0, B) is equal to $\mathcal{X}_{\text{stab}}$ (use the characterization (4.22)). Consequently, the formula

$$\mathcal{V}^* + \mathcal{X}_{\text{stab}} = \mathcal{W}_g,$$

together with theorem 4.26 implies that $\mathcal{V}^*$ is outer-stabilizable with respect to the system (A_0, B). Hence there exists an $F : \mathcal{W}_g \to \mathcal{U}$ such that $(A_0 + BF)\mathcal{V}^* \subset \mathcal{V}^*$ and

$$\sigma(A_0 + BF \mid \mathcal{W}_g/\mathcal{V}^*) \subset \mathbb{C}_g.$$

Extend this F to a map on $\mathcal{X}$ in an arbitrary way. Since A_0 and A coincide on $\mathcal{W}_g$, we obtain $\sigma(A_F \mid \mathcal{W}_g/\mathcal{V}^*) \subset \mathbb{C}_g$. ■

Theorem 4.39 *Consider the system (4.25). There exists a linear map $F : \mathcal{X} \to \mathcal{U}$ such that $G_F(s)$ is stable if and only if* $\operatorname{im} E \subset \mathcal{V}^*(\ker H) + \mathcal{X}_{\text{stab}}$.

Proof : ($\Rightarrow$) This was already proven. ($\Leftarrow$) This is an application of lemma 4.38: Let F be such that $A_F\mathcal{V}^* \subset \mathcal{V}^*$ and $\sigma(A_F \mid \mathcal{W}_g/\mathcal{V}^*) \subset \mathbb{C}_g$ ($\mathcal{W}_g$ is automatically A_F-invariant). Since $\mathcal{V}^* \subset \mathcal{W}_g$, $\mathcal{V}^* \subset \ker H$ and $\operatorname{im} E \subset \mathcal{W}_g$, we may conclude from lemma 4.35 that $G_F(s)$ is stable. ■

By using theorem 4.26, we see that the subspace inclusion of theorem 4.39 can in principle be verified computationally. Indeed, $\mathcal{X}_{\text{stab}} = \mathcal{X}_g(A) + \langle A \mid \operatorname{im} B\rangle$ so $\mathcal{X}_{\text{stab}}$ can be calculated from first principles. In section 4.3 we gave an algorithm to compute $\mathcal{V}^*(\ker H)$. Of course, again we do not address numerical issues here.

4.9 Exercises

4.1 (Output null-controllability.) Consider the system

$$\dot{x}(t) = Ax(t) + Bu(t), \qquad z(t) = Hx(t).$$

If $x_0 \in \mathcal{X}$ and u is an input function, then the corresponding output is denoted by $z_u(t, x_0) := Hx_u(t, x_0)$. A point x_0 is called *output null-controllable* if there is a $T > 0$ and an input u such that $z_u(t, x_0) = 0$ for all $t \geqslant T$. The subspace of all output null-controllable points is denoted by $\mathcal{S}$. Prove that

$$\mathcal{S} = \mathcal{V}^*(\ker H) + \langle A \mid \operatorname{im} B\rangle.$$

Hint: $x_u(T, x_0) = e^{AT}x_0 + x_u(T, 0)$. Use the facts that

$$x_u(T, x_0) \in \mathcal{V}^*(\ker H),$$
$$x_u(T, 0) \in \langle A \mid \operatorname{im} B\rangle$$

and that

$$\mathcal{V}^*(\ker H) + \langle A \mid \operatorname{im} B\rangle$$

is A-invariant.

4.2 (Strong invariance.) Consider the system $\dot{x}(t) = Ax(t) + Bu(t)$. A subspace $\mathcal{V}$ of $\mathcal{X}$ is called *strongly invariant* if for each $x_0 \in \mathcal{V}$ and for each input function u we have $x_u(t, x_0) \in \mathcal{V}$ for all $t \geqslant 0$. Show that

a. The reachable subspace $\langle A \mid \operatorname{im} B \rangle$ is strongly invariant.

b. $\mathcal{V}$ is strongly invariant if and only if $\mathcal{V}$ is controlled invariant and $\langle A \mid \operatorname{im} B \rangle \subset \mathcal{V}$.

c. If $\mathcal{V}$ is strongly invariant then $(A + BF)\mathcal{V} \subset \mathcal{V}$ for all F.

d. $\mathcal{V}$ is strongly invariant if and only if $A\mathcal{V} + \operatorname{im} B \subset \mathcal{V}$.

4.3 Consider the system (A, B). Let $F : \mathcal{X} \to \mathcal{U}$ be a linear map and let $G : \mathcal{U} \to \mathcal{U}$ be an isomorphism. Show that the classes of (A, B)-invariant subspaces and $(A + BF, BG)$-invariant subspaces coincide.

4.4 Consider the system $\dot{x}(t) = Ax(t) + Bu(t)$. Let $x_0 \in \mathcal{X}$, let u be an input function, and let $x_u(\cdot, x_0)$ be the resulting state trajectory. Let $\mathcal{V}$ denote the linear span of the vectors $\{x_u(t, x_0) \mid t \geqslant 0\}$. Show that $\mathcal{V}$ is a controlled invariant subspace.

4.5 (The model matching problem) In this exercise we study the connection between DDP and the solvability of a rational matrix equation. Consider the system (4.5). Define $R_1(s) := H(Is - A)^{-1}B$ and $R_2(s) := H(Is - A)^{-1}E$. R_1 and R_2 are strictly proper real rational matrices of dimensions $q \times m$ and $q \times r$, respectively. We consider the equation

$$R_1 Q = R_2$$

in the unknown Q, which is required to be a $(m \times r)$ strictly proper real rational matrix. A more systemic interpretation of this equation is the following: given a system Σ_1 with transfer matrix $R_1(s)$ and a system Σ_2 with transfer matrix $R_2(s)$, find a system Σ_m such that the cascade (= parallel) connection of Σ_m and Σ_1 is equal to Σ_2

The above is called the problem of *exact model matching*: given the system Σ_1 (plant) and a 'desired' system Σ_2, find a 'precompensator' Σ_m for Σ_1 such that the resulting cascade connection has exactly the same input/output behaviour as the given system Σ_2.

a. Show that the equation $R_1 Q = R_2$ has a strictly proper real rational solution Q if and only if there exists an $(m \times r)$ matrix Bohl function U such that

$$\int_0^t He^{A(t-\tau)}BU(\tau)\,d\tau = He^{At} \text{ for all } t \geqslant 0.$$

b. Show that if there exists U such that the equation in (i) holds, then for each $x_0 \in \operatorname{im} E$ there exists an input function u for the system $\dot{x}(t) = Ax(t) + Bu(t)$ such that $Hx_u(t, x_0) = 0$ for all $t \geqslant 0$.

c. Show that if $F \in \underline{F}(\mathcal{V}^*(\ker H))$ then for each $x_0 \in \mathcal{V}^*(\ker H)$ we have

$$-R_1(s)F(Is - A - BF)^{-1}x_0 = H(Is - A)^{-1}x_0.$$

d. Conclude that the equation $R_1 Q = R_2$ has a strictly proper real rational solution Q if and only if $\operatorname{im} E \subset \mathcal{V}^*(\ker H)$.

4.6 (Disturbance decoupling with feedforward.) Consider the system (4.5). In the previous section we studied the problem of disturbance decoupling by state feedback. Sometimes, instead of restricting ourselves to feedback control laws of the form $u(t) = Fx(t)$, we want to allow the use of control laws of the form $u(t) = Fx(t) + Nd(t)$. If such a control law is connected to our system, then the closed-loop system is given by the equation

$$\dot{x}(t) = (A + BF)x(t) + (BN + E)d(t), \; z(t) = Hx(t).$$

Thus we may pose the problem of disturbance decoupling by state feedback *with feedforward*: find linear maps $F : \mathcal{X} \to \mathcal{U}$ and $N : \mathcal{D} \to \mathcal{U}$ such that the given closed-loop system is disturbance decoupled.

a. Show that the closed-loop system is disturbance decoupled if and only if there exists an $(A + BF)$-invariant subspace $\mathcal{V}$ such that $\operatorname{im}(BN + E) \subset \mathcal{V} \subset \ker H$.

b. Let $N : \mathcal{D} \to \mathcal{U}$ be given. Show that there exists $F : \mathcal{X} \to \mathcal{U}$ such that the closed-loop system is disturbance decoupled if and only if $\operatorname{im}(BN + E) \subset \mathcal{V}^*(\ker H)$.

c. Show that there exist F and N such that the closed-loop system is disturbance decoupled if and only if $\operatorname{im} E \subset \mathcal{V}^*(\ker H) + \operatorname{im} B$.

4.7 Consider the system $\dot{x}(t) = Ax(t) + Bu(t)$. A subspace $\mathcal{V}$ of $\mathcal{X}$ is called a *reachability subspace* if for every $x_1 \in \mathcal{V}$ there exists $T > 0$ and an input function u such that $x_u(t, 0) \in \mathcal{V}$ for all $0 \leqslant t \leqslant T$ and $x_u(T, 0) = x_1$, i.e., if every point in the subspace can be reached from the origin in finite time along a trajectory that does not leave the subspace. Show that:

a. every reachability subspace is controlled invariant,

b. a subspace $\mathcal{V}$ is a reachability subspace if and only if it is a controllability subspace,

c. a subspace $\mathcal{V}$ is a controllability subspace if and only if it has the property that for any pair of points $x_0, x_1 \in \mathcal{V}$ there exists $T > 0$ and an input function u such that $x_u(t, x_0) \in \mathcal{V}$ for all $0 \leqslant t \leqslant T$ and $x_u(T, x_0) = x_1$.

4.8 Consider the system $\dot{x}(t) = Ax(t) + Bu(t)$. A subspace $\mathcal{V}$ of $\mathcal{X}$ is called a *coasting subspace* if for each $x_0 \in \mathcal{V}$ there is *exactly one* input function u such that $x_u(t, x_0) \in \mathcal{V}$ for all $t \geqslant 0$. Show that the following three conditions are equivalent:

a. $\mathcal{V}$ is a coasting subspace,

b. $\mathcal{V}$ is controlled invariant, $\mathcal{R}^*(\mathcal{V}) = 0$ and B is injective,

c. $\underline{F}(\mathcal{V}) \neq \varnothing$ and if $F_1, F_2 \in \underline{F}(\mathcal{V})$ then $F_1 \mid \mathcal{V} = F_2 \mid \mathcal{V}$.

4.9 Consider the single-input system $\dot{x}(t) = Ax(t) + bu(t)$. Assume that (A, b) is controllable.

a. Find all controllability subspaces associated with the system (A, b).

b. Show that every controlled invariant subspace $\mathcal{V}$ with $\mathcal{V} \neq \mathcal{X}$ is a coasting subspace.

Let $\mathcal{K}$ be a subspace of $\mathcal{X}$ with $\mathcal{K} \neq \mathcal{X}$. Assume that $x_0 \in \mathcal{K}$ and let u be such that $x_u(t, x_0) \in \mathcal{K}$ for all $t \geqslant 0$.

c. Show that u is given by a state feedback control law, i.e., there is a linear map $f : \mathcal{X} \to \mathcal{U}$ such that $u = fx$.

4.10 (Output regulation by state feedback.) Consider the system $\dot{x}(t) = Ax(t) + Bu(t) + Ed(t)$, $z(t) = Hx(t)$. For a given feedback control law $u(t) = Fx(t)$, let $z_F(t, x_0, d)$ denote the output of the closed-loop system corresponding to the initial condition x_0 and disturbance d. In this exercise we study the problem of *output regulation by state feedback.* We will say that F *achieves output regulation* if $z_F(t, x_0, d) \to 0$ $(t \to \infty)$ for all $x_0 \in \mathcal{X}$ and every disturbance d.

a. Show that F achieves output regulation if and only if $He^{A_F t}E = 0$ for all t and $He^{A_F t} \to 0$ $(t \to \infty)$.

b. Let $\mathcal{X}_{\text{stab}}$ denote the stabilizable subspace of the pair (A, B) with respect to the stability set $\mathbb{C}^- = \{s \in \mathbb{C} \mid \Re e\, s < 0\}$. Show that there exists F such that $He^{A_F t} \to 0$ $(t \to \infty)$ if and only if $\mathcal{X} = \mathcal{V}^*(\ker H) + \mathcal{X}_{\text{stab}}$.

c. Show that there exists a map F that achieves output regulation if and only if $\mathcal{V}^*(\ker H)$ is outer-stabilizable and $\operatorname{im} E \subset \mathcal{V}^*(\ker H)$.

4.11 (Input/output stabilization with feedforward.) Again consider the system (4.5). Suppose that $\mathbb{C}_g$ is a stability domain. The problem of input/output stabilization by state feedback *with feedforward* is to find a control law $u(t) = Fx(t) + Nd(t)$ such that the transfer function of the resulting closed-loop system, i.e.

$$G_{F,N}(s) := H(Is - A_F)^{-1}(BN + E),$$

is stable (see also exercise 4.6). Show that there exists a control law $u(t) = Fx(t) + Nd(t)$ such that $G_{F,N}(s)$ is stable if and only if $\operatorname{im} E \subset \mathcal{V}^*(\ker H) + \mathcal{X}_{\text{stab}}$. Conclude that allowing feedforward of the disturbance input does not enlarge the class of systems that can be made input/output stable.

4.12 Give a proof of theorem 4.25.

4.13 (Disturbance decoupling by state feedback with pole placement.) In addition to the ordinary disturbance decoupling problem, DDP, and the disturbance decoupling problem with stability, DDPS, we can also consider the disturbance decoupling problem with pole placement, DDPPP. Here, the question is to find conditions under which for *any* stability domain $\mathbb{C}_g$, there exists a map $F : \mathcal{X} \to \mathcal{U}$ such that $\sigma(A + BF) \subset \mathbb{C}_g$ and $T_F = 0$ (where, as usual, T_F denotes the closed loop impulse response from d to z). In this exercise we derive necessary and sufficient conditions for this to hold. Denote $\mathcal{V}^*(\mathcal{K})$ by $\mathcal{V}^*$, $\mathcal{R}^*(\mathcal{K})$ by $\mathcal{R}^*$, and $\mathcal{V}_g^*(\mathcal{K})$ by $\mathcal{V}_g^*$.

a. Observe that if for any stability domain $\mathbb{C}_g$ there exists a map $F : \mathcal{X} \to \mathcal{U}$ such that $\sigma(A + BF) \subset \mathbb{C}_g$ and $T_F = 0$, then for any stability domain $\mathbb{C}_g$ we have $\operatorname{im} E \subset \mathcal{V}_g^*$.

b. For $F \in \underline{F}(\mathcal{V}^*)$, let τ denote the fixed spectrum $\sigma(A + BF \mid \mathcal{V}^*/\mathcal{R}^*)$. Show that if $\mathbb{C}_g$ is a stability domain with the property that $\tau \cap \mathbb{C}_g = \varnothing$, then $\mathcal{V}_g^* = \mathcal{R}^*$.

c. Show that if (A, B) is controllable, then for any pair of real monic polynomials (p_1, p_2) such that $\deg p_1 = \dim \mathcal{R}^*$ and $\deg p_2 = n - \dim \mathcal{R}^*$, there exist $F \in \underline{F}(\mathcal{R}^*)$ such that $\chi_{A_F|\mathcal{R}^*} = p_1$ and $\chi_{A_F|\mathcal{X}/\mathcal{R}^*} = p_2$. (Hint: apply theorem 4.18 to $\mathcal{V} = \mathcal{R}^*$).

d. Show that if (A, B) is controllable and $\operatorname{im} E \subset \mathcal{R}^*$, then for any real monic polynomial p of degree n such that $p = p_1 p_2$, with p_1 and p_2 monic polynomials and $\deg p_1 = \dim \mathcal{R}^*$, there exists $F : \mathcal{X} \to \mathcal{U}$ such that $\chi_{A_F} = p$ and $T_F = 0$.

e. Show that for any stability domain $\mathbb{C}_g$ there exists a map $F : \mathcal{X} \to \mathcal{U}$ such that $\sigma(A + BF) \subset \mathbb{C}_g$ and $T_F = 0$ if and only if (A, B) is controllable and $\operatorname{im} E \subset \mathcal{R}^*$.

4.10 Notes and references

Controlled invariant subspaces were introduced independently by Basile and Marro [10,11] and Wonham and Morse [224]. An extensive treatment, including the disturbance decoupling problem by state feedback, can also be found in Wonham's classical textbook [223], and in the textbook [14] by Basile and Marro. A characterization of controlled invariant subspaces in terms of vectors of rational functions was given by Hautus in [72]. A study of controlled invariant subspaces in the context of polynomial models can be found in the work of Fuhrmann and Willems [51].

Alternative conditions for the existence of disturbance decoupling state feedback control laws, in terms of the open loop control input-to-output, and disturbance input-to-output transfer matrices, were obtained by Bhattacharyya in [19]. Robustness issues in the context of design of disturbance decoupling state feedback controllers were studied by Bhattacharyya, Del Nero Gomez and Howze in [22], and by Bhattacharyya in [21]. Extensions to characterize the freedom in placing the closed loop poles under the constraint of achieving disturbance decoupling has been studied by Chen, Saberi, Sannuti and Shamash [29] and later also by Malabre, Martinez-Garcia and Del-Muro-Cuellar [120].

For additional information on the invariant subspace algorithm we refer to Wonham and Morse [224]. In [17], Bhattacharyya derived a simplified algorithm, valid for a special class of systems, to compute supremal controlled invariant subspaces. A standard reference for the numerical computation of controlled invariant subspaces is the work of Moore and Laub [127], see also [128].

Controllability subspaces were introduced by Wonham and Morse in [224]. The basic properties of these subspaces were also discussed by these authors in [132] and [131]. A treatment of controlled invariant subspaces and reachability subspaces in terms of polynomial models can be found in the work of Emre and Hautus [43]. Relations between controllability subspaces and feedback simulation are described by Heymann in [78]. The spectral assignability properties of controllability subspaces were already described by Wonham in [223], section 5.2. Our section 4.5 provides an extension of these results. Most of the material of section 4.5 is based on the work of Schumacher [167], see also [168].

Stabilizability subspaces were introduced by Wonham in section 5.6 of [223]. There, also the disturbance decoupling problem with internal stability was studied in full detail. A characterization of stabilizability subspaces in terms of stable rational vectors was given by Hautus in [72]. The problem of external stabilization by state feedback, treated in section 4.8, was introduced and resolved by Hautus in [72].

An early reference for the model matching problem, studied in exercise 4.5, is the work of Wang and Davison [209]. The connection between the model matching problem and the problem of disturbance decoupling was discussed already in Morse [130] and in Emre and Hautus [43]. Additional information can be found in the work of Anderson and Scott [7]. Exercise 4.10 is a generalization of the output stabilization problem, OSP, which can be found in Wonham's book [223], section 4.4., see also the work of Bhattacharyya, Pearson and Wonham [23]. Related material on this problem can be found in Bhattacharyya [16].

Important extensions of the theory of controlled invariant subspaces and their application to disturbance decoupling problems are the notions of *almost* controlled invariant subspaces and *almost* disturbance decoupling. Here, the aim is not to make the closed loop transfer matrix exactly equal to zero, but to make it approximately equal to zero. These ideas were introduced by Willems in [214] and [215], and extended by Trentelman in [196], [194], [195] and by Trentelman and Willems in [199]. A characterization of almost controlled invariant subspaces in terms of rational vectors can be found in the work of Schumacher [170]. Almost stabilizability subspaces were discussed in Schumacher [173]. Finally, yet another extension of the existing notion of controlled invariant subspace was given by Basile and Marro in [12], where the notion of self-bounded controlled invariant subspace was introduced.

Chapter 5

Conditioned invariant subspaces

In this chapter, we introduce conditioned invariance (also called (C, A)-invariance) and the notion of detectability subspace. These concepts are closely connected with maintaining and recovering information on the state vector of an observed linear system. It is shown that conditioned invariance and controlled invariance are, in fact, dual concepts. This fact makes it possible to 'translate' many of the results obtained in the previous chapter on controlled invariant subspaces into results on conditioned invariant subspaces. In the final section of this chapter we discuss the problem of estimation in the presence of disturbances.

5.1 Conditioned invariance

In this section we introduce the notion of conditioned invariant subspace. Consider the controlled and observed system Σ given by the equations

$$\begin{aligned} \dot{x}(t) &= Ax(t) + Bu(t), \\ y(t) &= Cx(t). \end{aligned} \tag{5.1}$$

As before, in these equations u is the control input, taking its values in $\mathcal{U}$, and y is the output that we measure, taking its values in $\mathcal{Y}$. The state x takes its values in $\mathcal{X}$. We take the point of view that, at any time instant t, both the values $u(t)$ as well as $y(t)$ are known to us. These values are called the observation at time t. The state trajectory x in (5.1) is unknown. The idea is that we want to use the observations on u and y to maintain, in some sense, information about the unknown state trajectory x.

To be more concrete, let us first discuss what we mean by 'information on the unknown vector $x_0 \in \mathcal{X}$'. Assume that $\mathcal{S}$ is a subspace of the state space $\mathcal{X}$ and assume that ξ_0 is a (known) vector in $\mathcal{X}$. Consider the statement

$$x_0 \text{ lies in the hyperplane } \xi_0 + \mathcal{S}.$$

Clearly, this statement provides information on the unknown vector x_0. It says that $x_0 - \xi_0 \in \mathcal{S}$ or, equivalently, that the equivalence classes $x_0/\mathcal{S}$ and $\xi_0/\mathcal{S}$ are equal. Thus, the equivalence class $x_0/\mathcal{S}$ is known: x_0 is known modulo $\mathcal{S}$.

Let us now return to the system (5.1). Recall that the underlying idea was that x is unknown. However, assume that, by some external mechanism, we do know $x(0)/\mathcal{S}$. Then we can ask ourselves: is it possible by using the observations $u(\tau),\ y(\tau)$ $(0 \leqslant \tau < t)$ to obtain exact knowledge of $x(t)/\mathcal{S}$? Stated differently: does there exist a mechanism that, using exact knowledge of $x(0)/\mathcal{S}$ together with the observations $u(\tau),\ y(\tau)$ on the interval $[0, t)$ provides $x(t)/\mathcal{S}$? Such mechanism in a sense *maintains information modulo* $\mathcal{S}$: we started with exact knowledge of $x(t)/\mathcal{S}$ for $t = 0$ and the information modulo $\mathcal{S}$ is maintained in the sense that the mechanism gives us $x(t)/\mathcal{S}$ for all $t \geqslant 0$. A mechanism as described here is called an observer for $x/\mathcal{S}$. We restrict ourselves to the case that these are given by *finite-dimensional linear time-invariant* systems with u and y as input:

Definition 5.1 *Consider the system (5.1). Let $\mathcal{S}$ be a subspace of $\mathcal{X}$. An* observer for *$x/\mathcal{S}$ is a system Ω*

$$\dot{w}(t) = Pw(t) + Qu(t) + Ry(t), \quad \zeta(t) = Sw(t) \tag{5.2}$$

with finite-dimensional state space $\mathcal{W}$ and output space $\mathcal{X}/\mathcal{S}$ such that for each pair of initial conditions $(x(0), w(0))$ and any input function u we have: $\zeta(0) = x(0)/\mathcal{S}$ implies $\zeta(t) = x(t)/\mathcal{S}$ for all $t \geqslant 0$.

We may now ask ourselves: if $\mathcal{S}$ is a subspace of $\mathcal{X}$, does there always exist an observer for $x/\mathcal{S}$? The answer is: no. In fact, if for a given subspace $\mathcal{S}$ such an observer exists we call it conditioned invariant:

Definition 5.2 *A subspace $\mathcal{S}$ of $\mathcal{X}$ is called* conditioned invariant *if there exists an observer for $x/\mathcal{S}$.*

Example 5.3 Consider the system (5.1) and let $\mathcal{S} = 0$, the zero subspace. Define a system (5.2) as follows: take the state space to be equal to $\mathcal{X}$, $P = A$, $Q = B$, $R = 0$ and $S = I$. Clearly, if $\zeta(0) = x(0)$ then for any input function u we have $\zeta(t) = x(t)$ for $t \geqslant 0$. Hence the system $\dot{w}(t) = Aw(t) + Bu(t)$, $\zeta(t) = w(t)$ is an observer for x and consequently 0 is conditioned invariant. Note that a system (5.2) is an observer for $x (= x/0)$ if and only if it is a state observer (see definition 3.34).

Example 5.4 Again consider (5.1). Let $\mathcal{S}$ be an A-invariant subspace of $\mathcal{X}$. Define a system (5.2) as follows: let the state space be equal to $\mathcal{X}/\mathcal{S}$, take P equal to $A \mid \mathcal{X}/\mathcal{S}$, $R = 0$ and S the identity map on $\mathcal{X}/\mathcal{S}$. Let $\Pi : \mathcal{X} \to \mathcal{X}/\mathcal{S}$ be the canonical projection. Define $Q := \Pi B$. We claim that $\dot{w}(t) = Pw(t) + Qu(t)$, $\zeta(t) = w(t)$ is an observer for $x/\mathcal{S}$. Indeed, set $e(t) := w(t) - x(t)/\mathcal{V}$. Then $\dot{e}(t) =$

$Pw(t) - \Pi Ax(t)$. Since $P\Pi = \Pi A$ we obtain $\dot{e}(t) = Pe(t)$. If we assume now that $\zeta(0) = x(0)/\mathcal{S}$, then $e(0) = 0$ and hence $e(t) = 0$ for all t. We conclude that any A-invariant subspace is conditioned invariant.

The following result provides several equivalent characterizations of conditioned invariance:

Theorem 5.5 *Consider the system (5.1). Let $\mathcal{S}$ be a subspace of $\mathcal{X}$. The following statements are equivalent:*

(i) *$\mathcal{S}$ is conditioned invariant,*

(ii) $A(\mathcal{S} \cap \ker C) \subset \mathcal{S}$,

(iii) *there exists a linear map $G : Y \to \mathcal{X}$ such that $(A + GC)\mathcal{S} \subset \mathcal{S}$.*

Proof: (i) $\Rightarrow$ (ii) Assume $\mathcal{S}$ is conditioned invariant and let Ω be an observer for $x/\mathcal{S}$. Let $x_0 \in \mathcal{S} \cap \ker C$. Take as initial states $x(0) = x_0$ and $w(0) = 0$ and take $u = 0$. We check that $\zeta(0) = x(0)/\mathcal{S}$. Indeed, on the one hand $\zeta(0) = Sw(0) = 0$ and on the other hand $x(0)/\mathcal{S} = 0$ since $x_0 \in \mathcal{S}$. We may thus conclude that $\zeta(t) = x(t)/\mathcal{S}$ for all $t \geqslant 0$. This implies that $\dot{\zeta}(0^+) = \dot{x}(0^+)/\mathcal{S}$. Now,

$$\dot{\zeta}(0^+) = S\dot{w}(0^+) = SPw(0) + SRCx_0 = 0.$$

Consequently,

$$(Ax_0)/\mathcal{S} = \dot{x}(0^+)/\mathcal{S} = 0.$$

This however implies that $Ax_0 \in \mathcal{S}$.

(ii) $\Rightarrow$ (iii) Choose a basis $x_1, \ldots, x_k$ for $\mathcal{S}$ such that $x_1, \ldots, x_l$ $(l \leqslant k)$ is a basis for $\mathcal{S} \cap \ker C$. It is easily seen that the vectors $Cx_{l+1}, \ldots, Cx_k$ are independent. Let $G : Y \to \mathcal{X}$ be a map such that $GCx_i = -Ax_i (i = l + 1, \ldots, k)$. Then we have $(A + GC)x_i = Ax_i \in \mathcal{S}$ for $i = 1, \ldots, l$ and $(A + GC)x_i = 0$ for $i = l + 1, \ldots, k$. Hence $(A + GC)\mathcal{S} \subset \mathcal{S}$.

(iii) $\Rightarrow$ (i) Let $\Pi : \mathcal{X} \to \mathcal{X}/\mathcal{S}$ be the canonical projection. Define $P := (A + GC) \mid \mathcal{X}/\mathcal{S}$, $Q := \Pi B$, $R := -\Pi G$ and let S be the identity map on $\mathcal{X}/\mathcal{S}$. We claim that the system thus defined is an observer for $x/\mathcal{S}$. Indeed, take an arbitrary input function u and assume that $\zeta(0) = x(0)/\mathcal{S}$. Then $w(0) = \Pi x(0)$. Let $w(t)$ be the solution of $\dot{w}(t) = Pw(t) + Qu(t) + Ry(t),\ w(0) = \Pi x(0)$. Since

$$\begin{aligned}\frac{\mathrm{d}}{\mathrm{d}t}\Pi x(t) &= \Pi(A + GC)x(t) + \Pi Bu(t) - \Pi GCx(t)\\ &= P\Pi x(t) + Qu(t) + Ry(t),\end{aligned}$$

we must have $\Pi x(t) = w(t)$ for all $t \geqslant 0$ (uniqueness of solutions). We conclude that $\zeta(t) = x(t)/\mathcal{S}$ for all $t \geqslant 0$. ■

In the above, the characterization (ii) is a geometric characterization of conditioned invariance. From this characterization it is seen that the property of conditioned invariance only depends on the maps A and C. In order to display this dependence, conditioned invariant subspaces are often called (C, A)-invariant subspaces. Also, it is easily seen that the intersection of any (finite or infinite) number of conditioned invariant subspaces is again a conditioned invariant subspace. A map $G : Y \to \mathcal{X}$ (from the output space into the state space) is sometimes called an *output injection*. The characterization (iii) is called the output injection characterization of conditioned invariance. If $G : Y \to \mathcal{X}$ is a linear map and if T is an isomorphism of the output space Y then the classes of (C, A)-invariant subspaces and $(TC, A+GC)$-invariant subspaces coincide. This can be expressed by saying that the property of conditioned invariance is invariant under output injection and isomorphisms of the output space.

In the sequel, we often denote $A + GC$ by A_G. The set of all maps $G : Y \to \mathcal{X}$ such that $\mathcal{S}$ is invariant under A_G is denoted by $\underline{G}(\mathcal{S})$.

It turns out that there is a duality between the concepts of controlled invariance and conditioned invariance. In fact, the subspace $\mathcal{S}$ is conditioned invariant with respect to the system $\dot{x}(t) = Ax(t)$, $y(t) = Cx(t)$ if and only if its orthogonal complement $\mathcal{S}^\perp$ is controlled invariant with respect to the dual system $\dot{x}(t) = A^{\mathrm{T}}x(t) + C^{\mathrm{T}}u(t)$:

Theorem 5.6 *Consider the system (C, A). Let $\mathcal{S}$ be a subspace of $\mathcal{X}$. Then $\mathcal{S}$ is (C, A)-invariant if and only if $\mathcal{S}^\perp$ is $(A^{\mathrm{T}}, C^{\mathrm{T}})$-invariant.*

Proof : This follows immediately from the observation that $\mathcal{S}$ is invariant under $A + GC$ if and only if $\mathcal{S}^\perp$ is invariant under $(A + GC)^{\mathrm{T}} = A^{\mathrm{T}} + C^{\mathrm{T}}G^{\mathrm{T}}$. ∎

The duality between (A, B)-invariant subspaces and (C, A)-invariant subspaces can be used to translate properties of (A, B)-invariant subspaces into properties of (C, A)-invariant subspaces. An example of this is the following:

Theorem 5.7 *Consider the system (5.1). Let $\mathcal{E}$ be a subspace of $\mathcal{X}$. There exists a smallest conditioned invariant subspace containing $\mathcal{E}$, i.e. a subspace $\mathcal{S}^*(\mathcal{E})$ such that*

(i) $\mathcal{S}^*(\mathcal{E})$ is conditioned invariant,

(ii) $\mathcal{E} \subset \mathcal{S}^*(\mathcal{E})$,

(iii) if $\mathcal{S}$ is a conditioned invariant subspace such that $\mathcal{E} \subset \mathcal{S}$ then we have $\mathcal{S}^*(\mathcal{E}) \subset \mathcal{S}$.

Proof : Denote $\mathcal{V}^* := \mathcal{V}^*(\mathcal{E}^\perp, A^{\mathrm{T}}, C^{\mathrm{T}})$, the largest $(A^{\mathrm{T}}, C^{\mathrm{T}})$-invariant subspace contained in $\mathcal{E}^\perp$. Define $\mathcal{S}^*(\mathcal{E}) := \mathcal{V}^{*\perp}$. Then, by the previous theorem, $\mathcal{S}^*(\mathcal{E})$ is

(C, A)-invariant. Moreover, since $\mathcal{V}^* \subset \mathcal{E}^\perp$, we have $\mathcal{E} \subset \mathcal{S}^*(\mathcal{E})$. Finally, assume that $\mathcal{S}$ is a (C, A)-invariant subspace containing $\mathcal{E}$. Then $\mathcal{S}^\perp$ is $(A^{\mathrm{T}}, C^{\mathrm{T}})$-invariant and is contained in $\mathcal{E}^\perp$. Consequently, $\mathcal{S}^\perp \subset \mathcal{V}^*$. It follows that $\mathcal{S}^*(\mathcal{E}) \subset \mathcal{S}$. ■

If we denote the smallest (C, A)-invariant subspace containing a given subspace $\mathcal{E}$ by $\mathcal{S}^*(\mathcal{E}, C, A)$ then, by the above, we have the following relation:

$$\mathcal{S}^*(\mathcal{E}, C, A) = \mathcal{V}^*(\mathcal{E}^\perp, A^{\mathrm{T}}, C^{\mathrm{T}})^\perp. \tag{5.3}$$

In section 4.3, we showed that the largest (A, B)-invariant subspace contained in a given subspace $\mathcal{K}$ can be computed using the invariant subspace algorithm. By using the duality relation (5.3) it is possible to establish a recurrence relation to compute the smallest (C, A)-invariant subspace containing a given subspace $\mathcal{E}$. Indeed, if we define a sequence of subspaces $\mathcal{V}_0, \mathcal{V}_1, \mathcal{V}_2, \ldots$ by the recurrence relation

$$\mathcal{V}_0 = \mathcal{E}^\perp, \quad \mathcal{V}_{t+1} = \mathcal{E}^\perp \cap (A^{\mathrm{T}})^{-1}(\mathcal{V}_t + \operatorname{im} C^{\mathrm{T}}), \tag{5.4}$$

then according to theorem 4.10 the sequence $\{\mathcal{V}_t\}$ is monotonically decreasing and

$$\mathcal{V}^*(\mathcal{E}^\perp, A^{\mathrm{T}}, C^{\mathrm{T}}) = \mathcal{V}_k$$

for some integer $1 \leqslant k \leqslant \dim \mathcal{E}^\perp$. The subspace $\mathcal{S}^*(\mathcal{E}, C, A)$ is then equal to $\mathcal{V}_k^\perp$. A somewhat more elegant way to compute the subspace $\mathcal{S}^*$ is obtained from the observation that it is possible to derive from (5.4) a recurrence relation for the sequence $\{\mathcal{V}_t^\perp\}$. Indeed, $\mathcal{V}_0^\perp = \mathcal{E}$ and

$$\begin{aligned}\mathcal{V}_{t+1}^\perp &= \mathcal{E} + ((A^{\mathrm{T}})^{-1}(\mathcal{V}_t + \operatorname{im} C^{\mathrm{T}}))^\perp \\ &= \mathcal{E} + A(\mathcal{V}_t + \operatorname{im} C^{\mathrm{T}})^\perp \\ &= \mathcal{E} + A(\mathcal{V}_t^\perp \cap \ker C).\end{aligned}$$

(see section 2.3). Consequently, if we write down the recurrence relation

$$\mathcal{S}_0 = \mathcal{E}, \quad \mathcal{S}_{t+1} = \mathcal{E} + A(\mathcal{S}_t \cap \ker C), \tag{5.5}$$

then, in fact, $\mathcal{S}_t = \mathcal{V}_t^\perp$ for all t. The recurrence relation (5.5) is called the *conditioned invariant subspace algorithm*, CISA. The following result follows immediately from (4.10) and the above considerations:

Theorem 5.8 *Consider the system (5.1). Let $\mathcal{E}$ be a subspace of $\mathcal{X}$. Let $\mathcal{S}_t$, $t = 0, 1, 2, \ldots$, be defined by the algorithm (5.5). Then we have*

(i) $\mathcal{S}_0 \subset \mathcal{S}_1 \subset \mathcal{S}_2 \subset \cdots$,

(ii) *There exists $k \leqslant n - \dim \mathcal{E}$ such that $\mathcal{S}_k = \mathcal{S}_{k+1}$,*

(iii) *If $\mathcal{S}_k = \mathcal{S}_{k+1}$ then $\mathcal{S}_k = \mathcal{S}_t$ for all $t \geqslant k$,*

(iv) *If $\mathcal{S}_k = \mathcal{S}_{k+1}$ then $\mathcal{S}^*(\mathcal{E}) = \mathcal{S}_k$.*

5.2 Detectability subspaces

Until now, we have not talked about stability issues in our considerations on observers and conditioned invariant subspaces. In this section we define what we mean by stable observers. After we have done this, we give a definition of the notion of detectability subspace. This will be a subspace of the state space with the property that there exists a stable observer for the state vector modulo that subspace.

Again consider the system (5.1). Let $\mathcal{S}$ be a subspace of $\mathcal{X}$. If $\mathcal{S}$ is conditioned invariant then there is an observer for $x/\mathcal{S}$. That is, there exists a finite-dimensional linear time-invariant system Ω:

$$\dot{w}(t) = Pw(t) + Qu(t) + Ry(t), \;\; \zeta(t) = Sw(t)$$

with the property that for any input function u and for each pair of initial conditions $(x(0), w(0))$ we have:

$$\zeta(0) = x(0)/\mathcal{S} \text{ implies } \zeta(t) = x(t)/\mathcal{S} \;\text{ for all }\; t \geqslant 0.$$

As explained in the previous section this can be interpreted by saying that information modulo $\mathcal{S}$ is maintained: if we assign to the observer output ζ the value $x(0)/\mathcal{S}$ at time $t = 0$ then on the basis of the observations $\{(u(\tau), y(\tau)) \mid 0 \leqslant \tau \leqslant t\}$ the observer output takes the value $x(t)/\mathcal{S}$ at time t.

Apart from this information-maintaining property (which it has by definition) an observer can have the property that it *recovers* information modulo $\mathcal{S}$. Assume that the value of $x(0)/\mathcal{S}$ is *not* known. Then a desired property of the observer would be that for any input function u and for any pair of initial conditions $(x(0), w(0))$ we have that $\zeta(t) - x(t)/\mathcal{S} \to 0$ as $t \to \infty$. This can be interpreted by saying that, on the basis of the observations $u(\tau), y(\tau)$ up to time t, the observer generates an estimate $\zeta(t)$ of the value $x(t)/\mathcal{S}$. The larger the interval on which these observations are gathered, the more accurate this estimate becomes: asymptotically the observer recovers information modulo $\mathcal{S}$.

If an observer for $x/\mathcal{S}$ has the latter property, we call it stable. Often, we want to be able to speak about stability while referring to more general stability domains. Again, this can be done in terms of the spectrum of Bohl functions. In the following, let $\mathbb{C}_g$ be a stability domain. Consider the system (5.1).

Definition 5.9 *Let $\mathcal{S}$ be a subspace of $\mathcal{X}$. An observer Ω for $x/\mathcal{S}$ is called* stable *if for any input function u and for any pair of initial conditions $(x(0), w(0))$ we have that $\zeta - x/\mathcal{S}$ is a stable Bohl function.*

Definition 5.10 *A subspace $\mathcal{S}$ of $\mathcal{X}$ is called a* detectability subspace *if there exists a stable observer for $x/\mathcal{S}$.*

Note that every detectability subspace is conditioned invariant. We have the following characterizations:

Theorem 5.11 *Consider the system (5.1). Let $\mathbb{C}_g$ be a stability domain. Let $\mathcal{S}$ be a subspace of $\mathcal{X}$. Then the following statements are equivalent:*

(i) *$\mathcal{S}$ is a detectability subspace,*

(ii) *for all $\lambda \in \mathbb{C}_b$ we have $(A - \lambda I)^{-1}\mathcal{S} \cap \ker C = \mathcal{S} \cap \ker C$,*

(iii) *there exists $G \in \underline{G}(\mathcal{S})$ such that $\sigma(A_G \mid \mathcal{X}/\mathcal{S}) \subset \mathbb{C}_g$.*

The proof of the above theorem is based on the following lemma. First, recall from the previous section that if $\mathcal{S}$ is a conditioned invariant subspace then it is in fact possible to find an observer for $x/\mathcal{S}$ with state space equal to $\mathcal{X}/\mathcal{S}$ and output map S equal to the identity map of $\mathcal{X}/\mathcal{S}$ (see the proof of theorem 5.5). In the next lemma we show that if $\mathcal{S}$ is a detectability subspace then such an observer can be found for which, in addition, the system map P is stable. In the following, let Π be the canonical projection onto $\mathcal{X}/\mathcal{S}$.

Lemma 5.12 *Assume that $\mathcal{S}$ is a detectability subspace. Then there exists an observer for $x/\mathcal{S}$ of the form*

$$\dot{w}(t) = Pw(t) + Qu(t) + Ry(t), \quad \zeta(t) = w(t)$$

such that $\sigma(P) \subset \mathbb{C}_g$ and

$$\Pi A - P\Pi = RC.$$

Proof : If $\mathcal{S}$ is a detectability subspace there exists a stable observer for $x/\mathcal{S}$ given by

$$\dot{w}(t) = Pw(t) + Qu(t) + Ry(t), \quad \zeta(t) = Sw(t). \tag{5.6}$$

Take $u = 0$ and take an initial condition pair $(x(0), w(0)) = (x_0, w_0)$ such that $\zeta(0) = x(0)/\mathcal{S}$ or, equivalently, $Sw_0 - \Pi x_0 = 0$. By definition we then have $\zeta(t) = x(t)/\mathcal{S}$ for all $t \geqslant 0$ or, equivalently, $Sw(t) - \Pi x(t) = 0$ for all $t \geqslant 0$. This implies $S\dot{w}(0^+) - \Pi\dot{x}(0^+) = 0$ whence

$$SPw_0 + (SRC - \Pi A)x_0 = 0.$$

Thus, we find that

$$\ker(S, -\Pi) \subset \ker(SP, SRC - \Pi A).$$

Consequently, there exists a map $L : \mathcal{X}/\mathcal{S} \to \mathcal{X}/\mathcal{S}$ such that $LS = SP$ and $-L\Pi = SRC - \Pi A$. Now, consider the system

$$\dot{w}_1(t) = Lw_1(t) + \Pi Bu(t) + SRy(t), \quad \zeta_1(t) = w_1(t). \tag{5.7}$$

We claim that this is an observer for $x/\mathcal{S}$. Indeed, let $e_1(t) := \zeta_1(t) - x(t)/\mathcal{S}$. Then we have

$$\begin{aligned}\dot{e}_1(t) &= Lw_1(t) + \Pi Bu(t) + SRCx(t) - \Pi Ax(t) - \Pi Bu(t)\\ &= Lw_1(t) - L\Pi x(t)\\ &= Le_1(t).\end{aligned}$$

Thus, $e_1(t)$ satisfies a first order linear autonomous differential equation and consequently $e_1(0) = 0$ implies $e_1(t) = 0$ for all $t \geqslant 0$. We conclude that (5.7) defines an observer for $x/\mathcal{S}$ of the required form. We now show that, in fact, L is stable. To prove this, we use the fact that the observer we started with is stable. Take the input function $u = 0$ and let $e(t) := \zeta(t) - \Pi x(t)$. Then we have

$$\begin{aligned}\dot{e}(t) &= \dot{\zeta}(t) - \Pi\dot{x}(t)\\ &= SPw(t) + SRCx(t) - \Pi Ax(t)\\ &= LSw(t) - L\Pi x(t)\\ &= Le(t).\end{aligned}$$

Now let $e(0)$ be arbitrary. Since Π is surjective there exists $x_0 \in \mathcal{X}$ such that $e(0) = -\Pi x_0$. Let $w_0 = 0$. With the initial condition pair $(x(0), w(0)) = (x_0, w_0)$ we must have $\zeta - x/\mathcal{S}$ stable, or equivalently, e stable. Thus for any $e(0)$, the solution of $\dot{e}(t) = Le(t)$ is stable. It follows that $\sigma(L) \subset \mathbb{C}_g$. This completes the proof of the lemma. ■

Proof of theorem 5.11 : (i) ⇒ (ii) Let $\mathcal{S}$ be a detectability subspace. According to the previous lemma we can find an observer for $x/\mathcal{S}$ of the form

$$\dot{w}(t) = Pw(t) + Qu(t) + Ry(t), \quad \zeta(t) = w(t)$$

with $\Pi A - P\Pi = RC$ and $\sigma(P) \subset \mathbb{C}_g$. Now assume that $\lambda \in \mathbb{C}_b$ and $(A - \lambda I)x_0 \in \mathcal{S}$, $Cx_0 = 0$. We want to show that $x_0 \in \mathcal{S}$ or, equivalently, that $\Pi x_0 = 0$. We have $\Pi Ax_0 = \lambda\Pi x_0$ and hence $(RC + P\Pi)x_0 = \lambda\Pi x_0$. This implies $P\Pi x_0 = \lambda\Pi x_0$. If Πx_0 were unequal to 0 this would imply $\lambda \in \sigma(P)$, which is a contradiction. Conversely, if $x_0 \in \mathcal{S} \cap \ker C$ then by the fact that $\mathcal{S}$ is conditioned invariant we have $Ax_0 \in \mathcal{S}$. Thus, for all $\lambda \in \mathbb{C}$ we have $(A - \lambda I)x_0 \in \mathcal{S}$. We conclude that $x_0 \in (A - \lambda I)^{-1}\mathcal{S} \cap \ker C$.

(ii) ⇒ (iii) We first show that $\mathcal{S}$ is (C, A)-invariant. Let $x_0 \in \mathcal{S} \cap \ker C$. Then $(A - \lambda I)x_0 \in \mathcal{S}$ for all $\lambda \in \mathbb{C}_b$ and hence $Ax_0 \in \mathcal{S}$. Now let $G \in \underline{G}(\mathcal{S})$. It is easily seen that (ii) is equivalent to:

$$(A + GC - \lambda I)^{-1}\mathcal{S} \cap \ker C = \mathcal{S} \cap \ker C \quad \text{for all } \lambda \in \mathbb{C}_b. \tag{5.8}$$

In turn, the latter is equivalent to

$$(A + GC - \lambda I)^{-1}\mathcal{S} \cap (\ker C + \mathcal{S}) = \mathcal{S} \quad \text{for all } \lambda \in \mathbb{C}_b. \tag{5.9}$$

Indeed, (5.9) follows from (5.8) by adding $\mathcal{S}$ to both sides of the equation and using the fact that $\mathcal{S} \subset (A_G - \lambda I)^{-1}\mathcal{S}$ for all $\lambda \in \mathbb{C}_b$. The fact that (5.9) implies (5.8) is verified immediately.

Now, obviously we have $\ker C \subset \ker C + \mathcal{S}$. Let $M : Y \to Y$ be a map such that $\ker MC = \ker C + \mathcal{S}$. Define a map $A_0 : \mathcal{X}/\mathcal{S} \to \mathcal{X}/\mathcal{S}$ by $A_0 := A_G \mid \mathcal{X}/\mathcal{S}$. Let $C_0 : \mathcal{X}/\mathcal{S} \to Y$ be a map such that $C_0\Pi = MC$ (such map exists since $\mathcal{S} \subset \ker MC$, see section 2.4). We contend that (5.9) is equivalent to

$$\ker(A_0 - \lambda I) \cap \ker C_0 = 0 \quad \text{for all} \quad \lambda \in \mathbb{C}_b. \tag{5.10}$$

Indeed, $(A_0 - \lambda I)\Pi x = 0$ and $C_0\Pi x = 0$ if and only if $\Pi(A_G - \lambda I)x = 0$ and $MCx = 0$. The latter is equivalent to $(A_G - \lambda I)x \in \mathcal{S}$ and $x \in \ker C + \mathcal{S}$. Thus, for all $\lambda \in \mathbb{C}$ we have

$$\Pi^{-1}(\ker(A_0 - \lambda I) \cap \ker C_0) = (A_G - \lambda I)^{-1}\mathcal{S} \cap (\ker C + \mathcal{S}),$$

which proves the equivalence of (5.9) and (5.10). The condition (5.10) states that all (C_0, A_0)-unobservable eigenvalues are in $\mathbb{C}_g$ and by theorem 3.38 this is equivalent to detectability of the system (C_0, A_0). Consequently, there is a linear map $G_0 : Y \to \mathcal{X}/\mathcal{S}$ such that $\sigma(A_0 + G_0C_0) \subset \mathbb{C}_g$. Choose a map $G_1 : Y \to \mathcal{X}$ such that $\Pi G_1 = G_0$ (such map exists since Π is surjective so $\operatorname{im} G_0 \subset \operatorname{im} \Pi$). Define then $G_2 := G + G_1M$. Since $\mathcal{S} \subset \ker MC$ we have that $(A + G_2C)\mathcal{S} \subset \mathcal{S}$. Moreover,

$$\begin{aligned}
\Pi(A + G_2C) &= \Pi(A + GC) + \Pi G_1MC \\
&= A_0\Pi + G_0C_0\Pi \\
&= (A_0 + G_0C_0)\Pi
\end{aligned}$$

and hence $(A + G_2C) \mid \mathcal{X}/\mathcal{S} = A_0 + G_0C_0$ is stable.

(iii) $\Rightarrow$ (i) Let $G \in \underline{G}(\mathcal{S})$ be such that $A_G \mid \mathcal{X}/\mathcal{S}$ is stable. Furthermore, define $P := A_G \mid \mathcal{X}/\mathcal{S}$, $Q := \Pi B$ and $R := -\Pi G$. We claim that the system

$$\dot{w}(t) = Pw(t) + Qu(t) + Ry(t), \quad \zeta(t) = w(t)$$

(with state space $\mathcal{X}/\mathcal{S}$) is a stable observer for $x/\mathcal{S}$. Indeed, if we define $e(t) := \zeta(t) - \Pi x(t)$ then

$$\begin{aligned}
\dot{e}(t) &= Pw(t) + Qu(t) + RCx(t) - \Pi Ax(t) - \Pi Bu(t) \\
&= Pw(t) - \Pi(A + GC)x(t) \\
&= Pw(t) - P\Pi x(t) \\
&= Pe(t).
\end{aligned}$$

Consequently, for any input function u, $e(0) = 0$ implies $e(t) = 0$ for all $t \geqslant 0$. Also, since $\sigma(P) \subset \mathbb{C}_g$, the difference $e = \zeta - x/\mathcal{S}$ is a stable Bohl function for any initial condition pair $(x(0), w(0))$ and any input function u. ■

As noted in the proof of the above theorem, for any $G \in \underline{G}(\mathcal{S})$ and any map $M : Y \to Y$ such that $\mathcal{S} + \ker C = \ker MC$, the condition (ii) above is equivalent to saying that the factor system (C_0, A_0) (with $A_0 : A_G \mid \mathcal{X}/\mathcal{S}$ and C_0 such that $C_0\Pi = MC$) is detectable. Accordingly, a detectability subspace can be considered as a subspace by which, after suitable output injection and enlargement of the nullspace of the output map, the system can be factored out such that the resulting system is detectable (compare this with the remarks following theorem 4.22).

Also note that the result of lemma 5.12 becomes more or less obvious once we know that (i) and (iii) are equivalent: an observer of the form as described in lemma 5.12 is obtained by defining $P := A_G \mid \mathcal{X}/\mathcal{S}$ and $R := -\Pi G$ for any $G \in \underline{G}(\mathcal{S})$ and $Q := \Pi B$. However, we stress that lemma 5.12 was used to prove the equivalence of (i) and (iii) above.

It follows immediately from theorem 5.11 (iii) that the class of detectability subspaces is closed under output injection maps and isomorphisms of the output space: if $G : \mathcal{Y} \to \mathcal{X}$ is a linear map and T is an isomorphism of $\mathcal{Y}$ then the classes of detectability subspaces of the systems (C, A) and $(TC, A + GC)$, respectively, coincide. From theorem 5.11 (ii) it is easily seen that the intersection of any (finite or infinite) number of detectability subspaces is a detectability subspace.

We now show that the concepts of stabilizability subspace and detectability subspace are dual:

Theorem 5.13 *Let $\mathcal{S}$ be a subspace of $\mathcal{X}$. Then $\mathcal{S}$ is a detectability subspace with respect to the system (C, A) if and only if $\mathcal{S}^\perp$ is a stabilizability subspace with respect to the system (A^T, C^T).*

Proof : The subspace $\mathcal{V}$ is $(A + GC)$-invariant if and only if the orthogonal complement $\mathcal{V}^\perp$ is $(A^T + C^TG^T)$-invariant. In addition, using (2.9), we have $\sigma(A + GC \mid \mathcal{X}/\mathcal{V}) = \sigma(A^T + C^TG^T \mid \mathcal{V}^\perp)$. The claim of the theorem then follows from theorem 4.22. ■

Again, the duality established above makes it possible to translate most of the results that were obtained for stabilizability subspaces to results for detectability subspaces. Several examples of this are collected in the exercises following this chapter.

As was the case for conditioned invariant subspaces, for an arbitrary subspace of the state space there always exists a smallest detectability subspace that contains this subspace:

Theorem 5.14 *Consider the system (5.1). Let $\mathcal{E}$ be a subspace of $\mathcal{X}$. There exists a smallest detectability subspace containing $\mathcal{E}$, i.e., a subspace $\mathcal{S}^*_g(\mathcal{E})$ such that*

(i) $\mathcal{S}^*_g(\mathcal{E})$ *is a detectability subspace,*

(ii) $\mathcal{E} \subset \mathcal{S}^*_g(\mathcal{E})$,

(iii) if $\mathcal{S}$ is a detectability subspace such that $\mathcal{E} \subset \mathcal{S}$ then $\mathcal{S}^*_g(\mathcal{E}) \subset \mathcal{S}$.

Proof : This is proven completely analogously to theorem 5.7. We define $\mathcal{S}_g^*(\mathcal{E})$ to be equal to $\mathcal{V}_g^*(\mathcal{E}^\perp, A^{\mathrm{T}}, C^{\mathrm{T}})^\perp$, the orthogonal complement of the largest stabilizability subspace in $\mathcal{E}^\perp$ with respect to the dual system. This subspace satisfies the above requirements. ■

Sometimes we denote $\mathcal{S}_g^*(\mathcal{E})$ by $\mathcal{S}_g^*(\mathcal{E}, C, A)$. The following duality relation thus holds by definition:

$$\mathcal{S}_g^*(\mathcal{E}, C, A) = \mathcal{V}_g^*(\mathcal{E}^\perp, A^{\mathrm{T}}, C^{\mathrm{T}})^\perp.$$

For a given subspace $\mathcal{E}$ the following inclusions hold:

$$\mathcal{E} \subset \mathcal{S}^*(\mathcal{E}) \subset \mathcal{S}_g^*(\mathcal{E}).$$

We leave it as an exercise to the reader to check that, in fact, $\mathcal{S}_g^*(\mathcal{S}^*(\mathcal{E})) = \mathcal{S}_g^*(\mathcal{E})$. An important role is played by $\mathcal{S}_g^*(0)$, the smallest detectability subspace of the system (5.1). This subspace will be called the *undetectable subspace* of (C, A) and is denoted by $\mathcal{X}_{\mathrm{det}}$ or $\mathcal{X}_{\mathrm{det}}(C, A)$ (note that this subspace still depends on the underlying stability domain $\mathbb{C}_g$). In terms of recovery of information on the system $\dot{x}(t) = Ax(t) + Bu(t),\ \ y(t) = Cx(t)$, $\mathcal{X}_{\mathrm{det}}$ can be characterized as the smallest subspace $\mathcal{S}$ such that there exists a stable observer for $x/\mathcal{S}$. Indeed, since $\mathcal{X}_{\mathrm{det}}$ is a detectability subspace, a stable observer of $x/\mathcal{X}_{\mathrm{det}}$ exists. If $\mathcal{S}$ is a subspace such that a stable observer for $x/\mathcal{S}$ exists then by definition it is a detectability subspace and hence $\mathcal{X}_{\mathrm{det}} \subset \mathcal{S}$. The above can be interpreted by saying that, for each t, $x(t)/\mathcal{X}_{\mathrm{det}}$ is the 'maximal amount' of information that can be recovered on the state $x(t)$ of the system $\dot{x}(t) = Ax(t) + Bu(t),\ \ y(t) = Cx(t)$ using observations $\{(u(\tau), y(\tau)); 0 \leqslant \tau \leqslant t\}$. (If $\mathcal{S}$ is a subspace and x some unknown vector then knowledge of $x/\mathcal{S}$ is the same as knowing that $x \in \xi + \mathcal{S}$ for some (known) vector ξ. The smaller $\mathcal{S}$ is, the more information this provides).

From section 4.6, recall that $\mathcal{X}_{\mathrm{stab}}(A^{\mathrm{T}}, C^{\mathrm{T}})$, the stabilizable subspace of $(A^{\mathrm{T}}, C^{\mathrm{T}})$, is defined as the largest stabilizability subspace of the system $(A^{\mathrm{T}}, C^{\mathrm{T}})$. Thus, by definition, we have $\mathcal{X}_{\mathrm{det}}(C, A) = \mathcal{X}_{\mathrm{stab}}(A^{\mathrm{T}}, C^{\mathrm{T}})^\perp$. Using theorem 4.26 and using the fact that $\langle A^{\mathrm{T}} \mid \operatorname{im} C^{\mathrm{T}}\rangle^\perp = \langle \ker C \mid A\rangle$, we thus obtain:

Theorem 5.15 $\mathcal{X}_{\mathrm{det}} = \mathcal{X}_b(A) \cap \langle \ker C \mid A\rangle$.

According to theorem 5.11, a detectability subspace is a subspace $\mathcal{S}$ for which there is a map $G : Y \to \mathcal{X}$ such that $(A + GC)\mathcal{S} \subset \mathcal{S}$ and $\sigma(A + GC \mid \mathcal{X}/\mathcal{S}) \subset \mathbb{C}_g$. Sometimes we are interested rather in $\sigma(A + GC \mid \mathcal{S})$. A conditioned invariant subspace $\mathcal{S}$ will be called *inner-detectable* if there exists $G \in \underline{G}(\mathcal{S})$ such that $\sigma(A + GC \mid \mathcal{S}) \subset \mathbb{C}_g$. Correspondingly, detectability subspaces are sometimes called *outer-detectable* conditioned invariant subspaces.

From section 2.3, recall that if $A : \mathcal{X} \to \mathcal{X}$ is a linear map and if $\mathcal{S}$ is a subspace of $\mathcal{X}$ then $\mathcal{S}$ is A-invariant if and only if $\mathcal{S}^\perp$ is A^{T}-invariant. Moreover, in that case we have $\sigma(A \mid \mathcal{S}) = \sigma(A^{\mathrm{T}} \mid \mathcal{X}/\mathcal{S}^\perp)$. Using this, it can be verified immediately that a subspace $\mathcal{S}$ of $\mathcal{X}$ is an inner-detectable (C, A)-invariant subspace if and only if $\mathcal{S}^\perp$ is an outer-stabilizable $(A^{\mathrm{T}}, C^{\mathrm{T}})$-invariant subspace. It then follows from theorem 4.29 that a conditioned invariant subspace $\mathcal{S}$ is inner-detectable if and only if $\mathcal{S} \cap \mathcal{X}_{\mathrm{det}} = 0$. By definition, the pair (C, A) is detectable if and only if the state space $\mathcal{X}$ is inner-detectable. Consequently, (C, A) is detectable if and only if $\mathcal{X}_{\mathrm{det}} = 0$. Recalling the observer interpretation of $\mathcal{X}_{\mathrm{det}}$ we thus find: (C, A) is detectable if and only if there exists a stable observer for $x (= x/0)$. In case that $\mathbb{C}_g = \mathbb{C}^-$ this means that detectable systems $\dot{x}(t) = Ax(t) + Bu(t), \;\; y(t) = Cx(t)$ are characterized by the property that the entire state x can be recovered asymptotically by means of an observer (compare this with theorem 3.38).

To conclude this section we state the following theorem which can be obtained directly from theorem 4.30 by dualization:

Theorem 5.16 *The following statements are equivalent:*

(*i*) (C, A) *is detectable,*

(*ii*) $\sigma(A \mid \langle \ker C \mid A \rangle) \subset \mathbb{C}_g$,

(*iii*) $\mathcal{X}_b(A) \cap \langle \ker C \mid A \rangle = 0$,

(*iv*) $\langle \ker C \mid A \rangle \subset \mathcal{X}_g(A)$.

5.3 Estimation in the presence of disturbances

Consider the following system Σ:

$$\begin{aligned} \dot{x}(t) &= Ax(t) \;\; + Ed(t), \\ y(t) &= Cx(t), \\ z(t) &= Hx(t). \end{aligned} \tag{5.11}$$

Here, d represents an unknown disturbance which is assumed to be any element of some function space $\mathbf{D}$. The exact specification of the function space $\mathbf{D}$ is not important. One can, for example, take $\mathbf{D}$ to be the class of all piecewise continuous functions, or the class of all locally integrable functions. The variable y represents an output that can be measured and the variable z is an output that we want to estimate on the basis of the output y. To this end, we want to construct a finite-dimensional linear time-invariant system Ω:

$$\begin{aligned} \dot{w}(t) &= Pw(t) + Qy(t), \\ \zeta(t) &= Rw(t) + \;Sy(t), \end{aligned} \tag{5.12}$$

such that ζ is an estimate of z, in the following sense. Let $e := z - \zeta$, the error between z and ζ. Let $\mathcal{W}$ be the state space of the system Ω. The interconnection of Σ and Ω is a system with state space $\mathcal{X} \times \mathcal{W}$, described by the equations

$$\begin{aligned} \dot{x}_e(t) &= A_e x_e(t) \;\; + E_e d(t), \\ e(t) &= H_e x_e(t), \end{aligned} \tag{5.13}$$

where we have introduced the following notation:

$$A_e := \begin{pmatrix} A & 0 \\ QC & P \end{pmatrix}, \quad E_e := \begin{pmatrix} E \\ 0 \end{pmatrix}, \quad H_e := \begin{pmatrix} H - SC & -R \end{pmatrix},$$

$$x_e(t) := \begin{pmatrix} x(t) \\ w(t) \end{pmatrix}.$$

We want to find a system Ω such that for all initial conditions $x(0) = x_0$ and $w(0) = w_0$, and for all disturbances d, the error $e(t)$ converges to zero as $t \to \infty$. That is, we want to construct a system $\Omega = (P, Q, R, S)$ that for all initial conditions, whatever disturbance d might occur, produces an estimate ζ of z in the sense that the error $e(t) = z(t) - \zeta(t) \to 0$ as $t \to \infty$. This situation is depicted in the following diagram:

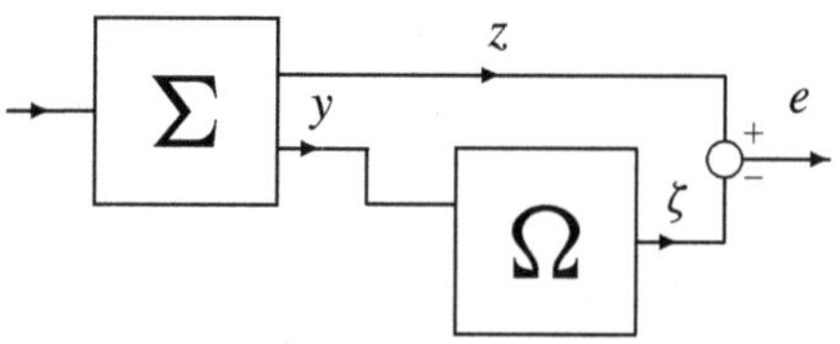

Denote $\mathcal{X}_e := \mathcal{X} \times \mathcal{W}$. For a given $x_{e,0} \in \mathcal{X}_e$, and a given disturbance input function d, denote the estimation error corresponding to $x_e(0) = x_{e,0}$ and disturbance input d by $e(t, x_{e,0}, d)$.

Definition 5.17 *Consider the system Σ given by (5.11). The system Ω given by (5.12) is called an* estimator *for Σ if $e(t, x_{e,0}, d) \to 0$ $(t \to \infty)$ for all $x_{e,0} \in \mathcal{X}_e$ and all disturbance functions d.* The problem of estimation in the presence of disturbances *is to find an estimator for Σ.*

Let us denote $T_e(t) = H_e e^{A_e t} E_e$ and $W_e(t) = H_e e^{A_e t}$. It turns out that the error converges to zero for all initial conditions and all disturbances if and only if the impulse response matrix T_e between the disturbance input and the error is equal to zero and W_e is stable with respect to $\mathbb{C}^-$, the open left half complex plane:

Lemma 5.18 *The system Ω is an estimator for Σ if and only if $T_e = 0$ and W_e is stable with respect to $\mathbb{C}^-$.*

Proof : ($\Rightarrow$) By taking d equal to zero and $x_e(0) = x_{e,0}$, we find that $W_e(t)x_{e,0} \to 0$ $(t \to \infty)$ for all $x_{e,0} \in \mathcal{X}_e$. This implies that W_e is stable with respect to $\mathbb{C}^-$. By taking $x_e(0) = 0$ we find that $e(t, 0, d) \to 0$ $(t \to \infty)$ for all d. Thus, with initial state zero, the output e is bounded for every input function d. According to exercise 3.1, this implies that $T_e = 0$.

($\Leftarrow$) The converse implication is immediate. ■

The problem of estimation in the presence of disturbances can thus be reformulated equivalently as: find a system $\Omega = (P, Q, R, S)$ (that is, find a finite-dimensional linear space $\mathcal{W}$ and maps $P : \mathcal{W} \to \mathcal{W}$, $Q : \mathcal{Y} \to \mathcal{W}$, $R : \mathcal{W} \to \mathcal{Z}$ and $S : \mathcal{Y} \to \mathcal{Z}$ such that $T_e = 0$, and W_e is stable with respect to $\mathbb{C}^-$. Of course, we might as well pose this problem in terms of an arbitrary stability domain $\mathbb{C}_g$. Our first result gives necessary and sufficient conditions for the existence of suitable maps P, Q and R in the case that S is already given:

Lemma 5.19 *Let $\mathbb{C}_g$ be a stability domain and let $S : \mathcal{Y} \to \mathcal{Z}$ be a map. Then there exist a linear space $\mathcal{W}$ and maps $P : \mathcal{W} \to \mathcal{W}$, $Q : \mathcal{Y} \to \mathcal{W}$ and $R : \mathcal{W} \to \mathcal{Z}$ such that the system $\Omega := (P, Q, R, S)$ yields $T_e = 0$ and W_e stable if and only if there exists a (C, A)-detectability subspace $\mathcal{S}$ such that*

$$\operatorname{im} E \subset \mathcal{S} \subset \ker(H - SC).$$

Proof : ($\Rightarrow$) If $T_e = 0$ then the system (5.13) is disturbance decoupled and hence, by theorem 4.6, there is an A_e-invariant subspace $\mathcal{S}_e$ such that $\operatorname{im} E_e \subset \mathcal{S}_e \subset \ker H_e$. Let $\mathcal{N}_e$ be the unobservable subspace of the pair (H_e, A_e), i.e., $\mathcal{N}_e := \langle \ker H_e \mid A_e \rangle$. By section 3.3, $\mathcal{N}_e$ is the largest A_e-invariant subspace contained in $\ker H_e$ and consequently we have $\operatorname{im} E_e \subset \mathcal{N}_e \subset \ker H_e$. Define a subspace $\mathcal{S}$ of $\mathcal{X}$ by

$$\mathcal{S} := \{x \in \mathcal{X} \mid \begin{pmatrix} x \\ 0 \end{pmatrix} \in \mathcal{N}_e\}.$$

(Here, 0 stands for the zero element of $\mathcal{W}$, the state space of the system Ω). It is verified immediately that $\operatorname{im} E \subset \mathcal{S} \subset \ker(H - SC)$. We contend that $\mathcal{S}$ is a detectability subspace with respect to (C, A). To prove this, let $x \in \mathcal{S} \cap \ker C$. Then $(x^{\mathrm{T}}, 0)^{\mathrm{T}} \in \mathcal{N}_e$ and $Cx = 0$. Since $\mathcal{N}_e$ is invariant under A_e we have

$$\begin{pmatrix} Ax \\ 0 \end{pmatrix} = A_e \begin{pmatrix} x \\ 0 \end{pmatrix} \in \mathcal{N}_e$$

and therefore $Ax \in \mathcal{S}$. It follows that $(A - \lambda I)x \in \mathcal{S}$ for all $\lambda \in \mathbb{C}$ and hence $x \in (A - \lambda I)^{-1}\mathcal{S} \cap \ker C$. Conversely, if $(A - \lambda I)x \in \mathcal{S}$ for some $\lambda \in \mathbb{C}_b$ and $Cx = 0$ then

$$(A_e - \lambda I) \begin{pmatrix} x \\ 0 \end{pmatrix} = \begin{pmatrix} (A - \lambda I)x \\ 0 \end{pmatrix} \in \mathcal{N}_e. \tag{5.14}$$

Let Π_e be the canonical projection onto $\mathcal{X}_e/\mathcal{N}_e$. Let $\bar{A}_e$ and $\bar{H}_e$ be the quotient maps defined by $\Pi_e A_e = \bar{A}_e \Pi_e$ and $\bar{H}_e \Pi_e = H_e$. Then the system $(\bar{H}_e, \bar{A}_e)$ is observable (see exercise 3.7). Also define $\bar{W}_e(t) := \bar{H}_e e^{\bar{A}_e t}$. Then we have $\bar{W}_e \Pi_e = W_e$ and hence (since Π_e is surjective) $\bar{W}_e$ is stable. By theorems 3.21 and 3.23 this implies $\sigma(\bar{A}_e) \subset \mathbb{C}_g$. On the other hand, since $\ker \Pi_e = \mathcal{N}_e$, it follows from (5.14) that

$$\bar{A}_e \Pi_e \begin{pmatrix} x \\ 0 \end{pmatrix} = \lambda \Pi_e \begin{pmatrix} x \\ 0 \end{pmatrix}.$$

Since $\lambda \in \mathbb{C}_b$, the latter immediately implies that $(x^{\mathrm{T}}, 0)^{\mathrm{T}} \in \mathcal{N}_e$. It then follows that $x \in \mathcal{S}$ and hence $x \in \mathcal{S} \cap \ker C$. By applying theorem 5.11 we may then conclude that $\mathcal{S}$ is a detectability subspace.

($\Leftarrow$) Assume $\mathcal{S}$ is a detectability subspace with $\operatorname{im} E \subset \mathcal{S} \subset \ker(H - SC)$. Let $\Pi : \mathcal{X} \to \mathcal{X}/\mathcal{S}$ be the canonical projection. Then $\Pi E = 0$ and there exists a map $\bar{H} : \mathcal{X}/\mathcal{V} \to \mathcal{Z}$ such that $\bar{H}\Pi = H - SC$ (see section 2.4). Let $G \in \underline{G}(\mathcal{S})$ be such that $\sigma(A + GC \mid \mathcal{X}/\mathcal{S}) \subset \mathbb{C}_g$. Define maps P, Q and R as follows. Let $\mathcal{W} := \mathcal{X}/\mathcal{S}$. Define $P := A_G \mid \mathcal{X}/\mathcal{S}$, $Q := -\Pi G$ and $R := \bar{H}$. We claim that with this choice $T_e = 0$ and W_e is stable. Indeed, if $\dot{w}(t) = Pw(t) + Qy(t)$, $\zeta(t) = Rw(t) + Sy(t)$ then $e = z - \zeta = \bar{H}(\Pi x - w)$. Define $e_1 := \Pi x - w$. Then

$$\begin{aligned}
\dot{e}_1(t) &= \Pi\dot{x}(t) - \dot{w}(t) \\
&= \Pi Ax(t) - Pw(t) - QCx(t) \\
&= \Pi(A + GC)x(t) - Pw(t) \\
&= Pe_1(t).
\end{aligned}$$

Consequently, $e(t) = \bar{H}e_1(t) = \bar{H}e^{Pt}e_1(0)$. From this we see that e is independent of d so $T_e = 0$. Also, since $\sigma(P) \subset \mathbb{C}_g$, e is stable for all initial states $x(0)$ and $w(0)$. Hence W_e must be stable. This completes the proof of the lemma. ■

As an immediate consequence of the previous lemma we have:

Corollary 5.20 *Let $\mathbb{C}_g$ be a stability domain and let $S : \mathcal{Y} \to \mathcal{Z}$ be a map. There exist a linear space $\mathcal{W}$ and maps $P : \mathcal{W} \to \mathcal{W}$, $Q : \mathcal{Y} \to \mathcal{W}$ and $R : \mathcal{W} \to \mathcal{Z}$ such that the system $\Omega := (P, Q, R, S)$ yields $T_e = 0$ and W_e stable if and only if $\mathcal{S}_g^*(\operatorname{im} E) \subset \ker(H - SC)$.*

The following lemma deals with the existence of a suitable map S, representing the direct feedthrough map of the system Ω to be designed:

Lemma 5.21 *Let $\mathcal{X}$, $\mathcal{Y}$ and $\mathcal{Z}$ be linear spaces and $C : \mathcal{X} \to \mathcal{Y}$ and $H : \mathcal{X} \to \mathcal{Z}$ be maps. Suppose that $\mathcal{V}$ is a subspace of $\mathcal{X}$. There exists a map $S : \mathcal{Y} \to \mathcal{Z}$ such that $\mathcal{V} \subset \ker(H - SC)$ if and only if $\mathcal{V} \cap \ker C \subset \ker H$.*

Proof : ($\Rightarrow$) This implication is immediate.

($\Leftarrow$) Let V be a map such that $\ker V = \mathcal{V}$. Then $\ker(V^{\mathrm{T}}, C^{\mathrm{T}})^{\mathrm{T}} \subset \ker H$ so there exists a map L such that $L(V^{\mathrm{T}}, C^{\mathrm{T}})^{\mathrm{T}} = H$. Partition $L = (L_1 \;\; L_2)$ to obtain $L_1V + L_2C = H$. Define the map S that we are looking for by $S := L_2$. Then $L_1V = H - SC$. This implies that $\ker V \subset \ker(H - SC)$. ■

By simply combining the above we obtain

Corollary 5.22 *Let $\mathbb{C}_g$ be a stability domain. There exists a system $\Omega = (P, Q.R, S)$ such that $T_e = 0$ and W_e is stable if and only if*

$$\mathcal{S}_g^*(\operatorname{im} E) \cap \ker C \subset \ker H.$$

In order to apply the previous result to our original estimation problem, we set $\mathbb{C}_g = \mathbb{C}^-$. Accordingly, denote the smallest detectability subspace containing im E, corresponding to the stability domain $\mathbb{C}^-$ by $\mathcal{S}^*_-(\text{im } E)$. Then we obtain:

Corollary 5.23 *There exists an estimator Ω for the system Σ if and only if*

$$\mathcal{S}^*_-(\text{im } E) \cap \ker C \subset \ker H.$$

5.4 Exercises

5.1 Using definition 5.1, give a direct proof of the fact that the intersection of two conditioned invariant subspaces is again conditioned invariant.

5.2 Consider the system (C, A). Let $T : Y \to Y$ be an isomorphism and let $G : Y \to \mathcal{X}$ be a linear map. Prove immediately from definition 5.2 that the classes of (C, A)-invariant subspaces and $(TC, A + GC)$-invariant subspaces coincide.

5.3 Consider the system (5.1). Let $\mathcal{S}$ be a conditioned invariant subspace. Show that $\mathcal{S}$ is a detectability subspace if and only if $(A - \lambda I)^{-1}\mathcal{S} \cap \ker C \subset \mathcal{S}$ for all $\lambda \in \mathbb{C}_b$.

5.4 Consider the system $\dot{x}(t) = Ax(t)+Bu(t)$, $y(t) = Cx(t)$. Assume that $p = m$, i.e. the dimension of the output space $\mathcal{Y}$ is equal to the dimension of the input space $\mathcal{U}$. Let $\mathbb{C}_g$ be a stability domain. Assume that CB is non-singular.

a. Show that $\mathcal{X} = \ker C \oplus \text{im } B$.

b. Let $P : \mathcal{X} \to \mathcal{X}$ be the projection onto $\ker C$ along im B. Assume that $\sigma(PA \mid \ker C) \subset \mathbb{C}_g$. Show that im B is a detectability subspace.

c. Let $\mathcal{V}^* := \mathcal{V}^*(\ker C, A, B)$ and $\mathcal{R}^* := \mathcal{R}^*(\ker C, A, B)$. Show that $\mathcal{V}^* = \ker C$ and $\mathcal{R}^* = 0$

d. Define $F := -(CB)^{-1}CA$. Show that $F \in \underline{F}(\mathcal{V}^*)$.

e. Show that the fixed spectrum $\sigma(A + BF \mid \mathcal{V}^*/R^*)$ is equal to $\sigma(PA \mid \ker C)$.

5.5 Consider (5.1). Let $\mathbb{C}_g$ be a stability domain and $\mathcal{S}$ a subspace of $\mathcal{X}$.

a. Let $\mathcal{S}$ be conditioned invariant. Show that $\mathcal{S}$ is inner-detectable if and only if

$$\sigma(A \mid \langle \ker C \mid A \rangle \cap \mathcal{S}) \subset \mathbb{C}_g.$$

b. Assume that (C, A) is detectable. Show that $\mathcal{S}$ is a detectability subspace if and only if there exists $G \in \underline{G}(\mathcal{S})$ such that $\sigma(A + GC) \subset \mathbb{C}_g$.

5.6 Give a direct proof of theorem 5.8, based on the geometric characterization of (C, A)-invariance.

5.7 Let $\mathbb{C}_g$ be a stability domain, and let $\mathcal{E}$ be a subspace of $\mathcal{X}$. Prove that

$$\mathcal{S}_g^*(\mathcal{S}^*(\mathcal{E})) = \mathcal{S}_g^*(\mathcal{E}).$$

5.8 (Observability subspaces.) Again consider the system $\dot{x}(t) = Ax(t) + Bu(t)$, $y(t) = Cx(t)$. A subspace $\mathcal{N}$ of $\mathcal{X}$ is called an *observability subspace* if for *each* stability domain $\mathbb{C}_g$ there exists a $(\mathbb{C}_g)$-stable observer Ω for $x/\mathcal{N}$.

- **a.** Show that if $\mathcal{N}$ is an observability subspace then for each stability domain $\mathbb{C}_g$ there exists an observer Ω for $x/\mathcal{N}$ of the form $\dot{w}(t) = Pw(t) + Ry(t)$, $\xi(t) = w(t)$, with $\sigma(P) \subset \mathbb{C}_g$ and $\Pi A - P\Pi = RC$ (here, Π is the canonical projection onto $\mathcal{X}/\mathcal{N}$).
- **b.** Prove that the following statements are equivalent:
  1. $\mathcal{N}$ is an observability subspace,
  2. $(A - \lambda I)^{-1}\mathcal{N} \cap \ker C = \mathcal{N} \cap \ker C$ for all $\lambda \in \mathbb{C}$,
  3. for each real monic polynomial p with $\deg p = n - \dim \mathcal{N}$ there exists $G \in \underline{G}(\mathcal{N})$ such that the characteristic polynomial of $A_G \mid \mathcal{X}/\mathcal{N}$ is equal to p.

5.9 Let $\mathcal{N}$ be a subspace of $\mathcal{X}$. Show that $\mathcal{N}$ is an observability subspace with respect to (C, A) if and only if $\mathcal{N}^{\perp}$ is a controllability subspace with respect to $(A^{\mathrm{T}}, C^{\mathrm{T}})$.

5.10 Consider the system (C, A). Let $\mathcal{E}$ be a subspace of $\mathcal{X}$. Let $\mathbb{C}_g$ be a stability domain.

- **a.** Show that there exists a smallest observability subspace, $\mathcal{N}^*(\mathcal{E})$, containing $\mathcal{E}$.
- **b.** Show that $\mathcal{E} \subset \mathcal{S}^*(\mathcal{E}) \subset \mathcal{S}_g^*(\mathcal{E}) \subset \mathcal{N}^*(\mathcal{E})$.
- **c.** Denote $\mathcal{N}^* := \mathcal{N}^*(\mathcal{E})$ and $\mathcal{S}^* := \mathcal{S}^*(\mathcal{E})$. Show that $\underline{G}(\mathcal{S}^*) \subset \underline{G}(\mathcal{N}^*)$ and that for all $G \in \underline{G}(\mathcal{S}^*)$ we have

$$\mathcal{N}^* = \langle \mathcal{S}^* + \ker C \mid A + GC \rangle.$$

- **d.** Determine $\mathcal{N}^*(0)$, the smallest observability subspace of (C, A).

5.11 (Computation of the smallest detectability subspace containing a given subspace.) Consider the system (C, A). Let $\mathcal{E}$ be a subspace of $\mathcal{X}$ and let $\mathbb{C}_g$ be a stability domain. Show that for all $G \in \underline{G}(\mathcal{S}^*(\mathcal{E}))$ we have

$$\mathcal{S}^*(\mathcal{E}) = (\mathcal{X}_b(A + GC) \cap \mathcal{S}^*(\mathcal{E})) \cap \mathcal{N}^*(\mathcal{E}).$$

5.12 (Pole placement by output injection under invariance constraints.) Consider the system (5.1). Let $\mathcal{S} \subset \mathcal{X}$ be a conditioned invariant subspace. In this problem we investigate the freedom in assigning the eigenvalues of $A + GC$ if $G : \mathcal{Y} \to \mathcal{X}$ is restricted to satisfy $(A + GC)\mathcal{S} \subset \mathcal{S}$. Let $\mathcal{N} := \mathcal{N}^*(\mathcal{S})$ be the smallest observability subspace containing $\mathcal{S}$. Furthermore, let $\mathcal{T} := \mathcal{S} \cap \langle \ker C \mid A \rangle$, the intersection of $\mathcal{S}$ with the unobservable subspace of (5.1).

a. Show that $\underline{G}(\mathcal{S}) \subset \underline{G}(\mathcal{N}) \cap \underline{G}(\mathcal{T})$.

b. Show that for any pair of real monic polynomials (p_1, p_2), with $\deg p_1 = \dim \mathcal{S} - \dim \mathcal{T}$ and $\deg p_2 = n - \dim \mathcal{N}$, there exist $G \in \underline{G}(\mathcal{S})$ such that $\chi_{A+GC|\mathcal{S}/\mathcal{T}} = p_1$ and $\chi_{A+GC|\mathcal{X}/\mathcal{N}} = p_2$.

c. Show that for $G_1, G_2 \in \underline{G}(\mathcal{S})$ we have $\chi_{A+G_1C|\mathcal{N}/\mathcal{S}} = \chi_{A+G_2C|\mathcal{N}/\mathcal{S}}$, i.e., $\chi_{A+GC|\mathcal{N}/\mathcal{S}}$ is fixed for all $G \in \underline{G}(\mathcal{S})$.

d. Show that $\mathcal{T}$ is $(A + GC)$-invariant for all $G : \mathcal{Y} \to \mathcal{X}$ and that for all $G : \mathcal{Y} \to \mathcal{X}$ we have $A + GC \mid \mathcal{T} = A \mid \mathcal{T}$.

e. Draw a lattice diagram that illustrates which parts of the spectrum of $A + GC$ are fixed, and which parts are free, under the restriction $G \in \underline{G}(\mathcal{S})$.

f. Let $\mathcal{S}$ be an observability subspace and assume that (C, A) is observable. Show that for any real monic polynomial p of degree n such that $p = p_1 p_2$, with p_1 and p_2 real monic polynomials, and $\deg p_1 = \dim \mathcal{S}$, there exists $G \in \underline{G}(\mathcal{S})$ such that $\chi_{A+GC} = p$.

5.5 Notes and references

Conditioned invariant subspaces were introduced by Basile and Marro in [10], and applied to the problem of estimation in the presence of unknown disturbance inputs by the same authors in [11], see also the more recent textbook [14]. Conditioned invariant subspaces were studied in terms of their duality properties with respect to controlled invariant subspaces by Morse in [129], see also the textbook by Wonham [223], page 127. The point of view of defining conditioned invariant subspaces in terms of the existence of observers originates from Willems [218].

Detectability subspaces were introduced at approximately the same time by Schumacher in [169] and Willems and Commault in [218]. The 'hybrid' characterization of detectability subspaces of theorem 5.11 is due to Schumacher [169]. The definition of observability subspace, see exercise 5.8, appeared for the first time in Willems and Commault [218].

The design of observers for systems with unknown inputs was studied by Bhattacharyya in [18], and by Hautus in [74]. A version of the problem was also studied in Willems and Commault [218]. More recent material on the design of estimators in the presence of unknown disturbances can be found in the work of Hou, Pugh and Müller [83], and in Tsui [201]. The connection between robust observer design and the problem of observer design with unknown inputs was studied by Bhattacharyya in [20].

Extensions to *almost* conditioned invariant subspaces and, related, the design of approximate observers and estimators, and PID observers, can be found in the work of Willems [216].

Chapter 6

(C, A, B)-pairs and dynamic feedback

In chapter 4 we have considered several feedback design problems that require the design of static state feedback control laws. Often, it is more realistic to assume that only *part* of the state vector is available for feedback. This can be modelled by specifying a linear function of the state vector, called the measurement output. Instead of the entire state vector, we then only allow the use of this measurement output for feedback. Instead of static feedback, however, we then allow *dynamic* feedback.

A central role in this chapter is played by the notion of *(C, A, B)-pair of subspaces*. We use this concept here to study the dynamic feedback versions of the disturbance decoupling problem, the disturbance decoupling problem with internal stability, and, finally, the problem of external stabilization.

6.1 (C, A, B)-pairs

In this section we introduce the notion of (C, A, B)-pair of subspaces. Consider the controlled and observed system Σ:

$$\begin{aligned} \dot{x}(t) &= Ax(t) + Bu(t), \\ y(t) &= Cx(t). \end{aligned} \tag{6.1}$$

In section 3.9 it was explained that if we control this system using the finite-dimensional linear time-invariant controller Γ:

$$\begin{aligned} \dot{w}(t) &= Kw(t) + Ly(t), \\ u(t) &= Mw(t) + Ny(t), \end{aligned} \tag{6.2}$$

then the resulting closed loop system is described by the autonomous linear differential equation

$$\begin{pmatrix} \dot{x}(t) \\ \dot{w}(t) \end{pmatrix} = \begin{pmatrix} A+BNC & BM \\ LC & K \end{pmatrix} \begin{pmatrix} x(t) \\ w(t) \end{pmatrix}. \tag{6.3}$$

The controller (6.2) takes the observations y as its input and generates the control function u as its output. The state space of the controller is denoted by $\mathcal{W}$ and is assumed to be a finite-dimensional real linear space. K, L, M and N are assumed to be linear maps from $\mathcal{W} \to \mathcal{W}$, $\mathcal{Y} \to \mathcal{W}$, $\mathcal{W} \to \mathcal{U}$ and $\mathcal{Y} \to \mathcal{U}$, respectively. The control action of interconnecting the controller Γ with the system (6.1) is called *dynamic feedback*. The state space of the closed loop system (6.3) is called the *extended state space* and is equal to the Cartesian product $\mathcal{X} \times \mathcal{W}$. The extended state space is often denoted by $\mathcal{X}_e$. The system mapping of the closed loop system (6.3) is called the *extended system mapping*. Thus, the extended system mapping corresponding to the controller $\Gamma = (K, L, M, N)$ is equal to

$$A_e := \begin{pmatrix} A+BNC & BM \\ LC & K \end{pmatrix}. \tag{6.4}$$

It will turn out that the control problems that are formulated and studied in this chapter essentially amount to finding linear maps K, L, M and N such that the extended system mapping A_e has certain geometric properties. It is therefore important to study the relationship between the geometric properties of the maps A, B and C (defining the system (6.1)) and those of the extended system maps (6.4). For describing this relationship it is convenient to introduce the notion of (C, A, B)-pair of subspaces. This notion is defined as follows:

Definition 6.1 *Consider the system (6.1). A pair of subspaces $(\mathcal{S}, \mathcal{V})$ of $\mathcal{X}$ is called a (C, A, B)-pair if $\mathcal{S} \subset \mathcal{V}$, $\mathcal{S}$ is (C, A)-invariant and $\mathcal{V}$ is (A, B)-invariant.*

For a given subspace $\mathcal{V}_e$ of the extended state space $\mathcal{X}_e = \mathcal{X} \times \mathcal{W}$ we consider the following two subspaces of the original state space $\mathcal{X}$:

$$p(\mathcal{V}_e) := \left\{ x \in \mathcal{X} \;\middle|\; \exists\, w \in \mathcal{W} \text{ such that } \begin{pmatrix} x \\ w \end{pmatrix} \in \mathcal{V}_e \right\} \tag{6.5}$$

and

$$i(\mathcal{V}_e) := \left\{ x \in \mathcal{X} \;\middle|\; \begin{pmatrix} x \\ 0 \end{pmatrix} \in \mathcal{V}_e \right\}. \tag{6.6}$$

Note that the first subspace is the *projection* of $\mathcal{V}_e$ onto the $\mathcal{X}$-plane, whereas the second is the *intersection* of $\mathcal{V}_e$ with this plane. Both spaces are subspaces of the original state space $\mathcal{X}$. It turns out that, starting with a subspace $\mathcal{V}_e$ that is invariant under some extended system mapping A_e, the pair of subspaces obtained by taking the intersection and projection forms a (C, A, B)-pair.

Theorem 6.2 *Let $\mathcal{V}_e$ be a subspace of $\mathcal{X}_e$ that is invariant under A_e. Then*

$$(i(\mathcal{V}_e), p(\mathcal{V}_e))$$

is a (C, A, B)-pair.

Proof : Obviously, $i(\mathcal{V}_e) \subset p(\mathcal{V}_e)$. Let $x \in i(\mathcal{V}_e) \cap \ker C$. Then we have

$$\begin{pmatrix} Ax \\ 0 \end{pmatrix} = A_e \begin{pmatrix} x \\ 0 \end{pmatrix} \in \mathcal{V}_e$$

and hence $Ax \in i(\mathcal{V}_e)$. It follows that $i(\mathcal{V}_e)$ is (C, A)-invariant. Next, let $x \in p(\mathcal{V}_e)$. Then there exists a vector $w \in \mathcal{W}$ such that $(x^{\mathrm{T}}, w^{\mathrm{T}})^{\mathrm{T}} \in \mathcal{V}_e$. Consequently,

$$\begin{pmatrix} Ax + B(NCx + Mw) \\ LCx + Kw \end{pmatrix} = A_e \begin{pmatrix} x \\ w \end{pmatrix} \in \mathcal{V}_e.$$

Since $LCx + Kw \in \mathcal{W}$ it follows that $Ax + B(NCx + Mw) \in p(\mathcal{V}_e)$. We conclude that $Ax \in p(\mathcal{V}_e) + \operatorname{im} B$. This implies that $p(\mathcal{V}_e)$ is (A, B)-invariant. ∎

The above shows that A_e-invariant subspaces of an extended state space give rise to (C, A, B)-pairs. In the sequel we also need to have some sort of converse of theorem 6.2, stating that if we start with a (C, A, B)-pair $(\mathcal{S}, \mathcal{V})$, there exists a controller $\Gamma = (K, L, M, N)$ and a subspace $\mathcal{V}_e$ of the corresponding extended state space such that $A_e\mathcal{V}_e \subset \mathcal{V}_e$, $p(\mathcal{V}_e) = \mathcal{V}$ and $i(\mathcal{V}_e) = \mathcal{S}$. This will be investigated now. Our first step is to construct the mapping N:

Lemma 6.3 *Let $(\mathcal{S}, \mathcal{V})$ be a (C, A, B)-pair. There exists a linear mapping $N : \mathcal{Y} \to \mathcal{U}$ such that $(A + BNC)\mathcal{S} \subset \mathcal{V}$.*

Proof : Let $q_1, \ldots, q_j$ be a basis of $\mathcal{S} \cap \ker C$ and extend this to a basis $q_1, \ldots, q_k$ of $\mathcal{S}$. Since $\mathcal{S} \subset \mathcal{V}$ and $\mathcal{V}$ is (A, B)-invariant, there exist $u_i \in \mathcal{U}$ and $v_i \in \mathcal{V}$ such that

$$Aq_i = v_i + Bu_i \quad (i = 1, 2, \ldots, k).$$

Furthermore, the vectors $Cq_{j+1}, \ldots, Cq_k$ are linearly independent: suppose

$$\alpha_{j+1}Cq_{j+1} + \cdots + \alpha_k Cq_k = 0.$$

Then

$$\tilde{q} := \alpha_{j+1}q_{j+1} + \cdots + \alpha_k q_k \in \mathcal{S} \cap \ker C$$

and hence $\tilde{q}$ is a linear combination of $q_1, \ldots, q_j$, say $\tilde{q} = \alpha_1 q_1 + \cdots + \alpha_j q_j$. It follows that

$$\alpha_1 q_1 + \cdots + \alpha_j q_j - \alpha_{j+1}q_{j+1} - \cdots - \alpha_k q_k = 0.$$

Since $q_1, \ldots, q_k$ are linearly independent it follows that $\alpha_1 = \alpha_2 = \cdots = \alpha_k = 0$. This proves the assertion on the linear independence. Therefore, there exists a linear mapping $N : \mathcal{Y} \to \mathcal{U}$ such that $NCq_i = -u_i$ $(i = j + 1, \ldots, k)$. We then have

$$(A + BNC)q_i = v_i \in \mathcal{V} \quad (i = j + 1, \ldots, k)$$

and

$$(A + BNC)q_i = Aq_i \in \mathcal{S} \subset \mathcal{V} \quad (i = 1, \ldots, j).$$

The latter follows from the facts that for $i = 1, \ldots, j$ the vectors q_i are in $\mathcal{S} \cap \ker C$ and that $\mathcal{S}$ is (C, A)-invariant. This completes the proof of the lemma. ■

We now state and prove the converse of theorem 6.2 as announced above.

Theorem 6.4 *Let $(\mathcal{S}, \mathcal{V})$ be a (C, A, B)-pair. Then there exists a controller (6.2) and an A_e-invariant subspace $\mathcal{V}_e$ of the extended state space $\mathcal{X}_e$ such that $\mathcal{S} = i(\mathcal{V}_e)$ and $\mathcal{V} = p(\mathcal{V}_e)$. Specifically, for any $N : \mathcal{Y} \to \mathcal{U}$ such that $(A + BNC)\mathcal{S} \subset \mathcal{V}$, and for any $F \in \underline{F}(\mathcal{V})$ and $G \in \underline{G}(\mathcal{S})$, the controller*

$$\begin{aligned} \dot{w}(t) &= (A + BF + GC - BNC)w(t) + (BN - G)y(t), \\ u(t) &= (F - NC)w(t) + Ny(t), \end{aligned} \tag{6.7}$$

with state space $\mathcal{W} = \mathcal{X}$, and the subspace

$$\mathcal{V}_e := \left\{ \begin{pmatrix} x_1 \\ 0 \end{pmatrix} + \begin{pmatrix} x_2 \\ x_2 \end{pmatrix} \in \mathcal{X}_e \,\middle|\, x_1 \in \mathcal{S},\ x_2 \in \mathcal{V} \right\}$$

satisfy the desired properties.

Before we give a proof of this theorem we want to make some remarks. First note that the state space of the controller (6.7) is equal to $\mathcal{X}$, the state space of our original system (6.1). In general, the dimension of the state space of a system is called its *dynamic order*. Hence, the dynamic order of the controller (6.7) is equal to the dynamic order of the system.

The controller defined by (6.7) can be interpreted as being the combination of a state observer Ω, static state feedback and static output feedback. To be specific, consider the system Ω defined by

$$\begin{aligned} \dot{w}(t) &= (A + GC)w(t) + Bu(t) - Gy(t), \\ \xi(t) &= w(t). \end{aligned} \tag{6.8}$$

We claim that Ω is a state observer for (6.1) (see definition 3.34). Indeed, if we define $e(t) := \xi(t) - x(t)$ then $e(t)$ can be seen to satisfy the differential equation $\dot{e}(t) = (A + GC)e(t)$. Hence $\xi(0) = x(0)$ implies $\xi(t) = x(t)$ for $t \geqslant 0$. If, in addition to the equations (6.8), we consider the output equation

$$u(t) = (F - NC)\xi(t) + Ny(t), \tag{6.9}$$

then the resulting system coincides with the controller (6.7). Thus (6.7) can be given the interpretation of a state observer together with a static feedback part (6.9). The static feedback part generates the input value $u(t)$ on the basis of the state estimate $\xi(t)$ together with the measurement output value $y(t)$. This internal structure of the controller (6.7) is depicted in the following diagram:

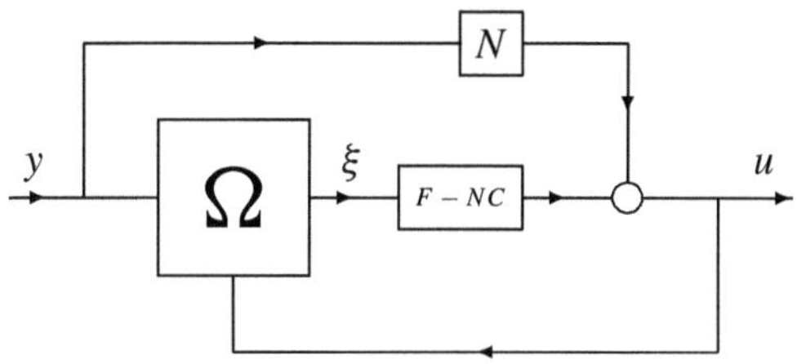

Proof of theorem 6.4 : The extended system mapping resulting from the controller (6.7) is equal to

$$A_e = \begin{pmatrix} A+BNC & B(F-NC) \\ (BN-G)C & A+BF+GC-BNC \end{pmatrix}. \tag{6.10}$$

First, we show that the subspace $\mathcal{V}_e$, as defined in theorem 6.4, is invariant under A_e. If $x_1 \in \mathcal{S}$ then

$$A_e \begin{pmatrix} x_1 \\ 0 \end{pmatrix} = \begin{pmatrix} (A+GC)x_1 \\ 0 \end{pmatrix} + \begin{pmatrix} (A+BNC)x_1 \\ (A+BNC)x_1 \end{pmatrix} - \begin{pmatrix} (A+GC)x_1 \\ (A+GC)x_1 \end{pmatrix}.$$

Since $(A+GC)x_1 \in \mathcal{S}$, $(A+BNC)x_1 \in \mathcal{V}$ and $\mathcal{S} \subset \mathcal{V}$ the latter sum is an element of $\mathcal{V}_e$ again. Furthermore, if $x_2 \in \mathcal{V}$ then we have

$$A_e \begin{pmatrix} x_2 \\ x_2 \end{pmatrix} = \begin{pmatrix} (A+BF)x_2 \\ (A+BF)x_2 \end{pmatrix} \in \mathcal{V}_e,$$

since $(A+BF)x_2 \in \mathcal{V}$. This proves that $\mathcal{V}_e$ is A_e-invariant. Next, we show that $\mathcal{S} = i(\mathcal{V}_e)$. Assume $x_1 \in \mathcal{S}$. Then by definition $(x_1^{\mathrm{T}}, 0)^{\mathrm{T}} \in \mathcal{V}_e$. Thus $x_1 \in i(\mathcal{V}_e)$. Conversely, if $x_1 \in i(\mathcal{V}_e)$ then $(x_1^{\mathrm{T}}, 0)^{\mathrm{T}} \in \mathcal{V}_e$. According to the definition of $\mathcal{V}_e$, there is $a \in \mathcal{S}$ and $b \in \mathcal{V}$ such that

$$\begin{pmatrix} a \\ 0 \end{pmatrix} + \begin{pmatrix} b \\ b \end{pmatrix} = \begin{pmatrix} x_1 \\ 0 \end{pmatrix}.$$

Hence, $b = 0$ and $x_1 = a \in \mathcal{S}$. Finally, we show that $\mathcal{V} = p(\mathcal{V}_e)$. If $x_2 \in \mathcal{V}$ then $(x_2^{\mathrm{T}}, x_2^{\mathrm{T}})^{\mathrm{T}} \in \mathcal{V}_e$. Thus $x_2 \in p(\mathcal{V}_e)$. Conversely, if $x_2 \in p(\mathcal{V}_e)$ then there is w such that $(x_2^{\mathrm{T}}, w^{\mathrm{T}})^{\mathrm{T}} \in \mathcal{V}_e$. This implies that there are $a \in \mathcal{S}$, $b \in \mathcal{V}$ such that

$$\begin{pmatrix} x_2 \\ w \end{pmatrix} = \begin{pmatrix} a \\ 0 \end{pmatrix} + \begin{pmatrix} b \\ b \end{pmatrix}.$$

Hence, $w = b \in \mathcal{V}$ and $x_2 = a + b \in \mathcal{S} + \mathcal{V} = \mathcal{V}$. ■

6.2 Disturbance decoupling by measurement feedback

Consider the control system

$$\begin{aligned} \dot{x}(t) &= Ax(t) + Bu(t) + Ed(t), \\ y(t) &= Cx(t), \\ z(t) &= Hx(t). \end{aligned} \tag{6.11}$$

Again, d represents a disturbance input. The functions d that can occur are assumed to be elements of a given function space. The variable z represents a to-be-controlled output. We want to control the system such that in the closed loop system the output z does not depend on the disturbance input d. In section 4.2 we considered this problem for the case that the control input $u(t)$ is allowed to depend on the current state $x(t)$ via some static state feedback control law $u(t) = Fx(t)$. Implicitly, this assumes that the entire state vector x can be used for control purposes, at any time instant $t \geqslant 0$. Often, it is more realistic to assume that we have access only to *part* of the state vector x. This can be formalized by including the additional output equation $y(t) = Cx(t)$ in the set of equations of the control system (with C a linear mapping from the state space $\mathcal{X}$ into some finite-dimensional linear space $\mathcal{Y}$), and to require the feedback mechanism that generates u to be driven by the output y (the *measurement output*). Here, we will allow these feedback mechanisms to be given by controllers Γ of the form (6.2). If the control system (6.11) and the controller (6.2) are interconnected, then the resulting closed loop system is described by the equations

$$\begin{aligned} \begin{pmatrix} \dot{x}(t) \\ \dot{w}(t) \end{pmatrix} &= \begin{pmatrix} A + BNC & BM \\ LC & K \end{pmatrix} \begin{pmatrix} x(t) \\ w(t) \end{pmatrix} + \begin{pmatrix} E \\ 0 \end{pmatrix} d(t), \\ z(t) &= \begin{pmatrix} H & 0 \end{pmatrix} \begin{pmatrix} x(t) \\ w(t) \end{pmatrix}. \end{aligned} \tag{6.12}$$

In addition to the compact notation (6.4) for the system mapping A_e of (6.12), we introduce

$$E_e := \begin{pmatrix} E \\ 0 \end{pmatrix} \text{ and } H_e := \begin{pmatrix} H & 0 \end{pmatrix}.$$

For a given controller $\Gamma = (K, L, M, N)$, the corresponding closed loop impulse response matrix between d and z is denoted by T_Γ and is equal to

$$T_\Gamma(t) := H_e e^{A_e t} E_e.$$

The corresponding transfer function $H_e(Is - A_e)_e^{-1} E_e$ is denoted by $G_\Gamma(s)$. The problem of disturbance decoupling by measurement feedback is to find a controller Γ such that the closed loop system (6.12) is disturbance decoupled (see section 4.2). We stress that this amounts to identifying a suitable finite-dimensional linear space $\mathcal{W}$ (the state space of the controller), together with four maps K, L, M and N:

Definition 6.5 *Consider the system (6.11).* The problem of disturbance decoupling by measurement feedback, DDPM, *is to find a controller* $\Gamma = (K, L, M, N)$ *such that* $T_\Gamma = 0$ *(or, equivalently, such that* $G_\Gamma = 0$*).*

Our aim is now to establish necessary and sufficient conditions for the existence of a controller that makes the closed loop system disturbance decoupled. We want these conditions to be entirely in terms of the original control system (6.11), that is, in terms only of the maps A, B, C, E and H. It turns out that the previous section provides most of the ingredients to establish such conditions:

Theorem 6.6 *There exists a controller* Γ *such that* $T_\Gamma = 0$ *if and only if there exists a* (C, A, B)*-pair* $(\mathcal{S}, \mathcal{V})$ *such that* $\operatorname{im} E \subset \mathcal{S} \subset \mathcal{V} \subset \ker H$.

Proof : ($\Rightarrow$) Assume that (6.12) is disturbance decoupled. According to theorem 4.6 there exists an A_e-invariant subspace $\mathcal{V}_e$ such that $\operatorname{im} E_e \subset \mathcal{V}_e \subset \ker H_e$. Define $\mathcal{S} := i(\mathcal{V}_e)$ and $\mathcal{V} := p(\mathcal{V}_e)$. According to theorem 6.2, $(\mathcal{S}, \mathcal{V})$ is a (C, A, B)-pair. Let $x \in \operatorname{im} E$. Then $(x^{\mathsf{T}}, 0)^{\mathsf{T}} \in \operatorname{im} E_e \subset \mathcal{V}_e$, so $x \in i(\mathcal{V}_e) = \mathcal{S}$. Let $x \in \mathcal{V} = p(\mathcal{V}_e)$. Then there exists $w \in \mathcal{W}$ (the state space of the controller Γ) such that $(x^{\mathsf{T}}, w^{\mathsf{T}})^{\mathsf{T}} \in \mathcal{V}_e \subset \ker H_e$. Hence $Hx = H_e(x^{\mathsf{T}}, w^{\mathsf{T}})^{\mathsf{T}} = 0$, so $\mathcal{V} \subset \ker H$.

($\Leftarrow$) Conversely, let $(\mathcal{S}, \mathcal{V})$ be a (C, A, B)-pair between $\operatorname{im} E$ and $\ker H$. According to theorem 6.4 there exists a controller Γ and an A_e-invariant subspace $\mathcal{V}_e$ of the extended state space such that $\mathcal{S} = i(\mathcal{V}_e)$ and $\mathcal{V} = p(\mathcal{V}_e)$. We claim that

$$\operatorname{im} E_e \subset \mathcal{V}_e \subset \ker H_e.$$

Indeed, $(x^{\mathsf{T}}, w^{\mathsf{T}})^{\mathsf{T}} \in \operatorname{im} E_e$ implies $w = 0$ and $x \in \operatorname{im} E \subset \mathcal{S}$. Consequently

$$\begin{pmatrix} x \\ w \end{pmatrix} = \begin{pmatrix} x \\ 0 \end{pmatrix} \in \mathcal{V}_e.$$

Furthermore, $(x^{\mathsf{T}}, w^{\mathsf{T}})^{\mathsf{T}} \in \mathcal{V}_e$ implies that $x \in \mathcal{V} \subset \ker H$. Hence $H_e(x^{\mathsf{T}}, w^{\mathsf{T}})^{\mathsf{T}} = Hx = 0$. This proves the claim. It then follows from theorem 4.6 that the closed loop system (6.12) is disturbance decoupled. ∎

Let $\mathcal{S}^*(\operatorname{im} E)$ be the smallest (C, A)-invariant subspace containing $\operatorname{im} E$ (see theorem 5.7) and let $\mathcal{V}^*(\ker H)$ be the largest (A, B)-invariant subspace contained in $\ker H$ (see theorem 4.5). If the condition $\mathcal{S}^*(\operatorname{im} E) \subset \mathcal{V}^*(\ker H)$ holds then obviously $(\mathcal{S}^*(\operatorname{im} E), \mathcal{V}^*(\ker H))$ is a (C, A, B)-pair between $\operatorname{im} E$ and $\ker H$. Conversely, if $(\mathcal{S}, \mathcal{V})$ is a (C, A, B)-pair such that $\operatorname{im} E \subset \mathcal{S}$ and $\mathcal{V} \subset \ker H$ then we must have $\mathcal{S}^*(\operatorname{im} E) \subset \mathcal{S}$ and $\mathcal{V} \subset \mathcal{V}^*(\ker H)$. Thus, we obtain the following corollary:

Corollary 6.7 *There exists a controller* Γ *such that* $T_\Gamma = 0$ *if and only if*

$$\mathcal{S}^*(\operatorname{im} E) \subset \mathcal{V}^*(\ker H).$$

The condition of corollary 6.7 can be checked computationally by means of the invariant subspace algorithm, ISA, described in section 4.3 (see theorem 4.10) and the conditioned invariant subspace algorithm, CISA, described in section 5.1 (see theorem 5.8). If the subspace inclusion of corollary 6.7 holds then a procedure to calculate an actual controller Γ can immediately be obtained from theorem 6.4. First, we calculate a mapping $F : \mathcal{X} \to \mathcal{U}$ such that $A_F\mathcal{V}^* \subset \mathcal{V}^*$, a mapping $G : \mathcal{Y} \to \mathcal{X}$ such that $A_G\mathcal{S}^* \subset \mathcal{S}^*$ and a mapping $N : \mathcal{Y} \to \mathcal{U}$ such that $(A + BNC)\mathcal{S}^* \subset \mathcal{V}^*$. Next we put $K := A+BF+GC-BNC,\ L := BN-G$ and $M := F-NC$. The controller Γ defined by $\dot{w}(t) = Kw(t) + Ly(t),\ u(t) = Mw(t) + Ny(t)$ then achieves the desired disturbance decoupling. The state space of Γ is equal to $\mathcal{W} = \mathcal{X}$, so its dynamic order is equal to the dynamic order of the underlying control system.

6.3 (C, A, B)-pairs and internal stability

In section 6.1 it was shown how a (C, A, B)-pair leads to an extended system mapping and an invariant subspace of the extended state space such that certain properties hold. In this section we study the connection between the spectrum of the extended system mapping and the stabilizability and detectability properties of the underlying (C, A, B)-pair.

Recall from theorem 6.2 that if we have an extended system mapping A_e of the form (6.4) working on some extended state space $\mathcal{X}_e$, and if $\mathcal{V}_e$ is a subspace of $\mathcal{X}_e$ that is A_e-invariant, then $(i(\mathcal{V}_e), p(\mathcal{V}_e))$ constitutes a (C, A, B)-pair. The following result is more specific:

Theorem 6.8 *Let $\mathcal{V}_e$ be an A_e-invariant subspace of $\mathcal{X}_e$. Then we have:*

(i) if $\mathcal{V}_e$ is inner-stable, then $i(\mathcal{V}_e)$ is inner-detectable and $p(\mathcal{V}_e)$ is inner-stabilizable,

(ii) if $\mathcal{V}_e$ is outer-stable, then $i(\mathcal{V}_e)$ is outer-detectable and $p(\mathcal{V}_e)$ is outer-stabilizable.

Proof : (i) Since $\mathcal{V}_e$ is inner-stable, it follows from theorem 2.17 that

$$(\lambda I - A_e)\mathcal{V}_e = \mathcal{V}_e$$

for all $\lambda \in \mathbb{C}_b$. We first show that $p(\mathcal{V}_e)$ satisfies the condition (ii) of theorem 4.22. By theorem 6.2, $p(\mathcal{V}_e)$ is controlled invariant so for all $\lambda \in \mathbb{C}$ we have

$$(\lambda I - A)p(\mathcal{V}_e) + \operatorname{im} B \subset p(\mathcal{V}_e) + \operatorname{im} B.$$

Let $\lambda \in \mathbb{C}_b$ and let $x \in p(\mathcal{V}_e)$. We have to show that x can be written as $(\lambda I - A)\tilde{x} + Bu$, for some $\tilde{x} \in p(\mathcal{V}_e)$ and $u \in \mathcal{U}$. Since $x \in p(\mathcal{V}_e)$, there exists $w \in \mathcal{W}$ such that $x_e := (x^{\mathrm{T}}, w^{\mathrm{T}})^{\mathrm{T}} \in \mathcal{V}_e$. There exists $\tilde{x}_e \in \mathcal{V}_e$, say $\tilde{x}_e = (\tilde{x}^{\mathrm{T}}, \tilde{w}^{\mathrm{T}})^{\mathrm{T}}$, such that $(\lambda I - A_e)\tilde{x}_e = x_e$. This however implies

$$(\lambda I - A - BNC)\tilde{x} - BM\tilde{w} = x.$$

If we define $u := -(NC\tilde{x} + M\tilde{w})$ then this yields the desired representation of x (note that $\tilde{x} \in p(\mathcal{V}_e)$). We conclude that $p(\mathcal{V}_e)$ is a stabilizability subspace. Next, we show that $i(\mathcal{V}_e)$ is inner-detectable. It was already shown that $i(\mathcal{V}_e)$ is conditioned invariant. Thus, we have to show that

$$\sigma(A \mid i(\mathcal{V}_e) \cap \langle \ker C \mid A \rangle) \subset \mathbb{C}_g$$

(see exercise 5.5). Let λ and x be such that $Ax = \lambda x$ and such that $x \in \langle \ker C \mid A \rangle \cap i(\mathcal{V}_e)$, $x \neq 0$. Since $Cx = 0$ we obtain

$$A_e \begin{pmatrix} x \\ 0 \end{pmatrix} = \begin{pmatrix} Ax \\ 0 \end{pmatrix} = \lambda \begin{pmatrix} x \\ 0 \end{pmatrix}.$$

Using the fact that $x \in i(\mathcal{V}_e)$ we have $(x^{\mathrm{T}}, 0)^{\mathrm{T}} \in \mathcal{V}_e$. Hence $\lambda \in \sigma(A_e \mid \mathcal{V}_e) \subset \mathbb{C}_g$.

(ii) If $\mathcal{V}_e$ is outer-stable then $\mathcal{V}_e^{\perp}$ is inner-stable with respect to A_e^T. Since

$$A_e^{\mathrm{T}} = \begin{pmatrix} A^{\mathrm{T}} + C^{\mathrm{T}} N^{\mathrm{T}} B^{\mathrm{T}} & C^{\mathrm{T}} L^{\mathrm{T}} \\ M^{\mathrm{T}} B^{\mathrm{T}} & K^{\mathrm{T}} \end{pmatrix}$$

we can conclude from (i) that $i(\mathcal{V}_e^{\perp})$ is inner-detectable with respect to $(B^{\mathrm{T}}, A^{\mathrm{T}})$ and that $p(\mathcal{V}_e^{\perp})$ is inner-stabilizable with respect to $(A^{\mathrm{T}}, C^{\mathrm{T}})$. By exercise 6.1, we have $i(\mathcal{V}_e^{\perp}) = p(\mathcal{V}_e)^{\perp}$ and $p(\mathcal{V}_e^{\perp}) = i(\mathcal{V}_e)^{\perp}$. Thus, we conclude that $p(\mathcal{V}_e)$ is outer-stabilizable and that $i(\mathcal{V}_e)$ is a detectability subspace (see theorem 5.13). ■

Our next result is, in a sense, the converse of theorem 6.8. As we already saw in section 6.1, if $(\mathcal{S}, \mathcal{V})$ is a (C, A, B)-pair then we can construct a controller Γ and an A_e-invariant subspace $\mathcal{V}_e$ of the associated extended state space $\mathcal{X}_e$ such that $i(\mathcal{V}_e) = \mathcal{S}$ and $p(\mathcal{V}_e) = \mathcal{V}$. The following theorem shows how the spectrum of A_e depends on the spectra of $A + BF$ and $A + GC$, if the linear maps F and G are chosen from $\underline{F}(\mathcal{V})$ and $\underline{G}(\mathcal{S})$, respectively.

Theorem 6.9 *Let $(\mathcal{S}, \mathcal{V})$ be a (C, A, B)-pair. Assume that $F \in \underline{F}(\mathcal{V})$ and $G \in \underline{G}(\mathcal{S})$. Then there exists a controller Γ and an A_e-invariant subspace $\mathcal{V}_e$ of the extended state space $\mathcal{X}_e$ such that $i(\mathcal{V}_e) = \mathcal{S}$, $p(\mathcal{V}_e) = \mathcal{V}$ and*

$$\sigma(A_e \mid \mathcal{V}_e) = \sigma(A_F \mid \mathcal{V}) \cup \sigma(A_G \mid \mathcal{S}), \tag{6.13}$$

$$\sigma(A_e \mid \mathcal{X}_e/\mathcal{V}_e) = \sigma(A_F \mid \mathcal{X}/\mathcal{V}) \cup \sigma(A_G \mid \mathcal{X}/\mathcal{S}). \tag{6.14}$$

Specifically, if $N : \mathcal{Y} \to \mathcal{U}$ is a linear mapping such that $(A + BNC)\mathcal{S} \subset \mathcal{V}$ then the controller

$$\begin{aligned} \dot{w}(t) &= (A + BF + GC - BNC)w(t) + (BN - G)y(t), \\ u(t) &= (F - NC)w(t) + Ny(t), \end{aligned} \tag{6.15}$$

with state space $\mathcal{W} = \mathcal{X}$, and the subspace

$$\mathcal{V}_e := \left\{ \begin{pmatrix} x_1 \\ 0 \end{pmatrix} + \begin{pmatrix} x_2 \\ x_2 \end{pmatrix} \,\middle|\, x_1 \in \mathcal{S},\ x_2 \in \mathcal{V} \right\} \tag{6.16}$$

satisfy the desired properties.

Proof : The facts that the controller (6.15) and the subspace $\mathcal{V}_e$ given by (6.16) have the properties that $i(\mathcal{V}_e) = \mathcal{S}$, $p(\mathcal{V}_e) = \mathcal{V}$ and $\mathcal{V}_e$ is A_e-invariant were already proven in theorem 6.4. It remains to show (6.13) and (6.14). We first prove (6.13). Let $x_e \in \mathcal{V}_e$ be an eigenvector of A_e with eigenvalue λ. Write

$$x_e = \begin{pmatrix} x_1 + x_2 \\ x_2 \end{pmatrix}$$

with $x_1 \in \mathcal{S}$ and $x_2 \in \mathcal{V}$. Using (6.10), the expression $A_e x_e = \lambda x_e$ becomes

$$(A + BNC)x_1 + (A + BF)x_2 = \lambda(x_1 + x_2), \tag{6.17}$$

$$(BNC - GC)x_1 + (A + BF)x_2 = \lambda x_2. \tag{6.18}$$

Subtracting (6.18) from (6.17) yields

$$(A + GC)x_1 = \lambda x_1.$$

First assume $x_1 \neq 0$. Then we may conclude $\lambda \in \sigma(A_G \mid \mathcal{S})$. If, on the other hand, $x_1 = 0$ then (6.18) becomes

$$(A + BF)x_2 = \lambda x_2.$$

Since in that case $x_2 \neq 0$ ($x_e \neq 0$!) the latter implies that $\lambda \in \sigma(A_G \mid \mathcal{V})$. We now prove the converse inclusion in (6.13). Assume $0 \neq x_2 \in \mathcal{V}$ such that $A_F x_2 = \lambda x_2$. Define $x_e := (x_2^{\mathrm{T}}, x_2^{\mathrm{T}})^{\mathrm{T}}$. Then $x_e \in \mathcal{V}_e$ and $A_e x_e = \lambda x_e$. Thus $\lambda \in \sigma(A_e \mid \mathcal{V}_e)$. Next, assume that $0 \neq x_1 \in \mathcal{S}$ with $A_G x_1 = \lambda x_1$. Since $(x_1^{\mathrm{T}}, 0)^{\mathrm{T}} \in \mathcal{V}_e$, since $\mathcal{V}_e$ is A_e- invariant and since

$$A_e \begin{pmatrix} x_1 \\ 0 \end{pmatrix} = \begin{pmatrix} (A + BNC)x_1 \\ (BNC - GC)x_1 \end{pmatrix},$$

the vector $\tilde{x}_1 := (BNC - GC)x_1$ lies in $\mathcal{V}$. Now, we may assume that $\lambda \notin \sigma(A_F \mid \mathcal{V})$. In that case the mapping $(\lambda I - A_F) \mid \mathcal{V}$ is invertible and the equation

$$\tilde{x}_1 + (\lambda I - A_F)x_2 = 0$$

has a (unique) solution $x_2 \in \mathcal{V}$. Define then

$$x_e := \begin{pmatrix} x_1 + x_2 \\ x_2 \end{pmatrix} \in \mathcal{V}_e.$$

It can then be verified that

$$A_e x_e = \begin{pmatrix} A_G x_1 + \tilde{x}_1 + (\lambda I - A_F)x_2 + \lambda x_2 \\ \tilde{x}_1 + (\lambda I - A_F)x_2 + \lambda x_2 \end{pmatrix} = \lambda \begin{pmatrix} x_1 + x_2 \\ x_2 \end{pmatrix} = \lambda x_e.$$

Since $x_1 \neq 0$, also $x_e \neq 0$. Hence $\lambda \in \sigma(A_e \mid \mathcal{V}_e)$. Thus we have proven (6.13). We now prove (6.14). We do this by dualization of the first part of this theorem. Define a subspace of the extended state space by

$$\tilde{\mathcal{V}}_e := \left\{ \begin{pmatrix} x_1 \\ 0 \end{pmatrix} + \begin{pmatrix} x_2 \\ -x_2 \end{pmatrix} \,\middle|\, x_1 \in \mathcal{V}^{\perp},\ x_2 \in \mathcal{S}^{\perp} \right\}.$$

In exercise 6.1 we show that $\tilde{\mathcal{V}}_e = \mathcal{V}_e^{\perp}$ and that consequently $\tilde{\mathcal{V}}_e$ is A_e^{T}-invariant. Note that $(A^{\mathrm{T}} + F^{\mathrm{T}}B^{\mathrm{T}})\mathcal{V}^{\perp} \subset \mathcal{V}^{\perp}$ and that $(A^{\mathrm{T}} + G^{\mathrm{T}}C^{\mathrm{T}})\mathcal{S}^{\perp} \subset \mathcal{S}^{\perp}$. Now, it can be proven completely analogously to the proof of (6.13) above that

$$\sigma(A_e^{\mathrm{T}} \mid \tilde{\mathcal{V}}_e) = \sigma(A^{\mathrm{T}} + G^{\mathrm{T}}C^{\mathrm{T}} \mid \mathcal{S}^{\perp}) \cup \sigma(A^{\mathrm{T}} + F^{\mathrm{T}}B^{\mathrm{T}} \mid \mathcal{V}^{\perp}).$$

The details are left as an exercise to the reader. Using then that $\tilde{\mathcal{V}}_e = \mathcal{V}_e^{\perp}$, the latter equality immediately implies that (6.14) holds. ■

The relation between the spectra of $A + GC$ and $A + BF$, and the spectrum of A_e is depicted in the lattice diagram in Figure 6.1.

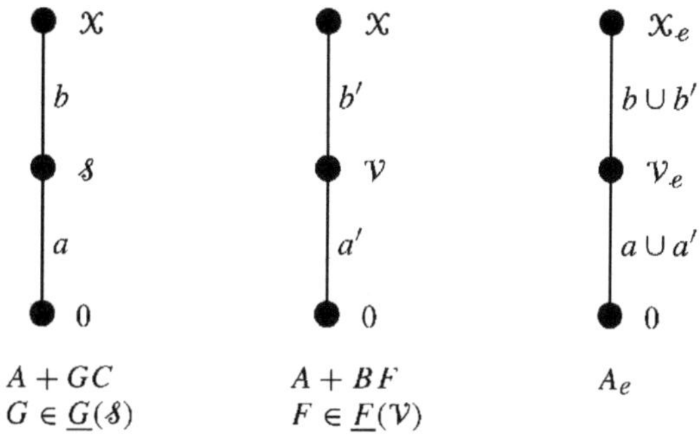

Figure 6.1

Corollary 6.10 *Let $(\mathcal{S}, \mathcal{V})$ be a (C, A, B)-pair. Assume that $F \in \underline{F}(\mathcal{V})$ and $G \in \underline{G}(\mathcal{S})$. Then there exists a controller Γ and an A_e-invariant subspace $\mathcal{V}_e$ of the extended state space $\mathcal{X}_e$ such that $i(\mathcal{V}_e) = \mathcal{S}$, $p(\mathcal{V}_e) = \mathcal{V}$ and $\sigma(A_e) = \sigma(A_F) \cup \sigma(A_G)$. Specifically, if $N : \mathcal{Y} \to \mathcal{U}$ is a linear mapping such that $(A + BNC)\mathcal{S} \subset \mathcal{V}$, then the controller (6.15) and the subspace (6.16) satisfy these properties.*

6.4 Disturbance decoupling with internal stability by measurement feedback

In section 6.2 we treated the problem of disturbance decoupling by dynamic measurement feedback. In the present section we study this problem under the additional constraint that the closed loop system should be internally stable. Again consider the control system (6.11). For a given controller Γ given by (6.2), let T_Γ be the closed loop impulse response matrix between d and z, let $G_\Gamma(s)$ be the closed loop transfer function and let A_e be the extended system mapping.

Definition 6.11 *Let $\mathbb{C}_g$ be a stability domain.* The problem of disturbance decoupling with internal stability by measurement feedback, DDPMS, *is to find a controller $\Gamma = (K, L, M, N)$ such that $T_\Gamma = 0$ (equivalently: $G_\Gamma = 0$) and $\sigma(A_e) \subset \mathbb{C}_g$.*

In order to establish necessary and sufficient conditions for the existence of a controller that makes the closed loop system disturbance decoupled and internally stable, we use the following lemma:

Lemma 6.12 *Let $(\mathcal{S}, \mathcal{V})$ be a (C, A, B)-pair, where $\mathcal{S}$ is a detectability subspace and $\mathcal{V}$ is a stabilizability subspace. Assume that (C, A) is detectable and (A, B) is stabilizable. Then there exists a controller Γ such that $\sigma(A_e) \subset \mathbb{C}_g$ and an A_e-invariant subspace $\mathcal{V}_e$ of the extended state space $\mathcal{X}_e$ such that $i(\mathcal{V}_e) = \mathcal{S}$ and $p(\mathcal{V}_e) = \mathcal{V}$.*

Proof : According to theorem 4.32 there exists $F \in \underline{F}(\mathcal{V})$ such that $\sigma(A_F) \subset \mathbb{C}_g$. By dualizing theorem 4.32, we find that there exists $G \in \underline{G}(\mathcal{S})$ such that $\sigma(A_G) \subset \mathbb{C}_g$. Finally, there exists a linear mapping $N : \mathcal{Y} \to \mathcal{U}$ such that $(A + BNC)\mathcal{S} \subset \mathcal{V}$. By applying corollary 6.10 we then find a controller Γ with the desired properties. ■

Theorem 6.13 *There exists a controller Γ such that $T_\Gamma = 0$ and $\sigma(A_e) \subset \mathbb{C}_g$ if and only if there exist a detectability subspace $\mathcal{S}$ and a stabilizability subspace $\mathcal{V}$ such that*

$$\operatorname{im} E \subset \mathcal{S} \subset \mathcal{V} \subset \ker H,$$

(C, A) is detectable, and (A, B) is stabilizable.

Proof : ($\Rightarrow$) If $T_\Gamma = 0$ then the closed loop system (6.12) is disturbance decoupled. By theorem 4.6, there is an A_e-invariant subspace $\mathcal{V}_e$ of the extended state space $\mathcal{X}_e$ such that $\operatorname{im} E_e \subset \mathcal{V}_e \subset \ker H_e$. Since, furthermore, $\sigma(A_e) \subset \mathbb{C}_g$, the subspace $\mathcal{V}_e$ must be inner-stable and outer-stable. According to theorem 6.8, this implies that $\mathcal{S} := i(\mathcal{V}_e)$ is a detectability subspace and that $\mathcal{V} := p(\mathcal{V}_e)$ is a stabilizability subspace. The claim that $\operatorname{im} E \subset \mathcal{S} \subset \mathcal{V} \subset \ker H$ was already proven in theorem 6.6. The facts that (C, A) is detectable and (A, B) is stabilizable follow from theorem 3.40 (note that Γ is a stabilizing controller).

($\Leftarrow$) By lemma 6.12, there is a controller Γ such that $\sigma(A_e) \subset \mathbb{C}_g$ and an A_e-invariant subspace $\mathcal{V}_e$ such that $i(\mathcal{V}_e) = \mathcal{S}$ and $p(\mathcal{V}_e) = \mathcal{V}$. We showed in theorem 6.6 that the latter implies that $\operatorname{im} E_e \subset \mathcal{V}_e \subset \ker H_e$. According to theorem 4.6 we must therefore have that the closed loop system (6.12) is disturbance decoupled. ■

The previous result immediately yields the following corollary which provides conditions that can be checked computationally:

Corollary 6.14 *There exists a controller Γ such that $T_\Gamma = 0$ and $\sigma(A_e) \subset \mathbb{C}_g$ if and only if*

$$\mathcal{S}_g^*(\operatorname{im} E) \subset \mathcal{V}_g^*(\ker H),$$

(C, A) is detectable, and (A, B) is stabilizable.

Proof : The proof of this is left as an exercise to the reader. ■

If (C, A) is detectable, (A, B) is stabilizable, and the subspace inclusion of corollary 6.14 holds, then a suitable controller can be calculated as follows. First, take a mapping $F : \mathcal{X} \to \mathcal{U}$ such that $A_F \mathcal{V}_g^* \subset \mathcal{V}_g^*$ and $\sigma(A_F) \subset \mathbb{C}_g$ (see theorem 4.32). Find a mapping $G : \mathcal{Y} \to \mathcal{X}$ such that $A_G \mathcal{S}_g^* \subset \mathcal{S}_g^*$ and $\sigma(A_G) \subset \mathbb{C}_g$ (dual of theorem 4.32). Next, let $N : \mathcal{Y} \to \mathcal{U}$ be such that $(A + BNC)\mathcal{S}_g^* \subset \mathcal{V}_g^*$. Put $K := A + BF + GC - BNC$, $L := BN - G$ and $M := F - NC$. Then the controller $\Gamma = (K, L, M, N)$ with state space $\mathcal{W} = \mathcal{X}$ will achieve disturbance decoupling and internal stability.

6.5 Pairs of (C, A, B)-pairs

In the previous sections we have seen how (C, A, B)-pairs can be used to construct controllers. We showed that a (C, A, B)-pair between $\operatorname{im} E$ and $\ker H$ immediately gives rise to a controller that achieves disturbance decoupling. In the present section we will rather work with *pairs* of (C, A, B)-pairs. These will be applied to extend the results of section 4.8 on the problem of external stabilization from the static state feedback case to the case of dynamic measurement feedback.

First consider the situation that we have two (A, B)-invariant subspaces $\mathcal{V}_1$ and $\mathcal{V}_2$ such that $\mathcal{V}_1 \subset \mathcal{V}_2$. We may then ask ourselves the question: does there exists a single linear mapping $F : \mathcal{X} \to \mathcal{U}$ such that both $\mathcal{V}_1$ and $\mathcal{V}_2$ are invariant under $A + BF$? The answer is yes:

Lemma 6.15 *Let $\mathcal{V}_1 \subset \mathcal{V}_2$ be (A, B)-invariant subspaces. Then there exists a linear mapping $F : \mathcal{X} \to \mathcal{U}$ such that $(A + BF)\mathcal{V}_i \subset \mathcal{V}_i$ $(i = 1, 2)$.*

Proof : Let $F_1 \in \underline{F}(\mathcal{V}_1)$ and $F_2 \in \underline{F}(\mathcal{V}_2)$. Choose a basis for $\mathcal{X}$ as follows. First choose a basis $q_1, \ldots, q_k$ for $\mathcal{V}_1$. Extend this to a basis $q_1, \ldots, q_l$ for $\mathcal{V}_2$ and subsequently to a basis $q_1, \ldots, q_n$ for $\mathcal{X}$. Let F be a linear mapping $\mathcal{X} \to \mathcal{U}$ such that $Fq_i = F_1 q_i$ $(i = 1, \ldots, k)$ and $Fq_i = F_2 q_i$ $(i = k + 1, \ldots, l)$. It is easily verified that $F \in \underline{F}(\mathcal{V}_1) \cap \underline{F}(\mathcal{V}_2)$. ■

Of course, this result can immediately be dualized to obtain the corresponding result for conditioned invariant subspaces:

Lemma 6.16 *Let $\mathcal{S}_1 \subset \mathcal{S}_2$ be (C, A)-invariant subspaces. There exists a linear mapping $G : \mathcal{Y} \to \mathcal{X}$ such that $(A + GC)\mathcal{S}_i \subset \mathcal{S}_i$ $(i = 1, 2)$.*

Proof : The proof of this lemma is left as an exercise to the reader (use theorem 5.6). ■

In section 6.1 it was shown that if $(\mathcal{S}, \mathcal{V})$ is a (C, A, B)-pair then there exists an output-feedback mapping $N : \mathcal{Y} \to \mathcal{U}$ such that $(A + BNC)\mathcal{S} \subset \mathcal{V}$. The following lemma is a generalization of this result to the case that we have two (C, A, B)-pairs with, in a suitable sense, the first one contained in the other.

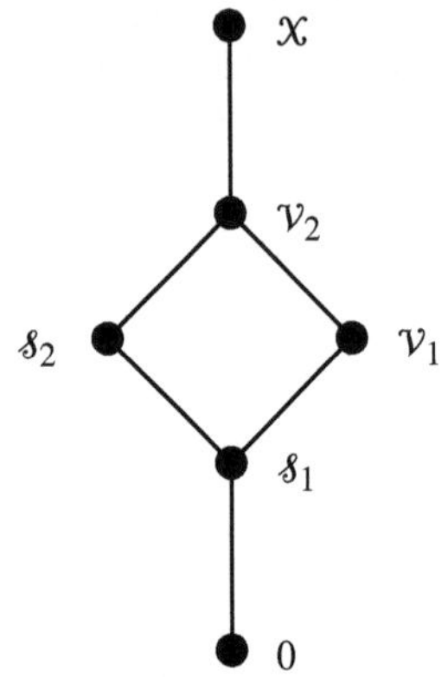

Lemma 6.17 *Let $(\mathcal{S}_1, \mathcal{V}_1)$ and $(\mathcal{S}_2, \mathcal{V}_2)$ be (C, A, B)-pairs such that $\mathcal{S}_1 \subset \mathcal{S}_2$ and $\mathcal{V}_1 \subset \mathcal{V}_2$. There exists a linear mapping $N : \mathcal{Y} \to \mathcal{U}$ such that $(A + BNC)\mathcal{S}_i \subset \mathcal{V}_i$ $(i = 1, 2)$.*

Proof : Let $q_1, \ldots, q_l$ be a basis for $\mathcal{S}_1$ such that $q_1, \ldots, q_k$ is a basis for $\mathcal{S}_1 \cap \ker C$ (where $k \leqslant l$). Since $\mathcal{S}_1 \cap \ker C \subset \mathcal{S}_2 \cap \ker C$, this basis $q_1, \ldots, q_k$ can be extended to a basis for $\mathcal{S}_2 \cap \ker C$ by adding vectors, say, $q_{l+1}, \ldots, q_r$. $(r \geqslant l)$. We claim that the vectors $q_1, \ldots, q_r$ are linearly independent. Indeed, assume $\alpha_1 q_1 + \cdots + \alpha_r q_r = 0$. Then $\tilde{x} := \alpha_{k+1} q_{k+1} + \cdots + \alpha_l q_l$ can be written as a linear combination of $q_1, \ldots, q_k, q_{l+1}, \ldots, q_r$. Since these vectors lie in $\ker C$ we find that $\tilde{x} \in \mathcal{S}_1 \cap \ker C$. This implies that $\tilde{x} = 0$ and hence that $\alpha_{k+1} = \cdots = \alpha_l = 0$. We thus find that $\alpha_1 q_1 + \cdots + \alpha_k q_k + \alpha_{l+1} q_{l+1} + \cdots + \alpha_r q_r = 0$. However, $q_1, \ldots, q_k, q_{l+1}, \ldots, q_r$ forms a basis for $\mathcal{S}_2 \cap \ker C$ and hence $\alpha_1 = \cdots = \alpha_k = \alpha_{l+1} = \cdots = \alpha_r = 0$. Extend $q_1, \ldots, q_r$ to a basis $q_1, \ldots, q_s$ for $\mathcal{S}_2$ $(s \geqslant r)$. We then claim that the vectors $Cq_{k+1}, \ldots, Cq_l, Cq_{r+1}, \ldots, Cq_s$ are linearly independent: suppose that

$$\beta_{k+1} C q_{k+1} + \cdots + \beta_l C q_l + \beta_{r+1} C q_{r+1} + \cdots + \beta_s C q_s = 0.$$

Then the vector $\beta_{k+1} q_{k+1} + \cdots + \beta_l q_l + \beta_{r+1} q_{r+1} + \cdots + \beta_s q_s$ lies in $\ker C$. Since it also lies in $\mathcal{S}_2$, it can be written as a linear combination of $q_1, \ldots, q_k, q_{l+1}, \ldots, q_r$. This then yields a linear combination of $q_1, \ldots, q_s$ that is equal to zero. It follows that $\beta_{k+1} = \cdots = \beta_l = \beta_{r+1} = \cdots = \beta_s = 0$. Since $\mathcal{S}_1 \subset \mathcal{V}_1$ and $\mathcal{S}_2 \subset \mathcal{V}_2$ and since $\mathcal{V}_1$ and $\mathcal{V}_2$ are controlled invariant, there exist $v_{k+1}, \ldots, v_l \in \mathcal{V}_1, v_{r+1}, \ldots, v_s \in \mathcal{V}_2$ and $u_{k+1}, \ldots, u_l, u_{r+1}, \ldots, u_s \in \mathcal{U}$ such that

$$Aq_i = v_i + Bu_i \; (i = k + 1, \ldots, l \text{ and } i = r + 1, \ldots, s).$$

Since $Cq_{k+1}, \ldots, Cq_l, Cq_{r+1}, \ldots, Cq_s$ are linearly independent, there exists a linear mapping $N : \mathcal{Y} \to \mathcal{U}$ such that $NCq_i = -u_i$ ($i = k+1, \ldots, l$ and $i = r+1, \ldots, s$). We then have

$$\begin{aligned} (A + BNC)q_i &= Aq_i \in \mathcal{S}_1 \subset \mathcal{V}_1 \quad (i = 1, \ldots, k), \\ (A + BNC)q_i &= v_i \in \mathcal{V}_1 \quad (i = k+1, \ldots, l), \end{aligned}$$

which implies $(A + BNC)\mathcal{S}_1 \subset \mathcal{V}_1$. In the same way we find that

$$\begin{aligned} (A + BNC)q_i &= Aq_i \in \mathcal{S}_2 \subset \mathcal{V}_2 \quad (i = 1, \ldots, r), \\ (A + BNC)q_i &= v_i \in \mathcal{V}_2 (i = r+1, \ldots, s), \end{aligned}$$

which implies that $(A + BNC)\mathcal{S}_2 \subset \mathcal{V}_2$. ∎

By combining the previous three lemmas with theorem 6.4 and corollary 6.10, we arrive at the following theorem:

Theorem 6.18 *Let $(\mathcal{S}_1, \mathcal{V}_1)$ and $(\mathcal{S}_2, \mathcal{V}_2)$ be (C, A, B)-pairs such that $\mathcal{S}_1 \subset \mathcal{S}_2$ and $\mathcal{V}_1 \subset \mathcal{V}_2$. Let $F \in \underline{F}(\mathcal{V}_1) \cap \underline{F}(\mathcal{V}_2)$, and let $G \in \underline{G}(\mathcal{S}_1) \cap \underline{G}(\mathcal{S}_2)$. Then there exists a controller Γ and A_e-invariant subspaces $\mathcal{V}_{e,1} \subset \mathcal{V}_{e,2}$ of the extended state space $\mathcal{X}_e$ such that*

$$i(\mathcal{V}_{e,1}) = \mathcal{S}_1, \quad p(\mathcal{V}_{e,1}) = \mathcal{V}_1, \tag{6.19}$$

$$i(\mathcal{V}_{e,2}) = \mathcal{S}_2, \quad p(\mathcal{V}_{e,2}) = \mathcal{V}_2, \tag{6.20}$$

$$\sigma(A_e) = \sigma(A_F) \cup \sigma(A_G), \tag{6.21}$$

$$\sigma(A_e \mid \mathcal{V}_{e,2}/\mathcal{V}_{e,1}) = \sigma(A_F \mid \mathcal{V}_2/\mathcal{V}_1) \cup \sigma(A_G \mid \mathcal{S}_2/\mathcal{S}_1). \tag{6.22}$$

Specifically, for any linear mapping $N : \mathcal{Y} \to \mathcal{U}$ such that $(A + BNC)\mathcal{S}_i \subset \mathcal{V}_i$ ($i = 1, 2$) the controller Γ given by

$$\begin{aligned} \dot{w}(t) &= (A + BF + GC - BNC)w(t) + (BN - G)y(t), \\ u(t) &= (F - NC)w(t) + Ny(t) \end{aligned} \tag{6.23}$$

with state space $\mathcal{W} = \mathcal{X}$, and the subspaces

$$\mathcal{V}_{e,i} := \left\{ \begin{pmatrix} x_1 \\ 0 \end{pmatrix} + \begin{pmatrix} x_2 \\ x_2 \end{pmatrix} \in \mathcal{X}_e \;\middle|\; x_1 \in \mathcal{S}_i, \; x_2 \in \mathcal{V}_i \right\} \quad (i = 1, 2) \tag{6.24}$$

satisfy these desired properties.

Proof : The claim that the controller Γ given by (6.23) yields an extended system mapping A_e under which the subspaces (6.24) are invariant follows from theorem 6.4. Also (6.19) and (6.20) follow from theorem 6.4. Property (6.21) is a consequence of corollary 6.10. The fact that $\mathcal{V}_{e,1} \subset \mathcal{V}_{e,2}$ follows from the definition. It remains to prove (6.22). Let $\Pi : \mathcal{V}_{e,2} \to \mathcal{V}_{e,2}/\mathcal{V}_{e,1}$ be the canonical projection. Denote the mapping $A_e \mid \mathcal{V}_{e,2}/\mathcal{V}_{e,1}$ by A_{e0}. Let $\lambda \in \sigma(A_{e0})$ and let Πx_e be an associated eigenvector ($x_e \in \mathcal{V}_{e,2}$). Since $\ker \Pi = \mathcal{V}_{e,1}$ we have $x_e \notin \mathcal{V}_{e,1}$. By the fact that $A_{e0}\Pi = \Pi A_e$ we have $\Pi A_e x_e = \lambda \Pi x_e$. Hence there exists a vector $\tilde{x}_e \in \mathcal{V}_{e,1}$ such that $A_e x_e = \lambda x_e + \tilde{x}_e$. From the definition of $\mathcal{V}_{e,1}$ and $\mathcal{V}_{e,2}$ we find that there are $\tilde{x}_1 \in \mathcal{S}_1$, $\tilde{x}_2 \in \mathcal{V}_1$, $x_1 \in \mathcal{S}_2$ and $x_2 \in \mathcal{V}_2$ such that

$$\tilde{x}_e = \begin{pmatrix} \tilde{x}_1 + \tilde{x}_2 \\ \tilde{x}_2 \end{pmatrix} \text{ and } x_e = \begin{pmatrix} x_1 + x_2 \\ x_2 \end{pmatrix}.$$

Using the representation (6.10) for A_e this yields

$$(A + BNC)(x_1 + x_2) + (BF - BNC)x_2 = \lambda(x_1 + x_2) + \tilde{x}_1 + \tilde{x}_2, \tag{6.25}$$

$$(BNC - GC)(x_1 + x_2) + (A + BF + GC - BNC)x_2 = \lambda x_2 + \tilde{x}_2. \tag{6.26}$$

This, in turn, yields

$$(A + BNC)x_1 + (A + BF)x_2 = \lambda(x_1 + x_2) + \tilde{x}_1 + \tilde{x}_2, \tag{6.27}$$

$$(BNC - GC)x_1 + (A + BF)x_2 = \lambda x_2 + \tilde{x}_2. \tag{6.28}$$

By subtracting (6.28) from (6.27) we obtain

$$(A + GC)x_1 = \lambda x_1 + \tilde{x}_1. \tag{6.29}$$

Let $\Pi_1 : \mathcal{V}_2 \to \mathcal{V}_2/\mathcal{V}_1$ and $\Pi_2 : \mathcal{S}_2 \to \mathcal{S}_2/\mathcal{S}_1$ be the canonical projections. Denote $A_F \mid \mathcal{V}_2/\mathcal{V}_1$ by A_1 and $A_G \mid \mathcal{S}_2/\mathcal{S}_1$ by A_2. We now distinguish two cases, the case that $x_1 \notin \mathcal{S}_1$ and the case that $x_1 \in \mathcal{S}_1$. First assume that $x_1 \notin \mathcal{S}_1$. Then $\Pi_2 x_1 \neq 0$. Since $\tilde{x}_1 \in \mathcal{S}_1$ it follows from (6.29) that

$$A_2 \Pi_2 x_1 = \Pi_2 A_G x_1 = \lambda \Pi_2 x_1.$$

Hence, $\Pi_2 x_1$ is an eigenvector of A_2 with eigenvalue λ so $\lambda \in \sigma(A_G \mid \mathcal{S}_2/\mathcal{S}_1)$. Next, assume $x_1 \in \mathcal{S}_1$. Since $\tilde{x}_1, \tilde{x}_2 \in \mathcal{V}_1$ and $(A + BNC)x_1 \in \mathcal{V}_1$ it then follows from (6.27) that

$$A_1 \Pi_1 x_2 = \Pi_1 A_F x_2 = \lambda \Pi_1 x_2.$$

We claim that $\Pi_1 x_2 \neq 0$. Indeed, $x_2 \in \mathcal{V}_1$ would, together with $x_1 \in \mathcal{S}_1$, imply that $x_e \in \mathcal{V}_{1,e}$. This is a contradiction. Hence, $\Pi_1 x_2$ is an eigenvector of A_1 with eigenvalue λ so $\lambda \in \sigma(A_F \mid \mathcal{V}_2/\mathcal{V}_1)$. We have now proven that in (6.22) we have inclusion from the left to the right.

To prove the converse inclusion, let $\lambda \in \sigma(A_1)$. There is $x_2 \in \mathcal{V}_2$, $x_2 \notin \mathcal{V}_1$ such that

$$A_1 \Pi_1 x_2 = \lambda \Pi_1 x_2.$$

Consequently, there is a vector $\tilde{x}_2 \in \mathcal{V}_1$ such that

$$A_F x_2 = \lambda x_2 + \tilde{x}_2.$$

Define

$$x_e := \begin{pmatrix} x_2 \\ x_2 \end{pmatrix} \quad \text{and} \quad \tilde{x}_e := \begin{pmatrix} \tilde{x}_2 \\ \tilde{x}_2 \end{pmatrix}.$$

Then $x_2 \in \mathcal{V}_{e,2}$ and $\tilde{x}_e \in \mathcal{V}_{e,1}$. It is straightforward to verify that

$$A_e x_e = \lambda x_e + \tilde{x}_e$$

and hence $A_{e0}\Pi x_e = \Pi A_e x_e = \lambda \Pi x_e$. Since $x_2 \notin \mathcal{V}_1$ we must have $\Pi x_e \neq 0$ and hence Πx_e is an eigenvector of A_{e0} with eigenvalue λ. We conclude that $\lambda \in \sigma(A_e \mid \mathcal{V}_{e,2}/\mathcal{V}_{e,1})$. To prove the fact that also the spectrum of $A_G \mid \mathcal{S}_2/\mathcal{S}_1$ is contained in the spectrum of $A_e \mid \mathcal{V}_{e,2}/\mathcal{V}_{e,1}$, assume that $\lambda \in \sigma(A_2)$. There is $x_1 \in \mathcal{S}_2,\ x_1 \notin \mathcal{S}_1$ such that

$$A_2 \Pi_2 x_1 = \lambda \Pi_2 x_1.$$

Consequently, there exists a vector $\tilde{x}_1 \in \mathcal{S}_1$ such that

$$A_G x_1 = \lambda x_1 + \tilde{x}_1.$$

The vector $\left(\begin{smallmatrix} x_1 \\ 0 \end{smallmatrix}\right)$ lies in $\mathcal{V}_{e,2}$ and since this subspace is A_e-invariant we have

$$A_e \begin{pmatrix} x_1 \\ 0 \end{pmatrix} = \begin{pmatrix} (A + BNC)x_1 \\ (BNC - GC)x_1 \end{pmatrix} \in \mathcal{V}_{e,2}.$$

This immediately implies that $(BNC - GC)x_1 \in \mathcal{V}_2$. Now, assume that $\lambda \notin \sigma(A_F \mid \mathcal{V}_2/\mathcal{V}_1)$. Then $(A_1 - \lambda I)^{-1}$ exists. Let $x_2 \in \mathcal{V}_2$ be such that

$$(A_1 - \lambda I)^{-1}\Pi_1(BNC - GC)x_1 = -\Pi_1 x_2.$$

The latter implies that

$$\Pi_1(BNC - GC)x_1 = -\Pi_1(A_F - \lambda I)x_2$$

and therefore

$$\tilde{x}_2 := (BNC - GC)x_1 + (A_F - \lambda I)x_2 \in \mathcal{V}_1.$$

Define now $x_e \in \mathcal{V}_{2,e}$ and $\tilde{x}_e \in \mathcal{V}_{1,e}$ by

$$x_e := \begin{pmatrix} x_1 + x_2 \\ x_2 \end{pmatrix}, \quad \tilde{x}_e := \begin{pmatrix} \tilde{x}_2 \\ \tilde{x}_2 \end{pmatrix}.$$

Since $x_1 \notin \mathcal{S}_1$ we must have $x_e \notin \mathcal{V}_{1,e}$. Hence $\Pi x_e \neq 0$. It is now easily verified that

$$A_e x_e = \begin{pmatrix} (A + BNC)x_1 + A_F x_1 \\ (BNC - GC)x_1 + A_F x_2 \end{pmatrix} = \begin{pmatrix} A_G x_1 + \lambda x_2 + \tilde{x}_2 \\ \lambda x_2 + \tilde{x}_2 \end{pmatrix} = \lambda x_e + \tilde{x}_e.$$

From this we obtain that

$$A_{e0}\Pi x_e = \Pi A_e x_e = \lambda \Pi x_e,$$

and we conclude that $\lambda \in \sigma(A_{e0})$. This completes the proof of the theorem. ■

The relation between the spectra of $A + GC$ and $A + BF$, and the spectrum of A_e is depicted in the lattice diagram (6.2).

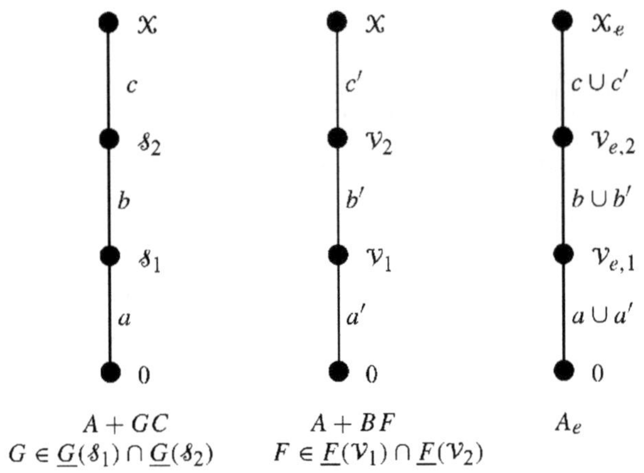

Figure 6.2

6.6 External stabilization by measurement feedback

Again consider the control system (6.11). In section 4.8 we discussed the problem of external stabilization by (static) state feedback. In the present section we consider the dynamic measurement feedback version of this problem. More concretely, we consider the problem of finding a controller for the system (6.11) such that the closed loop system (6.12) is externally stable:

Definition 6.19 *Consider the system (6.11). Let $\mathbb{C}_g$ be a stability domain.* The problem of external stabilization by measurement feedback, *ESPM, is to find a controller Γ such that the closed loop transfer function $G_\Gamma(s)$ is stable.*

Let us take a look at this. If Γ is a controller that makes the transfer function $G_\Gamma(s)$ stable, then for all $x_0 \in \operatorname{im} E$ the Bohl function

$$H_e e^{A_e t}\begin{pmatrix} x_0 \\ 0 \end{pmatrix}$$

is stable. Thus, the state trajectory $x_e(t) = (x(t)^T, w(t)^T)^T$ of the extended system (6.3), with initial state $x_e(0) = (x_0^T, 0)^T$, has the property that Hx is stable. Now, x satisfies the equation

$$\dot{x}(t) = Ax(t) + B(NCx(t) + Mw(t)),$$

with, simultaneously,

$$\dot{w}(t) = LCx(t) + Kw(t).$$

Define $v(t) := NCx(t) + Mw(t)$. Then v is a Bohl function and x satisfies $\dot{x}(t) = Ax(t) + Bv(t)$, $x(0) = x_0$. We conclude that, in fact, $x(t) = x_v(t, x_0)$, i.e., x is the state trajectory of the system (A, B) resulting from the initial state $x(0) = x_0$ and the control input v. Since Hx is stable, this immediately implies that x_0 is an element of the subspace $\mathcal{W}_g(\ker H)$ defined by (4.28). Since this holds for any $x_0 \in \operatorname{im} E$, after applying theorem 4.37 we obtain as a necessary condition for the existence of an externally stabilizing controller Γ that

$$\operatorname{im} E \subset \mathcal{V}^*(\ker H) + \mathcal{X}_{\text{stab}}. \tag{6.30}$$

We can say even more. For, if Γ is such that $G_\Gamma(s)$ is stable then also the transposed transfer matrix

$$G_\Gamma^{\mathrm{T}}(s) = E_e^{\mathrm{T}}(Is - A_e^{\mathrm{T}})^{-1} H_e^{\mathrm{T}}$$

is stable. Since

$$A_e^{\mathrm{T}} = \begin{pmatrix} A^{\mathrm{T}} + C^{\mathrm{T}} N^{\mathrm{T}} B^{\mathrm{T}} & C^{\mathrm{T}} L^{\mathrm{T}} \\ M^{\mathrm{T}} B^{\mathrm{T}} & K^{\mathrm{T}} \end{pmatrix}, \quad H_e^{\mathrm{T}} = \begin{pmatrix} H^{\mathrm{T}} \\ 0 \end{pmatrix},$$
$$E_e^{\mathrm{T}} = \begin{pmatrix} E^{\mathrm{T}} & 0 \end{pmatrix},$$

this implies that the controller $\tilde{\Gamma} = (K^{\mathrm{T}}, M^{\mathrm{T}}, L^{\mathrm{T}}, N^{\mathrm{T}})$ achieves external stability for the dual control system

$$\begin{aligned} \dot{x}(t) &= A^{\mathrm{T}} x(t) + C^{\mathrm{T}} u(t) + H^{\mathrm{T}} d(t), \\ y(t) &= B^{\mathrm{T}} x(t), \\ z(t) &= E^{\mathrm{T}} x(t). \end{aligned} \tag{6.31}$$

This immediately implies that

$$\operatorname{im} H^{\mathrm{T}} \subset \mathcal{V}^*(E^{\mathrm{T}}, A^{\mathrm{T}}, C^{\mathrm{T}}) + \mathcal{X}_{\text{stab}}(A^{\mathrm{T}}, C^{\mathrm{T}}).$$

By taking orthogonal complements this yields

$$\mathcal{V}^*(E^{\mathrm{T}}, A^{\mathrm{T}}, C^{\mathrm{T}})^{\perp} \cap \mathcal{X}_{\text{stab}}(A^{\mathrm{T}}, C^{\mathrm{T}})^{\perp} \subset \ker H,$$

which, by (5.3) and theorem 5.15, leads to the following necessary condition for the existence of an externally stabilizing controller Γ for the system (6.11):

$$\mathcal{S}^*(\operatorname{im} E) \cap \mathcal{X}_{\text{det}} \subset \ker H. \tag{6.32}$$

Here, $\mathcal{S}^*(\operatorname{im} E)$ and $\mathcal{X}_{\text{det}}$ are the smallest conditioned invariant subspace containing $\operatorname{im} E$, and the smallest detectability subspace, defined with respect to the system (C, A). We have now shown that the conditions (6.30) and (6.32) are both necessary for the existence of a controller that achieves external stability.

In the remainder of this section we show that the pair of conditions (6.30) and (6.32) is also sufficient for the existence of such a controller. The idea is to use the conditions (6.30) and (6.32) to obtain a pair of (C, A, B)-pairs with certain desired properties and then to apply theorem 6.18 of the previous section to these (C, A, B)-pairs to obtain a suitable controller. Denote $\mathcal{W}_g(\ker H)$ by $\mathcal{W}_g$, and define

$$\mathcal{T}_g := \mathcal{S}^*(\operatorname{im} E) \cap \mathcal{X}_{\text{det}}. \tag{6.33}$$

In order to be able to describe the duality between the subspaces $\mathcal{W}_g$ and $\mathcal{T}_g$, let us be a bit more precise on the maps in terms of which these subspaces are defined. Thus, denote $\mathcal{W}_g = \mathcal{W}_g(H, A, B)$ and $\mathcal{T}_g = \mathcal{T}_g(E, C, A)$. As already noted above, we have

$$\mathcal{T}_g(E, C, A) = \mathcal{W}_g(E^{\mathrm{T}}, A^{\mathrm{T}}, C^{\mathrm{T}})^{\perp}. \tag{6.34}$$

This duality can be used to obtain the following result:

Lemma 6.20 *The subspace $\mathcal{T}_g$ is invariant under $A + GC$ for any linear mapping $G : \mathcal{Y} \to \mathcal{X}$. There exists $G \in \underline{G}(\mathcal{S}^*(\operatorname{im} E))$ such that $\sigma(A_G \mid \mathcal{S}^*(\operatorname{im} E)/\mathcal{T}_g) \subset \mathbb{C}_g$.*

Proof : The subspace $\mathcal{T}_g$ is (C, A)-invariant since it is the intersection of (C, A)-invariant subspaces. Since $\mathcal{X}_{\text{det}} \subset \ker C$ we have $\mathcal{T}_g \subset \ker C$. Hence, for any mapping $G : \mathcal{Y} \to \mathcal{X}$ we have

$$(A + GC)\mathcal{T}_g = A\mathcal{T}_g = A(\mathcal{T}_g \cap \ker C) \subset T_g.$$

The second assertion can be obtained by dualizing theorem 4.38 and using (5.3) and (6.34). The details are left as an exercise to the reader. (Use the fact that if $\mathcal{V}$ and $\mathcal{W}$ are A-invariant subspaces and $\mathcal{V} \subset \mathcal{W}$, then $\mathcal{V}^{\perp}$ and $\mathcal{W}^{\perp}$ are A^{T}-invariant and $\sigma(A \mid \mathcal{W}/\mathcal{V}) = \sigma(A^{\mathrm{T}} \mid \mathcal{V}^{\perp}/\mathcal{W}^{\perp})$). ■

We are now ready to prove the following crucial instrument:

Lemma 6.21 *Consider the system (6.11). Assume that* $\operatorname{im} E \subset \mathcal{W}_g$ *and* $\mathcal{T}_g \subset \ker H$. *Then there exists a controller Γ and A_e-invariant subspaces $\mathcal{V}_{e,1} \subset \mathcal{V}_{e,2}$ of the extended state space $\mathcal{X}_e$ such that*

$$\operatorname{im} E_e \subset \mathcal{V}_{e,2} \text{ and } \mathcal{V}_{e,1} \subset \ker H_e \tag{6.35}$$

and

$$\sigma(A_e \mid \mathcal{V}_{e,2}/\mathcal{V}_{e,1}) \subset \mathbb{C}_g. \tag{6.36}$$

Specifically, if $F \in \underline{F}(\mathcal{V}^*(\ker H))$ and $G \in \underline{G}(\mathcal{S}^*(\operatorname{im} E))$ are such that

$$\sigma(A_F \mid \mathcal{W}_g/\mathcal{V}^*(\ker H)) \subset \mathbb{C}_g \tag{6.37}$$

and

$$\sigma(A_G \mid \mathcal{S}^*(\operatorname{im} E)/\mathcal{T}_g) \subset \mathbb{C}_g \tag{6.38}$$

then the controller Γ given by

$$\begin{aligned} \dot{w}(t) &= (A + BF + GC)w(t) - Gy(t), \\ u(t) &= Fw(t) \end{aligned} \tag{6.39}$$

with state space $\mathcal{W} = \mathcal{X}$, and the subspaces

$$\mathcal{V}_{e,1} := \left\{ \begin{pmatrix} x_1 \\ 0 \end{pmatrix} + \begin{pmatrix} x_2 \\ x_2 \end{pmatrix} \in \mathcal{X}_e \;\middle|\; x_1 \in \mathcal{T}_g,\ x_2 \in \mathcal{V}^*(\ker H) \right\} \tag{6.40}$$

and

$$\mathcal{V}_{e,2} := \left\{ \begin{pmatrix} x_1 \\ 0 \end{pmatrix} + \begin{pmatrix} x_2 \\ x_2 \end{pmatrix} \in \mathcal{X}_e \;\middle|\; x_1 \in \mathcal{S}^*(\operatorname{im} E),\ x_2 \in \mathcal{W}_g \right\} \tag{6.41}$$

satisfy these properties.

Proof : Since $\mathcal{W}_g$ is A-invariant, it is also (C, A)-invariant. Hence, the condition $\operatorname{im} E \subset \mathcal{W}_g$ is equivalent to $\mathcal{S}^*(\operatorname{im} E) \subset \mathcal{W}_g$. Dually, the subspace $\mathcal{T}_g$ is A-invariant so also (A, B)-invariant. Hence, the condition $\mathcal{T}_g \subset \ker H$ is equivalent to $\mathcal{T}_g \subset \mathcal{V}^*(\ker H)$. Denote $\mathcal{S}^* = \mathcal{S}^*(\operatorname{im} E)$ and $\mathcal{V}^* = \mathcal{V}^*(\ker H)$. Consider now the pair of (C, A, B)-pairs

$$(\mathcal{T}_g, \mathcal{V}^*), \quad (\mathcal{S}^*, \mathcal{W}_g).$$

These pairs are related via $\mathcal{T}_g \subset \mathcal{S}^*$ and $\mathcal{V}^* \subset \mathcal{W}_g$. This brings us in the situation of theorem 6.18. We claim that $(A + BNC)\mathcal{T}_g \subset \mathcal{V}^*$ and $(A + BNC)\mathcal{S}^* \subset \mathcal{W}_g$ if we take $N = 0$. Indeed, $\mathcal{T}_g$ and $\mathcal{W}_g$ are both A-invariant and hence

$$A\mathcal{T}_g \subset \mathcal{T}_g \subset \mathcal{V}^*$$

and

$$A\mathcal{S}^* \subset A\mathcal{W}_g \subset \mathcal{W}_g.$$

Let F and G be such that (6.37) and (6.38) hold. Define $\mathcal{V}_{e,1}$ and $\mathcal{V}_{e,2}$ by (6.40) and (6.41). Then it follows from theorem 6.18 that $\mathcal{V}_{e,1} \subset \mathcal{V}_{e,2}$ and that (6.36) holds. Moreover, $i(\mathcal{V}_{e,2}) = \mathcal{S}^*$ and $p(\mathcal{V}_{e,1}) = \mathcal{V}^*$. These two conditions imply (6.35). ■

Using the previous lemma we can now conclude that the subspace inclusions (6.30) and (6.32) are necessary and sufficient conditions for the existence of an externally stabilizing controller.

Corollary 6.22 *Consider the system (6.11). There exists a controller Γ such that $G_\Gamma(s)$ is stable if and only if*

$$\operatorname{im} E \subset \mathcal{V}^*(\ker H) + \mathcal{X}_{\text{stab}}$$

and

$$\mathcal{S}^*(\operatorname{im} E) \cap \mathcal{X}_{\text{det}} \subset \ker H.$$

Moreover, if these conditions hold then for any $F \in \underline{F}(\mathcal{V}^*(\ker H))$ such that

$$\sigma(A_F \mid (\mathcal{V}^*(\ker H) + \mathcal{X}_{\text{stab}})/\mathcal{V}^*(\ker H)) \subset \mathbb{C}_g$$

and any $G \in \underline{G}(\mathcal{S}^*(\operatorname{im} E))$ *such that*

$$\sigma(A_G \mid \mathcal{S}^*(\operatorname{im} E)/(\mathcal{S}^*(\operatorname{im} E) \cap \mathcal{X}_{\mathrm{det}})) \subset \mathbb{C}_g,$$

the controller Γ *defined by*

$$\begin{aligned} \dot{w}(t) &= (A + BF + GC)w(t) - Gy(t), \\ u(t) &= Fw(t) \end{aligned} \tag{6.42}$$

with state space $\mathcal{W} = \mathcal{X}$, *achieves external stability.*

Proof : ($\Rightarrow$) This was already proven.

($\Leftarrow$) This is an application of lemma 4.35: by the previous lemma, the controller defined by (6.42) yields A_e-invariant subspaces $\mathcal{V}_{e,1} \subset \mathcal{V}_{e,2}$ of the extended state space $\mathcal{X}_e$ such that (6.35) and (6.36) hold. It is then an immediate consequence of lemma 4.35 that the transfer function $G_\Gamma(s)$ is stable. ■

6.7 Exercises

6.1 Let $\mathcal{S} \subset \mathcal{V}$ be subspaces of the state space $\mathcal{X}$. Define subspaces of the Cartesian product $\mathcal{X}_e := \mathcal{X} \times \mathcal{X}$ by

$$\mathcal{V}_e := \left\{ \begin{pmatrix} x_1 \\ 0 \end{pmatrix} + \begin{pmatrix} x_2 \\ x_2 \end{pmatrix} \middle| \; x_1 \in \mathcal{S},\; x_2 \in \mathcal{V} \right\}$$
$$\tilde{\mathcal{V}}_e := \left\{ \begin{pmatrix} x_1 \\ 0 \end{pmatrix} + \begin{pmatrix} x_2 \\ -x_2 \end{pmatrix} \middle| \; x_1 \in \mathcal{V}^\perp,\; x_2 \in \mathcal{S}^\perp \right\}.$$

a. Show that $\dim \mathcal{V}_e = \dim \mathcal{S} + \dim \mathcal{V}$.

b. $\mathcal{V}_e^\perp = \tilde{\mathcal{V}}_e$.

c. If $A_e : \mathcal{X}_e \to \mathcal{X}_e$ is a linear mapping such that $\mathcal{V}_e$ is invariant under A_e, then $\tilde{\mathcal{V}}_e$ is invariant under A_e^{T}.

d. $i(\mathcal{V}_e)^\perp = p(\mathcal{V}_e^\perp)$ and $p(\mathcal{V}_e)^\perp = i(\mathcal{V}_e^\perp)$.

6.2 According to (6.6), there exists a controller Γ such that $T_\Gamma = 0$ if and only if there exists a (C, A, B)-pair $(\mathcal{S}, \mathcal{V})$ with $\operatorname{im} E \subset \mathcal{S} \subset \mathcal{V} \subset \ker H$. In that case, the controller given by (6.7) does the job. As noted in section 6.1, the dynamic order of this controller is equal to n, the dynamic order of the system to-be-controlled. The purpose of this exercise is to show that in general one can construct a controller with dynamic order *less* than n. Throughout this exercise, we assume that there exists a (C, A, B)-pair $(\mathcal{S}, \mathcal{V})$ with $\operatorname{im} E \subset \mathcal{S} \subset \mathcal{V} \subset \ker H$. We will construct a controller Γ with dynamic order equal to $\dim \mathcal{V} - \dim \mathcal{S}$.

First, we define $\mathcal{W}$, the state space of Γ, to be any real linear space of dimension $\dim \mathcal{V} - \dim \mathcal{S}$.

a. Show that there exists a surjective linear mapping $R : \mathcal{V} \to \mathcal{W}$ such that $\ker R = \mathcal{S}$.

b. Let $R^+ : \mathcal{W} \to \mathcal{V}$ be a right-inverse of R. Show that $\operatorname{im}(I - R^+R) \subset \mathcal{S}$.

c. Show that there exists a linear mapping $N : \mathcal{Y} \to \mathcal{U}$ such that

$$(A + BNC)\mathcal{S} \subset \mathcal{V}.$$

d. Show that there exists a linear mapping $F : \mathcal{X} \to \mathcal{U}$ such that

$$(A + BNC + BF)\mathcal{V} \subset \mathcal{V}$$

and $\mathcal{S} \subset \ker F$.

e. Show that there exists a linear mapping $G : \mathcal{Y} \to \mathcal{X}$ such that

$$(A + BNC + GC)\mathcal{S} \subset \mathcal{S}$$

and $\operatorname{im} G \subset \mathcal{V}$.

f. Define $K : \mathcal{W} \to \mathcal{W},\ L : \mathcal{Y} \to \mathcal{W}$ and $M : \mathcal{W} \to \mathcal{U}$ by

$$\begin{aligned} K &:= R(A + BNC + BF + GC)R^+, \\ L &:= -RG, \\ M &:= FR^+. \end{aligned}$$

Check that K, L and M are well-defined and prove that any choice of right-inverse R^+ yields the same K and M.

g. Define a subspace $\mathcal{V}_e$ of the extended state space $\mathcal{X} \times \mathcal{W}$ by

$$\mathcal{V}_e := \left\{ \begin{pmatrix} x \\ Rx \end{pmatrix} \middle| \ x \in \mathcal{V} \right\}.$$

Let A_e be the extended system mapping of the closed loop system corresponding to the controller $\Gamma = (K, L, M, N)$. Show that $\mathcal{V}_e$ is A_e-invariant.

h. Show that $T_\Gamma = 0$, i.e., the controller Γ achieves disturbance decoupling.

6.3 (Disturbance decoupling by measurement feedback with control feedthrough in the measurement output.) Consider the system Σ described by

$$\begin{aligned} \dot{x}(t) &= Ax(t) + Bu(t) + Ed(t), \\ y(t) &= Cx(t) + Du(t), \\ z(t) &= Hx(t). \end{aligned}$$

where $u(t) \in \mathbb{R}^m$ is a control input, $d(t) \in \mathbb{R}^r$ is a disturbance input, $y(t) \in \mathbb{R}^p$ a measurement output, and $z(t) \in \mathbb{R}^q$ an output to be controlled. Recall from section 3.13 that the controller Γ given by the equations

$$\begin{aligned} \dot{w}(t) &= Kw(t) + Ly(t), \\ u(t) &= Mw(t) + Ny(t) \end{aligned}$$

(simply denoted by $\Gamma = (K, L, M, N)$) makes the closed loop system $\Sigma \times \Gamma$ well posed if $I - DN$ is invertible.

a. Assume that $\Gamma = (K, L, M, N)$ makes the closed loop system well posed. Determine the equations for $\Sigma \times \Gamma$.

We say that $\Sigma \times \Gamma$ is disturbance decoupled if d has no influence on z. In this problem we want to find necessary and sufficient conditions for the existence of a controller $\Gamma = (K, L, M, N)$ such that: (i) $\Sigma \times \Gamma$ is well posed, and (ii) $\Sigma \times \Gamma$ is disturbance decoupled.

For this, it is convenient to consider the system $\tilde{\Sigma}$ given by

$$\begin{aligned} \dot{x}(t) &= Ax(t) + Bu(t) + Ed(t), \\ y(t) &= Cx(t), \\ z(t) &= Hx(t). \end{aligned}$$

b. Prove that if there exists a controller $\Gamma = (K, L, M, N)$ such that $\Sigma \times \Gamma$ is well posed and disturbance decoupled, then there exists a controller $\tilde{\Gamma} = (\tilde{K}, \tilde{L}, \tilde{M}, \tilde{N})$ such that $\tilde{\Sigma} \times \tilde{\Gamma}$ is disturbance decoupled and $I + D\tilde{N}$ is invertible.

c. Let $\tilde{\Gamma} = (\tilde{K}, \tilde{L}, \tilde{M}, \tilde{N})$ be such that $\tilde{\Sigma} \times \tilde{\Gamma}$ is disturbance decoupled. Show that

$$(A + B\tilde{N}C)\mathcal{S}^*(\operatorname{im} E) \subset \mathcal{V}^*(\ker H)$$

d. Show that if there exists a controller $\tilde{\Gamma} = (\tilde{K}, \tilde{L}, \tilde{M}, \tilde{N})$ such that $\tilde{\Sigma} \times \tilde{\Gamma}$ is disturbance decoupled and $I + D\tilde{N}$ is invertible, then there exists a controller $\Gamma = (K, L, M, N)$ such that $\Sigma \times \Gamma$ is well posed and disturbance decoupled.

e. Now prove the following: There exists a controller $\Gamma = (K, L, M, N)$ such that $\Sigma \times \Gamma$ is well posed and disturbance decoupled if and only if the following two conditions hold:

1. $\mathcal{S}^*(\operatorname{im} E) \subset \mathcal{V}^*(\ker H)$,

2. there exists a linear map $\tilde{N}$ such that $I + D\tilde{N}$ is invertible and

$$(A + B\tilde{N}C)\mathcal{S}^*(\operatorname{im} E) \subset \mathcal{V}^*(\ker H).$$

6.4 (Output regulation by measurement feedback.) Consider the system $\dot{x}(t) = Ax(t) + Bu(t) + Ed(t)$, $y(t) = Cx(t)$, $z(t) = Hx(t)$. For a given controller $\Gamma : \dot{w}(t) = Kw(t) + Ly(t)$, $u(t) = Mw(t) + Ny(t)$, let $z_\Gamma(t, x_0, w_0, d)$ denote the output of the closed loop system corresponding to the initial state $(x(0), w(0)) = (x_0, w_0)$ and disturbance d. We will say that Γ *achieves output regulation* if $z_\Gamma(t, x_0, w_0, d) \to 0$ $(t \to \infty)$ for all $(x_0, w_0) \in \mathcal{X} \times \mathcal{W}$ and for every disturbance function d.

a. Show that Γ achieves output regulation if and only if $H_e e^{A_e t} E_e = 0$ for all t, and $H_e e^{A_e t} \to 0$ $(t \to \infty)$.

b. Let $\mathcal{N}_e := \langle \ker H_e \mid A_e \rangle$ be the unobservable subspace of (H_e, A_e). Show that $H_e e^{A_e t} \to 0$ $(t \to \infty)$ if and only if $\mathcal{N}_e$ is outer-stable (with respect to $\mathbb{C}_g = \mathbb{C}^- = \{s \in \mathbb{C} \mid \Re e\, s < 0\}$).

c. Show that $H_e e^{A_e t} E_e = 0$ for all t and $H_e e^{A_e t} \to 0$ $(t \to \infty)$ if and only if there exists an outer-stable A_e-invariant subspace $\mathcal{V}_e$ such that $\text{im } E_e \subset \mathcal{V}_e \subset \ker H_e$ (Hint: for the "if" part, use theorem 4.35)

d. Prove the following: there exists a controller Γ that achieves output regulation if and only if there exists a (C, A, B)-pair $(\mathcal{S}, \mathcal{V})$ with $\mathcal{S}$ outer-detectable, $\mathcal{V}$ outer-stabilizable and $\text{im } E \subset \mathcal{S} \subset \mathcal{V} \subset \ker H$. (Hint: for the "if"-part apply (6.18) to the pair of (C, A, B)-pairs $(\mathcal{S}, \mathcal{V})$; $(\mathcal{X}, \mathcal{X})$).

e. Prove the following: there exists a controller Γ that achieves output regulation if and only if $\mathcal{V}^*(\ker H)$ is outer-stabilizable and $\mathcal{S}_g^*(\text{im } E) \subset \mathcal{V}^*(\ker H)$.

f. Describe how an actual output regulating controller Γ can be computed.

6.5 (Disturbance decoupling by measurement feedback with pole placement.) In this problem we investigate the disturbance decoupling problem *with pole placement*, DDPMPP. The question here is: given the control system (6.11), when does there exist, for *any* stability domain $\mathbb{C}_g$, a controller Γ such that $T_\Gamma = 0$ and $\sigma(A_e) \subset \mathbb{C}_g$? In this problem we denote $\mathcal{S}^* = \mathcal{S}^*(\text{im } E)$ and $\mathcal{S}_g^* = \mathcal{S}_g^*(\text{im } E)$. The smallest observability subspace containing im E (see problems 5.8, 5.9, 5.10 and 5.12)) is denoted by $\mathcal{N}^*$. Also, we denote $\mathcal{V}^* = \mathcal{V}^*(\ker H)$, $\mathcal{R}^* = \mathcal{R}^*(\ker H)$, and $\mathcal{V}_g^* = \mathcal{V}_g^*(\ker H)$.

a. Observe that if, for any stability domain $\mathbb{C}_g$, there exists Γ such that $T_\Gamma = 0$ and $\sigma(A_e) \subset \mathbb{C}_g$, then we have: (A, B) is $\mathbb{C}_g$-stabilizable, (C, A) is $\mathbb{C}_g$-detectable, and $\mathcal{S}_g^* \subset \mathcal{V}_g^*$ for any $\mathbb{C}_g$.

b. For $F \in \underline{F}(\mathcal{V}^*)$, let τ_1 denote the fixed spectrum $\sigma(A + BF \mid \mathcal{V}^*/\mathcal{R}^*)$. Show that if $\mathbb{C}_g$ is a stability domain with the property that $\tau_1 \cap \mathbb{C}_g = \varnothing$, then the corresponding stabilizability subspace $\mathcal{V}_g^*$ is equal to $\mathcal{R}^*$.

c. For $G \in \underline{G}(\mathcal{S}^*)$, let τ_2 denote the fixed spectrum $\sigma(A + GC \mid \mathcal{N}^*/\mathcal{S}^*)$ (see exercise 5.12). Show that if $\tau_2 \cap \mathbb{C}_g = \varnothing$, then the corresponding detectability subspace $\mathcal{S}_g^*$ is equal to $\mathcal{N}^*$.

d. Show that if for any $\mathbb{C}_g$ there exist Γ such that $T_\Gamma = 0$ and $\sigma(A_e) \subset \mathbb{C}_g$, then (A, B) is controllable, (C, A) is observable, and

$$\mathcal{N}^* \subset \mathcal{R}^*.$$

Thus we have obtained necessary conditions for DDPMPP. In the remainder of this problem we prove that these conditions are also sufficient. The idea is to apply corollary 6.10 to the (C, A, B)-pair $(\mathcal{N}^*, \mathcal{R}^*)$.

e. Show that if (A, B) is controllable, then for any real monic polynomial p of degree n such that $p = p_1 p_2$, with p_1 and p_2 real monic polynomials, and $\deg p_1 = \dim \mathcal{R}^*$, there exist $F \in \underline{F}(\mathcal{R}^*)$ such that $\chi_{A+BF} = p$.

f. Show that if (C, A) is observable, then for any real monic polynomial q of degree n such that $q = q_1 q_2$, with q_1 and q_2 real monic polynomials, and $\deg q_1 = \dim \mathcal{N}^*$, there exists $G \in \underline{G}(\mathcal{N}^*)$ such that $\chi_{A+GC} = q$.

g. Show that if (A, B) is controllable, (C, A) is observable, and $\mathcal{N}^* \subset \mathcal{R}^*$, then for any real monic polynomial r such that $r = p_1 p_2 q_1 q_2$, with p_1, p_2, q_1 and q_2 real monic polynomials, $\deg p_1 = \dim \mathcal{R}^*$, $\deg p_2 = n - \dim \mathcal{R}^*$, and $\deg q_1 = \dim \mathcal{N}^*$, $\deg q_2 = n - \dim \mathcal{N}^*$, there exists a controller Γ and an A_e-invariant subspace $\mathcal{V}_e$ of the extended state space $\mathcal{X}_e$ such that $i(\mathcal{V}_e) = \mathcal{N}^*$, $p(\mathcal{V}_e) = \mathcal{R}^*$ and $\chi_{A_e} = r$.

h. Prove that if (A, B) is controllable, (C, A) is observable, and $\mathcal{N}^* \subset \mathcal{R}^*$, then for any stability domain $\mathbb{C}_g$ there exists a controller Γ such that $T_\Gamma = 0$ and $\sigma(A_e) \subset \mathbb{C}_g$.

i. Indicate how such controller Γ can be computed.

6.6 Consider the system $\dot{x}(t) = Ax(t) + Bu(t) + Ed(t)$, $y(t) = Cx(t)$, $z(t) = Hx(t)$. A *static output feedback controller* is a controller of the form $u = Ky$, with $K : \mathcal{Y} \to \mathcal{X}$. In this problem we consider the problem of disturbance decoupling by static output feedback. The problem is to find a static output feedback controller $u = Kx$ such that the closed loop system is disturbance decoupled, equivalently, such that the closed loop transfer matrix $G_K(s) := H(Is - A - BKC)^{-1}E$ is equal to zero. Some thought reveals that the crucial concept here is that of (A, B, C)-invariance: A subspace $\mathcal{V}$ of the state space $\mathcal{X}$ is called (A, B, C)-invariant if there exists a map $K : \mathcal{Y} \to \mathcal{X}$ such that

$$(A + BKC)\mathcal{V} \subset \mathcal{V}.$$

a. Show that $\mathcal{V}$ is (A, B, C)-invariant if and only if it is (A, B)-invariant and (C, A)-invariant.

b. Show that there exists a static output feedback controller $u = Ky$ such that $G_K(s) = 0$ if and only if there exists a subspace $\mathcal{V}$ such that $\operatorname{im} E \subset \mathcal{V} \subset \ker H$, $\mathcal{V}$ is (A, B)-invariant, and $\mathcal{V}$ is (C, A)-invariant.

6.8 Notes and references

(C, A, B)-pairs were introduced by Schumacher in [166]. Also, this article is the earliest reference in which a complete solution of the disturbance decoupling problem by measurement feedback can be found. Around the same time, the disturbance decoupling problem by measurement feedback with internal stability was resolved independently by Willems and Commault [218], and by Imai and Akashi [85]. In [135], Ohm, Bhattacharyya and Howze obtained alternative necessary and sufficient conditions, in terms of solvability of rational matrix equations. A discussion of the disturbance decoupling problem by measurement feedback for discrete time systems can be found in Akashi and Imai [2]. More recently, Stoorvogel and van der Woude [191] treated the disturbance decoupling problem with measurement feedback for the more general case that there are nonzero direct feedthrough matrices from the control and disturbance inputs to the measured and to be controlled outputs (see also exercise 6.3). Extensions to characterize the freedom in placing the closed loop poles under the constraint of achieving disturbance decoupling has been studied in Saberi, Sannuti and Stoorvogel [158]. Related results also follow as a corollary of the work of Basile and Marro [13, 14]. In Schumacher [169], (C, A, B) -pairs are applied in tracking and regulation problems.

The problem of external stabilization by measurement feedback, ESPM, treated in section 6.6, is the measurement feedback version of the external stabilization problem, ESP, originally introduced by Hautus in [72]. The combination of disturbance decoupling and external stabilization with measurement feedback into one single synthesis problem, having ESPM as a special case, was treated by van der Woude in [203].

As extensions of the ideas treated in this chapter, we mention the work of van der Woude [204] on disturbance decoupling for structured systems, using graph theoretic methods. The natural extension of disturbance decoupling by measurement feedback to almost disturbance decoupling by measurement feedback was studied by Willems in [216], using almost controlled invariant and almost conditioned invariant subspaces. Almost disturbance decoupling by measurement feedback and internal stability was studied by Weiland and Willems in [211]. A complete solution of this problem can be found in the work of Ozcetin, Saberi, and Shamash [141] and in Saberi, Lin and Stoorvogel [157]. Finally, we mention more recent work by Otsuka [140] on generalized (C, A, B)-pairs applied to the problem of disturbance decoupling for uncertain systems.

Chapter 7

System zeros and the weakly unobservable subspace

In this chapter we first give a brief review of some elementary material on polynomial matrices and the Smith form. We then continue our study of the system Σ given by the equations

$$\begin{aligned}\dot{x}(t) &= Ax(t) + Bu(t),\\ y(t) &= Cx(t) + Du(t),\end{aligned}$$

and introduce an important polynomial matrix associated with this system, the system matrix of Σ. Using the system matrix, we introduce the concepts of transmission polynomials, and zeros of the system Σ. Next, we discuss the weakly unobservable subspace, and the controllable weakly unobservable subspace associated with Σ. The weakly unobservable subspace is used to give a geometric characterization of the property of strong observability. We conclude this chapter with a characterization of the transmission polynomials and the zeros of Σ in terms of the weakly unobservable and controllable weakly unobservable subspace.

7.1 Polynomial matrices and Smith form

Polynomial matrices play an important role in system and control theory. Here we discuss some of their relevant properties. A *polynomial matrix* is a matrix with polynomial entries, hence of the form

$$P(s) = \begin{pmatrix} p_{11}(s) & \cdots & p_{1m}(s)\\ \vdots & & \vdots\\ p_{n1}(s) & \cdots & p_{nm}(s)\end{pmatrix},$$

where the functions $p_{ij}(s)$ are polynomials with real coefficients. Alternatively, we may write such a matrix as a polynomial with matrix coefficients:

$$P(s) = P_0 s^k + P_1 s^{k-1} + \cdots + P_k.$$

Either interpretation can be useful. We say that a square polynomial matrix $P(s)$ is *nonsingular* if the polynomial $\det P(s)$ is not the zero polynomial, equivalently, if $P(\lambda)$ is nonsingular for at least one $\lambda \in \mathbb{C}$. A nonsingular polynomial matrix has an inverse which is a rational matrix. The *normal rank* of a (not necessarily square) polynomial matrix $P(s)$ is defined as $\text{normrank}\, P := \max\{\text{rank}\, P(\lambda) \mid \lambda \in \mathbb{C}\}$. Except for a finite number of points $\lambda \in \mathbb{C}$, one has $\text{rank}\, P(\lambda) = \text{normrank}\, P$. For example, if $P(s)$ is a nonsingular $n \times n$-polynomial matrix, then we have $\text{normrank}\, P = n$, and at points λ where $\det P(\lambda) = 0$, we have $\text{rank}\, P(\lambda) < n$. It is obvious that the normrank does not change when we pre- or postmultiply $P(s)$ with a nonsingular polynomial matrix.

Definition 7.1 *A square polynomial matrix $P(s)$ is called* unimodular *if it has a polynomial inverse. A not necessarily square polynomial matrix $P(s)$ is called* left unimodular *if it has a left polynomial inverse. Right unimodularity is defined similarly.*

Examples of unimodular polynomial matrices are for instance elementary-operation matrices. These are polynomial matrices that correspond to elementary row and column operations. The following provides a complete list of the elementary-operation matrices:

- The constant matrices that are obtained by interchanging two columns in the identity matrix. If the polynomial matrix $Q(s)$ is multiplied from the right (left) with the elementary-operation matrix obtained by interchanging in the identity matrix the ith and the jth column, then the effect will be an interchange of the corresponding two columns (rows) of $Q(s)$.

- The polynomial matrices that are obtained by replacing in the identity matrix one of the zero-entries by a polynomial. Multiplying a given polynomial matrix $Q(s)$ from the right (left) by the elementary-operation matrix obtained from the identity matrix by replacing the (i, j) zero-entry by the polynomial $\alpha(s)$, amounts to adding in $Q(s)$ $\alpha(s)$ times the ith column to the jth column (α times the jth row to the ith row).

- The constant matrices that are obtained by replacing in the identity matrix one of the one-entries by a non-zero constant. If the polynomial matrix $Q(s)$ is multiplied from the right (left) by the elementary operation matrix obtained by replacing the one in the (i, i) position by the constant $\alpha \neq 0$, then the ith column (row) of $Q(s)$ is multiplied by α.

These matrices are easily seen to be unimodular. On the other hand, it can be shown that these elementary operations can be used to bring an arbitrary polynomial matrix to a particular standard form. In order to be able to write the result in a convenient way, we introduce the following notation: if $\alpha_1, \ldots, \alpha_r$ are given numbers,

$\text{diag}_{m\times n}(\alpha_1, \alpha_2, \ldots, \alpha_r)$ denotes the $m \times n$ matrix, the entries a_{ij} of which are given by $a_{ii} := \alpha_i$ if $i \leqslant r$ and $a_{ij} := 0$ otherwise. Sometimes we omit the subscript $m \times n$ if the dimensions of the matrix are obvious.

Theorem 7.2 *For every $m \times n$ polynomial matrix $P(s)$ there exists a sequence of elementary (row and column) operations that brings $P(s)$ to the form*

$$\bar{P}(s) = \text{diag}_{m\times n}(\psi_1(s), \psi_2(s), \ldots, \psi_r(s)), \tag{7.1}$$

where $\psi_1, \psi_2, \ldots$ are monic non-zero polynomials satisfying $\psi_1 \mid \psi_2 \mid \psi_3 \cdots$ (i.e., each ψ divides the next one) and $r = \text{normrank}\, P$. Consequently, for every polynomial matrix $P(s)$ there exist unimodular matrices $U(s)$ and $V(s)$ such that $P(s) = U(s)\bar{P}(s)V(s)$, where $\bar{P}(s)$ has the form (7.1).

We will say that two polynomial matrices $P(s)$ and $Q(s)$ are *unimodularly equivalent* if there exist unimodular matrices $U(s)$ and $V(s)$ such that

$$P(s) = U(s)Q(s)V(s).$$

The polynomials

$$\psi_1, \ldots, \psi_r, 0, \ldots, 0$$

where the zero polynomial appears $\min(n, m) - r$ times, are called the *invariant factors* of the polynomial matrix $P(s)$. The non-zero polynomials $\psi_1, \ldots, \psi_r$ in this list are called the *non-trivial* invariant factors of $P(s)$. The total number of invariant factors is always equal to the number $\min(n, m)$. Invariant factors are counted with multiplicity, i.e., the same polynomial can appear in the list of invariant factors more than once. The invariant factors can be shown to be uniquely defined and independent of the way in which $P(s)$ is transformed to the form (7.1). Also, they do not change when the matrix is pre- or postmultiplied with a unimodular matrix, or equivalently, when an elementary row or column operation is applied to $P(s)$. The matrix in (7.1) is called the *Smith form* of $P(s)$. The proof of theorem 7.2 will be omitted.

Corollary 7.3 *Two $m \times n$ polynomial matrices $P(s)$ and $Q(s)$ are unimodularly equivalent if and only if they have the same invariant factors (or, equivalently, the same Smith form).*

If $P(s)$ is a nonsingular $m \times m$ polynomial matrix, it follows that

$$\det P(s) = \alpha \prod_{i=1}^{m} \psi_i(s), \tag{7.2}$$

for some constant α, hence the zeros of the invariant factors are the points λ at which $P(\lambda)$ is singular. More generally, the zeros of the invariant factors of a polynomial matrix $P(s)$ are the point λ for which $\text{rank}\, P(\lambda) < \text{normrank}\, P$. These values are commonly referred to as the *zeros of the polynomial matrix* $P(s)$.

Corollary 7.4 *Let $P(s)$ be an $m \times m$ polynomial matrix. Then the following statements are equivalent:*

(i) *$P(s)$ is unimodular,*

(ii) *$P(s)$ is the product of elementary-operation matrices,*

(iii) *the Smith form of $P(s)$ is I,*

(iv) *all invariant factors of $P(s)$ are equal to 1,*

(v) det $P(s)$ *is a nonzero constant,*

(vi) *$P(\lambda)$ is invertible for every $\lambda \in \mathbb{C}$.*

Proof : (i) $\Rightarrow$ (vi) If $P(s)$ has a polynomial inverse then $P(\lambda)$ has an inverse for every $\lambda \in \mathbb{C}$.

(vi) $\Rightarrow$ (v) If $P(\lambda)$ is invertible for every λ then det $P(\lambda)$ has no zeros. A polynomial without zeros is a nonzero constant.

(v) $\Rightarrow$ (iv) This follows from (7.2) and the fact that the invariant factors are monic.

(iv) $\Rightarrow$ (iii), (iii) $\Rightarrow$ (ii) and (ii) $\Rightarrow$ (i) are trivial. ■

For nonsquare polynomial matrices we have the following results:

Corollary 7.5 *Let $P(s)$ be a polynomial matrix. The following statements are equivalent:*

(i) *$P(s)$ is left unimodular,*

(ii) *there exists a polynomial matrix $Q(s)$ such that $\begin{pmatrix} P(s) & Q(s) \end{pmatrix}$ is a unimodular matrix,*

(iii) *the Smith form of $P(s)$ is of the form $\begin{pmatrix} I \\ 0 \end{pmatrix}$,*

(iv) *the number of columns is not more than the number of rows and the invariant factors are all equal to 1.*

(v) *$P(\lambda)$ has full column rank for every $\lambda \in \mathbb{C}$.*

Proof : (i) $\Rightarrow$ (v) $P(\lambda)$ has a left inverse for every $\lambda \in \mathbb{C}$.

(v) $\Rightarrow$ (iv) A matrix with full column rank cannot have more columns than rows. If any of the invariant factors would have a zero λ, the Smith form, and hence the matrix itself, would not have full column rank at λ. Hence the invariant factors must be constant.

(iv) $\Rightarrow$ (iii) is obvious.

(iii) $\Rightarrow$ (ii) We have

$$P(s) = U(s)\bar{P}(s)V(s) = U(s)\begin{pmatrix} V(s) \\ 0 \end{pmatrix}$$

where $\bar{P}(s)$ is the Smith form of $P(s)$. We can extend $\bar{P}(s)$ (by the addition of columns) to the unit matrix. We obtain:

$$P_e(s) := U(s)\begin{pmatrix} V(s) & 0 \\ 0 & I \end{pmatrix}$$

as the desired extension of $P(s)$.

(ii) $\Rightarrow$ (i) If $Q(s)$ is such that $(P(s) \quad Q(s))$ is unimodular, this matrix has an inverse $U(s)$. Decomposing U according to the decomposition of $(P \quad Q)$ into

$$U = \begin{pmatrix} U_1 \\ U_2 \end{pmatrix},$$

we find that $U_1 P = I$. ∎

Next we discuss substitution of matrices into polynomial matrices. Suppose we are given an $m \times n$ polynomial matrix

$$P(s) = P_0 + P_1 s + \cdots + P_k s^k.$$

If A is an $n \times n$ matrix, we may substitute A from the right into $P(s)$. The result will be:

$$P_r(A) := P_0 + P_1 A + \cdots + P_k A^k.$$

If A is an $m \times m$ matrix, left substitution is possible:

$$P_\ell(A) := P_0 + AP_1 + \cdots + A^k P_k.$$

Even if both expressions are defined (i.e., $m = n$), they will not be the same, unless A commutes with all the coefficient matrices P_i. So it is important to distinguish between the two concepts. Also, familiar formulas for the substitution of a matrix into a scalar polynomial (see (2.10)) are not valid in this more general setting. Again this is due to the lack of commutativity of matrix multiplication. So, in general, even if all expressions are defined, $P_r(A)Q_r(A)$ will not be equal to $(PQ)_r(A)$. However, the following result is easily verified:

Theorem 7.6 *If $Q(s)$ and A commute, i.e., if A commutes with every coefficient matrix of $Q(s)$, then*

$$P_r(A)Q_r(A) = (PQ)_r(A).$$

for every polynomial matrix P

This simple result has powerful applications. As an example, we mention the Cayley-Hamilton theorem: starting point is Cramer's rule: if A is invertible, $A^{-1} = B/\det A$, where B is the adjoint matrix of A. Hence $BA = I \det A$. Replacing A by $sI - A$, we obtain $B(s)(sI - A) = Ip(s)$, where p is the characteristic polynomial of A and $B(s)$ a suitable polynomial matrix. Substituting $s = A$ from the right, we obtain $B_r(A)(AI - A) = Ip(A)$, where $p(A)$ is the matrix obtained by substituting A in the scalar polynomial $p(s)$ as defined in section 2.5. Since the left-hand side is obviously zero, we obtain $p(A) = 0$.

7.2 System matrix, transmission polynomials, and zeros

In this section we introduce the concepts of transmission polynomials and zeros of a system. Consider the system Σ given by the equations

$$\begin{aligned} \dot{x}(t) &= Ax(t) + Bu(t), \\ y(t) &= Cx(t) + Du(t), \end{aligned} \tag{7.3}$$

where, again, u takes its values in $\mathcal{U}$, x takes its values in $\mathcal{X}$, and y takes its values in $\mathcal{Y}$. As before, instead of writing down these equations explicitly, we will often denote Σ by (A, B, C, D). By formally taking the Laplace transforms of the equations (7.3) and denoting by $\hat{u}(s)$, $\hat{x}(s)$ and $\hat{y}(s)$ the Laplace transforms of $u(t)$, $x(t)$ and $y(t)$, respectively, (7.3) transforms to the system equations in the frequency domain given by

$$\begin{aligned} (sI - A)\hat{x}(s) - B\hat{u}(s) &= x_0, \\ C\hat{x}(s) + D\hat{u}(s) &= \hat{y}(s), \end{aligned}$$

This can be written equivalently as

$$P_\Sigma(s)\begin{pmatrix} \hat{x}(s) \\ \hat{u}(s) \end{pmatrix} = \begin{pmatrix} x_0 \\ \hat{y}(s) \end{pmatrix},$$

where $P_\Sigma(s)$ is defined by

$$P_\Sigma(s) := \begin{pmatrix} sI - A & -B \\ C & D \end{pmatrix}. \tag{7.4}$$

$P_\Sigma(s)$ is called the *system matrix* of the system Σ. It is an $(n + p) \times (n + m)$ polynomial matrix, containing the relevant information of the system.

The invariant factors of the polynomial matrix $P_\Sigma(s)$ are called *the transmission polynomials of* Σ. A transmission polynomial is called *non-trivial* if it is unequal to zero. The product of the non-trivial transmission polynomials of Σ is called *the zero polynomial* of the system. Any complex root of the zero polynomial is called a *zero* of the system Σ. We will say that a zero has *multiplicity* r if it is an r-fold root of the zero polynomial.

The systemic relevance of these zeros are most easily given under the assumption that $P_\Sigma(s)$ is left invertible, i.e., normrank $P_\Sigma = n + m$. In this case all transmission polynomials of Σ are unequal to zero. Suppose that λ is a zero of Σ. Then rank $P_\Sigma(\lambda) < n + m$. Hence there exists a nonzero vector $(x_0^{\mathrm{T}}, u_0^{\mathrm{T}})^{\mathrm{T}}$ such that $P_\Sigma(\lambda)(x_0^{\mathrm{T}}, u_0^{\mathrm{T}})^{\mathrm{T}} = 0$. Therefore, if we have the input $u(t) = e^{\lambda t}u_0$, the function $x(t) := e^{\lambda t}x_0$ satisfies the differential equation for x. The corresponding output will be zero. Consequently, if $P_\Sigma(s)$ has normal rank $n + m$, a zero of the system is a frequency λ with the property that there exists an input (viz. $u = e^{\lambda t}u_0$) and an initial state x_0, not both zero, for which the output is identically zero.

The following clarifies the concept of a zero as a systemtheoretic property:

Lemma 7.7 *Let $\lambda \in \mathbb{C}$. Suppose that $x_0 \in \mathcal{X}$ and $u_0 \in \mathcal{U}$ are such that*

$$P_\Sigma(\lambda)\begin{pmatrix} x_0 \\ u_0 \end{pmatrix} = 0. \tag{7.5}$$

Let $u(t) := e^{\lambda t}u_0$. Then the output resulting from the initial state x_0 and input function u is zero, i.e. $y_u(t, x_0) = 0$ for all $t \geqslant 0$.

Proof : It follows from (7.5) that $x(t) := e^{\lambda t}x_0$ satisfies $\dot{x} = Ax + Bu$, $x(0) = x_0$. Hence $x_u(t, x_0) = e^{\lambda t}x_0$. Again by (7.5) this yields $y_u(t, x_0) = 0$ for all $t \geqslant 0$. ■

7.3 The weakly unobservable subspace

In the previous section, we have seen that the zeros of the system Σ are associated with initial states that, by choosing an appropriate input, yield zero output. In general a point in the state space of Σ for which this property holds, is called a weakly unobservable point:

Definition 7.8 *A point $x_0 \in \mathcal{X}$ is called* weakly unobservable *if there exists an input function u such that the corresponding output function satisfies $y_u(t, x_0) = 0$ for all $t \geqslant 0$. The set of all weakly unobservable points of Σ is denoted by $\mathcal{V}(\Sigma)$ and is called the* weakly unobservable subspace *of Σ.*

It is easily seen that the set of all weakly unobservable points of Σ is indeed a linear subspace of $\mathcal{X}$. Thus, the above terminology is justified. Let $\langle \ker C \mid A \rangle$ be the unobservable subspace of the pair (C, A). In section 3.3 this subspace was characterized as the set of points x_0 for which the output resulting from the input $u = 0$ is equal to zero. Thus we have $\langle \ker C \mid A \rangle \subset \mathcal{V}(\Sigma)$. As an immediate consequence of definition 4.4, we see that if the direct feedthrough map D is equal to zero, then $\mathcal{V}(\Sigma) = \mathcal{V}^*(\ker C)$, the largest controlled invariant subspace contained in $\ker C$. However, also if $D \neq 0$ the weakly unobservable subspace is controlled invariant:

Lemma 7.9 *Let $x_0 \in \mathcal{V}(\Sigma)$ and let u be an input function such that the corresponding output function satisfies $y_u(t, x_0) = 0$ for all $t \geqslant 0$. Then the associated state satisfies $x_u(t, x_0) \in \mathcal{V}(\Sigma)$ for all $t \geqslant 0$.*

Proof : The proof of this is analogous to the corresponding part of the proof of theorem 4.5 and is left as an exercise. ■

Let $x_0 \in \mathcal{V}(\Sigma)$ and let u be such that $y_u(t, x_0) = 0$ for all $t \geqslant 0$. According to the previous lemma we have $x_u(t, x_0) \in \mathcal{V}(\Sigma)$ for all $t \geqslant 0$. Since $\mathcal{V}(\Sigma)$ is a linear subspace of $\mathcal{X}$, also

$$\dot{x}(0^+) := \lim_{t \downarrow 0} \frac{1}{t}(x_u(t, x_0) - x_0) \in \mathcal{V}(\Sigma).$$

Since $\dot{x}(0^+) = Ax_0 + Bu(0)$ and $Cx_0 + Du(0) = 0$, we see that for any given $x_0 \in \mathcal{V}(\Sigma)$ there exists a vector $u_0 \in \mathcal{U}$ (take $u_0 := u(0)$) such that $Ax_0 + Bu_0 \in \mathcal{V}(\Sigma)$ and $Cx_0 + Du_0 = 0$. Equivalently, the subspace $\mathcal{V} = \mathcal{V}(\Sigma)$ satisfies

$$\begin{pmatrix} A \\ C \end{pmatrix} \mathcal{V} \subset (\mathcal{V} \times 0) + \operatorname{im} \begin{pmatrix} B \\ D \end{pmatrix}. \tag{7.6}$$

Now, let $\mathcal{V}$ be any subspace of $\mathcal{X}$ with the property (7.6). Choose a basis $x_1, \ldots, x_n$ for $\mathcal{X}$ such that $x_1, \ldots, x_r$ is a basis for $\mathcal{V}$ ($r \leqslant n$). By (7.6) there are vectors $u_i \in \mathcal{U}$ such that for $i = 1, 2, \ldots, r$ we have $Ax_i + Bu_i \in \mathcal{V}$ and $Cx_i + Du_i = 0$. Let $F : \mathcal{X} \to \mathcal{U}$ be any linear map such that $Fx_i = u_i$ ($i = 1, \ldots, r$). Then we have $(A+BF)x_i \in \mathcal{V}$ and $(C+DF)x_i = 0$. Since $x_1, \ldots, x_r$ is a basis for $\mathcal{V}$, we conclude that there exists a map $F : \mathcal{X} \to \mathcal{U}$ such that

$$(A + BF)\mathcal{V} \subset \mathcal{V} \text{ and } (C + DF)\mathcal{V} = 0. \tag{7.7}$$

In fact, we even have the following:

Theorem 7.10

(i) *$\mathcal{V}(\Sigma)$ is the largest subspace $\mathcal{V}$ of $\mathcal{X}$ for which (7.6) holds.*

(ii) *$\mathcal{V}(\Sigma)$ is the largest subspace $\mathcal{V}$ of $\mathcal{X}$ for which there exists a linear map $F : \mathcal{X} \to \mathcal{U}$ such that (7.7) holds.*

Proof : (i) We have already shown that $\mathcal{V} = \mathcal{V}(\Sigma)$ satisfies (7.6). Let $\mathcal{V}$ be an arbitrary subspace that satisfies (7.6). According to the above, there is an F such that (7.7) holds. Let $x_0 \in \mathcal{V}$ and apply the feedback control $u(t) = Fx(t)$. The resulting trajectory $x_u(t, x_0)$ then remains in $\mathcal{V}$ for all $t \geqslant 0$. Hence, $y_u(t, x_0) = (C+DF)x_u(t, x_0) = 0$ for all $t \geqslant 0$. We conclude that $x_0 \in \mathcal{V}(\Sigma)$. Thus $\mathcal{V} \subset \mathcal{V}(\Sigma)$.

(ii) We already showed that there exists an F such that (7.7) holds with $\mathcal{V} = \mathcal{V}(\Sigma)$. Let $\mathcal{V}$ be any subspace such that (7.7) holds for some F. It can then be seen immediately that $\mathcal{V}$ satisfies (7.6). According to part (i) of this theorem this implies $\mathcal{V} \subset \mathcal{V}(\Sigma)$. ■

In the sequel, we often denote $C + DF$ by C_F (and of course as before, $A + BF$ by A_F). The following result is a generalization of theorem 4.3:

Theorem 7.11 *Let $F : \mathcal{X} \to \mathcal{U}$ be a linear map such that $A_F\mathcal{V}(\Sigma) \subset \mathcal{V}(\Sigma)$ and $C_F\mathcal{V}(\Sigma) = 0$. Let L be a linear map such that $\operatorname{im} L = \ker D \cap B^{-1}\mathcal{V}(\Sigma)$. Let $x_0 \in \mathcal{V}(\Sigma)$ and u be an input function. Then the output resulting from u and x_0 is zero if and only if u has the form $u(t) = Fx(t) + Lw(t)$ for some function w.*

Proof : ($\Rightarrow$) Assume that $y_u(t, x_0) = 0$ for all $t \geqslant 0$. According to lemma 7.9, $x_u(t, x_0) \in \mathcal{V}(\Sigma)$ for all $t \geqslant 0$. Define an input v by $v(t) := u(t) - Fx_u(t, x_0)$. Then

$$\dot{x}_u(t, x_0) = A_F x_u(t, x_0) + Bv(t)$$

and

$$C_F x_u(t, x_0) + Dv(t) = 0.$$

Since $A_F x_u(t, x_0)$ and $\dot{x}_u(t, x_0)$ lie in $\mathcal{V}(\Sigma)$ for all $t \geqslant 0$ we find that $Bv(t) \in \mathcal{V}(\Sigma)$ and $Dv(t) = 0$ for all t. Thus, $v(t) \in \ker D \cap B^{-1}\mathcal{V}(\Sigma)$ for all $t \geqslant 0$. It follows that there exists a function w such that $v(t) = Lw(t)$ for all t.

($\Leftarrow$) Assume that $u(t) = Fx(t) + Lw(t)$. Then we have

$$x_u(t, x_0) = e^{A_F t}x_0 + \int_0^t e^{A_F(t-\tau)}BLw(\tau)\,d\tau.$$

Since $\mathcal{V}(\Sigma)$ is A_F-invariant, $x_0 \in \mathcal{V}(\Sigma)$ and

$$\operatorname{im} BL \subset \mathcal{V}(\Sigma),$$

we must have $x_u(t, x_0) \in \mathcal{V}(\Sigma)$ for all $t \geqslant 0$. The resulting output is equal to

$$y_u(t, x_0) = C_F x_u(t, x_0) + DLw(t).$$

Since $C_F\mathcal{V}(\Sigma) = 0$ and $\operatorname{im} L \subset \ker D$, the latter is equal to zero for all $t \geqslant 0$. ■

We conclude this section by giving an algorithm to calculate for a given system Σ the weakly unobservable subspace $\mathcal{V}(\Sigma)$. As in section 4.3, for this it is convenient to think in terms of the discrete time version of the system (7.3):

$$\begin{aligned} x_{t+1} &= Ax_t + Bu_t, \\ y_t &= Cx_t + Du_t, \quad t = 0, 1, 2, \ldots. \end{aligned} \tag{7.8}$$

Given an input sequence $u = (u_0, u_1, u_2, \ldots)$ and an initial state x_0, let

$$x = (x_0, x_1, x_2, \ldots) \text{ and } y = (y_0, y_1, \ldots)$$

be the resulting state trajectory and output. The discrete time analogue of $\mathcal{V}(\Sigma)$ is denoted by $\mathcal{V}_d(\Sigma)$ and is defined by

$$\mathcal{V}_d(\Sigma) := \{x_0 \in \mathcal{X} \mid \text{there is an input sequence } u \text{ such that } y = 0\}.$$

Define a sequence of subspaces $\mathcal{V}_0, \mathcal{V}_1, \mathcal{V}_2, \ldots$ by $\mathcal{V}_0 := \mathcal{X}$ and for $t = 1, 2, \ldots$ by

$$\mathcal{V}_t := \{x_0 \in \mathcal{X} \mid \text{there is an input sequence } u \text{ such that} \\ y_k = 0 \text{ for } k = 0, 1, \ldots, t-1\}.$$

Clearly, the subspaces $\mathcal{V}_t$ form a chain, that is, they satisfy the inclusion relation $\mathcal{V}_0 \supset \mathcal{V}_1 \supset \mathcal{V}_2 \supset \cdots$. We derive a recurrence relation for $\mathcal{V}_t$. It follows from the definition of $\mathcal{V}_t$ that $x_0 \in \mathcal{V}_{t+1}$ if and only if there exists a vector $u_0 \in \mathcal{U}$ such that $Ax_0 + Bu_0 \in \mathcal{V}_t$ and $Cx_0 + Du_0 = 0$. Thus, $x_0 \in \mathcal{V}_{t+1}$ if and only if there is a $u_0 \in \mathcal{U}$ such that

$$\begin{pmatrix} A \\ C \end{pmatrix} x_0 + \begin{pmatrix} B \\ D \end{pmatrix} u_0 \in \mathcal{V}_t \times 0.$$

We conclude that the sequence $(\mathcal{V}_t)$ is generated by the recurrence relation

$$\mathcal{V}_0 = \mathcal{X}, \ \mathcal{V}_{t+1} = \begin{pmatrix} A \\ C \end{pmatrix}^{-1} \left[(\mathcal{V}_t \times 0) + \operatorname{im} \begin{pmatrix} B \\ D \end{pmatrix} \right]. \tag{7.9}$$

From this, it follows that if for some integer k we have $\mathcal{V}_k = \mathcal{V}_{k+1}$ then $\mathcal{V}_t = \mathcal{V}_k$ for all $t \geqslant k$. Consequently, the inclusion chain for $\mathcal{V}_t$ must have the form

$$\mathcal{V}_0 \supset \mathcal{V}_1 \supset \cdots \supset \mathcal{V}_k = \mathcal{V}_{k+1} = \mathcal{V}_{k+2} = \cdots,$$

for some integer $k \leqslant n$. Here, $\supset$ denotes *strict* inclusion.We contend that $\mathcal{V}_d(\Sigma) = \mathcal{V}_k$. The proof of this is similar to the proof of the corresponding result in section 4.3. In the following theorem we summarize the properties of the sequence $\{\mathcal{V}_t\}$ as derived above and show that (7.9) leads to the subspace $\mathcal{V}(\Sigma)$ in a finite number of recursion steps:

Theorem 7.12 *Let $\mathcal{V}_t$, $t = 0, 1, 2, \ldots$ be defined by the algorithm (7.9). Then we have*

(i) $\mathcal{V}_0 \supset \mathcal{V}_1 \supset \mathcal{V}_2 \supset \cdots$,

(ii) *there exists $k \leqslant n$ such that $\mathcal{V}_k = \mathcal{V}_{k+1}$,*

(iii) *if $\mathcal{V}_k = \mathcal{V}_{k+1}$ then $\mathcal{V}_k = \mathcal{V}_t$ for all $t \geqslant k$,*

(iv) *if $\mathcal{V}_k = \mathcal{V}_{k+1}$ then $\mathcal{V}(\Sigma) = \mathcal{V}_k$.*

Proof : The properties (i), (ii) and (iii) have already been proven. We now prove (iv). Assume $\mathcal{V}_k = \mathcal{V}_{k+1}$. Using (7.9) this implies

$$\begin{pmatrix} A \\ C \end{pmatrix} \mathcal{V}_k \subset (\mathcal{V}_k \times 0) + \operatorname{im} \begin{pmatrix} B \\ D \end{pmatrix}.$$

According to theorem 7.10, $\mathcal{V}(\Sigma)$ is the largest subspace of $\mathcal{X}$ for which this inclusion holds. Consequently, $\mathcal{V}_k \subset \mathcal{V}(\Sigma)$. The converse inclusion can be proven analogously as in theorem 4.10. This proof is left as an exercise to the reader. ■

7.4 Controllable weakly unobservable points

In this section we consider the subspace of $\mathcal{V}(\Sigma)$ consisting of all points in the state space for which there exists an input u such that the resulting output is zero, while, at the same time, the state trajectory is steered to the zero state in finite time:

Definition 7.13 *A point $x_0 \in \mathcal{X}$ is called* controllable weakly unobservable *if there exists an input function u, and $T > 0$ such that $y_u(t, x_0) = 0$ for all $t \in [0, T]$ and $x_u(T, x_0) = 0$. The set of all such points is denoted by $\mathcal{R}(\Sigma)$ and is called the* controllable weakly unobservable subspace of Σ.

Again, it is straightforward to verify that the set $\mathcal{R}(\Sigma)$ is indeed a linear subspace of $\mathcal{X}$. Also, $\mathcal{R}(\Sigma) \subset \mathcal{V}(\Sigma)$. Clearly, if $D = 0$ then $\mathcal{R}(\Sigma) = \mathcal{R}^*(\ker C)$, the largest controllability subspace contained in $\ker C$ (see theorem 4.15). Analogously to the definition of $\underline{F}$ in section 4.1 we denote by $\underline{F}(\mathcal{V}(\Sigma))$ the set of all linear maps $F : \mathcal{X} \to \mathcal{U}$ such that $(A + BF)\mathcal{V}(\Sigma) \subset \mathcal{V}(\Sigma)$ and $(C + DF)\mathcal{V}(\Sigma) = 0$. The following result generalizes theorem 4.17:

Theorem 7.14 *Let $F \in \underline{F}(\mathcal{V}(\Sigma))$. Then we have*

$$\mathcal{R}(\Sigma) = \langle A + BF \mid \mathcal{V}(\Sigma) \cap B \ker D \rangle.$$

In particular, for any $F \in \underline{F}(\mathcal{V}(\Sigma))$ we have $A_F \mathcal{R}(\Sigma) \subset \mathcal{R}(\Sigma)$ and $C_F \mathcal{R}(\Sigma) = 0$.

Proof : Let L be such that $\operatorname{im} L = B^{-1}\mathcal{V}(\Sigma) \cap \ker D$. Assume that $x_0 \in \mathcal{R}(\Sigma)$ and let u and $T > 0$ be such that $y_u(t, x_0) = 0$ for all $t \geqslant 0$ and $x_u(T, x_0) = 0$. By theorem 7.11, u can be written as $u(t) = Fx(t) + Lw(t)$ for some w. Thus, $x_u(\cdot, x_0)$ is a state trajectory of the system $\dot{x}(t) = A_F x(t) + BLw(t)$ with state space $\mathcal{V}(\Sigma)$. Along this trajectory, x_0 is steered to 0 at time $t = T$. It follows that $x_0 \in \langle A_F \mid \operatorname{im} BL \rangle$ (see section 3.2). Conversely, let $x_0 \in \langle A_F \mid \operatorname{im} BL \rangle$. Consider the system $\dot{x}(t) = A_F x(t) + BLw(t)$ with state space $\mathcal{V}(\Sigma)$. Since $\langle A_F \mid \operatorname{im} BL \rangle$ is the reachable subspace of this system, there exists w and $T > 0$ such that the trajectory $x(t)$ resulting from x_0 and w satisfies $x(T) = 0$. Of course, $x(t) \in \mathcal{V}(\Sigma)$ for all t. Define u by $u(t) := Fx(t) + Lw(t)$. Then $x(t) = x_u(t, x_0)$ (where the latter denotes the trajectory of our original system (A, B)). We have $x_u(T, x_0) = 0$ and

$$y_u(t, x_0) = C_F x_u(t, x_0) + DLw(t).$$

Since $C_F \mathcal{V}(\Sigma) = 0$ and $\operatorname{im} L \subset \ker D$ the latter is equal to zero for all $t \geqslant 0$. We conclude that $x_0 \in \mathcal{R}(\Sigma)$. The second assertion of the theorem is an immediate consequence of the foregoing. ■

Analogously to the case that $D = 0$, the following interpretation can be given of the above theorem. By taking F from $\underline{F}(\mathcal{V}(\Sigma))$ and by taking L such that $\operatorname{im} L = B^{-1}\mathcal{V}(\Sigma) \cap B \ker D$, we obtain a new system $\dot{x}(t) = A_F x(t) + BLw(t)$, with state space $\mathcal{V}(\Sigma)$. This system can be considered to be obtained from the original one by restricting the trajectories to the subspace $\mathcal{V}(\Sigma)$ and by restricting the input functions to take their values in $B^{-1}\mathcal{V}(\Sigma) \cap \ker D$. Since $\operatorname{im} BL = \mathcal{V}(\Sigma) \cap B \ker D$, theorem 7.14 says that $\mathcal{R}(\Sigma)$ is the reachable subspace of the restricted system.

7.5 Strong observability

In section 3.3 we have considered observability of the system Σ. It was shown that Σ is observable if and only if for each initial state x_0 we have that if the output, with input set to zero, satisfies $y_0(t, x_0) = 0$ for all $t \geqslant 0$, then $x_0 = 0$. That is, if the output of the uncontrolled system is zero, then the initial state must be zero. The system Σ is called strongly observable if this property holds when we use an arbitrary input function:

Definition 7.15 *Σ is called* strongly observable *if for all $x_0 \in \mathcal{X}$ and for every input function u, the following holds: $y_u(t, x_0) = 0$ for all $t \geqslant 0$ implies $x_0 = 0$.*

It follows immediately from definition 7.8 that Σ is strongly observable if and only if $\mathcal{V}(\Sigma) = 0$. We claim that Σ is strongly observable if and only if for *every* linear map $F : \mathcal{X} \to \mathcal{U}$, the pair $(C + DF, A + BF)$ is observable. To show this, let us denote

$$\Sigma_F := (A_F, B, C_F, D).$$

It is easily seen that the weakly unobservable subspaces of Σ and Σ_F coincide, i.e. $\mathcal{V}(\Sigma_F) = \mathcal{V}(\Sigma)$ (see also exercise 7.1). It was noted in section 7.3 that we always have $\langle \ker C \mid A \rangle \subset \mathcal{V}(\Sigma)$. Consequently, for any F we have

$$\langle \ker C_F \mid A_F \rangle \subset \mathcal{V}(\Sigma_F) = \mathcal{V}(\Sigma).$$

This shows that if $\mathcal{V}(\Sigma) = 0$ then for any F the pair (C_F, A_F) is observable. Conversely, according to exercise 7.2, for all $F \in \underline{F}(\mathcal{V}(\Sigma))$ we have $\mathcal{V}(\Sigma) = \langle \ker C_F \mid A_F \rangle$. Consequently, if (C_F, A_F) is observable for all F then we must have $\mathcal{V}(\Sigma) = 0$. We have thus proven the following:

Theorem 7.16 *The following statements are equivalent:*

(*i*) Σ *is strongly observable,*

(*ii*) $\mathcal{V}(\Sigma) = 0$,

(*iii*) $(C + DF, A + BF)$ *is observable for all* F.

It is also possible to connect strong observability of Σ with properties of its system matrix P_Σ. Recall that the system matrix of Σ is defined as the real polynomial matrix

$$P_\Sigma(s) = \begin{pmatrix} Is - A & -B \\ C & D \end{pmatrix}.$$

Theorem 7.17 *The following statements are equivalent:*

(*i*) Σ *is strongly observable,*

(*ii*) $\operatorname{rank} P_\Sigma(\lambda) = n + \operatorname{rank} \binom{B}{D}$ *for all* $\lambda \in \mathbb{C}$,

(*iii*) *the Smith form of* P_Σ *is equal to the constant matrix*

$$Q := \begin{pmatrix} I & 0 \\ 0 & 0 \end{pmatrix}.$$

Here, I denotes the $(n + r) \times (n + r)$ *identity matrix with* $r := \operatorname{rank} \binom{B}{D}$.

Proof : Assume that $\operatorname{rank} \binom{B}{D} = r$. There exists an isomorphism $T : \mathcal{U} \to \mathcal{U}$ of the input space such that

$$\begin{pmatrix} B \\ D \end{pmatrix} T = \begin{pmatrix} B_1 & 0 \\ D_1 & 0 \end{pmatrix},$$

with $(B_1^{\mathrm{T}}, D_1^{\mathrm{T}})^{\mathrm{T}}$ injective. The number of columns of $(B_1^{\mathrm{T}}, D_1^{\mathrm{T}})^{\mathrm{T}}$ is equal to r. Define a system Σ_T by $\Sigma_T := (A, B_1, C, D_1)$. It is easily seen from definition 7.8 that $\mathcal{V}(\Sigma_T) = \mathcal{V}(\Sigma)$. The system matrix of Σ_T has $n + r$ columns and is equal to

$$P_{\Sigma_T}(s) = \begin{pmatrix} Is - A & -B_1 \\ C & D_1 \end{pmatrix}.$$

Of course, $\operatorname{rank} P_{\Sigma_T}(\lambda) = \operatorname{rank} P_\Sigma(\lambda)$ for all $\lambda \in \mathbb{C}$.

(i) $\Rightarrow$ (ii) Assume (ii) does not hold. Then there exists λ such that

$$\operatorname{rank} P_{\Sigma_T}(\lambda) < n + r.$$

Consequently, $P_{\Sigma_T}(\lambda)$ is not injective so we can find x_0 and u_0, not both equal to zero, such that

$$P_{\Sigma_T}(\lambda) \begin{pmatrix} x_0 \\ u_0 \end{pmatrix} = 0. \tag{7.10}$$

We claim that $x_0 \neq 0$. Indeed, $x_0 = 0$ would imply $B_1 u_0 = 0$ and $D_1 u_0 = 0$, which contradicts injectivity of $\binom{B_1}{D_1}$. Now apply lemma 7.7: the output resulting from $u(t) := e^{\lambda t} u_0$ and initial condition x_0 is zero. Since $x_0 \neq 0$, this contradicts the assumption that Σ is strongly observable.

(ii) $\Rightarrow$ (i) Assume that (i) does not hold. According to theorem 7.16 we then have $\mathcal{V}(\Sigma_T) \neq 0$. Let $F \in \underline{F}(\mathcal{V}(\Sigma_T))$. Let λ be an eigenvalue of $A + B_1 F \mid \mathcal{V}(\Sigma_T)$ and let $x_0 \in \mathcal{V}(\Sigma_T)$ be the corresponding eigenvector. Let $u_0 := F x_0$. Then (7.10) holds. Consequently, rank $P_{\Sigma_T}(\lambda) < n + r$. This contradicts (ii).

(ii) $\Leftrightarrow$ (iii) Note that P_Σ is unimodularly equivalent with $(P_{\Sigma_T} \quad 0)$ (with 0 the $(n + p) \times (n + m - r)$ zero matrix). If (ii) holds then, according to corollary 7.5, P_{Σ_T} has Smith form $(I, 0)^{\mathrm{T}}$. This implies (iii). Conversely, if (iii) holds then P_{Σ_T} has Smith form $\binom{I}{0}$. Again by corollary 7.5, this yields (ii). ■

7.6 Transmission polynomials and zeros in state space

In this section we show that the non-trivial transmission polynomials and the zeros of Σ can be characterized as the non-trivial invariant factors and the eigenvalues, respectively, of a linear map associated with the weakly unobservable subspace of Σ. Recall from theorem 7.14 that if $F \in \underline{F}(\mathcal{V}(\Sigma))$ then we automatically have

$$A_F \mathcal{R}(\Sigma) \subset \mathcal{R}(\Sigma), \ C_F \mathcal{R}(\Sigma) = 0.$$

Thus, for any $F \in \underline{F}(\mathcal{V}(\Sigma))$ the map $A_F \mid \mathcal{V}(\Sigma)/\mathcal{R}(\Sigma)$ is well-defined. We show that this map is independent of F for $F \in \underline{F}(\mathcal{V}(\Sigma))$:

Theorem 7.18 *Let $F_1, F_2 \in \underline{F}(\mathcal{V}(\Sigma))$. Then*

$$A_{F_1} \mid \mathcal{V}(\Sigma)/\mathcal{R}(\Sigma) = A_{F_2} \mid \mathcal{V}(\Sigma)/\mathcal{R}(\Sigma).$$

Proof : According to exercise 7.3 we have

$$(F_1 - F_2)\mathcal{V}(\Sigma) \subset B^{-1}\mathcal{V}(\Sigma) \cap \ker D.$$

Consequently, by theorem 7.14 we have

$$B(F_1 - F_2)\mathcal{V}(\Sigma) \subset \mathcal{R}(\Sigma).$$

The remainder of the proof is similar to the proof of theorem 4.18 (iii) and is left as an exercise. ■

The map $A_F \mid \mathcal{V}(\Sigma)/\mathcal{R}(\Sigma)$ for $F \in \underline{F}(\mathcal{V}(\Sigma))$ is denoted by M_Σ. The map M_Σ is a linear map from $\mathcal{V}(\Sigma)/\mathcal{R}(\Sigma)$ into itself. We will show that the transmission polynomials and the zeros, with their multiplicities, of the system Σ are completely determined by the spectral properties of the quotient map M_Σ.

In the following let $\mu_1, \ldots, \mu_k$ be the non-trivial invariant factors of the map M_Σ and let χ_Σ be the characteristic polynomial of M_Σ. Furthermore, let $\tau_1, \ldots, \tau_\ell$ be the non-trivial transmission polynomials of Σ. Let ζ_Σ be the zero polynomial of Σ. It turns out that the non-trivial transmission polynomials of Σ coincide with the non-trivial invariant factors of M_Σ. Thus, the zero polynomial of Σ is equal to the characteristic polynomial of M_Σ and the set of zeros of Σ must be equal to the spectrum of M_Σ:

Theorem 7.19 *$k = \ell$ and for $i = 1, 2, \ldots, k$ we have $\tau_i = \mu_i$. Also, $\zeta_\Sigma = \chi_\Sigma$.*

Proof : It suffices to show that the polynomial matrix P_Σ is unimodularly equivalent with a polynomial matrix of the form

$$\begin{pmatrix} I' & 0 & 0 \\ 0 & Is - M_\Sigma & 0 \\ 0 & 0 & 0 \end{pmatrix}, \tag{7.11}$$

where I' denotes the identity map of a suitable dimension. Indeed, the non-trivial invariant factors of (7.11) are equal to $\mu_1, \ldots, \mu_k$. On the other hand, the non-trivial invariant factors of P_Σ are $\tau_1, \ldots, \tau_\ell$. Now, if P_Σ is equivalent to (7.11) then these non-trivial invariant factors must coincide and we must have $k = \ell$ and $\tau_i = \mu_i$ for $i = 1, 2, \ldots, k$.

We now show that P_Σ is indeed unimodularly equivalent with a polynomial matrix of the form (7.11). To begin with, we apply multiplication with the following types of transformations:

$$U_1 := \begin{pmatrix} I & 0 \\ F & I \end{pmatrix}, \tag{7.12}$$

$$U_2 := \begin{pmatrix} S & 0 \\ 0 & T \end{pmatrix}, \quad U_3 := \begin{pmatrix} S^{-1} & 0 \\ 0 & I \end{pmatrix}. \tag{7.13}$$

Here, F is a linear map from $\mathcal{X}$ to $\mathcal{U}$, S is an isomorphism of $\mathcal{X}$ and T is an isomorphism of $\mathcal{U}$. The identity matrices appearing in these transformations of course act on distinct spaces, but for convenience we all denote them by I. U_1, U_2 and U_3 are then invertible real matrices and hence also unimodular polynomial matrices. It is easy to see that

$$U_3 P_\Sigma U_1 U_2 = \begin{pmatrix} Is - S^{-1}A_F S & S^{-1}BT \\ C_F S & DT \end{pmatrix}. \tag{7.14}$$

We now specify the maps F, S and T appearing in the above. For F we take any element of $\underline{F}(\mathcal{V}(\Sigma))$. The maps S and T are defined as the transformations in $\mathcal{X}$ and $\mathcal{U}$, respectively, corresponding to the following choice of bases (see exercise 7.7):

- choose a basis of $\mathcal{X}$ adapted to $\mathcal{R}(\Sigma)$ and $\mathcal{V}(\Sigma)$,
- choose a basis of $\mathcal{U}$ adapted to $B^{-1}\mathcal{V}(\Sigma) \cap \ker D$.

By combining exercise 7.7 with (7.14) we then see that P_Σ is unimodularly equivalent with

$$\begin{pmatrix} Is - A_{11} & 0 - A_{12} & -A_{13} & -B_{11} & -B_{12} \\ 0 & Is - A_{22} & -A_{23} & 0 & -B_{22} \\ 0 & 0 & Is - A_{33} & 0 & -B_{32} \\ 0 & 0 & C_3 & 0 & D_2 \end{pmatrix}. \tag{7.15}$$

An important observation here is that A_{22} is a matrix of the map M_Σ. By permutation of columns (which also represents unimodular transformation) we see that (7.15) is equivalent with

$$\begin{pmatrix} Is - A_{11} & -B_{11} & -A_{12} & -A_{13} & -B_{12} \\ 0 & 0 & Is - A_{22} & -A_{23} & -B_{22} \\ 0 & 0 & 0 & Is - A_{33} & -B_{32} \\ 0 & 0 & 0 & C_3 & D_2 \end{pmatrix}. \tag{7.16}$$

According to exercise 7.7, the system (A_{11}, B_{11}) is controllable. Consequently (see corollary 7.5) the polynomial matrix $(Is - A_{11} \;\; B_{11})$ has Smith form $(I \;\; 0)$. Also, the system $(A_{33}, B_{32}, C_3, D_2)$ is strongly observable and $(-B_{32}^{\mathrm{T}} \;\; D_2^{\mathrm{T}})^{\mathrm{T}}$ is injective. Hence, by theorem 7.17,

$$\begin{pmatrix} Is - A_{33} & -B_{32} \\ C_3 & D_2 \end{pmatrix}$$

has Smith form $(I \;\; 0)^{\mathrm{T}}$. Here, the dimension of the identity matrix I is equal to the number of columns of $(A_{33} \;\; B_{32})$. After applying a number of elementary column and row operations we then find that (7.16) is equivalent with

$$\begin{pmatrix} I & 0 & 0 & 0 \\ 0 & 0 & Is - A_{22} & 0 \\ 0 & 0 & 0 & I \\ 0 & 0 & 0 & 0 \end{pmatrix}. \tag{7.17}$$

As already noted, A_{22} is a matrix of the map M_Σ. It is then obvious that (7.17) is unimodularly equivalent with a matrix of the form (7.11). This completes the proof of the theorem. ■

7.7 Exercises

7.1 Consider the system $\Sigma = (A, B, C, D)$. Let $F : \mathcal{X} \to \mathcal{U}$ and $G : \mathcal{Y} \to \mathcal{X}$ be linear maps. Define

$$\Sigma_{F,G} := (A + BF + GC + GDF, B + GD, C + DF, D).$$

a. Show that $\mathcal{V}(\Sigma_{F,G}) = \mathcal{V}(\Sigma)$, i.e. the weakly unobservable subspace is invariant under state feedback and output injection.

b. Show that $\mathcal{V}(\Sigma)$ is $(A + GC, B + GD)$-invariant for any linear map $G : \mathcal{Y} \to \mathcal{X}$.

7.2 Show that for all $F \in \underline{F}(\mathcal{V}(\Sigma))$ the weakly unobservable subspace $\mathcal{V}(\Sigma)$ is equal to $\langle \ker C_F \mid A_F \rangle$, the unobservable subspace of the system (C_F, A_F).

7.3 Let $F_0 \in \underline{F}(\mathcal{V}(\Sigma))$. Show that a linear map $F : \mathcal{X} \to \mathcal{U}$ is an element of $\underline{F}(\mathcal{V}(\Sigma))$ if and only if $(F_0 - F)\mathcal{V}(\Sigma) \subset B^{-1}\mathcal{V}(\Sigma) \cap \ker D$.

7.4 (Output stabilizability.) Consider the system $\dot{x}(t) = Ax(t) + Bu(t)$, $y(t) = Cx(t) + Du(t)$, with state space $\mathcal{X}$. Let $\mathcal{X}_{\text{stab}}$ be the stabilizable subspace with respect to the stability domain $\mathbb{C}^-$. For a given feedback law $u(t) = Fx(t)$, let $y_F(t, x_0)$ denote the output of the closed loop system, corresponding to the initial state x_0. The system Σ is called *output stabilizable* if there exists F such that $y_F(t, x_0) \to 0$ $(t \to \infty)$ for all x_0. Show that that the following statements are equivalent:

a. Σ is output stabilizable,

b. For all x_0 there exists a Bohl function u such that $y_u(\cdot, x_0)$ is $\mathbb{C}^-$-stable,

c. $\mathcal{X} = \mathcal{V}(\Sigma) + \mathcal{X}_{\text{stab}}$.

7.5 Consider the system $\Sigma = (A, B, C, D)$ and assume that D is injective.

a. Show that for each $x_0 \in \mathcal{V}(\Sigma)$ there is exactly one input function u such that $y_u(t, x_0) = 0$ for all $t \geqslant 0$, and this input satisfies

$$u(t) = -(D^{\mathrm{T}}D)^{-1}D^{\mathrm{T}}Cx_u(t, x_0).$$

b. Show that $\mathcal{V}(\Sigma)$ is equal to the subspace

$$\langle \ker(C - D(D^{\mathrm{T}}D)^{-1}D^{T}C) \mid A - B(D^{\mathrm{T}}D)^{-1}D^{\mathrm{T}}C \rangle,$$

the unobservable subspace of the system (C_F, A_F) obtained by applying the feedback

$$u(t) = -(D^{\mathrm{T}}D)^{-1}D^{\mathrm{T}}Cx(t).$$

c. Let $F : \mathcal{X} \to \mathcal{U}$ be a linear map. Show that $F \in \underline{F}(\mathcal{V}(\Sigma))$ if and only if

$$F \mid \mathcal{V}(\Sigma) = -(D^{\mathrm{T}}D)^{-1}D^{\mathrm{T}}C \mid \mathcal{V}(\Sigma).$$

7.6 Show that $x_1 \in \mathcal{R}(\Sigma)$ if and only if there exist $u \in \mathbf{U}$ and $T > 0$ such that $x_u(T, 0) = x_1$ and $y_u(t, 0) = 0$ for all $t \geqslant 0$.

7.7 Consider the system Σ. Define $\mathcal{U}_1 := B^{-1}\mathcal{V}(\Sigma) \cap \ker D$. Choose a basis of the input space $\mathcal{U}$ adapted to $\mathcal{U}_1$. Choose a basis of the state space $\mathcal{X}$ adapted to $\mathcal{R}(\Sigma)$ and $\mathcal{V}(\Sigma)$. Choose any basis in $\mathcal{Y}$. Now let $F \in \underline{F}(\mathcal{V}(\Sigma))$.

a. Show that the matrices of A_F, B, C_F and D with respect to the given bases have the form

$$\begin{pmatrix} A_{11} & A_{12} & A_{13} \\ 0 & A_{22} & A_{23} \\ 0 & 0 & A_{33} \end{pmatrix}, \begin{pmatrix} B_{11} & B_{12} \\ 0 & B_{22} \\ 0 & B_{32} \end{pmatrix},$$

$$\begin{pmatrix} 0 & 0 & C_3 \end{pmatrix}, \begin{pmatrix} 0 & D_2 \end{pmatrix},$$

respectively.

b. Show that $\begin{pmatrix} B_{32} \\ D_2 \end{pmatrix}$ is injective.

c. Show that the pair (A_{11}, B_{11}) is controllable and that the system $(A_{33}, B_{32}, C_3, D_2)$ is strongly observable.

d. Show that

$$\begin{pmatrix} sI - A_{33} & -B_{32} \\ C_3 & D_2 \end{pmatrix}$$

has Smith form of the form $\begin{pmatrix} I \\ 0 \end{pmatrix}$.

7.8 Show that for each real monic polynomial $q(s)$ of degree $\dim \mathcal{R}(\Sigma)$ there exists $F \in \underline{F}(\mathcal{V}(\Sigma))$ such that the characteristic polynomial of the map $(A + BF) \mid \mathcal{R}(\Sigma)$ is equal to q.

7.9 (Strong detectability.) The system Σ is called *strongly detectable* if for all $x_0 \in \mathcal{X}$ and $u \in \mathbf{U}$ the following implication holds:

$$\{y_u(t, x_0) = 0 \text{ for all } t \geqslant 0\} \Rightarrow \left\{\lim_{t\to\infty} x_u(t, x_0) = 0\right\}.$$

If all zeros of Σ are contained in $\mathbb{C}^-$, then the system Σ is called a *minimum phase system*. This terminology is borrowed from the electrical engineering literature, where a SISO transfer function $g(s) = \frac{p(s)}{q(s)}$ is called minimum phase if all roots of p are contained in $\mathbb{C}^-$. Show that the following statements are equivalent:

a. Σ is strongly detectable,

b. $\mathcal{R}(\Sigma) = 0$ and Σ is minimum phase,

c. $(C + DF, A + BF)$ is $\mathbb{C}^-$-detectable for all F.

7.10 Assume that $\mathcal{R}(\Sigma) = 0$. Let $F \in \underline{F}(\mathcal{V}(\Sigma))$. Show that the zeros of Σ are the unobservable eigenvalues of (C_F, A_F).

7.11 Let $F : \mathcal{X} \to \mathcal{U}$ and $G : Y \to \mathcal{X}$ be linear maps. Show that the transmission polynomials of the systems Σ and $\Sigma_{F,G}$ (see exercise 7.1) coincide. Conclude that the zeros of Σ, including their multiplicities, are invariant under state feedback and output injection transformations.

7.12 Let $\Sigma^{\mathrm{T}} = (A^{\mathrm{T}}, C^{\mathrm{T}}, B^{\mathrm{T}}, D^{\mathrm{T}})$ be the dual of $\Sigma = (A, B, C, D)$. Show that the transmission polynomials and the zeros with their respective multiplicities of Σ and Σ^{T} coincide.

7.13 Give a complete proof of theorem 7.12.

7.14 Give a complete proof of theorem 7.18.

7.15 (Disturbance decoupling by state feedback with direct control feedthrough.) Consider the control system with disturbances, given by $\dot{x} = Ax + Bu + Ed$, $z = Cx + Du$. Here, u is the control input, and d represents an unknown disturbance. We want to find a state feedback control law $u = Fx$ such that the controlled system is disturbance decoupled, i.e., the transfer matrix $G_F(s) := (C + DF)(sI - A - BF)^{-1}E$ is equal to zero.

a. Show that the control law $u = Fx$ makes the closed loop system disturbance decoupled if and only if there exists a subspace $\mathcal{V}$ of $\mathcal{X}$ such that $(A + BF)\mathcal{V} \subset \mathcal{V}$, $(C + DF)\mathcal{V} = 0$, and $\operatorname{im} E \subset \mathcal{V}$.

b. Show that there exists $F : \mathcal{X} \to \mathcal{U}$ such that $u = Fx$ makes the controlled system disturbance decoupled if and only if $\operatorname{im} E \subset \mathcal{V}(\Sigma)$.

c. Assume that the map D is injective. Show that the feedback law $u = -(DD^{\mathrm{T}})^{-1}D^{\mathrm{T}}Cx$ makes the controlled system disturbance decoupled.

7.16 (The stabilizable weakly unobservable subspace.) Consider the system $\Sigma = (A, B, C, D)$, and let $\mathbb{C}_g$ be a stability domain. We define the *stabilizable weakly unobservable subspace* as the linear subspace of $\mathcal{X}$ given by

$$\begin{aligned}\mathcal{V}_g(\Sigma) := \{x_0 \in \mathcal{X} \mid \; & \text{there exists a Bohl function } u \text{ such that} \\ & y_u(t, x_0) = 0 \text{ for all } t \geqslant 0, \text{ and } x_u(\cdot, x_0) \text{ is } \mathbb{C}_g - \text{stable}\},\end{aligned}$$

i.e., the subspace of all initial states for which there exists a Bohl input such that the output is zero and the state is stable.

a. Show that $\mathcal{R}(\Sigma) \subset \mathcal{V}_g(\Sigma) \subset \mathcal{V}(\Sigma)$.

b. Show that if $x_0 \in \mathcal{V}_g(\Sigma)$ and u is a Bohl function such that $y_u(t, x_0) = 0$ for $t \geqslant 0$ and $x_u(\cdot, x_0)$ is stable, then $x_u(t, x_0) \in \mathcal{V}_g(\Sigma)$ for all $t \geqslant 0$.

c. Use (b) to show that there exists a map $F : \mathcal{X} \to \mathcal{U}$ such that $(A + BF)\mathcal{V}_g(\Sigma) \subset \mathcal{V}_g(\Sigma)$ and $(C + DF)\mathcal{V}_g(\Sigma) = 0$.

Let $F_0 : \mathcal{X} \to \mathcal{U}$ be such that $(A + BF_0)\mathcal{V}_g(\Sigma) \subset \mathcal{V}_g(\Sigma)$ and $(C + DF_0)\mathcal{V}_g(\Sigma) = 0$. Also, let L be a map such that $\operatorname{im} L = \ker D \cap B^{-1}\mathcal{V}_g(\Sigma)$.

d. Show that if $x_0 \in \mathcal{V}_g(\Sigma)$ and u is a Bohl function, then $y_u(t, x_0) = 0$ for all $t \geqslant 0$ if and only if there exists a Bohl function w such that u can be written as $u(t) = Fx_u(t, x_0) + Lw(t)$.

e. Consider now the system restricted to $\mathcal{V}_g(\Sigma)$:

$$\dot{x}(t) = (A + BF_0)x(t) + BLw(t),$$

with state space $\mathcal{V}_g(\Sigma)$. Show that for each $x_0 \in \mathcal{V}_g(\Sigma)$ there exists a Bohl function w such that the resulting state trajectory x is $\mathbb{C}_g$-stable. (Hint: use (d).)

f. Conclude from (e) that the restricted system is $\mathbb{C}_g$-stabilizable.

g. Prove that there exists a map $F : \mathcal{X} \to \mathcal{U}$ such that $(A + BF)\mathcal{V}_g(\Sigma) \subset \mathcal{V}_g(\Sigma)$, $(C + DF)\mathcal{V}_g(\Sigma) = 0$, and $\sigma(A + BF \mid \mathcal{V}_g(\Sigma)) \subset \mathbb{C}_g$.

h. Show that $\mathcal{V}_g(\Sigma)$ is the largest subspace of $\mathcal{X}$ for which there exists a map F satisfying the three properties under (g).

7.17 Consider again the system $\Sigma = (A, B, C, D)$. Let $F : \mathcal{X} \to \mathcal{U}$ be a map such that $(A + BF)\mathcal{V}(\Sigma) \subset \mathcal{V}(\Sigma)$ and $(C + DF)\mathcal{V}(\Sigma) = 0$. Let L be a map such that $\operatorname{im} L = \ker D \cap B^{-1}\mathcal{V}(\Sigma)$.

a. Let $x_0 \in \mathcal{V}(\Sigma)$ and let u be a Bohl function. Show that $y_u(t, x_0) = 0$ for all $t \geqslant 0$ if and only if there exists a Bohl function w such that $u(t) = Fx_u(t, x_0) + Lw(t)$.

b. Consider now the restricted system $\dot{x}(t) = (A + BF)x(t) + BLw(t)$ with state space $\mathcal{V}(\Sigma)$. Let $\mathbb{C}_g$ be a stability domain. Show that $\mathcal{V}_g(\Sigma)$ (the stabilizable weakly unobservable subspace, see problem 7.16) is equal to the stabilizable subspace of this restricted system. (Hint: use (a).)

c. Conclude that for all $F \in \underline{F}(\mathcal{V}(\Sigma))$ we have

$$\mathcal{V}_g(\Sigma) = \mathcal{R}(\Sigma) + \mathcal{X}_g(A + BF \mid \mathcal{V}(\Sigma))$$

7.8 Notes and references

General material on polynomial matrices can be found in the textbooks by Wedderburn [210] and MacDuffee [122]. For a general treatment of the the Smith form, we refer to the book by Gantmacher [55]. A more recent textbook containing an excellent overview of most of the relevant material on polynomial matrices is Kailath [90]. Additional material can be found in the book by Vardulakis [206].

The notion of system matrix is due to Rosenbrock [155]. Zeros of linear multivariable systems were introduced by Rosenbrock in terms of the Smith McMillan form of the system transfer matrix in [155]. Additional work on system zeros can be found in Desoer and Schulman [37], MacFarlane and Karcanias [117], and Pugh [151].

The weakly unobservable subspace was introduced for discrete-time systems by Silverman in [178]. Additional information can be found in the work of Molinari

[124] and [126], where the recurrence relation 7.9 was introduced. For continuous-time systems, an extensive treatment of the weakly unobservable subspace can be found in Hautus and Silverman [76], see also Willems, Kitapçi and Silverman [219]. Of course, for the case that $D = 0$, the weakly unobservable subspace coincides with the largest controlled invariant subspace contained in the kernel of the map C, as studied extensively in Wonham [223].

The connection between the spectral properties of the weakly unobservable subspace and the system zeros was also studied in the book by Wonham [223], section 5.5. More information can be found in Hosoe [82], and in Anderson [4]. Related material on the connection between the system zeros and the state space geometric structure of the system, can be found in Morse [129], Aling and Schumacher [3], Malabre [119] and, more recently, in Chen [27].

Chapter 8

System invertibility and the strongly reachable subspace

In this chapter we will extend the class of inputs to include distributions. To do this, we need to give suitable meaning to the notion of initial state in case the input is a distribution. We will do this using the distributional set-up discussed in the Appendix on distributions. We will derive the exact formulas for the state trajectory and output corresponding to a given initial state and impulsive-smooth distributional input.

Using the distributional set-up, we introduce the notions of left-invertibility, right-invertibility, and invertibility of systems. It is shown that these properties of the system can be characterized in terms of corresponding properties of the transfer matrix of the system, and in terms of properties of the system matrix.

The distributional set-up also gives rise to a couple of new relevant subspaces of the state space, the strongly reachable subspace, and the distributionally weakly unobservable subspace. We study the connection of these new subspaces with the ordinary weakly unobservable subspace, and the controllable weakly unobservable subspace introduced in chapter 7. Finally, we give a state space characterization of the invertibility properties of the system in terms of these new subspaces.

8.1 Distributions as inputs

Up to now we have only considered functions as inputs. In the present section we extend the class of inputs to include distributions.

Again consider the system Σ given by

$$\begin{aligned} \dot{x} &= Ax + Bu, \\ y &= Cx + Du. \end{aligned} \tag{8.1}$$

In the differential equation above, if u is an element of the distribution space $\mathfrak{D}'^{m}_{+}$ (see appendix A), instead of simply a function taking its values in the input space $\mathcal{U}$

(to be identified with $\mathbb{R}^m$), we run into a number of conceptual difficulties as soon as we try to define the concept of solution of the differential equation starting in a given point $x_0 \in \mathcal{X}$ (to be identified with $\mathcal{R}^n$). Indeed, this is a non-trivial matter, since distributions do not have a well-defined value at a particular time instant t_0.

Example 8.1 Consider the (scalar) initial value problem $\dot{x} = 0,\ x(0) = x_0$, with $x_0 \in \mathbb{R}$. First let us interpret $\dot{x}$ as ordinary differentiation. Then the solution on $\mathbb{R}^+ = [0, \infty)$ is given by $x(t) = x_0,\ t \geqslant 0$. This solution can be identified with the distribution $x_0 h$, the scalar multiple x_0 times the Heaviside distribution. Next, let us interpret $\dot{x}$ as distributional derivative. Of course, the condition '$x(0) = x_0$' has no meaning if we are looking for a distribution x as a solution. The question is now: how should '$\dot{x} = 0,\ x(0) = x_0$' be interpreted in order to obtain $x = x_0 h$ as its solution. The answer is that we should incorporate the 'initial condition' in the differential equation and replace '$\dot{x} = 0,\ x(0) = x_0$' by $\dot{x} = x_0\delta$. (Here, δ stands for the Dirac distribution). Indeed, by example A.4 the solution over the distribution space $\mathfrak{D}'^{\,n}_+$ is unique, and is equal to $x = x_0 h$.

Example 8.2 Consider the scalar initial value problem $\dot{x} = ax,\ x(0) = x_0$. Again, if we interpret this in the ordinary sense, then the solution x on $\mathbb{R}^+$ is given by $x(t) = e^{at}x_0$ $(t \geqslant 0)$. This solution can be identified with the distribution corresponding to the function $e^{at}i_{\mathbb{R}^+}(t)x_0$ (where $i_{\mathbb{R}^+}$ denotes the indicator function of $\mathbb{R}^+$), which is smooth on $\mathbb{R}^+$. If we interpret $\dot{x}$ as distributional derivative then the initial value problem should be written as $\dot{x} = ax + x_0\delta$. Indeed, by example A.8 the smooth distribution corresponding to the function $e^{at}i_{\mathbb{R}^+}(t)x_0$ is the unique solution to this equation: its derivative is equal to the sum of x_0 (its jump at $t = 0$) times the Dirac distribution δ and the distribution corresponding to the function $ae^{at}i_{\mathbb{R}^+}(t)x_0$.

In general, in order to be able to give a useful meaning to the concept 'solution of $\dot{x} = Ax + Bu,\ x(0) = x_0$' if u is a distribution, we restrict u to the subclass $\mathfrak{D}_0^m$ of $\mathfrak{D}'^{\,m}_+$ of *impulsive-smooth distributions* (see definition A.6).

Definition 8.3 *Assume that u is an m-vector of impulsive-smooth distributions, i.e. $u \in \mathfrak{D}_0^m$. Then the solution of the initial value problem $\dot{x} = Ax + Bu,\ x(0) = x_0$ is defined as the solution over the distribution space $\mathfrak{D}'^{\,n}_+$ of the differential equation*

$$\dot{x} = Ax + Bu + x_0\delta. \tag{8.2}$$

We stress that $\dot{x}$ stands for distributional derivative. Of course, the first question that should be answered is: does the equation (8.2) indeed have a solution and, if so, is this solution unique? The answer to both questions is: yes. A proof of this is given as follows. Recall that differentiation of a distribution x is the same as taking the convolution of x with $I\dot{\delta}$, the product of the identity matrix I and the distribution $\dot{\delta}$ (the derivative of δ, see example A.4). Thus (8.2) is equivalent with

$$I\dot{\delta} * x = Ax + Bu + x_0\delta,$$

which, since $\delta * x = x$, is equivalent with

$$(I\dot{\delta} - A\delta) * x = Bu + x_0\delta.$$

Since the impulsive distribution $I\dot{\delta} - A\delta$ is invertible, with inverse $(I\dot{\delta} - A\delta)^{-1} = T(\dot{\delta})$, where $T(s) := (sI - A)^{-1}$ (see example A.15), we see that

$$x_{u,x_0} := T(\dot{\delta}) * (Bu + x_0\delta) = T(\dot{\delta}) * Bu + T(\dot{\delta})x_0 \tag{8.3}$$

is the unique solution of (8.2).

If, in (8.3), u is smooth on $\mathbb{R}^+$, i.e. if u corresponds to a function $\tilde{u}(t)i_{\mathbb{R}^+}(t)$ with $\tilde{u} \in \mathfrak{C}^\infty(\mathbb{R}, \mathbb{R}^m)$, then one would of course hope that (8.3) reduces to the variations of constant formula (3.2). We show that this is indeed the case. Define $\tilde{T}(t) := e^{At}i_{\mathbb{R}^+}(t)$. By example A.12, $(I\dot{\delta} - A\delta)^{-1}$ is smooth on $\mathbb{R}^+$, and corresponds to the function $\tilde{T}(t)$. Thus x_{u,x_0} is equal to the smooth distribution corresponding to the function $\tilde{x}$ given by

$$\tilde{x}(t) = \tilde{T}(t)x_0 + (\tilde{T} * B\tilde{u}i_{\mathbb{R}^+})(t). \tag{8.4}$$

(here, $*$ means ordinary convolution of functions). It is easily verified that $\tilde{x}(t) = 0$ for $t < 0$. For $t \geqslant 0$, (8.4) becomes

$$\tilde{x}(t) = e^{At}x_0 + \int_0^t e^{A(t-\tau)}Bu(\tau)\,\mathrm{d}\tau$$

(check this!). We conclude that for inputs u that are smooth on $\mathbb{R}^+$, (8.3) indeed reduces to (3.2).

We will now show that if $u \in \mathfrak{D}_0^m$, then for each x_0 the solution x_{u,x_0} of (8.2) is an element of $\mathfrak{D}_0^n$, i.e. is also impulsive-smooth. We have already shown that if u is smooth on $\mathbb{R}^+$ then x_{u,x_0} is smooth on $\mathbb{R}^+$. Now assume that $u = u_1 + u_2$ with u_1 impulsive and u_2 smooth on $\mathbb{R}^+$. Then of course

$$x_{u,x_0} = T(\dot{\delta}) * Bu_1 + x_{u_2,x_0}.$$

It therefore suffices to show that, if u is impulsive, then $x_{u,0}$ is impulsive-smooth. Let $u = u_0\delta + u_1\delta^{(1)} + \cdots + u_p\delta^{(k)}$, with $u_i \in \mathcal{U}$. Then

$$\begin{aligned} x_{u,0} &= T(\dot{\delta}) * Bu \\ &= \sum_{i=0}^{k} T(\dot{\delta}) * Bu_i\delta^{(i)}. \end{aligned}$$

Convolution with $\delta^{(i)}$ is the same as taking the i-th derivative. Hence, the ith term in the sum above is equal to the i-th derivative of $T(\dot{\delta})Bu_i$ which, by example A.13, is equal to

$$\sum_{j=0}^{i-1} A^j Bu_i\delta^{(i-1-j)} + T(\dot{\delta})A^i Bu_i.$$

We conclude that

$$x_{u,0} = \sum_{i=0}^{k}\sum_{j=0}^{i-1} A^j Bu_i\delta^{(i-1-j)} + T(\dot{\delta})\sum_{i=0}^{k} A^i Bu_i.$$

This is indeed an impulsive-smooth distribution. Note that the regular part of $x_{u,0}$ corresponds to the function $e^{At} i_{\mathbb{R}^+}(t) \sum_{i=0}^{k} A^i B u_i$. Thus, if $u = \sum_{i=0}^{k} u_i \delta^{(i)}$, then $x_{u,0}$ consists of an impulsive part followed by a regular 'free motion' starting in the point $\sum_{i=0}^{k} A^i B u_i$. We summarize the above in the following theorem:

Theorem 8.4 *Let $x_0 \in \mathcal{X}$ and $u \in \mathfrak{D}_0^m$. Let $u = u_1 + u_2$ with $u_1 = \sum_{i=0}^{k} u_i \delta^{(i)}$ and where u_2 is smooth on $\mathbb{R}^+$. Then the differential equation (8.2) has a unique solution x_{u,x_0} in ${\mathfrak{D}'_+}^n$. This solution is impulsive-smooth, i.e. $x_{u,x_0} \in \mathfrak{D}_0^n$. In fact, $x_{u,x_0} = x_1 + x_2$ with*

$$x_1 = \sum_{i=0}^{k} \sum_{j=0}^{i-1} A^j B u_i \delta^{(i-1-j)}$$

the impulsive part, and

$$x_2 = T(\dot{\delta}) \left(x_0 + \sum_{i=0}^{k} A^i B u_i \right) + T(\dot{\delta}) B * u_2$$

the smooth part.

Note that if the smooth part u_2 corresponds to the function $\tilde{u}_2(t) i_{\mathbb{R}^+}(t)$, then the smooth part x_2 of x_{u,x_0} corresponds to the function $\tilde{x}_2$, given for $t < 0$ by $\tilde{x}_2(t) = 0$ and for $t \geqslant 0$ by

$$\tilde{x}_2(t) = e^{At} \left(x_0 + \sum_{i=0}^{k} A^i B u_i \right) + \int_0^t e^{A(t-\tau)} B \tilde{u}_2(\tau) \, d\tau.$$

Thus we see that x_{u,x_0} consists of an impulsive part, 'followed by' a smooth part starting in the point $x_0 + \sum_{i=0}^{k} A^i B u_i$, and driven only by $\tilde{u}_2$. We define

$$x_{u,x_0}(0^+) := x_0 + \sum_{i=0}^{k} A^i B u_i. \tag{8.5}$$

Apparently, application of $u \in \mathfrak{D}_0^m$ results in an *instantaneous jump* from x_0 to $x_{u,x_0}(0^+)$. This jump is completely determined by the impulsive part u_1 of u. Note that for $t \geqslant 0$ we have $\tilde{x}_2(t) = x_{\tilde{u}_2}(t, x_0 + \sum_{i=0}^{k} A^i B u_i)$, the 'ordinary' state trajectory of the system $\dot{x}(t) = Ax(t) + Bu(t)$ with initial state $x(0) = \sum_{i=0}^{k} A^i B u_i$, and input function $u = \tilde{u}_2$.

The coefficient vectors of the impulsive part of x_{u,x_0} turn out to be generated by a recurrence relation:

Theorem 8.5 *Let $u = u_1 + u_2$, where $u_1 = \sum_{i=0}^{k} u_i \delta^{(i)}$, and where u_2 is smooth on $\mathbb{R}^+$. Then the impulsive part x_1 of x_{u,x_0} is equal to*

$$x_1 = \sum_{i=0}^{k-1} \xi_i \delta^{(i)},$$

where the coefficient vectors ξ_i *are generated by the backward recursion*

$$\xi_k := 0, \quad \xi_{i-1} = A\xi_i + Bu_i \quad (i = k, \ldots, 0).$$

Moreover,

$$x_{u,x_0}(0^+) = x_0 + \xi_{-1}.$$

Proof : This follows immediately from theorem 8.4. The details are left to the reader. ■

When we compare the distributional set-up for linear system with the conventional interpretation, we observe that a number of new concepts arises, not present for systems with only regular inputs. In particular, we can introduce the space of instantaneously reachable points. A point $x_1 \in \mathcal{X}$ is called *instantaneously reachable* if there exists an input $u \in \mathfrak{D}_0^m$ such that $x_{u,0}(0^+) = x_1$. That is, if there exists an input u that causes an instantaneous jump from the zero initial state to the point x_1. By the previous theorem, x_1 is instantaneously reachable if and only if there exist vectors $u_0, \ldots, u_k \in \mathcal{U}$ such that $x_1 = \sum_{i=0}^{k} A^i B u_i$. From this it follows immediately that the space of instantaneously reachable points is equal to $\langle A \mid \operatorname{im} B \rangle$, the ordinary reachable subspace of the system (A, B) (see section 3.2).

Finally, let us take the output equation $y = Cx + Du$ into account. If $x_0 \in \mathcal{X}$ and if $u \in \mathfrak{D}_0^m$, then the output will of course be given by to

$$y_{u,x_0} = Cx_{u,x_0} + Du.$$

Obviously, also y_{u,x_0} is an impulsive-smooth distribution. In the following, the transfer matrix of Σ will be denoted by $G(s) = C(Is-A)^{-1}B+D$. We have the following result:

Theorem 8.6 *Let* $u = u_1 + u_2$, *where* $u_1 = \sum_{i=0}^{k} u_i \delta^{(i)}$, *and where* u_2 *is smooth on* $\mathbb{R}^+$. *Then* y_{u,x_0} *is the impulsive-smooth distribution given by*

$$y_{u,x_0} = CT(\dot{\delta})x_0 + G(\dot{\delta}) * u.$$

The impulsive part of y_{u,x_0} *is given by*

$$y_1 = \sum_{i=0}^{k-1} y_i \delta^{(i)},$$

where

$$y_i = C\xi_i + Du_i \quad (i = k, \ldots, 0), \tag{8.6}$$

with ξ_i *generated by the backward recursion*

$$\xi_k = 0, \quad \xi_{i-1} = A\xi_i + Bu_i \quad (i = k, \ldots, 0). \tag{8.7}$$

The regular part of y_{u,x_0} *is given by*

$$y_2 = CT(\dot{\delta})(x_0 + \xi_{-1}) + G(\dot{\delta}) * u_2.$$

Proof : This follows immediately from theorems 8.4 and 8.5 (see also example A.16). ■

8.2 System invertibility

In this section we discuss the properties of left-invertibility, right-invertibility, and invertibility of systems. We relate these properties with properties of the system's transfer matrix, and of the system matrix.

Consider the system $\Sigma = (A, B, C, D)$. In the distributional set-up as introduced in section 8.1, if we take $x(0) = 0$ then for an impulsive-smooth distribution $u \in \mathfrak{D}_0^m$ the corresponding output is given by

$$y_{u,0} = G(\dot{\delta}) * u, \tag{8.8}$$

where $G(\dot{\delta})$ is the matrix distribution associated with the transfer matrix $G(s) = C(Is - A)^{-1}B + D$ (see appendix A). According to example A.16 this distribution is equal to $C(I\dot{\delta} - A\delta)^{-1}B + D\delta$. By the assignment $u \mapsto y_{u,0}$, the system Σ defines an operator that maps $\mathfrak{D}_0^m$ to $\mathfrak{D}_0^p$. Of course, this operator is *linear*. If it is injective, then we call the system Σ left-invertible:

Definition 8.7 *Σ is called* left-invertible *if for all $u_1, u_2 \in \mathfrak{D}_0^m$ the following holds: $y_{u_1,0} = y_{u_2,0}$ implies $u_1 = u_2$.*

In other words, Σ is left-invertible if, with initial state $x(0) = 0$, no pair of distinct inputs gives rise to one and the same output. By linearity, Σ is left-invertible if and only if for all $u \in \mathfrak{D}_0^m$ we have that $y_{u,0} = 0$ implies $u = 0$. As expected, left-invertibility of Σ can be expressed in terms of properties of the transfer matrix $G(s)$:

Theorem 8.8 *Σ is left-invertible if and only if $G(s)$ is a left-invertible rational matrix.*

Proof : ($\Rightarrow$) If $G(s)$ is not left-invertible, then there exists a rational vector $q(s) \neq 0$ such that $G(s)q(s) = 0$. Without loss of generality, assume that $q(s)$ is a polynomial vector. Define $u \in \mathfrak{D}_0^m$ by $u := q(\dot{\delta})$. Then $u \neq 0$ and according to example A.14 we have

$$y_{u,0} = G(\dot{\delta}) * q(\dot{\delta}) = (Gq)(\dot{\delta}) = 0. \tag{8.9}$$

This contradicts the assumption that Σ is left-invertible.

($\Leftarrow$) If $G(s)$ is left-invertible, then it has a left-inverse, i.e. there exists a rational matrix $G_L(s)$ such that $G_L(s)G(s) = I$. Let $u \in \mathfrak{D}_0^m$ be such that $G(\dot{\delta}) * u = 0$. Then we obtain

$$u = (G_L G)(\dot{\delta}) * u = G_L(\dot{\delta}) * G(\dot{\delta}) * u = 0.$$

Thus, Σ is left-invertible. ■

It is also possible to connect left-invertibility of Σ with properties of the polynomial matrix $P_\Sigma(s)$, the system matrix of Σ. Note that the transfer matrix $G(s)$ is left-invertible if and only if $\operatorname{normrank} G = m$. It turns out that the normal ranks of $G(s)$ and $P_\Sigma(s)$ are related as follows:

Lemma 8.9 $\operatorname{normrank} P_\Sigma = n + \operatorname{normrank} G$.

Proof : Let $R(s) := Is - A$. Then $R(s)$ is a rational matrix with normal rank n. Let I, $I_{\mathcal{U}}$ and $I_{\mathcal{Y}}$ be the identity matrices of dimensions n, m and p, respectively. We have

$$P_\Sigma(s) = \begin{pmatrix} I & 0 \\ CR(s)^{-1} & I_{\mathcal{U}} \end{pmatrix} \begin{pmatrix} R(s) & 0 \\ 0 & G(s) \end{pmatrix} \begin{pmatrix} I & -R^{-1}(s)B \\ 0 & I_{\mathcal{Y}} \end{pmatrix}.$$

The first and third factor in this product are invertible rational matrices. Consequently,

$$\operatorname{normrank} P_\Sigma = \operatorname{normrank} \begin{pmatrix} R & 0 \\ 0 & G \end{pmatrix} = n + \operatorname{normrank} G. \qquad ■$$

An immediate consequence of this is the following:

Corollary 8.10 *Σ is left-invertible if and only if* $\operatorname{rank} P_\Sigma(\lambda) = n + m$ *for all but finitely many $\lambda \in \mathbb{C}$.*

Note that, if Σ is left-invertible, then those $\lambda \in \mathbb{C}$ for which $\operatorname{rank} P_\Sigma(\lambda) < n + m$ are exactly the zeros of the system Σ.

We now turn to right-invertibility. We will call Σ right-invertible if the operator $u \mapsto y_{u,0}$ from $\mathfrak{D}_0^m$ to $\mathfrak{D}_0^p$ is surjective:

Definition 8.11 *Σ is called* right-invertible *if for every $y \in \mathfrak{D}_0^p$ there exists $u \in \mathfrak{D}_0^m$ such that $y = y_{u,0}$.*

In other words, Σ is right-invertible if for every impulsive-smooth distribution y there exists an impulsive smooth distribution u such that y is the output corresponding to initial state zero and input u.

Example 8.12 Consider the system Σ with transfer function $G(s) = s^{-1}$. This system is right-invertible: if y is impulsive-smooth, then $u := \dot{\delta} * y$ is impulsive-smooth. Moreover, $G(\dot{\delta}) * u = \dot{\delta}^{-1} * \dot{\delta} * y = y$.

The following theorem characterizes right-invertibility in terms of properties of the transfer matrix $G(s)$ and the system matrix $P_\Sigma(s)$.

Theorem 8.13 *The following statements are equivalent:*

(*i*) Σ *is right-invertible,*

(*ii*) $G(s)$ *is a right-invertible rational matrix,*

(*iii*) $\operatorname{rank} P_\Sigma(\lambda) = n + p$ *for all but finitely many* $\lambda \in \mathbb{C}$.

Proof : (i) $\Rightarrow$ (ii) If $G(s)$ is not right-invertible, there exists a polynomial row-vector $q(s) \neq 0$ such that $q(s)G(s) = 0$. Let y_0 be any vector in $\mathcal{Y}$ and define $y := y_0\delta$. Then y is impulsive-smooth. There exists an input u such that $y = G(\dot{\delta}) * u$. This implies that $q(\dot{\delta})y_0 = 0$. Since this holds for any y_0, we get $q(\dot{\delta}) = 0$, so $q(s) = 0$, which is a contradiction.

(ii) $\Leftarrow$ (i) Let $G_R(s)$ be a right-inverse of $G(s)$. Then $G_R(\dot{\delta})$ is a right-inverse of $G(\dot{\delta})$. Let $y \in \mathfrak{D}_0^p$. Define $u := G_R(\dot{\delta}) * y$. Then we have $u \in \mathfrak{D}_0^m$. Moreover, $y = G(\dot{\delta}) * u$.

(ii) $\Leftrightarrow$ (iii) This equivalence follows immediately from lemma 8.9. ■

If Σ is right-invertible, then those $\lambda \in \mathbb{C}$ for which $\operatorname{rank} P_\Sigma(\lambda) < n + p$ are the zeros of the system Σ.

Σ is called *invertible* if it is both right-invertible and left-invertible. Clearly, a system can only be invertible if $p = m$, i.e. the number of inputs and outputs are equal. For completeness, we formulate the following corollary:

Corollary 8.14 *The following statements are equivalent:*

(*i*) Σ *is invertible,*

(*ii*) $G(s)$ *is an invertible rational matrix.,*

(*iii*) $\operatorname{rank} P_\Sigma(\lambda) = n + m$ *for all but finitely many* $\lambda \in \mathbb{C}$.

8.3 The strongly reachable subspace

Consider the system Σ given by (8.1). Again assume that the inputs u are elements of the class $\mathfrak{D}_0'$ of impulsive-smooth distributions. In the previous section we have seen that if $x_0 \in \mathcal{X}$ and if $u \in \mathfrak{D}_0'$, then the resulting output y_{u,x_0} is impulsive-smooth. In the sequel, we are interested in those inputs u for which the resulting output is smooth on $\mathbb{R}^+$. For a given initial point x_0, let $\mathcal{U}_\Sigma(x_0)$ denote the subset of $\mathfrak{D}_0'$ consisting of those inputs u such that y_{u,x_0} is smooth on $\mathbb{R}^+$. We claim that this subset is, in fact, independent of x_0. Indeed, if $u \in \mathfrak{D}_0'$, then it follows from theorem 8.6 that

$$y_{u,x_0} = y_{u,0} + C(I\dot{\delta} - A\delta)^{-1}x_0. \tag{8.10}$$

Since the second term on the right of (8.10) is smooth on $\mathbb{R}^+$, we find that y_{u,x_0} is smooth on $\mathbb{R}^+$ if and only if $y_{u,0}$ is smooth on $\mathbb{R}^+$. Consequently, $\mathcal{U}_\Sigma(x_0) = \mathcal{U}_\Sigma(0)$ for all $x_0 \in \mathcal{X}$. We will denote this subset of $\mathfrak{D}'_0$ by $\mathcal{U}_\Sigma$.

A point in the state space is called strongly reachable if it is instantaneously reachable by means of an impulsive input u in $\mathfrak{U}_\Sigma$.

Definition 8.15 *$x_1 \in \mathcal{X}$ is called a* strongly reachable *point if there exists an impulsive $u \in \mathcal{U}_\Sigma$ such that $x_{u,0}(0^+) = x_1$. The set of all strongly reachable points of the system Σ is denoted by $\mathcal{T}(\Sigma)$ and is called the* strongly reachable subspace of Σ.

It is easily seen that $\mathcal{T}(\Sigma)$ is indeed a linear subspace of $\mathcal{X}$. It follows immediately from the definition that $\mathcal{T}(\Sigma)$ is contained in $\langle A \mid \text{im } B\rangle$, the space of (instantaneously) reachable points.

The strongly reachable subspace can be given an interpretation in terms of the recurrence relation (8.6) and (8.7). The condition that y_{u,x_0} should be regular translates to $y_i = 0$ $(i = k-1, \ldots, 0)$. Thus we find:

Theorem 8.16 *$\tilde{x} \in \mathcal{T}(\Sigma)$ if and only if there exists $k \in \mathbb{N}$ and vectors*

$$u_k, u_{k-1}, \ldots, u_0 \in \mathcal{U}$$

such that the vectors $\xi_k, \xi_{k-1}, \ldots, \xi_{-1}$ generated by the backward recursion

$$\xi_k = 0, \quad \xi_{i-1} = A\xi_i + Bu_i \quad (i = k, \ldots, 0) \tag{8.11}$$

satisfy $C\xi_i + Du_i = 0$ $(i = k, \ldots, 0)$ and $\xi_{-1} = \tilde{x}$.

Let $F : \mathcal{X} \to \mathcal{U}$ and $G : \mathcal{Y} \to \mathcal{X}$ be linear maps. Define the system $\Sigma_{F,G}$ by

$$\Sigma_{F,G} = (A + BF + GC + GDF, B + GD, C + DF, D).$$

We claim that the strongly reachable subspaces associated with the systems Σ and $\Sigma_{F,G}$, respectively, coincide. Stated differently: the strongly reachable subspace is invariant under state feedback and output injection (compare exercise 7.1).

Theorem 8.17 $\mathcal{T}(\Sigma_{F,G}) = \mathcal{T}(\Sigma)$.

Proof : Let $\tilde{x} \in \mathcal{T}(\Sigma_{F,G})$. According to theorem 8.16 there exists $k \in \mathbb{N}$ and vectors $v_k, v_{k-1}, \ldots, v_0 \in \mathcal{U}$ such that if ξ_i is generated by the recursion

$$\xi_k = 0, \quad \xi_{i-1} = (A + BF + GC + GDF) + (B + GD)v_i$$

then $(C + DF)\xi_i + Dv_i = 0$ $(i = k, \ldots, 0)$ and $\xi_{-1} = \tilde{x}$. Now, define $u_i := v_i + F\xi_i$. Then $C\xi_i + Du_i = 0$ $(i = k, \ldots, 0)$ and ξ_i satisfies the recursion (8.11). Thus we conclude that $\tilde{x} \in \mathcal{T}(\Sigma)$. The converse inclusion follows by noting that $\Sigma = (\Sigma_{F,G})_{-F,-G}$. ■

As another consequence of theorem 8.16 we have the following:

Lemma 8.18 *Let $\tilde{x} \in \mathcal{T}(\Sigma)$ and $\tilde{u} \in \mathcal{U}$ be such that $C\tilde{x} + D\tilde{u} = 0$. Then $A\tilde{x} + B\tilde{u} \in \mathcal{T}(\Sigma)$.*

Proof : If $\tilde{x} \in \mathcal{T}(\Sigma)$ then there are vectors $u_i \in \mathcal{U}$ $(i = k, \ldots, 0)$ such that the ξ_i defined by (8.11) satisfy $C\xi_i + Du_i = 0$ and $\xi_{-1} = \tilde{x}$. Define vectors $\bar{u}_{k+1}, \bar{u}_k, \ldots, \bar{u}_0 \in \mathcal{U}$ by $\bar{u}_{i+1} := u_i$ and $\bar{u}_0 := \tilde{u}$. Define $\bar{\xi}_{k+1}, \ldots, \bar{\xi}_0$ by $\bar{\xi}_{i+1} := \xi_i$ and $\bar{\xi}_{-1} := A\tilde{x} + B\tilde{u}$. Then $\bar{\xi}_i$ satisfies the recurrence relation

$$\bar{\xi}_{k+1} = 0, \;\; \xi_{i-1} = A\bar{\xi}_i + B\bar{u}_i \;\; (i = k+1, \ldots, 0)$$

and $C\bar{\xi}_i + D\bar{u}_i = 0$, $\bar{\xi}_{-1} = A\tilde{x} + B\tilde{u}$. We conclude $A\tilde{x} + B\tilde{u} \in \mathcal{T}(\Sigma)$. ■

The previous result can of course be restated in an alternative way as follows: for the subspace $\mathcal{V} = \mathcal{T}(\Sigma)$ the following inclusion holds

$$\begin{pmatrix} A & B \end{pmatrix} \left[(\mathcal{V} \times \mathcal{U}) \cap \ker \begin{pmatrix} C & D \end{pmatrix} \right] \subset \mathcal{V}. \tag{8.12}$$

Assume that $\mathcal{V}$ is an arbitrary subspace of $\mathcal{X}$ that satisfies (8.12). Choose a basis

$$\begin{pmatrix} x_1 \\ u_1 \end{pmatrix}, \begin{pmatrix} x_2 \\ u_2 \end{pmatrix}, \ldots, \begin{pmatrix} x_l \\ u_l \end{pmatrix} \tag{8.13}$$

of $\mathcal{V} \times \mathcal{U}$ such that the first r $(r \leqslant l)$ vectors form a basis of $(\mathcal{V} \times \mathcal{U}) \cap \ker(C \;\; D)$. Define $y_i \in \mathcal{Y}$ by

$$y_i := Cx_i + Du_i \;\; (i = 1, 2, \ldots, l),$$

Then $y_i = 0$ for $i = 1, \ldots, r$ and $y_{r+1}, \ldots, y_l$ are linearly independent. Let $G : \mathcal{Y} \to \mathcal{X}$ be a linear mapping such that

$$Gy_i = -Ax_i - Bu_i, \;\; (i = r+1, \ldots, l),$$

Then

$$w_i := \begin{pmatrix} A + GC & B + GD \end{pmatrix} \begin{pmatrix} x_i \\ u_i \end{pmatrix} = Ax_i + Bu_i + Gy_i.$$

We claim that $w_i \in \mathcal{V}$ for $i = 1, \ldots, l$. Indeed, for $i = 1, \ldots, r$ we have $w_i = Ax_i + Bu_i$ which, by (8.12), lies in $\mathcal{V}$. For $i = r+1, \ldots, l$ we have $w_i = 0$. It follows from this that

$$(A + GC)\mathcal{V} \subset \mathcal{V}, \;\; \operatorname{im}(B + GD) \subset \mathcal{V}. \tag{8.14}$$

Indeed, if $x \in \mathcal{V}$ then $(x^{\mathrm{T}}, 0)^{\mathrm{T}} \in \mathcal{V} \times \mathcal{U}$. By the above, the map $(A + GC \;\; B + GD)$ maps $\mathcal{V} \times \mathcal{U}$ into $\mathcal{V}$. Consequently, $(A + GC)x \in \mathcal{V}$. In the same way, if $u \in \mathcal{U}$ then $(0, u^{\mathrm{T}})^{\mathrm{T}} \in \mathcal{V} \times \mathcal{U}$ so $(B + GD)u \in \mathcal{V}$. This proves (8.14). We conclude that if a subspace $\mathcal{V}$ of $\mathcal{X}$ satisfies (8.12) then there exists an output injection mapping G such that (8.14) holds. In particular, there is an output injection mapping such that (8.14) holds for $\mathcal{V} = \mathcal{T}(\Sigma)$. In fact, we have:

Theorem 8.19

(i) $\mathcal{T}(\Sigma)$ *is the smallest subspace* $\mathcal{V}$ *of* $\mathcal{X}$ *for which*

$$\begin{pmatrix} A & B \end{pmatrix} (\mathcal{V} \times \mathcal{U}) \cap \ker \begin{pmatrix} C & D \end{pmatrix} \subset \mathcal{V}.$$

(ii) $\mathcal{T}(\Sigma)$ *is the smallest subspace* $\mathcal{V}$ *of* $\mathcal{X}$ *for which there exists a linear mapping* $G : \mathcal{Y} \rightarrow \mathcal{X}$ *such that*

$$(A + GC)\mathcal{V} \subset \mathcal{V}, \quad \text{im}(B + GD) \subset \mathcal{V}.$$

Proof : (i) We have already shown that $\mathcal{V} = \mathcal{T}(\Sigma)$ satisfies (8.12). Let $\mathcal{V}$ be an arbitrary subspace such that (8.12) holds. We are going to show that $\mathcal{T}(\Sigma) \subset \mathcal{V}$. Let $\tilde{x} \in \mathcal{T}(\Sigma)$. There are vectors u_i $(i = k, \ldots, 0)$ in $\mathcal{U}$ such that if ξ_i is generated by the recursion $\xi_k = 0, \xi_{i-1} = A\xi_i + Bu_i$ then we have $C\xi_i + Du_i = 0$ $(i = k, \ldots, 0)$ and $\xi_{-1} = \tilde{x}$. We claim that if for some i, $\xi_i \in \mathcal{V}$ then $\xi_{i-1} \in \mathcal{V}$. Indeed, if $\xi_i \in \mathcal{V}$ then we have

$$\begin{pmatrix} \xi_i \\ u_i \end{pmatrix} \in (\mathcal{V} \times \mathcal{U}) \cap \ \ker(C \quad D).$$

Since $\mathcal{V}$ satisfies (8.12), this implies $\xi_{i-1} = A\xi_i + Bu_i \in \mathcal{V}$. From the fact that $\xi_k = 0 \in \mathcal{V}$ we may now conclude that $\xi_i \in \mathcal{V}$ for all i. In particular $\tilde{x} = \xi_{-1} \in \mathcal{V}$.

(ii) It was already shown that there exists G such that (8.14) holds for $\mathcal{V} = \mathcal{T}(\Sigma)$. Let $\mathcal{V}$ be an arbitrary subspace such that (8.14) holds for some G. We claim that $\mathcal{V}$ then satisfies (8.12). Let $x \in \mathcal{V}$ and $u \in \mathcal{U}$ be such that $Cx + Du = 0$. We have $(A + GC)x \in \mathcal{V}$ and $(B + GD)u \in \mathcal{V}$. Consequently

$$Ax + Bu = (A + GC)x + (B + GD)u \in \mathcal{V}.$$

This proves our claim. It then follows from (i) above that $\mathcal{T}(\Sigma) \subset \mathcal{V}$. ■

Recall from theorem 5.5 that a subspace $\mathcal{V}$ of the state space is conditionally invariant or (C, A)-invariant if and only if there exists a linear mapping $G : \mathcal{Y} \rightarrow \mathcal{X}$ such that $(A + GC)\mathcal{V} \subset \mathcal{V}$. As a consequence of this, it follows from the previous theorem that the subspace $\mathcal{T}(\Sigma)$ is always (C, A)-invariant. As a matter of fact, since the strongly reachable subspace of Σ coincides with the strongly reachable subspace of the system $\Sigma_F := (A + BF, B, C + DF, D)$ (see theorem (8.17)), $\mathcal{T}(\Sigma)$ is $(C + DF, A + BF)$-invariant for *any* linear mapping F (compare exercise 7.1). This implies that for all F

$$A_F(\mathcal{T}(\Sigma) \cap \ker C_F) \subset \mathcal{T}(\Sigma). \tag{8.15}$$

According to theorem 8.19 (ii), if $D = 0$ then we even have $\mathcal{T}(\Sigma) = \mathcal{S}^*(\text{im}\, B)$, the smallest (C, A)-invariant subspace containing im B.

The set of all output injection mappings $G : Y \rightarrow \mathcal{X}$ such that $(A+GC)\mathcal{T}(\Sigma) \subset \mathcal{T}(\Sigma)$ and $\text{im}(B + GD) \subset \mathcal{T}(\Sigma)$ is denoted by $\underline{G}(\mathcal{T}(\Sigma))$. We recall that $A + GC$ is sometimes denoted by A_G. and $B + GD$ by B_G.

By combining theorems 7.10 and 8.19 it can be seen that the weakly unobservable subspace $\mathcal{V}(\Sigma)$ and the subspace $\mathcal{T}(\Sigma)$ are *dual concepts*. Specifically, if we define the dual of Σ by $\Sigma^{\mathrm{T}} := (A^{\mathrm{T}}, C^{\mathrm{T}}, B^{\mathrm{T}}, D^{\mathrm{T}})$ then we have

Theorem 8.20 $\mathcal{V}(\Sigma^{T}) = \mathcal{T}(\Sigma)^{\perp}$.

Proof : The proof of this is left as an exercise to the reader. ■

We now establish an algorithm to compute $\mathcal{T}(\Sigma)$. Define a sequence of subspaces $\mathcal{T}_0, \mathcal{T}_1, \mathcal{T}_2, \ldots$ of $\mathcal{T}(\Sigma)$ as follows: define $\mathcal{T}_0 := 0$ and for $t \geqslant 1$

$$\begin{aligned}\mathcal{T}_t := \{\tilde{x} \in \mathcal{X} \mid \text{there exist } & u_{t-1}, \ldots, u_0 \in \mathcal{U} \text{ such that } \xi_i \\ & \text{generated by } \xi_{t-1} := 0,\ \xi_{i-1} = A\xi_i + Bu_i \\ & \text{satisfies } C\xi_i + Du_i = 0\ (i = t-1, \ldots, 0) \text{ and } \xi_{-1} = \tilde{x}\}.\end{aligned}$$

It is easily verified that $\mathcal{T}_t$ is indeed a subspace of $\mathcal{T}(\Sigma)$. In fact, $\mathcal{T}_t$ consists exactly of the points in $\mathcal{X}$ that are strongly reachable my means of an impulsive input $u \in \mathcal{U}_{\Sigma}$ in which the highest order derivative of the Dirac distribution δ is less than or equal to $t-1$. Obviously, the sequence $\{\mathcal{T}_t\}$ is a chain in $\mathcal{X}$, i.e. $\mathcal{T}_0 \subset \mathcal{T}_1 \subset \mathcal{T}_2 \subset \cdots$. We show that the sequence $\mathcal{T}_t$ is generated by a recurrence relation. It follows from the definition that a point $\tilde{x}$ lies in $\mathcal{T}_{t+1}$ if and only if there exists $\bar{u} \in \mathcal{U}$ and $\bar{x} \in \mathcal{T}_t$ such that $\tilde{x} = A\bar{x} + B\bar{u}$ and $C\bar{x} + D\bar{u} = 0$. Equivalently $\tilde{x} \in \mathcal{T}_{t+1}$ if and only if there exists $(\bar{x}^{\mathrm{T}}, \bar{u}^{\mathrm{T}})^{\mathrm{T}} \in (\mathcal{T}_t \times \mathcal{U}) \cap \ker(C \quad D)$ such that $\tilde{x} = A\bar{x} + B\bar{u}$. Thus we see that the sequence $\{\mathcal{T}_t\}$ satisfies

$$\mathcal{T}_0 = 0, \quad \mathcal{T}_{t+1} = \begin{pmatrix} A & B \end{pmatrix} \left[(\mathcal{T}_t \times \mathcal{U}) \cap \ker \begin{pmatrix} C & D \end{pmatrix} \right]. \tag{8.16}$$

From this it follows immediately that if for some integer k we have $\mathcal{T}_k = \mathcal{T}_{k+1}$ then $\mathcal{T}_k = \mathcal{T}_t$ for all $t \geqslant k$. Consequently, the inclusion chain $\{\mathcal{T}_i\}$ has the form

$$\mathcal{T}_0 \subset \mathcal{T}_1 \subset \cdots \subset \mathcal{T}_k = \mathcal{T}_{k+1} = \cdots$$

for some $k \leqslant n$. In this formula, $\subset$ designates *strict* inclusion. We claim that $\mathcal{T}_k = \mathcal{T}(\Sigma)$. Indeed, we already know that $\mathcal{T}_k \subset \mathcal{T}(\Sigma)$. On the other hand, if $\mathcal{T}_k = \mathcal{T}_{k+1}$ then $\mathcal{V} = \mathcal{T}_k$ of course satisfies (8.12). Since $\mathcal{T}(\Sigma)$ is the smallest subspace satisfying (8.12) we find $\mathcal{T}(\Sigma) \subset \mathcal{T}_k$. We have now proven the following theorem:

Theorem 8.21 *Let $\mathcal{T}_t$, $t = 0, 1, 2, \ldots$ be defined by the algorithm (8.16). Then we have*

(i) $\mathcal{T}_0 \subset \mathcal{T}_1 \subset \mathcal{T}_2 \subset \cdots$.

(ii) *There exists $k \leqslant n$ such that $\mathcal{T}_k = \mathcal{T}_{k+1}$.*

(iii) *If $\mathcal{T}_k = \mathcal{T}_{k+1}$ then $\mathcal{T}_k = \mathcal{T}_t$ for all $t \geqslant k$.*

(iv) *If $\mathcal{T}_k = \mathcal{T}_{k+1}$ then $\mathcal{T}_k = \mathcal{T}(\Sigma)$.*

It is left as an exercise to the reader to check that there is a duality between the algorithm (7.9) to compute $\mathcal{V}(\Sigma)$ and the algorithm (8.16). Indeed, if we denote $\mathcal{V}_t$ in (7.9) by $\mathcal{V}_t(\Sigma)$ and $\mathcal{T}_t$ in (8.16) by $\mathcal{T}_t(\Sigma)$ then we have $\mathcal{V}_t(\Sigma^{\mathrm{T}}) = \mathcal{T}_t(\Sigma)^{\perp}$.

From definition 7.13, recall that a point $x_0 \in \mathcal{X}$ is called controllable weakly unobservable if there exists an input function u and $T > 0$ such that $y_u(t, x_0) = 0$, $t \in [0, T]$, and $x_u(T, x_0) = 0$. The controllable weakly unobservable subspace was denoted by $\mathcal{R}(\Sigma)$. As already noted, $\mathcal{R}(\Sigma)$ is a subspace of $\mathcal{V}(\Sigma)$. We will show that $\mathcal{R}(\Sigma)$ is also a subspace of $\mathcal{T}(\Sigma)$ and that it is, in fact, the intersection of the weakly unobservable subspace and the strongly reachable subspace:

Theorem 8.22 $\mathcal{R}(\Sigma) = \mathcal{V}(\Sigma) \cap \mathcal{T}(\Sigma)$.

Proof : ($\subset$) According to theorem 7.14, for any $F \in \underline{F}(\Sigma)$ we have $\mathcal{R}(\Sigma) = \langle A_F \mid \mathcal{V}(\Sigma) \cap B \ker D\rangle$. Thus, if $\tilde{x} \in \mathcal{R}(\Sigma)$ then there are vectors $u_0, \ldots, u_{n-1} \in \ker D$ such that $Bu_i \in \mathcal{V}(\Sigma)$ and

$$\tilde{x} = Bu_0 + A_F Bu_1 + \cdots + A_F^{n-1} Bu_{n-1}.$$

Let ξ_i be defined by the recursion

$$\xi_{n-1} := 0,\ \xi_{i-1} = A_F \xi_i + Bu_i\ (i = n-1, \ldots, 0).$$

Then we have $\xi_{-1} = \tilde{x}$. Moreover, since $A_F \mathcal{V}(\Sigma) \subset \mathcal{V}(\Sigma)$ we have $\xi_i \in \mathcal{V}(\Sigma)$ for all i. Since $C_F \mathcal{V}(\Sigma) = 0$ we therefore obtain $C_F \xi_i + Du_i = 0$ for all i. From theorem 8.16 we conclude that $\tilde{x} \in \mathcal{T}(\Sigma_F)$, with $\Sigma_F = (A_F, B, C_F, D)$. Using theorem 8.17 we then find that $\mathcal{R}(\Sigma) \subset \mathcal{T}(\Sigma)$.

($\supset$) Let $\mathcal{T}_t$, $t = 0, 1, 2, \ldots$ be defined by (8.16). We show that

$$\mathcal{V}(\Sigma) \cap \mathcal{T}_t \subset \mathcal{R}(\Sigma). \tag{8.17}$$

The proof is by induction. Assume $t = 1$. Since $\mathcal{T}_1 = B \ker D$, in this case (8.17) follows from theorem 7.14. Next, assume that (8.17) is true up to t. Let $x_1 \in \mathcal{V}(\Sigma) \cap \mathcal{T}_{t+1}$. There exists $x_0 \in \mathcal{T}_t$ and $u_0 \in \mathcal{U}$ such that $x_1 = Ax_0 + Bu_0$ and $Cx_0 + Du_0 = 0$. This implies that $Ax_0 \in \mathcal{V}(\Sigma) + \operatorname{im} B$ and $Cx_0 \in \operatorname{im} D$, which yields

$$\begin{pmatrix} A \\ C \end{pmatrix} x_0 \in (\mathcal{V}(\Sigma) \times 0) + \operatorname{im} \begin{pmatrix} B \\ D \end{pmatrix}.$$

Using theorem 7.10, the latter implies that $x_0 \in \mathcal{V}(\Sigma)$. Since also $x_0 \in \mathcal{T}_t$, by induction hypothesis we have $x_0 \in \mathcal{R}(\Sigma)$. Let $F \in \underline{F}(\mathcal{V}(\Sigma))$. We have

$$B(u_0 - Fx_0) = x_1 - (A + BF)x_0 \in \mathcal{V}(\Sigma),$$

$$D(u_0 - Fx_0) = -(C + DF)x_0 = 0,$$

and therefore

$$B(u_0 - Fx_0) \in \mathcal{V}(\Sigma) \cap B \ker D \subset \mathcal{R}(\Sigma).$$

Since $A_F\mathcal{R}(\Sigma) \subset R(\Sigma)$ (see (7.14)), we conclude that

$$x_1 = (A + BF)x_0 + B(u_0 - Fx_0) \in R(\Sigma). \qquad \blacksquare$$

8.4 The distributionally weakly unobservable subspace

In chapter 7 we have introduced the space of weakly unobservable points. A point in the state space of the system Σ was called weakly unobservable if there exists an input *function* such that the output resulting from that point and input is equal to zero. In this section we introduce distributionally weakly unobservable points. These will be points in the state space for which there exists a *distributional* input such that the resulting output is zero.

Example 8.23 Consider the system $\Sigma\ :\ \dot{x} = u,\ y = x$ with state space $\mathcal{X} = \mathbb{R}$. Clearly, $\mathcal{V}(\Sigma) = 0$. Let $x(0) = a \neq 0$. We claim that by choosing a suitable distribution u as input, the resulting output can be made zero. Indeed, for a given input u the corresponding state trajectory is obtained by solving $\dot{x} = u + a\delta$. Thus, if we take $u = -a\delta$ this yields $x = 0$ and hence $y = 0$.

Definition 8.24 $x_0 \in \mathcal{X}$ *is called a* distributionally weakly unobservable *point if there exists $u \in \mathcal{U}_\Sigma$ such that $y_{u,x_0} = 0$. The set of all distributionally weakly unobservable points of Σ is denoted by $\mathcal{W}(\Sigma)$ and is called the* distributionally weakly unobservable subspace *of* Σ.

Again, it is easily seen that the set $\mathcal{W}(\Sigma)$ is indeed a linear subspace of $\mathcal{X}$. Obviously, we have

$$\mathcal{V}(\Sigma) \subset \mathcal{W}(\Sigma). \tag{8.18}$$

From exercise 8.5 we have that for each $x_0 \in \mathcal{T}(\Sigma)$ there exists $u \in \mathcal{U}_\Sigma$ such that $x_{u,x_0}(0^+) = 0$ and $y_{u,x_0} = 0$. Consequently, also

$$\mathcal{T}(\Sigma) \subset \mathcal{W}(\Sigma). \tag{8.19}$$

In fact, it turns out that the sum of the weakly unobservable subspace and the strongly reachable subspace, is equal to the distributionally weakly unobservable subspace:

Theorem 8.25 $\mathcal{W}(\Sigma) = \mathcal{V}(\Sigma) + \mathcal{T}(\Sigma)$.

Proof : ($\supset$) This follows from the above.

($\subset$) Let $x_0 \in \mathcal{W}(\Sigma)$. There exists an input $u \in \mathcal{U}_\Sigma$ such that $y_{u,x_0} = 0$. The input u can be decomposed as $u = u_1 + u_2$ with u_1 impulsive and u_2 smooth on $\mathbb{R}^+$. Define $x_1 := x_{u,x_0}(0^+)$. It follows from theorem 8.5 that, in fact,

$$x_1 = x_{u_1,x_0}(0^+).$$

Also, it is easily seen that $u_1 \in \mathcal{U}_\Sigma$. Since

$$x_{u_1,x_0}(0^+) = x_0 + x_{u_1,0}(0^+),$$

we see that $x_1 - x_0 = x_{u_1,0}(0^+) \in \mathcal{T}(\Sigma)$. Since $x_0 = x_1 - (x_1 - x_0)$ it therefore suffices to show that $x_1 \in \mathcal{V}(\Sigma)$. Let the distribution u_2 correspond to the function $\tilde{u}_2(t)$ with $\tilde{u}_2(t) = 0$ for $t < 0$. According to theorem 8.6, the regular part of y_{u,x_0} corresponds to the function $\tilde{y}(t)$ with $\tilde{y}(t) = 0$ $(t < 0)$ and $\tilde{y}(t) = y_{\tilde{u}_2,x_1}(t)$ $(t \geqslant 0)$. Since $y_{u,x_0} = 0$, we must have $y_{\tilde{u}_2,x_1}(t) = 0$ $(t \geqslant 0)$. This however implies that $x_1 \in \mathcal{V}(\Sigma)$. This completes the proof of the theorem. ■

8.5 State space conditions for system invertibility

In this section we characterize the properties of left-invertibility, right-invertibility, and invertibility of the system Σ in terms of the various subspaces introduced in this chapter, and in chapter 7. Again consider the system $\Sigma = (A, B, C, D)$. We first characterize left-invertibility in terms of properties of the controllable weakly unobservable subspace of Σ.

Theorem 8.26 *The following statements are equivalent:*

(i) Σ *is left-invertible,*

(ii) $\mathcal{R}(\Sigma) = 0$ *and* $\binom{B}{D}$ *is injective,*

(iii) $\mathcal{V}(\Sigma) \cap B \ker D = 0$ *and* $\binom{B}{D}$ *is injective.*

Proof : (i) $\Rightarrow$ (ii) We first show that $\binom{B}{D}$ has full column rank. Let $P_\Sigma(s)$ be the system matrix of Σ. We have

$$\operatorname{normrank} P_\Sigma \leqslant n + \operatorname{rank} \binom{B}{D}.$$

Hence, if Σ is left-invertible then according to corollary 8.10 the matrix $(B^{\mathrm{T}}, D^{\mathrm{T}})$ has full rank m. Next, assume that $0 \neq x_1 \in \mathcal{R}(\Sigma)$. There exists an input function $\tilde{u}$ on $\mathbb{R}^+$ and $T > 0$ such that $x_{\tilde{u}}(T, 0) = x_1$ and $y_{\tilde{u}}(t, 0) = 0$ for all $t \geqslant 0$ (see exercise 7.6). Let $u \in \mathfrak{D}_0^m$ be the distribution corresponding to $\tilde{u}$. Then $u \neq 0$ while $y_{u,0} = 0$.

(ii) $\Rightarrow$ (i) If Σ is not left-invertible then the transfer matrix $G(s)$ is not left-invertible. Hence there exists a polynomial vector $q(s) \neq 0$ such that $G(s)q(s) = 0$.

Let $q(s) = q_0 + q_1 s + \cdots + q_k s^k$. Let $r(s) := s^{k+1}$. Define the rational vector $f(s)$ by $f(s) := q(s)/r(s)$. It is easily seen that the distribution $f(\dot{\delta})$ is smooth on $\mathbb{R}^+$ (recall that if $\psi(s) = s^\ell$ then $\psi(p)^{-1}$ is the ℓ-fold convolution product $h*h*\cdots*h$). Define as input $u := f(\dot{\delta})$. Clearly

$$\begin{aligned} y_{u,0} &= G(\dot{\delta}) * r(\dot{\delta})^{-1} * q(\dot{\delta}) \\ &= r(\dot{\delta})^{-1} * G(\dot{\delta}) * q(\dot{\delta}) = 0. \end{aligned}$$

Let $\tilde{u}$ be the smooth function, corresponding to u. Then $y_{\tilde{u}}(t, 0) = 0$ for all $t \geqslant 0$. By exercise 7.6, $x_{\tilde{u}}(t, 0) \in \mathcal{R}(\Sigma)$ for all $t \geqslant 0$. Consequently, $x_{\tilde{u}}(t, 0) = 0$ for all t and hence $\dot{x}_{\tilde{u}}(t, 0) = 0$ for all t. Since $x_{\tilde{u}}(\cdot, x_0)$ satisfies the differential equation $\dot{x} = Ax + B\tilde{u}(t)$ and $Cx_{\tilde{u}}(t, x_0) + D\tilde{u}(t) = 0$ for all $t \geqslant 0$, we find $B\tilde{u}(t) = 0$ and $D\tilde{u}(t) = 0$ for all $t \geqslant 0$. Since $\ker B \cap \ker D = 0$ we find $\tilde{u}(t) = 0$ for all $t \geqslant 0$. Hence $u = 0$. In turn, this implies $q(\dot{\delta}) = 0$. This contradicts $q(s) \neq 0$.

(ii) $\Leftrightarrow$ (iii) This equivalence is an immediate consequence of theorem 7.14. ■

We now turn to right-invertibility. Consider the dual system Σ^{T} with realization $(A^{\mathrm{T}}, C^{\mathrm{T}}, B^{\mathrm{T}}, D^{\mathrm{T}})$. Of course, the transfer matrix of Σ^{T} is equal to $G^{\mathrm{T}}(s)$, the transpose of $G(s)$. Since $G^{\mathrm{T}}(s)$ is a left-invertible rational matrix if and only if $G(s)$ is a right-invertible rational matrix, we find that Σ is right-invertible if and only if Σ^{T} is left-invertible. We use this observation to derive a state space characterization of right-invertibility. Recall that $\mathcal{V}(\Sigma)$ and $\mathcal{T}(\Sigma)$ are dual concepts, in the sense that $\mathcal{V}(\Sigma^{\mathrm{T}}) = \mathcal{T}(\Sigma)^{\perp}$ (see theorem 8.20). Using this, we see that also $\mathcal{R}(\Sigma)$ and $\mathcal{W}(\Sigma)$ are dual concepts, in the sense that $\mathcal{R}(\Sigma^{\mathrm{T}}) = \mathcal{W}(\Sigma)^{\perp}$. By applying theorem 8.26 to Σ^{T} we obtain:

Theorem 8.27 *The following statements are equivalent:*

(i) Σ *is right-invertible,*

(ii) $\mathcal{W}(\Sigma) = \mathcal{X}$ *and* $(C \quad D)$ *is surjective,*

(iii) $\mathcal{T}(\Sigma) + C^{-1}\operatorname{im} D = \mathcal{X}$ *and* $(C \quad D)$ *is surjective.*

To conclude this section we show that the notions of strong observability (see section 7.5) and strong controllability (see exercise 8.5) are related to the existence of *polynomial* left-inverses and right-inverses of $G(s)$, respectively. Recall that Σ is strongly observable if and only if $\mathcal{V}(\Sigma) = 0$. Thus, if Σ is strongly observable and if, in addition, $(B^{\mathrm{T}}, D^{\mathrm{T}})^{\mathrm{T}}$ is injective then Σ is left-invertible. Dually, Σ is strongly controllable if and only if $\mathcal{T}(\Sigma) = 0$. Thus, if Σ is strongly controllable and if $(C \quad D)$ is surjective, then Σ is right-invertible. We show that under these conditions the transfer matrix $G(s)$ has, in fact, a polynomial left-inverse and right-inverse, respectively.

Theorem 8.28

(i) If Σ *is strongly observable and* $(B^{\mathrm{T}}, D^{\mathrm{T}})^{\mathrm{T}}$ *is injective, then there exists a polynomial matrix* $P_L(s)$ *such that* $P_L(s)G(s) = I$.

(ii) If Σ is strongly controllable and $\begin{pmatrix} C & D \end{pmatrix}$ is surjective, then there exists a polynomial matrix $P_R(s)$ such that $G(s)P_R(s) = I$.

Proof : (i) According to theorem 7.17 the system matrix $P_\Sigma(s)$ has Smith form $\binom{I}{0}$. By corollary 7.5 this implies that $P_\Sigma(s)$ is left-unimodular, i.e. there exists a polynomial matrix $L(s)$ such that $L(s)P_\Sigma(s) = I$. Partition

$$L = \begin{pmatrix} L_{11}(s) & L_{12}(s) \\ L_{21}(s) & L_{22}(s) \end{pmatrix},$$

compatible with the partitioning of $P_\Sigma(s)$. It then follows that $L_{21}(s)(s)(Is - A) + L_{22}(s)C = 0$ and that $-L_{21}(s)B + L_{22}(s)D = 0$. From this it is easy to verify that $L_{22}(s)G(s) = I$.

(ii) The second assertion follows by a duality argument. ■

8.6 Exercises

8.1 Consider $\Sigma = (A, B, C, D)$. Show that $\mathcal{R}(\Sigma) = 0$ if and only if

$$\operatorname{rank} P_\Sigma(\lambda) = n + \operatorname{rank} \binom{B}{D} \tag{8.20}$$

for all but finitely may $\lambda \in \mathbb{C}$ (equivalently: if and only if normrank $P_\Sigma = n + \operatorname{rank}(B^{\mathrm{T}}, D^{\mathrm{T}})^{\mathrm{T}}$.

8.2 From exercise 7.9, recall the definitions of strong detectability and minimum phase system.

a. Show that Σ is strongly detectable if and only if (8.20) is satisfied for all $\lambda \in \mathbb{C}^+$.

b. Show that if Σ is strongly detectable and $(B^{\mathrm{T}}, D^{\mathrm{T}})^{\mathrm{T}}$ is injective, then all invariant factors of $P_\Sigma(s)$ are non-zero and Hurwitz (a non-zero polynomial is called *Hurwitz* if all its roots lie in $\mathbb{C}^-$).

c. Use (b) to show that if Σ is strongly detectable and $(B^{\mathrm{T}}, D^{\mathrm{T}})^{\mathrm{T}}$ is injective, then the transfer matrix $G(s)$ has a $\mathbb{C}^-$-stable left inverse (in general not proper).

d. Show that if Σ is an invertible minimum phase system then it has a $\mathbb{C}^-$-stable inverse (in general not proper).

8.3 Σ is called a *prime system* if it is strongly observable and strongly controllable. Show that if Σ is prime and if $(B^{\mathrm{T}}, D^{\mathrm{T}})^{\mathrm{T}}$ is injective and $(C \quad D)$ is surjective then the transfer matrix $G(s)$ has a polynomial inverse.

8.4 Consider the system Σ. A point $x_0 \in \mathcal{X}$ is called *strongly controllable* if there exists an impulsive input $u \in \mathcal{U}_\Sigma$ such that $x_{u,x_0}(0^+) = 0$.

a. Let $x_0 \in \mathcal{X}$ and $u \in \mathfrak{D}_0^m$. Show that $x_{u,x_0}(0^+) = x_0 + x_{u,0}(0^+)$.

b. Show that $x_0 \in \mathcal{X}$ is strongly controllable if and only if it is strongly reachable.

c. Show that for all $x_0 \in \mathcal{T}(\Sigma)$ there exists an impulsive $u \in \mathcal{U}_\Sigma$ such that $x_{u,x_0}(0^+) = 0$ and $y_{u,x_0} = 0$.

8.5 The system Σ is called *strongly controllable* if each $x_0 \in \mathcal{X}$ can be steered to the origin instantaneously, by means of an admissible input u. Formally: if for all $x_0 \in \mathcal{X}$ there exists an impulsive $u \in \mathcal{U}_\Sigma$ such that $x_{u,x_0}(0^+) = 0$. Show that the following statements are equivalent:

a. Σ is strongly controllable,

b. $\mathcal{T}(\Sigma) = \mathcal{X}$,

c. the dual system Σ^{T} is strongly observable,

d. $\operatorname{rank} P_\Sigma(\lambda) = n + \operatorname{rank}\begin{pmatrix} C & D \end{pmatrix}$ for all $\lambda \in \mathbb{C}$,

e. the Smith form of P_Σ is equal to the constant matrix

$$Q = \begin{pmatrix} I & 0 \\ 0 & 0 \end{pmatrix}.$$

Here I denotes the $(n+r) \times (n+r)$ identity matrix, $r := \operatorname{rank}\begin{pmatrix} C & D \end{pmatrix}$,

f. $(A + GC, B + GD)$ is controllable for all G.

8.6 Consider the system $\Sigma = (A, B, C, D)$ and let Σ^{T} be the dual of Σ.

a. Show that $\mathcal{R}(\Sigma)$ and $\mathcal{W}(\Sigma)$ are dual concepts, in the sense that $\mathcal{R}(\Sigma^{\mathrm{T}}) = \mathcal{W}(\Sigma)^\perp$.

b. Show that for all $G \in \underline{G}(\Sigma)$ we have

$$\mathcal{W}(\Sigma) = \langle \mathcal{T}(\Sigma) + C^{-1} \operatorname{im} D \mid A + GC \rangle,$$

(here $C^{-1} \operatorname{im} D := \{x \in \mathcal{X} \mid Cx \in \operatorname{im} D\}$).

8.7 Notes and references

The use of distributions in order to describe various subspaces of the state space of the system was suggested by Willems in [215], and by Hautus and Silverman in [76]. Additional material can be found in Willems, Kitapçi and Silverman [219].

Invertibility of linear multivariable systems was studied by Silverman in [176] and [177]. Conditions for invertibility in terms of the Toeplitz matrix associated with the system were obtained in Sain and Massey [161], see also the work of Wang and Davison [209]. In [179], Silverman and Payne derived conditions for system invertibility using the so-called structure algorithm.

The analogue of the strongly reachable subspace in a discrete-time framework already appeared in the work of Molinari [126]. In particular, the recursive algorithm 8.16 was introduced there. For continuous time systems, a detailed exposition of the strongly reachable subspace can be found in Hautus and Silverman [76]. Related material can be found in Willems, Kitapçi and Silverman [219]. The relation between system invertibility, and the strongly reachable and weakly unobservable subspace was already described in Wonham's textbook [223], ex. 5.17. A detailed treatment can be found in Hautus and Silverman [76]. In [171], Schumacher discussed strongly controllable systems and their invertibility properties. Finally, related material on the connection between invertibility properties of the system, and its geometric state space properties can be found in the work of Morse [129], and in the work of Aling and Schumacher [3] and Chen [27].

Chapter 9

Tracking and regulation

9.1 The regulator problem

An important feedback synthesis problem is to design for a given control system a dynamic feedback controller such that the output of the resulting closed-loop system *tracks* (i.e., converges to), some a priori given reference signal. This problem is known as the *servo problem*.

In the case that the reference signal is equal to a Bohl function from a certain time on (which covers the important special cases that the reference signal is, for example, a step function, a ramp function or a sinusoid), one way to approach the servo problem is to let the reference signal be generated by some dynamical model, more specifically, to set up some linear, time invariant, autonomous system that, for some appropriate initial state, has the reference signal as its output. Note that the frequencies of this reference signal are fixed by the dynamics of this autonomous system (to be called *the exosystem*) while phase and amplitude of the different frequencies is determined by the initial condition of this exosystem. One then incorporates the equations of this exosystem into the equations of the control system, and defines a new output as the difference between the outputs of the exosystem and the control system. The servo problem can then be reformulated as: design a dynamic feedback controller such that the output of the aggregated system converges to zero regardless of its initial state. In particular, by taking the appropriate initial state for the exosystem, the deviation of the output from the reference signal (called the *tracking error*) will then converge to zero as time tends to infinity.

A second important synthesis problem is the problem of *output regulation*. For a certain control system that is subjected to external disturbances, the problem is here to design a dynamic feedback controller such that the output of the closed-loop system converges to zero as time tends to infinity, regardless of the disturbance and the initial state. One way to approach this problem is to consider the disturbances to be completely unknown, but to be elements of some function class **D** (in fact, this setup is worked out in exercise 6.4). In this chapter, we will take an alternative point of view and assume the disturbances, albeit unknown, to be generated as outputs of some linear time-invariant autonomous system, again to be called the *exosystem*. This basically amounts to the fact that the function class **D** dictates that there are only a fixed set of frequencies in the disturbance signal. Each initial state of the exosystem then corresponds to one disturbance function and this initial state fixes the phase and amplitude of each frequency. One incorporates the equations of the exosystem into the equations of the control system, and requires the output of the new, aggregated, system to converge to zero (to be *regulated*), regardless of the initial state.

Of course, an even more general problem formulation is obtained by combining these two synthesis problems into a single one by requiring the design of a dynamic feedback controller such that the output of the closed loop system tracks a given reference signal, regardless of the disturbance and the initial state. It should be clear that this combined problem can be approached by combining the two exosystems into a single one and to require regulation of the tracking error.

As an illustration, consider a scalar control system whose output is required to track a sinusoid, in the presence of constant disturbances. Let the control system be given by

$$\dot{x}_1(t) = a_{11}x_1(t) + b_1u(t) + a_{14}d(t), \quad z_1(t) = x_1(t).$$

Suppose the reference signal is $r(t) = \sin \omega t$. This reference signal can be generated by the system

$$\begin{aligned} \dot{x}_2(t) &= x_3(t), \\ \dot{x}_3(t) &= -\omega^2 x_2(t), \\ r(t) &= x_2(t), \end{aligned}$$

by taking the initial conditions $x_2(0) = 0$ and $x_3(0) = \omega$. The tracking error is equal to $z_1(t) - r(t)$. Suppose that the disturbances d are known to be constant, but with unknown magnitude. This can be modelled by letting the disturbances be generated by

$$\begin{aligned} \dot{x}_4(t) &= 0, \\ d(t) &= x_4(t). \end{aligned}$$

Both reference signal and disturbance signals can be thought of as being generated by a single exosystem, obtained by combining the respective equations. The aggregated system is then given by

$$\begin{aligned} \begin{pmatrix} \dot{x}_1(t) \\ \dot{x}_2(t) \\ \dot{x}_3(t) \\ \dot{x}_4(t) \end{pmatrix} &= \begin{pmatrix} a_{11} & 0 & 0 & a_{14} \\ 0 & 0 & 1 & 0 \\ 0 & -\omega^2 & 0 & 0 \\ 0 & 0 & 0 & 0 \end{pmatrix} \begin{pmatrix} x_1(t) \\ x_2(t) \\ x_3(t) \\ x_4(t) \end{pmatrix} + \begin{pmatrix} b_1 \\ 0 \\ 0 \\ 0 \end{pmatrix} u(t), \\ z(t) &= \begin{pmatrix} 1 & -1 & 0 & 0 \end{pmatrix} \begin{pmatrix} x_1(t) \\ x_2(t) \\ x_3(t) \\ x_4(t) \end{pmatrix}. \end{aligned}$$

In addition to the requirements of tracking and regulation, a realistic design requires the property of internal stability. Obviously, one can not expect to be able to internally stabilize the aggregated system, since typically part of this system (the exosystem) can not be influenced by controls and will generally be unstable. Thus, in the present context the requirement of internal stabilization should be interpreted as internal stabilization of the interconnection of the original control system and the designed controller.

We will now make things more precise. Suppose that we are given a control system which is subject to a disturbance of a specified type, and whose output should track a given reference signal. This situation is modelled as the interconnection of two systems, Σ_1 and Σ_2, where Σ_2 denotes the control system and Σ_1 an autonomous system that generates the disturbances and the reference signal, called the *exosystem*.

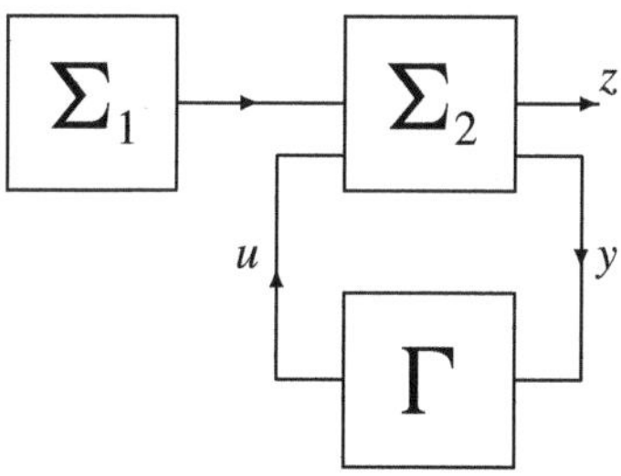

It is assumed that the control system also has a control input u and two outputs y and z, as in the previous chapters. Also, we assume that a stability domain $\mathbb{C}_g$ has been prescribed. The regulator problem then consists of finding a controller Γ such that for the resulting closed loop system the following properties hold:

- *the regulation property*: $z(t)$ is $\mathbb{C}_g$-stable for any initial state of the total closed loop system.
- *internal stability*: for zero initial state of the exosystem and any initial state of the control system and the controller, the combined state of the control system and the controller is $\mathbb{C}_g$-stable.

Again, we note that we can not hope to achieve internal stability of the total closed loop system, since the exosystem is completely uncontrollable and typically unstable. If it happens to be internally stable, the problem reduces to the classical stabilization problem, treated in section 3.12. In fact, one usually assumes that the exosystem is *antistable* (see definition 2.12).

Let us now specify the system considered. We shall assume that the exosystem Σ_1 is given by the equation

$$\dot{x}_1(t) = A_1 x_1(t), \tag{9.1}$$

while the plant Σ_2 is assumed to be given by the equations

$$\begin{aligned} \dot{x}_2(t) &= A_3x_1(t) + A_2x_2(t) + B_2u(t), \\ y(t) &= C_1x_1(t) + C_2x_2(t), \\ z(t) &= D_1x_1(t) + D_2x_2(t) + Eu(t). \end{aligned} \tag{9.2}$$

The state space $\mathcal{X}_2$ of Σ_2 is assumed to be n_2-dimensional. The output space $\mathcal{Z}$ is r-dimensional. The disturbance enters the plant via the term A_3x_1 in the state equation and via the terms C_1x_1 and D_1x_1 in the output equations. We allow for a direct feedthrough term Eu from the control input to the to-be-controlled variable. Such a term is omitted in the equation for y, because it would have been inconsequential for the present problem.

It is convenient to combine Σ_1 and Σ_2 according to equations (9.1) and (9.2) to one system Σ with state variable $x = (x_1^{\mathrm{T}}, x_2^{\mathrm{T}})^{\mathrm{T}}$ and coefficient matrices

$$\begin{aligned} &A := \begin{pmatrix} A_1 & 0 \\ A_3 & A_2 \end{pmatrix}, \quad B := \begin{pmatrix} 0 \\ B_2 \end{pmatrix}, \quad C := \begin{pmatrix} C_1 & C_2 \end{pmatrix}, \\ &D := \begin{pmatrix} D_1 & D_2 \end{pmatrix}, \end{aligned} \tag{9.3}$$

so that we have the following equations for Σ :

$$\begin{aligned} \dot{x}(t) &= Ax(t) + Bu(t), \\ y(t) &= Cx(t), \\ z(t) &= Dx(t) + Eu(t). \end{aligned} \tag{9.4}$$

Before investigating the existence of a controller with the desired properties, we describe what systems already satisfy the regulation property and the internal stability condition. We define these properties for systems without control inputs, equivalently, for $u = 0$. We say that Σ is *endostable* if Σ_2 is internally stable. This means that for $x_1(0) = 0$ and arbitrary $x_2(0)$, the state $x_2(t)$ is $\mathbb{C}_g$-stable. We say that Σ is (*output*) *regulated* if $z(t)$ is $\mathbb{C}_g$-stable for every initial state of Σ. These properties can be expressed in terms of the coefficient matrices. Obviously, Σ is endostable if and only if $\sigma(A_2) \subset \mathbb{C}_g$. For output regulation, we have the following result.

Lemma 9.1 *Assume that $\sigma(A_2) \subset \mathbb{C}_g$. Then the system Σ with input $u = 0$ is output regulated if the equations*

$$\begin{aligned} TA_1 - A_2T &= A_3, \\ D_2T + D_1 &= 0 \end{aligned} \tag{9.5}$$

in T are solvable. If A_1 is antistable, this condition is also necessary.

If A_1 is not antistable then we can delete the stable part since it does not effect regulation which is only an asymptotic condition. In this way we can basically reduce the general problem to the case when A_1 is antistable.

Proof : The main idea is that for large t, the state x_2 of the plant is close to Tx_1, for a suitable linear map T. So we introduce the variable $v := x_2 - Tx_1$, where we specify

T later on. In a straightforward calculation, one derives from the equations (9.1) and (9.2) that

$$\begin{aligned} \dot{v}(t) &= A_2 v(t) + (A_2 T - T A_1 + A_3) x_1(t)\,, \\ z(t) &= D_2 v(t) + (D_1 + D_2 T) x_1(t)\,. \end{aligned} \tag{9.6}$$

Now assume that T is a solution of the equations (9.5). Then the equations (9.6) reduce to

$$\begin{aligned} \dot{v}(t) &= A_2 v(t), \\ z(t) &= D_2 v(t). \end{aligned}$$

Since A_2 is $\mathbb{C}_g$-stable, theorem 3.23 implies that $z(t)$ is $\mathbb{C}_g$-stable.

Conversely, assume that A_1 is antistable and that the system Σ is regulated and endostable. Then it follows from Sylvester's Theorem (see section 9.3) that there exists a (unique) matrix T satisfying the first equation of (9.5). Substituting this into (9.6), we find that $v(t)$ is $\mathbb{C}_g$-stable. Since $z(t)$ is also $\mathbb{C}_g$-stable, this implies that $(D_2 T + D_1) x_1(t)$ is $\mathbb{C}_g$-stable. However, because x_1 is antistable for any initial condition, we must have that $D_2 T + D_1 = 0$. ■

Now we want to solve the regulator problem by constructing a controller such that the closed loop system satisfies the conditions of lemma 9.1. As usual, the controller Γ will be of the form

$$\begin{aligned} \dot{w}(t) &= K w(t) + L y(t), \\ u(t) &= M w(t) + N y(t). \end{aligned} \tag{9.7}$$

The closed loop system will be equal to the cascade connection Σ_{cl} of Σ_1 and $\Sigma_{2,cl}$, where $\Sigma_{2,cl}$ is the feedback interconnection of Σ_2 and Γ, given by

$$\begin{aligned} \dot{x}_{2e}(t) &= A_{2e} x_{2e}(t) + A_{3e} x_1(t), \\ z(t) &= D_{1e} x_1(t) + D_{2e} x_{2e}(t), \end{aligned}$$

where

$$\begin{aligned} A_{2e} &:= \begin{pmatrix} A_2 + B_2 N C_2 & B_2 M \\ L C_2 & K \end{pmatrix}, \quad & A_{3e} &:= \begin{pmatrix} A_3 + B_2 N C_1 \\ L C_1 \end{pmatrix}, \\ D_{2e} &:= \begin{pmatrix} D_2 + E N C_2 & E M \end{pmatrix}, \quad & D_{1e} &:= D_1 + E N C_1. \end{aligned} \tag{9.8}$$

We call Γ a *regulator* if Σ_{cl} is endostable and output regulated. The problem of finding a regulator will be called the *regulator problem.* It follows from lemma 9.1 that the regulator problem can be solved by finding $\Gamma = (K, L, M, N)$ such that A_{2e} is stable and the equations

$$T_e A_1 - A_{2e} T_e = A_{3e}\,, \quad D_{2e} T_e + D_{1e} = 0 \tag{9.9}$$

have a solution T_e. The existence of a solution T_e is necessary for the existence of a regulator if A_1 is antistable. In order to be able to solve this problem we shall make two assumptions:

- Σ_2 is stabilizable with u as input, i.e., (A_2, B_2) is stabilizable.
- Σ is detectable with y as output, i.e., (C, A) is detectable

Obviously, the stabilizability of Σ_2 is necessary for the existence of a regulator. Also the detectability of Σ_2 is necessary. However, here we impose a more restrictive condition on the system, viz. the detectability of the total system Σ. Note that there exists a standard reduction technique to solve the problem when Σ_2 is detectable but Σ is not detectable. In an appropriate manner, this technique deletes the undetectable modes and only requires us to design a suitable regulator for the remaining system which satisfies the stronger condition of detectability we impose in this chapter. For details we refer to [44, 159]. Then we have the following result:

Theorem 9.2 *Assume that (A_2, B_2) is stabilizable and (C, A) is detectable. Then there exists a regulator if the equations*

$$\begin{aligned} TA_1 - A_2T - B_2V &= A_3, \\ D_1 + D_2T + EV &= 0 \end{aligned} \tag{9.10}$$

have a solution (T, V). If A_1 is antistable, the solvability of (9.10) is necessary for the existence of a regulator. Specifically, if $G : \mathcal{Y} \to \mathcal{X}$ is such that $\sigma(A + GC) \subset \mathbb{C}_g$, $F_2 : \mathcal{X} \to \mathcal{U}$ is such that $\sigma(A_2 + B_2F_2) \subset \mathbb{C}_g$ and if $F_1 := -F_2T + V$, $F := (F_1 \;\; F_2)$, where (T, V) is a solution of (9.10), then a regulator is given by

$$\begin{aligned} \dot{w}(t) &= (A + GC + BF)w(t) - Gy(t), \\ u(t) &= Fw(t). \end{aligned} \tag{9.11}$$

Proof : Assume that A_1 is antistable and that a regulator exists. This regulator satisfies (9.9) for some T_e. We decompose T_e into $T_e = (T^{\mathrm{T}}, U^{\mathrm{T}})^{\mathrm{T}}$ and substitute (9.8) into (9.9). The first block row of the first of the resulting equations reads:

$$TA_1 - (A_2 + B_2NC_2)T - B_2MU = A_3 + B_2NC_1$$

and the second equation reads:

$$D_1 + (D_2 + ENC_2)T + EMU + ENC_1 = 0.$$

These relations show that $(T, NC_2T + MU + NC_1)$ is a solution of (9.10).

Conversely, assume that (T, V) satisfies (9.10). Define a controller Γ by

$$(K, L, M, N) := (A + GC + BF, -G, F, 0)$$

i.e., by (9.11) where F and G satisfy the conditions of theorem 9.2. We claim that Γ is a regulator. To show this, we have to prove that the resulting extended system is endostable, and that we have the regulation property. In order to prove endostability, we introduce $r := w - x$ and notice that x and r satisfy

$$\begin{aligned} \dot{x}(t) &= (A + BF)x(t) + BFr(t)\,, \\ \dot{r}(t) &= (A + GC)r(t)\,. \end{aligned}$$

Obviously, $r(t)$ is $\mathbb{C}_g$-stable. If $x_1(0) = 0$ (and hence $x_1(t) = 0$ for all t), the first equation reduces to

$$\dot{x}_2(t) = (A_2 + B_2F_2)x_2(t) + B_2Fr(t).$$

Then we also have $x_2(t)$ is $\mathbb{C}_g$-stable (compare theorem 2.7). Next we verify that Σ_e is output regulated. To this extent, we define $U := (I, T^{\mathrm{T}})^{\mathrm{T}}$ and we claim that $T_e := (T^{\mathrm{T}}, U^{\mathrm{T}})^{\mathrm{T}}$ satisfies (9.9). To show this, we substitute this and (9.8) into (9.9). Then for the first equation of (9.9) we have to prove that

$$\begin{pmatrix} T \\ U \end{pmatrix} A_1 - \begin{pmatrix} A_2 & B_2F \\ -GC_2 & A+GC+BF \end{pmatrix} \begin{pmatrix} T \\ U \end{pmatrix} = \begin{pmatrix} A_3 \\ -GC_1 \end{pmatrix}.$$

We notice that $FU = V$. Hence the first block equation is exactly the first equation of (9.10), viz. $TA_1 - A_2T - BV = A_3$. The second equation takes some more effort. It reads:

$$G(C_1 + C_2T - CU) + UA_1 - AU - BV = 0.$$

The expression between parentheses equals zero because of the definition of U. The remaining terms are decomposed according to (9.3):

$$\begin{pmatrix} I \\ T \end{pmatrix} A_1 - \begin{pmatrix} A_1 & 0 \\ A_3 & A_2 \end{pmatrix} \begin{pmatrix} I \\ T \end{pmatrix} - \begin{pmatrix} 0 \\ B_2 \end{pmatrix} V = \begin{pmatrix} A_1 - A_1 \\ TA_1 - A_3 - A_2T - B_2V \end{pmatrix} = 0,$$

where we have used the first equation of (9.10). This shows that the first equation of (9.9) is satisfied.

Next we consider the second equation of (9.9). It reads $D_1 + D_2T + EV = 0$, which is the same as the second equation of (9.10) and hence it is immediately clear that this is also satisfied. ■

9.2 Well-posedness of the regulator problem

A mathematical problem is called *well posed* if it is solvable and it remains solvable after a small perturbation of the data of the problem. The equation $x^2 + y^2 - ay + b = 0$, for example, is solvable (in $\mathbb{R}^2$) for $a = 2$ and $b = 1$, but it is not well posed for these values of a and b because the solvability is lost when b is replaced by $1 + \varepsilon$ for arbitrary $\varepsilon > 0$. The investigation of the well-posedness is easy for linear equations. As a matter of fact, we have

Lemma 9.3 *Let $\mathcal{X}$ and $\mathcal{Y}$ be finite-dimensional linear spaces and let $A : \mathcal{X} \to \mathcal{Y}$ be a linear map and $b \in \mathcal{Y}$. Then the equation $Ax = b$ in the variable x is well posed if and only if A is surjective.*

Proof : If A is surjective, any matrix representation of A will have a nonzero subdeterminant of dimension equal to the number of rows. Since this determinant is a continuous function of the entries of the matrix (in fact, a polynomial) it follows that it will remain nonzero when the entries are perturbed a little bit. Hence, A will remain surjective, after a small perturbation. Therefore, the equation $Ax = b$ remains solvable for small perturbations of A and b. Obviously, b can be perturbed in an arbitrary way and not just locally.

Conversely, if A is not surjective, im A is a proper subspace of $\mathcal{Y}$. The equation $Ax = b$ is solvable if and only if $b \in \operatorname{im} A$. However, b cannot be an interior point of $\operatorname{im} A$, since $\operatorname{im} A$ contains no interior points. Consequently, an arbitrary small perturbation of b may take it out of $\operatorname{im} A$ and hence destroy the solvability of the equation. ■

Remark 9.4 In concrete situations, it is of importance to specify more precisely what the 'data' of the problem is. Sometimes not all of the entries in a matrix are considered data and subject to perturbations. For example, if A is a companion matrix, only the last row is considered data. The remaining entries consist of 'hard' zeros and ones. It follows from the above proof that the necessity of the surjectivity of A still holds if only the vector b is considered data.

Remark 9.5 Usually, and in particular in the case of lemma 9.3, the well-posedness problem is easier to solve than the original equation. It is easier to verify the surjectivity of a map A than to examine the solvability of the equation $Ax = b$.

We want to apply the above result to the regulator problem. For the solvability of the regulator problem, a number of conditions are imposed. In the first place, it is assumed that Σ_2 is stabilizable and Σ detectable. It is not difficult to see that these properties are invariant under small perturbations. For instance, if (A, B) is stabilizable and F is a stabilizing feedback, then $A + BF$ is a stability matrix. Since the eigenvalues depend continuously on the matrix, $\bar{A} + \bar{B}F$ is also a stability matrix if $\bar{A}$ and $\bar{B}$ are close to A and B. The important part to check is the well-posedness of the matrix linear equation (9.10). For this, we apply the previous theorem. We find that the equation:

$$\begin{aligned} TA_1 - A_2T - B_2V &= A_3, \\ D_2T + EV &= -D_1 \end{aligned}$$

in the variables T and V is well posed if and only if the map

$$(T, V) \mapsto (TA_1 - A_2T - B_2V, D_2T + EV)$$

is surjective. In order to check this condition, one can give a matrix representation of this map using tensor products. We will however follow a different procedure, based on general considerations on the solvability of matrix equations. The advantage of the condition thus obtained will be that it can be interpreted in systemic terms, specifically, in terms of the zeros of a system.

9.3 Linear matrix equations

The subject of this section is the solvability of linear matrix equations of the form

$$\sum_{i=1}^{k} A_i X B_i = C, \tag{9.12}$$

where A_i, B_i and C are given matrices and X is unknown. We distinguish between universal and individual solvability of (9.12). We say that (9.12) is *universally* solvable if the equation has a solution for every C. Universal solvability thus is a condition on the matrices A_i and B_i. If we want to stress that the equation is solvable for the particular C given, we say that (9.12) is *individually* solvable. Conditions for solvability can be given by viewing the left-hand side of (9.12) as a linear map L acting on the matrix X. Then (9.12) is individually solvable if and only if $C \in \operatorname{im} L$, and (9.12) is universally solvable if and only if L is surjective. One can give explicit conditions for these properties using tensor (or Kronecker) products, but this will give rise to huge matrices and little insight. Rather, we would like to have results in the spirit of *Sylvester's theorem: the equation $AX - XB = C$, where A and B are square matrices, is universally solvable if and only if $\sigma(A) \cap \sigma(B) = \varnothing$*. It does not seem possible to obtain a similar result for the general equation (9.12). However, if we restrict ourselves to the case where the matrices B_i are of the form $B_i = q_i(B)$ for given polynomials q_i and a fixed matrix B, we can derive the following:

Theorem 9.6 *Let $A_i \in \mathbb{R}^{n\times m}$, $B \in \mathbb{R}^{p\times p}$, and let $q_i(s)$ be polynomials for $i = 1, \ldots, k$. Let*

$$A(s) := \sum_{i=1}^{k} A_i q_i(s). \tag{9.13}$$

Then the equation

$$\sum_{i=1}^{k} A_i X q_i(B) = C, \tag{9.14}$$

is universally solvable if and only if $\operatorname{rank} A(\lambda) = n$ *for all* $\lambda \in \sigma(B)$.

It is straightforward that Sylvester's theorem is a special case of this result.

Proof : ('only if':) Suppose that $\operatorname{rank} A(\lambda) < n$ for some $\lambda \in \sigma(B)$. Choose nonzero vectors v and w such that $Bv = \lambda v$ and $w^{\mathrm{T}} A(\lambda) = 0$. Then we have for any matrix X:

$$w^{\mathrm{T}} \sum A_i X q_i(B) v = w^{\mathrm{T}} \sum A_i X q_i(\lambda) v = w^{\mathrm{T}} A(\lambda) X v = 0.$$

Hence, if $w^{\mathrm{T}} C v \neq 0$ (e.g. if $C = wv^{\mathrm{T}}$), (9.14) does not have a solution.

('if':) The polynomial matrix $A(s)$ is obviously right invertible as a rational matrix. Hence there exists a polynomial matrix $D(s)$ and a scalar polynomial $a(s)$ such that $A(s)D(s) = a(s)I$, where $a(s)$ is the product of the invariant factors of $A(s)$. The assumption of the theorem implies that $a(\lambda) \neq 0$ for $\lambda \in \sigma(B)$, hence, by the spectral mapping theorem (see (2.11)), that $a(B)$ is nonsingular. Defining $C_1 := C(a(B))^{-1}$ and $E(s) := D(s)C_1$ we find

$$\sum_{i=1}^{k} A_i E(s) q_i(s) = A(s)E(s) = C_1 a(s).$$

Next we apply right substitution of $s = B$ into this equation and use theorem 7.6. This yields $\sum_{i=1}^{k} A_i E_r(B) q_i(B) = C_1 a(B) = C$, which shows that $X := E_r(B)$ (the index r indicates right substitution) is a solution of (9.14). ■

Next we investigate the individual solvability of equation (9.14). For the equations $AX - XB = C$ and $AX - YB = C$, such conditions were given by Roth in 1952, viz.

Theorem 9.7

(i) *Let A, B and C be matrices such that the equation $AX - XB = C$ is defined. Then this equation has a solution if and only if the matrices*

$$\begin{pmatrix} A & 0 \\ 0 & B \end{pmatrix} \quad \textit{and} \quad \begin{pmatrix} A & C \\ 0 & B \end{pmatrix}$$

are similar.

(ii) *If A, B and C are polynomial matrices of dimensions such that the equations $AX - YB = C$ makes sense, then this equation has a (polynomial matrix) solution if and only if*

$$\begin{pmatrix} A & 0 \\ 0 & B \end{pmatrix} \quad \textit{and} \quad \begin{pmatrix} A & C \\ 0 & B \end{pmatrix}$$

are unimodularly equivalent.

An elegant proof of these theorem can be found in [80, Theorem 4.4.22]. We want to generalize this result to the equation (9.14). A generalization in terms of similarity seems difficult. However, it is known that two matrices A and $\bar{A}$ are similar if and only if the polynomial matrices $sI - A$ and $sI - \bar{A}$ are unimodularly equivalent. Hence, according to Roth, the equation $AX - XB = C$ is solvable if and only if the polynomial matrices

$$\begin{pmatrix} sI - A & 0 \\ 0 & sI - B \end{pmatrix} \quad \text{and} \quad \begin{pmatrix} sI - A & C \\ 0 & sI - B \end{pmatrix}$$

are unimodularly equivalent. This formulation has a direct generalization:

Theorem 9.8 *Let A_i, B, $q_i(s)$ and $A(s)$ be as in theorem 9.6. Then equation (9.14) is (individually) solvable if and only if the matrices*

$$\begin{pmatrix} A(s) & 0 \\ 0 & sI - B \end{pmatrix} \quad \text{and} \quad \begin{pmatrix} A(s) & C \\ 0 & sI - B \end{pmatrix} \tag{9.15}$$

are unimodularly equivalent.

Proof : First we note that (9.14) has a solution if and only if the equation

$$A(s)P(s) + Q(s)(sI - B) = C \tag{9.16}$$

in the polynomial matrices $P(s)$, $Q(s)$ has a solution. In fact, if (9.16) has a solution, we apply right substitution of $s = B$ into this equation, yielding (9.14) with $X = P_r(B)$ (the index r indicates right substitution). Conversely, let X be a solution of (9.14). Then we write:

$$C - A(s)X = \sum_i A_i X(q_i(B) - q_i(s)I) = \sum_i A_i V_i(s)(sI - B)$$

for certain polynomial matrices $V_i(s)$. Hence $(P(s), Q(s)) := (X, \sum A_i V_i(s))$ is a solution of (9.16). Next we notice that equation (9.16) is an equation of the type given in theorem 9.7 (ii). Hence (9.16) is solvable if and only if the two matrices in (9.15) are unimodularly equivalent. ■

9.4 The regulator problem revisited

In theorem 9.2, we saw that subject to the assumptions that Σ_2 is stabilizable and Σ is detectable, a sufficient condition for the existence of a regulator is the solvability of the matrix equation (9.10). In the present section, we intend to apply the results of section 9.3. To this extent, we rewrite (9.10) to

$$\begin{pmatrix} -A_2 & -B_2 \\ D_2 & E \end{pmatrix} \begin{pmatrix} T \\ V \end{pmatrix} + \begin{pmatrix} I & 0 \\ 0 & 0 \end{pmatrix} \begin{pmatrix} T \\ V \end{pmatrix} A_1 = \begin{pmatrix} A_3 \\ -D_1 \end{pmatrix}. \tag{9.17}$$

This is an equation of the form (9.14). The solvability of this equation is not affected if the right-hand side is multiplied by -1. After this, the matrices defined in theorem 9.6 reduce to

$$A(s) = \begin{pmatrix} sI - A_2 & -B_2 \\ D_2 & E \end{pmatrix}, \quad B = A_1, \quad C = \begin{pmatrix} -A_3 \\ D_1 \end{pmatrix}. \tag{9.18}$$

Hence according to theorem 9.8, equation (9.17) has a solution if and only if the matrices

$$P(s) := \begin{pmatrix} sI - A_1 & 0 & 0 \\ -A_3 & sI - A_2 & -B_2 \\ D_1 & D_2 & E \end{pmatrix}$$

and

$$P_{\text{disc}}(s) := \begin{pmatrix} sI - A_1 & 0 & 0 \\ 0 & sI - A_2 & -B_2 \\ 0 & D_2 & E \end{pmatrix}$$

are unimodularly equivalent. Here we have applied an obvious row and column operation. Note that $P(s)$ is the system matrix as defined in section 7.2. Also $P_{\text{disc}}(s)$ can be interpreted as a system matrix, viz. the system matrix of the disconnected system Σ_{disc} obtained from Σ by disconnecting Σ_1 and Σ_2, i.e., by setting $A_3 = 0$, $D_1 = 0$. Recall that two polynomial matrices are unimodularly equivalent if and only if they have the same invariant factors (corollary 7.3). The invariant factors of the system matrix are defined to be the transmission polynomials of the system. Hence we have found the following:

Theorem 9.9 *Assume that (A_2, B_2) is stabilizable and that (C, A) is detectable. Then there exists a regulator for Σ if Σ and Σ_{disc} have the same transmission polynomials. If A_1 is antistable then this condition is also necessary.*

Now we investigate the well-posedness of the regulator problem. As was shown in section 9.2, this is guaranteed if the equation (9.10) or, equivalently, (9.17) is well posed (assuming that Σ_2 is stabilizable and Σ is detectable. Recall that these conditions are well posed). Hence, applying theorem 9.6, we find the following result:

Theorem 9.10 *Assume that (A_2, B_2) is stabilizable and (C, A) is detectable. Then the regulator problem is well posed if*

$$\operatorname{rank} \begin{pmatrix} \lambda I - A_2 & -B_2 \\ D_2 & E \end{pmatrix} = n_2 + r$$

(i.e., of full row rank) for every $\lambda \in \sigma(A_1)$. If A_1 is antistable this condition is also necessary.

In system-theoretic terms this condition requires that Σ_2 is *right-invertible and its zeros do not coincide with poles of* Σ_1 (for the notion of right-invertibility we refer to chapter 8). The necessary and sufficient conditions of this section are easily extended to the case where A_1 is not antistable. We omit the details. (See also exercise 9.2)

9.5 Exercises

9.1 Consider the system given as the interconnection of the exosystem

$$\begin{aligned} \dot{x}_1(t) &= -\omega x_2(t), \\ \dot{x}_2(t) &= \omega x_1(t), \end{aligned}$$

and the control system

$$\begin{aligned}
\dot{x}_3(t) &= -x_3(t) + x_5(t) + a x_1(t),\\
\dot{x}_4(t) &= x_5(t),\\
\dot{x}_5(t) &= x_3(t) + 3x_4(t) + 2x_5(t) + u(t),\\
y(t) &= \begin{pmatrix} x_2(t) \\ x_3(t) \\ x_4(t) \\ x_5(t) \end{pmatrix},\\
z(t) &= x_3(t) - x_1(t),
\end{aligned}$$

Assume that $\mathbb{C}_g = \mathbb{C}^-$.

a. For which values of a and ω is the regulator problem well posed?

b. Construct a regulator.

9.2 Consider the regulator problem without the assumption that A_1 is antistable. Show that if (A_2, B_2) is stabilizable and (C, A) is detectable, then the regulator problem is well posed if and only if

$$\operatorname{rank}\begin{pmatrix} \lambda I - A_2 & -B_2 \\ D_2 & E \end{pmatrix} = n_2.$$

for every $\lambda \in \sigma(A_1) \cap \mathbb{C}_b$.

9.3 Let $A \in \mathbb{R}^{n\times n}$ and $F(s)$ be an $n \times m$ polynomial matrix. Show that

$$\operatorname{rank}(\lambda I - A \quad F(\lambda)) = n$$

for all $\lambda \in \mathbb{C}$ if and only if $(A, F_\ell(A))$ is controllable. Here, $F_\ell(A)$ denotes the result of substituting $s = A$ into $F(s)$ from the left.

9.4 Show that (A, B) is controllable if and only if for every $n \times n$ matrix C there exist matrices X and U (of suitable dimensions) such that $XA - AX + BU = C$.

9.5 In this problem we consider the regulator problem. Let the exosystem be given by the equation

$$\Sigma_1 : \dot{x}_1 = \alpha_1 x_1,$$

with $x_1(t) \in \mathbb{R}, \alpha_1 \geqslant 0$. In addition, let the plant be given by

$$\Sigma_2 : \begin{array}{rl} \dot{x}_2 = & a_3 x_1 + A_2 x_2 + B_2 u, \\ y = z = & d_1 x_1 + D_2 x_2, \end{array}$$

with $x_2(t) \in \mathbb{R}^{n_2}$ (we write a_3, d_1 instead of A_3, D_1 to stress that these matrices consist of one column). Assume that (A_2, B_2) is $\mathbb{C}^-$-stabilizable and that (D, A) is $\mathbb{C}^-$-detectable where

$$A := \begin{pmatrix} \alpha_1 & 0 \\ a_3 & A_2 \end{pmatrix}; \; D := (d_1 \;\; D_2).$$

a. Show that if

$$\Gamma : \begin{array}{l} \dot{w} = Kw + Ly, \\ u = Mw + Ny, \end{array}$$

is a regulator, then there exists vectors t_0 and u_0 such that

$$\begin{array}{l} (\alpha_1 I - A_2)t_0 - B_2 M u_0 = a_3, \\ K u_0 = \alpha_1 u_0, \\ D_2 t_0 + d_1 = 0. \end{array}$$

b. Show that if Γ is a regulator then α_1 is an eigenvalue of K.

c. Now let the exosystem be given by

$$\Sigma_1 : \dot{x}_1 = A_1 x_1,$$

with $x_1(t) \in \mathbb{R}^{n_1}$, A_1 anti-stable. In addition, assume that the plant is given by

$$\Sigma_2 : \begin{array}{lll} \dot{x}_2 = & & A_2 x_2 + A_3 x_1 + B_2 u, \\ y & = z = & D_1 x_1 + D_2 x_2. \end{array}$$

Again assume that (A_2, B_2) is $\mathbb{C}^-$-stabilizable and that (D, A) is $\mathbb{C}^-$-detectable, with D and A defined as usual. Use the ideas from a) and b) to show that if Γ is a regulator, then we have $\sigma(A_1) \subset \sigma(K)$.

The phenomenon illustrated in this problem is an example of the famous *internal model principle*: the set of eigenvalues $\sigma(A_1)$ of the exosystem is contained in the set of eigenvalues $\sigma(K)$ of the regulator: in a sense, the regulator contains an internal model of the exosystem.

9.6 Notes and references

The regulator problem has been studied by many people. See for instance Davison in [35], Davison and Goldenberg in [36], Francis in [44], Francis and Wonham in [48] and Desoer and Wang in [38]. The theory has also been extended to for instance nonlinear systems by Isidori and Byrnes in [87]. Many results have recently been collected by Saberi, Stoorvogel and Sannuti in the book [159]. The regulator equations (9.9) were originally introduced by Francis in [45].

The techniques presented in section 9.3 can be found in the work of Hautus [73, 75]. The Sylvester equation is actually quite an old subject, and was originally introduced by Sylvester in [192].

Well-posedness was studied by Wonham in section 8.3 of [223] and by Hautus in [73]. Note that this basically still requires that if the system is perturbed then we need a new controller. In structural stability , we are looking for one controller which stabilizes a neighborhood of the given plant. This problem was studied in many variations and has been studied by Francis, Sebakhy and Wonham in [47], by Davison and Goldenberg in [36], by Desoer and Wang in [38], by Pearson, Shields and Staats in [142] and by Francis and Wonham in [48]. More recently the known results were extended by Saberi, Stoorvogel and Sannuti in the book [159].

The internal model principle studied by Wonham in [223] and by Francis and Wonham in [48] was only for the case that the to be regulated signal is equal to the measurement signal. Extensions to the general case can be found in the book [159] by Saberi, Stoorvogel and Sannuti. Note that the internal model principle is unrelated to well-posedness, structural stability (which is sometimes also referred to as robust regulation) which is sometimes alluded to in the literature. It is a basic property resulting directly from the fact that we achieve regulation.

Chapter 10

Linear quadratic optimal control

In the previous chapters we have been concerned with control problems that require the controlled system to satisfy specific, qualitative, properties, such as internal stability, the property of being disturbance decoupled, external stability, the regulation property, etc.. In the present chapter we will take into account *quantitative* aspects. Given a control system, we will express the performance of the controlled system in terms of a cost functional. The control problem will then be to find all optimal controllers i.e, all controllers that minimize the cost functional. Such controllers lead to a controlled system with optimal performance.

10.1 The linear quadratic regulator problem

Consider the control system Σ given by the equations

$$\begin{aligned} \dot{x}(t) &= Ax(t) + Bu(t), \\ z(t) &= Cx(t) + Du(t). \end{aligned} \tag{10.1}$$

Here, u is the control input, and z is the output to be controlled. We assume that u takes its values in $\mathcal{U}$, which we identify with $\mathbb{R}^m$, and z takes its values in $\mathcal{Z}$, to be identified with $\mathbb{R}^p$. The state x is assumed to take its values in $\mathcal{X}$, to be identified with $\mathbb{R}^n$.

We now explain what we mean by a disturbance. By a disturbance we mean the occurrence of an 'initial state' $x(0) = x_0$. This could alternatively be modelled by replacing the differential equation of Σ by $\dot{x} = Ax + Bu + \delta x_0$, where δx_0 is interpreted as a disturbance input.

Suppose that it is desired to keep all components of the output $z(t)$ as small as possible, for all t. In the ideal situation, the uncontrolled system ($u = 0$) evolves along the stationary solution $x = 0$. Of course, the output $z(t)$ will then also be equal to zero for all t. Suppose now that at time $t = 0$ the state of the system is perturbed to, say, $x(0) = x_0$. After the occurrence of this impulsive disturbance, the uncontrolled system will evolve along a state trajectory unequal to the stationary zero solution, and we will get $z(t) = Ce^{At}x_0$. If, however, from time $t = 0$ on, we apply an input function u, then for $t \geqslant 0$ the corresponding output becomes equal to $z(t) = Cx(t) + Du(t)$. Keeping in mind that we want the output $z(t)$ to be as small as possible for all $t \geqslant 0$, we measure its size by the quadratic cost functional

$$J(x_0, u) = \int_0^\infty \|z(t)\|^2 \, dt, \tag{10.2}$$

where $\|\cdot\|$ denotes the Euclidean norm. Our desire to keep the values of the output as small as possible can be expressed by requiring this integral to be as small as possible by suitable choice of input function u. In this way we arrive at the *linear quadratic regulator problem*:

Problem 10.1 *Consider the system* $\Sigma : \dot{x}(t) = Ax(t)+Bu(t)$, $z(t) = Cx(t)+Du(t)$. *Determine for every initial state* x_0 *an input* $u \in \mathbf{U}$ *(a space of functions* $[0, \infty) \to \mathcal{U}$*) such that*

$$J(x_0, u) := \int_0^\infty \|z(t)\|^2 \, dt \tag{10.3}$$

is minimal. Here $z(t)$ *denotes the output trajectory* $z_u(t, x_0)$ *of* Σ *corresponding to the initial state* x_0 *and input function* u.

Since the system is linear and the integrand in the cost functional is a quadratic function of z, the problem is called *linear quadratic*. Of course, $\|z\|^2 = x^{\mathrm{T}}C^{\mathrm{T}}Cx + 2u^{\mathrm{T}}D^{\mathrm{T}}Cx + u^{\mathrm{T}}D^{\mathrm{T}}Du$, so the integrand can also be considered as a quadratic function of (x, u).

In the problem formulation, we are not explicit about the exact nature of $\mathbf{U}$. In fact, like in the previous chapters, it is not very important what $\mathbf{U}$ is. One can take e.g. the space of piecewise continuous functions, the space of piecewise smooth (i.e. piecewise $\mathfrak{C}^\infty$) functions or the space of locally integrable functions. In the last case, one has to interpret the various formulas in a measure-theoretic sense (in particular, formulas containing derivatives have to be understood to hold 'almost everywhere').

Of course, the convergence of (10.3) is a point of concern. Therefore, one often considers the corresponding *finite horizon* problem in a preliminary investigation. In this problem, a final time T is given and one wants to minimize the integral

$$J(x_0, u, T) := \int_0^T \|z(t)\|^2 \, dt. \tag{10.4}$$

In contrast to this, the first problem above is sometimes called the *infinite horizon problem*.

An important issue is the convergence of the state. Obviously, convergence of the integral does not always imply the convergence to zero of the state. Therefore, distinction is made between the problem with zero and with free endpoint. Problem 10.1

as stated is referred to as the *problem with free endpoint*. If one restricts the inputs u in the problem to those for which the resulting state trajectory tends to zero, one speaks about the *problem with zero endpoint*. Specifically:

Problem 10.2 *In the situation of problem 10.1, determine for every initial state* x_0 *an input* $u \in \mathbf{U}$ *such that* $x(t) \to 0$ $(t \to \infty)$ *and such that under this condition,* $J(x_0, u)$ *is minimized.*

Remark 10.3 There is a special situation where the linear quadratic regulator problem (with free endpoint) is very easily solved. This is the case where $x_0 \in \mathcal{V}(\Sigma)$, the weakly unobservable subspace (see chapter 7). In fact, in this case, there exists a control $u \in \mathbf{U}$ such that the output is identically zero. Such a control is obviously optimal. This control can be given by the state feedback $u = Fx$, where $F := -(D^{\mathrm{T}}D)^{-1}D^{\mathrm{T}}C$. Likewise, the problem with zero endpoint is very easily solved if $x_0 \in \mathcal{V}_g(\Sigma)$, the stabilizable weakly unobservable subspace, with stability domain $\mathbb{C}_g$ equal to $\mathbb{C}^-$ (see exercise 7.16). In that case there exists a control such that the output is zero, and the state converges to zero as t tends to infinity. This control is optimal.

Various special cases of these problems have been considered in the literature. In the literature names have been associated to these special cases.

Definition 10.4 *Problems 10.1 and 10.2 are called* regular *if* D *is injective, equivalently,* $D^{\mathrm{T}}D > 0$. *The problem is said to be in* standard form *if* $C^{\mathrm{T}}D = 0$ *and* $D^{\mathrm{T}}D = I$.

In the standard case, the integrand in the cost functional reduces to $\|z\|^2 = x^{\mathrm{T}}C^{\mathrm{T}}Cx + u^{\mathrm{T}}u$. We often write $Q = C^{\mathrm{T}}C$.

One of the issues to be addressed is the existence of optimal controls. Since for any initial state and any input function u we have $J(x_0, u) \geqslant 0$, also $J^*(x_0) := \inf\{J(x_0, u) \mid u \in \mathbf{U}\} \geqslant 0$. Thus one is at least assured of the boundedness from below of the infimum of the integral. This, however, does not immediately imply the existence of an optimal control, since the infimum might not be attained. In general, solutions do not exist within the class of ordinary functions unless $D^{\mathrm{T}}D > 0$, i.e. the problem is regular. In the situation where this condition is not satisfied, one has to extend the set of admissible input functions to distributions. Henceforth, we restrict ourselves to the regular case.

The standard case is a special case, which is not essentially simpler than the general regular problem, but which gives rise to simpler formulas. The general regular problem can be reduced to the standard case by means of a feedback transformation, that is, by the introduction of a new control variable v which is related to u by $u = Fx + Gv$, where

$$F := -(D^{\mathrm{T}}D)^{-1}D^{\mathrm{T}}C \text{ and } G := -(D^{\mathrm{T}}D)^{-\frac{1}{2}}.$$

Indeed, after this preliminary feedback transformation and transformation of the input space, the original system is transformed to

$$\begin{aligned}\dot{x}(t) &= (A + BF)x(t) + BGv(t),\\ z(t) &= (C + DF)x(t) + DGv(t),\end{aligned}$$

which gives rise to a standard problem since $(DG)^{\mathrm{T}}(C + DF) = 0$ and $(DG)^{\mathrm{T}}DG = I$.

It is not very difficult to show that the infimum of J is a quadratic function of x_0, provided it is finite. We will not do this because it will be a result of our calculations anyway. However, the statement is made here because it will be a heuristic motivation of some of the steps in the treatment of the problems. It is very easy to see that, if $J^*(x_0) = \inf\{J(x_0, u) \mid u \in \mathbf{U}\}$ is finite, then $J^*(\lambda x_0) = \lambda^2 J^*(x_0)$, which makes it plausible that $J^*(x_0)$ is a quadratic function.

10.2 The finite-horizon problem

We start from the standard problem:

Problem 10.5 *Given the system $\dot{x}(t) = Ax(t) + Bu(t)$, a final time $T > 0$, and symmetric matrices N and Q such that $N \geqslant 0$ and $Q \geqslant 0$, determine for every initial state x_0 a piecewise continuous input function $u : [0, T] \to \mathcal{U}$ such that the integral*

$$J(x_0, u, T) := \int_0^T x(t)^{\mathrm{T}} Q x(t) + u(t)^{\mathrm{T}} u(t)\, dt + x(T)^{\mathrm{T}} N x(T) \tag{10.5}$$

is minimized.

In this problem, we have introduced a weight on the final state, using the matrix N. This generalization of the problem can be dealt with without any effort.

As remarked at the end of the previous section, the minimal value of J is expected to be a quadratic function of x_0, say $x_0^{\mathrm{T}} K x_0$. More generally, we expect that the minimum of the integral $\int_{t_0}^T x^{\mathrm{T}} Q x + u^{\mathrm{T}} u \,\mathrm{d}t$ is of the form $x_0^{\mathrm{T}} K(t_0) x_0$. We now assume that $K(t)$ is any symmetric-matrix-valued continuously differentiable function, defined on $[0, T]$ and we have in mind the interpretation of K given in the previous sentence. We consider the difference $J(x_0, u, T) - x_0^{\mathrm{T}} K(0) x_0$ for any admissible input u. Then we have

$$\begin{aligned} J(x_0, u, T) - x_0^{\mathrm{T}} K(0) x_0 &= \int_0^T x(t)^{\mathrm{T}} Q x(t) + u(t)^{\mathrm{T}} u(t) \,\mathrm{d}t \\ &\quad + \int_0^T \frac{\mathrm{d}}{\mathrm{d}t} x(t)^{\mathrm{T}} K(t) x(t) \,\mathrm{d}t + x(T)^{\mathrm{T}} (N - K(T)) x(T). \end{aligned}$$

We take the two integrals together and compute the second integral using the differential equation. The integrand will be (we omit the dependence on t):

$$\begin{aligned} x^{\mathrm{T}}Qx + u^{\mathrm{T}}u + \frac{\mathrm{d}}{\mathrm{d}t}x^{\mathrm{T}}Kx &= x^{\mathrm{T}}(Q + A^{\mathrm{T}}K + KA + \dot{K})x + 2u^{\mathrm{T}}B^{\mathrm{T}}Kx + u^{\mathrm{T}}u \\ &= x^{\mathrm{T}}KBB^{\mathrm{T}}Kx + 2u^{\mathrm{T}}B^{\mathrm{T}}Kx + u^{\mathrm{T}}u + x^{\mathrm{T}}Sx \\ &= \|u + B^{\mathrm{T}}Kx\|^2 + x^{\mathrm{T}}Sx, \end{aligned}$$

where

$$S := \dot{K} + A^{\mathrm{T}}K + KA - KBB^{\mathrm{T}}K + Q.$$

In the computations, we have completed the square. As a result, we get

$$\begin{aligned} J(x_0, u, T) = x_0^{\mathrm{T}}K(0)x_0 + \int_0^T [x(t)^{\mathrm{T}}S(t)x(t) + \|u(t) + B^{\mathrm{T}}K(t)x(t)\|^2]\,\mathrm{d}t \\ + x(T)^{\mathrm{T}}(N - K(T))x(T). \end{aligned} \quad (10.6)$$

This formula will play a crucial role in our further treatment. It will be referred to as the *completion of squares formula.* A primitive attempt to minimize the integral in (10.5) is to choose u such that the integrand is minimized for every value of t. This does not work, because the choice of u is going to affect the values of x for larger t, and it is difficult to predict what the total effect on J will be. A similar situation prevails when we try to minimize (10.6) in this way, unless we choose the matrix K in a particular way. In fact, if we are able to choose K such that $S(t)$ is identically zero and $K(T) = N$, the expression for J reduces to

$$J(x_0, u, T) = x_0^{\mathrm{T}}K(0)x_0 + \int_0^T \|u(t) + B^{\mathrm{T}}K(t)x(t)\|^2\,\mathrm{d}t. \quad (10.7)$$

It is obvious from this expression that $J(x_0, u, T) \geqslant x_0^{\mathrm{T}}K(0)x_0$ for all u and that equality is achieved if and only if $u(t) = -B^{\mathrm{T}}K(t)x(t)$ for all $t \in [0, T]$. Thus we have obtained the following result:

Theorem 10.6 *Let* $K : [0, T] \to \mathbb{R}^{n\times n}$ *be continuously differentiable and such that* $K(t)$ *is symmetric for every* $t \in [0, T]$*. If* K *is a solution of the* Riccati equation*:*

$$\dot{K}(t) = -A^{T}K(t) - K(t)A + K(t)BB^{T}K(t) - Q \quad (10.8)$$

with final condition $K(T) = N$*, then for each initial state* x_0 *we have*

$$J^*(x_0, T) := \inf\{J(x_0, u, T) \mid u \text{ piecewise continuous on } [0, T]\} = x_0^{T}K(0)x_0 \quad (10.9)$$

Furthermore, for a given initial state x_0*, an input function* u *is optimal, i.e.,* u *is piecewise continuous on* $[0, T]$ *and* $J(x_0, u, T) = J^*(x_0, T)$*, if and only if*

$$u(t) = -B^{T}K(t)x(t), \quad 0 \leqslant t \leqslant T. \quad (10.10)$$

Assuming for the moment that K satisfying the desired properties can be found, we obtain an optimal control as a time-varying feedback (see (10.10)). For a given initial state, we can construct an *open-loop* control function by substituting (10.10) into the differential equation. This yields a time varying linear autonomous equation: $\dot{x}(t) = (A - BB^{\mathrm{T}}K(t))x(t)$ with initial state $x(0) = x_0$. Such an initial value problem is well known to have a unique solution on any time interval. If the solution is denoted with $\xi(t)$ then the open-loop formula for u is $u(t) = -B^{\mathrm{T}}K(t)\xi(t)$. Thus we see that for each initial state x_0 there is a unique optimal control input.

It follows that the optimization problem is solved when the Riccati equation is solved. This equation is a nonlinear initial (or rather final) value problem. The existence of a solution of such a problem is not always guaranteed, as follows from the example $\dot{x} = x^2 + 1$, which has $x(t) = \tan(t + c)$ as a general solution. For no c, a solution exists over an interval larger than π. The general theory of ordinary differential equations only guarantees the existence of a *local* solution, defined over an interval whose length cannot be prescribed. Specifically:

Theorem 10.7 *Let $T > 0$, $x_0 \in \mathbb{R}^n$, and let $f : \mathbb{R}^n \times [0, T] \to \mathbb{R}^n$ be continuously differentiable. Then we have:*

(i) *There exists T_1 satisfying $0 < T_1 < T$ such that the differential equation with initial condition*

$$\dot{x}(t) = f(x(t), t), \quad x(0) = x_0 \tag{10.11}$$

has a solution on $[0, T_1]$.

(ii) *If (10.11) has a solution on some interval $[0, T_1]$, it is unique.*

(iii) *If the solution on each interval $[0, T_1]$, on which it exists, is bounded with a bound independent of T_1, then there exists a solution of (10.11) on $[0, T]$.*

We apply this theorem to the Riccati equation and with the time reversed. Note that this equation can be viewed as an equation of the type considered in theorem 10.7. According to theorem 10.7, there exists a unique solution K of

$$\dot{K}(t) = -A^{\mathrm{T}}K(t) - K(t)A + K(t)BB^{\mathrm{T}}K(t) - Q, \quad K(T) = N. \tag{10.12}$$

on some interval $[T_1, T]$. If we want to emphasize the dependence of this solution on the final time T we will write this solution as $K(t, T)$. Note also that, since N is symmetric, the matrix function $K^{\mathrm{T}}(t, T)$ will satisfy (10.12). By uniqueness, it follows that $K^{\mathrm{T}}(t, T) = K(t, T)$ for $T_1 \leqslant t \leqslant T$. Using the special interpretation of the solution of the Riccati equation as minimal value of the integral, one can actually derive global existence of the solution:

Theorem 10.8 *The Riccati equation with initial value, (10.12), has a unique solution on the interval $[0, T]$ for every symmetric N. This solution is symmetric.*

Proof : The only thing we have to prove is that $K(t, T)$ is uniformly bounded on each interval of its existence. Let $x_0 \in \mathcal{X}$ be arbitrary and suppose that $K(t, T)$ exists on $[T_1, T]$. Choose the input u identically zero on $[0, T]$. Take t_0 any element of $[T_1, T]$. By time invariance, the function $L(t) := K(t + t_0, T)$ is a solution of the Riccati equation on $[0, T - t_0]$. Hence by theorem 10.6, $x_0^{\mathrm{T}} L(0) x_0$ is the minimum of $J(x_0, u, T - t_0)$ over all u. In particular,

$$x_0^{\mathrm{T}} K(t_0, T) x_0 = x_0^{\mathrm{T}} L(0) x_0 \leqslant J(x_0, 0, T - t_0).$$

Clearly,

$$\begin{aligned} J(x_0, 0, T - t_0) &= \int_0^{T-t_0} x_0^{\mathrm{T}} e^{A^{\mathrm{T}} t} Q e^{At} x_0 \, \mathrm{d}t + x_0^{\mathrm{T}} e^{A^{\mathrm{T}}(T-t_0)} N e^{A(T-t_0)} x_0 \\ &\leqslant \int_0^{T} x_0^{\mathrm{T}} e^{A^{\mathrm{T}} t} Q e^{At} x_0 \, \mathrm{d}t + x_0^{\mathrm{T}} e^{A^{\mathrm{T}}(T-t_0)} N e^{A(T-t_0)} x_0. \end{aligned}$$

Now, $t \mapsto x_0^{\mathrm{T}} e^{A(T-t)} N e^{A(T-t)} x_0$ is a continuous function on the compact interval $[0, T]$ and is therefore bounded from above by a constant, say M. Thus we conclude that $x_0^{\mathrm{T}} K(t_0, T) x_0$ is bounded from above by a constant, independent of t_0 and T_1. This implies that the matrix function $K(t, T)$ itself is bounded with a bound independent of T_1 (see also exercise 10.2). Hence according to theorem 10.7, the Riccati equation has a solution on the total interval. ■

The dependence of the solution $K(t, T)$ on both t and T is inconvenient if we want to use the results of this section for the infinite horizon case. Fortunately, K is actually a function of one variable, specifically, $T - t$. In fact, if we define $P(t) := K(T - t, T)$, then P satisfies the initial-value problem:

$$\dot{P}(t) = A^{\mathrm{T}} P(t) + P(t) A - P(t) B B^{\mathrm{T}} P(t) + Q, \quad P(0) = N. \tag{10.13}$$

Obviously, the solution of this problem does not depend on T. We reformulate the results of this section in terms of $P(t)$:

Theorem 10.9 *Consider the system $\dot{x}(t) = Ax(t) + Bu(t)$, together with the cost functional*

$$J(x_0, u, T) := \int_0^T x(t)^T Q x(t) + u(t)^T u(t) \, dt + x(T)^T N x(T), \tag{10.14}$$

where $Q \geqslant 0$ and $N \geqslant 0$, and furthermore, $T > 0$. Then we have the following properties:

(*i*) *The Riccati equation with initial value*

$$\dot{P}(t) = A^T P(t) + P(t) A - P(t) B B^T P(t) + Q, \quad P(0) = N, \tag{10.15}$$

has a unique solution on $[0, \infty)$. This solution is symmetric and positive semidefinite for all $t \geqslant 0$.

(*ii*) *For each* x_0 *there is exactly one optimal input function, i.e., a piecewise continuous function* u^* *on* $[0, T]$ *such that* $J(x_0, u^*, T) = J^*(x_0, T)$. *This optimal input function* u^* *is generated by the time-varying feedback control law*

$$u(t) = -B^T P(T - t)x(t) \quad (0 \leqslant t \leqslant T). \tag{10.16}$$

(*iii*) *For each* x_0, *the minimal value of the cost functional equals*

$$J^*(x_0, T) = x_0^T P(T)x_0.$$

(*iv*) *If* $N = 0$, *then the function* $t \mapsto P(t)$ *is an increasing function in the sense that* $P(t) - P(s)$ *is positive semidefinite for* $t \geqslant s$.

The monotonicity statement on P is an easy consequence of the interpretation of P as the matrix representing the minimal cost of the performance criterion (see (iii)).

Remark 10.10 There is a second monotonicity statement we can make. If we consider two optimal control problems with the same system and the same Q matrix, but with two matrices N_1 and N_2, satisfying $N_1 \geqslant N_2$, then for every x_0, the function J associated with N_1 will be larger than the function J corresponding to N_2. There will therefore be a similar relation between the minimal costs. Hence $P_{N_1}(t) \geqslant P_{N_2}(t)$ for all $t \geqslant 0$.

10.3 The infinite-horizon problem, standard case

We consider the situation as described in theorem 10.9 with $N = 0$. An obvious conjecture is that $x_0^T P(T)x_0$ converges to the minimal cost of the infinite-horizon problem as $T \to \infty$. The convergence of $x_0^T P(T)x_0$ for all x_0 is equivalent to the convergence of the matrix $P(T)$ for $T \to \infty$ to some matrix P^- (see exercise 10.2). Such a convergence does not always take place. For instance, if we consider the system $\dot{x}(t) = 0$ with state space $\mathcal{X} = \mathbb{R}$, and $Q = 1$, we have $\dot{P}(t) = 1$, $P(0) = 0$, with solution $P(t) = t$. This is not surprising because in this case, if $x(0) = c \neq 0$, the integral to be minimized equals $\int_0^\infty x^2(t) + u^2(t)\,dt \geqslant \int_0^\infty c^2\,dt = \infty$ for all inputs u. In order to achieve convergence, we make the following assumption: for every x_0, there exists an input u for which the integral

$$J(x_0, u) := \int_0^\infty x(t)^T Qx(t) + u(t)^T u(t)\,dt \tag{10.17}$$

converges, i.e., for which the cost $J(x_0, u)$ is finite. Obviously, for the problem to make sense for all x_0, this condition is necessary. It is easily seen that the stabilizability of (A, B) is a sufficient condition for the above assumption to hold (not necessary, take e.g. $Q = 0$). Take an arbitrary initial state x_0 and assume that $\bar{u}$ is a function such that the integral (10.17) is finite. We have for every $T > 0$ that

$$x_0^T P(T)x_0 \leqslant J(x_0, \bar{u}, T) \leqslant J(x_0, \bar{u}),$$

which implies that for every x_0, the expression $x_0^{\mathrm{T}} P(T) x_0$ is bounded. This implies that $P(T)$ is bounded (see exercise 10.2). Since $P(T)$ is increasing with respect to T, it follows that $P^- := \lim_{T\to\infty} P(T)$ exists. Since P satisfies the differential equation (10.15), it follows that also $\dot{P}(t)$ has a limit as $t \to \infty$. It is easily seen that this latter limit must be zero. Hence $P = P^-$ satisfies the following equation:

$$A^{\mathrm{T}} P + PA - PBB^{\mathrm{T}} P + Q = 0. \tag{10.18}$$

This is is called the *Algebraic Riccati Equation* (ARE). Note that the solutions of this equation are exactly the constant solutions of the Riccati differential equation. The previous consideration shows that the ARE has a positive semidefinite solution P^-. The solution is not necessarily unique, not even with the extra condition that $P \geqslant 0$.

We would like to use the previous results in order to solve the infinite horizon problem. For this purpose, an obvious method would be to find a positive semidefinite solution P of (10.18) and to use the control $u = -B^{\mathrm{T}} P x$, obtained from (10.16) by taking the limit for $T \to \infty$. Here we expect the optimal cost to be $x_0^{\mathrm{T}} P x_0$. There are two points that have to be clarified, however. In the first place, because of the lack of uniqueness of P, we do not know which solution of (10.18) to take. In the second place, even if we have found the correct solution of the ARE, it is not obvious that the proposed control actually is optimal. Both points are already settled in the following lemma:

Lemma 10.11 *Suppose that for every x_0 there exists an input $u \in \mathbf{U}$ such that $J(x_0, u) < \infty$. Then we have:*

(i) *$P^- := \lim_{T\to\infty} P(T)$ exists, where P is the solution of Riccati differential equation (10.15) with $N = 0$.*

(ii) *P^- is the smallest real symmetric positive semidefinite solution of the ARE (10.18). Hence, for every real symmetric $P \geqslant 0$ satisfying ARE, we have $P \geqslant P^-$.*

(iii) *For every $u \in \mathbf{U}$, the following holds*

$$J(x_0, u) = x_0^T P^- x_0 + \int_0^\infty \|u(t) + B^T P^- x(t)\|^2 \, dt. \tag{10.19}$$

(iv) *For every x_0, $J^*(x_0) := \inf\{J(x_0, u) \mid u \in \mathbf{U}\} = x_0^T P^- x_0$. Furthermore, for every x_0, there is exactly one optimal input function, i.e., a function $u^* \in U$ such that $J(x_0, u^*) = J^*(x_0)$. This optimal input is generated by the time-invariant feedback law $u(t) = -B^T P^- x(t)$.*

Proof : We have seen already that (i) holds and that P^- satisfies ARE. Assume that $P^* \geqslant 0$ is any solution of the ARE. Then P^* is a constant solution of the Riccati differential equation. Also $P(t)$ is a solution and $P(0) = 0 \leqslant P^*$. Hence, $P(t) \leqslant P^*$ for all $t \geqslant 0$ (see remark 10.10), and consequently $P^- = \lim_{t\to\infty} P(t) \leqslant P^*$.

In order to prove (iii), we apply the completion of squares formula (10.6) with $K(t) := P(T-t)$, $N := 0$ and u arbitrary. Then

$$J(x_0, u, T) = x_0^{\mathrm{T}} P(T) x_0 + \int_0^T \|u(t) + B^{\mathrm{T}} P(T-t) x(t)\|^2 \, \mathrm{d}t.$$

Choose a fixed $T_0 > 0$. Then we have for $T > T_0$ that

$$\begin{aligned} &\int_0^T \|u(t) + B^{\mathrm{T}} P(T-t) x(t)\|^2 \, \mathrm{d}t \\ &\qquad \geqslant \int_0^{T_0} \|u(t) + B^{\mathrm{T}} P(T-t) x(t)\|^2 \, \mathrm{d}t \to \int_0^{T_0} \|u(t) + B^{\mathrm{T}} P^- x(t)\|^2 \, \mathrm{d}t \end{aligned}$$

for $T \to \infty$. Hence

$$\liminf_{T\to\infty} \int_0^T \|u(t) + B^{\mathrm{T}} P(T-t) x(t)\|^2 \, \mathrm{d}t \geqslant \int_0^{T_0} \|u(t) + B^{\mathrm{T}} P^- x(t)\|^2 \, \mathrm{d}t$$

for all T_0. This implies

$$\liminf_{T\to\infty} J(x_0, u, T) \geqslant x_0^{\mathrm{T}} P^- x_0 + \int_0^\infty \|u(t) + B^{\mathrm{T}} P^- x(t)\|^2 \, \mathrm{d}t.$$

Conversely, if in the completion of squares formula we take $K(t) = P^-$, $N = 0$, and u arbitrary, then

$$\begin{aligned} J(x_0, u, T) &= x_0^{\mathrm{T}} P^- x_0 + \int_0^T \|u(t) + B^{\mathrm{T}} P^- x(t)\|^2 \, \mathrm{d}t - x(T)^{\mathrm{T}} P^- x(T) \\ &\leqslant x_0^{\mathrm{T}} P^- x_0 + \int_0^\infty \|u(t) + B^{\mathrm{T}} P^- x(t)\|^2 \, \mathrm{d}t. \end{aligned} \tag{10.20}$$

Combination of these inequalities yields (10.19).

Finally, (iv) is a direct consequence of (iii). ■

Notice that the minimality condition uniquely determines the matrix P^-.

Remark 10.12 It follows from the previous proof that $x(T)^{\mathrm{T}} P^- x(T) \to 0$ whenever $J(x_0, u)$ is finite. In particular, in the case that P^- is positive definite, we infer that $x(t) \to 0$ for $t \to \infty$, whenever $J(x_0, u) < \infty$.

Although the lemma provides a solution to the infinite horizon problem, it is unsatisfactory because it starts from the assumption that for each initial state there is at least one control input such that the corresponding cost is finite. In the remainder of this section, we establish necessary and sufficient conditions, in terms of the matrices A, B and Q, under which the finite cost assumption holds.

Factorize $Q = C^{\mathrm{T}} C$ and introduce the output

$$z(t) = \begin{pmatrix} C \\ 0 \end{pmatrix} x(t) + \begin{pmatrix} 0 \\ I \end{pmatrix} u(t). \tag{10.21}$$

Our cost can then be written as $J(x_0, u) = \int_0^\infty \|z\|^2 \, dt$.

Now, from exercise 7.4, recall the concept of output stabilizability. In general, a system $\Sigma : \dot{x}(t) = Ax(t) + Bu(t), z(t) = Cx(t) + Du(t)$ is called *output stabilizable* if there exists a state feedback control law $u(t) = Fx(t)$ such that for each initial state x_0 the corresponding output $z_F(t, x_0)$ converges to zero as t tends to infinity. According to exercise 7.4, the system $\Sigma = (A, B, C, D)$ is output stabilizable if and only if

$$\mathcal{X} = \mathcal{V}(\Sigma) + \mathcal{X}_{\text{stab}}, \tag{10.22}$$

where $\mathcal{X}$ is the state space of Σ, $\mathcal{V}(\Sigma)$ the weakly unobservable subspace, and $\mathcal{X}_{\text{stab}}$ the stabilizable subspace of (A, B) with respect to the stability domain $\mathbb{C}^-$.

We will apply this to the system Σ given by $\dot{x}(t) = Ax(t) + Bu(t)$, together with the output equation (10.21). It is easily seen that for this special case we have $\mathcal{V}(\Sigma) = \langle \ker C \mid A \rangle$ (see exercise 7.5), so the system Σ is output stabilizable if and only if

$$\mathcal{X} = \langle \ker C \mid A \rangle + \mathcal{X}_{\text{stab}}.$$

It is clear that if Σ is output stabilizable then for any initial state x_0 there is a control input u such that $J(x_0, u) < \infty$. It follows from lemma 10.11 that also the converse holds. Indeed, if for all x_0 there exists an input u such that $J(x_0, u) < \infty$, then for all u (10.19) holds. This implies that the feedback law $u(t) = Fx(t)$ with $F := -B^{\mathrm{T}} P^-$ yields $\int_0^\infty \|z_F(t, x_0)\|^2 \, dt < \infty$, for any initial state x_0. This in turn implies that $z_F(t, x_0) \to 0$ $(t \to \infty)$ for all x_0, so the system Σ is output stabilizable. We conclude that the finite cost assumption of lemma 10.11 is equivalent to output stabilizability of Σ.

It also follows from lemma 10.11 that the finite cost assumption implies the existence of a real symmetric positive semidefinite solution of the ARE (10.18). We will now show that also the converse holds. Indeed, assume P is a positive semidefinite solution of the ARE. Then (10.20) with P^- replaced by P implies

$$J(x_0, u, T) \leqslant x_0^{\mathrm{T}} P x_0 + \int_0^\infty \|u(t) + B^{\mathrm{T}} P x(t)\|^2 \, dt,$$

for all T. By taking $u(t) = Fx(t)$ with $F := -B^{\mathrm{T}} P$ we obtain $J(x_0, u, T) \leqslant x_0^{\mathrm{T}} P x_0$ for all T and hence $J(x_0, u) < \infty$.

Summarizing the above we obtain the following theorem:

Theorem 10.13 *Consider the system $\dot{x}(t) = Ax(t) + Bu(t)$ together with the cost functional*

$$J(x_0, u) := \int_0^\infty x(t)^T Q x(t) + u(t)^T u(t) \, dt,$$

with $Q \geqslant 0$. Factorize $Q = C^T C$. Then the following statements are equivalent:

(i) *For every* $x_0 \in \mathcal{X}$ *there exists* $u \in \mathbf{U}$ *such that* $J(x_0, u) < \infty$.

(ii) *The ARE (10.18) has a real symmetric positive semidefinite solution* P,

(iii) *The system*

$$\Sigma = \left(A, B, \begin{pmatrix} C \\ 0 \end{pmatrix}, \begin{pmatrix} 0 \\ I \end{pmatrix}\right)$$

is output stabilizable,

(iv) $\langle \ker C \mid A \rangle + \mathcal{X}_{\text{stab}} = \mathcal{X}$.

Assume that one of the above conditions holds. Then there exists a smallest real symmetric positive semidefinite solution of the ARE, i.e., there exists a real symmetric solution $P^- \geqslant 0$ *such that for every real symmetric solution* $P \geqslant 0$ *we have* $P^- \leqslant P$. *For every* x_0 *we have*

$$J^*(x_0) := \inf\{J(x_0, u) \mid u \in \mathbf{U}\} = x_0^T P^- x_0.$$

Furthermore, for every x_0, *there is exactly one optimal input function, i.e., a function* $u^* \in \mathbf{U}$ *such that* $J(x_0, u^*) = J^*(x_0)$. *This optimal input is generated by the time-invariant feedback law*

$$u(t) = -B^T P^- x(t).$$

10.4 The infinite horizon problem with zero endpoint

In this section we are going to impose the condition $x(t) \to 0$ $(t \to \infty)$, which we will briefly write as $x(\infty) = 0$. Obviously, for the existence of an input u, for arbitrary x_0, such that $x(\infty) = 0$, the pair (A, B) has to be stabilizable. This is going to be a standing assumption in this section. It implies in particular any of the equivalent conditions in the statement of theorem 10.13. We apply the completion of squares formula with $K(t) = P$, any real symmetric solution of the ARE and u any control such that the corresponding state converges to zero. (We assume the initial state x_0 to be fixed). The result is

$$J(x_0, u, T) = x_0^T P x_0 + \int_0^T \|u(t) + B^T P x(t)\|^2 \, dt - x(T)^T P x(T).$$

This implies that

$$J(x_0, u) = x_0^T P x_0 + \int_0^\infty \|u(t) + B^T P x(t)\|^2 \, dt \tag{10.23}$$

for every u such that $x(\infty) = 0$. Let

$$J_0^*(x_0) := \inf\{J(x_0, u) \mid u \in \mathbf{U},\ x(\infty) = 0\}.$$

Then (10.23) implies that for every real symmetric solution P of the ARE, we have

$$x_0^{\mathrm{T}} P x_0 \leqslant J_0^*(x_0). \tag{10.24}$$

It also follows from (10.23) that if we choose the feedback control law $u(t) = -B^{\mathrm{T}} P x(t)$ and if the resulting state converges to zero, then for every initial state x_0 this will give the optimal control and the optimal value will be $x_0^{\mathrm{T}} P x_0$. As a consequence, we observe that there exists at most one real symmetric solution P of the ARE for which $A - BB^{\mathrm{T}}P$ is a stability matrix, and that this matrix P, if it exists, must be the maximal solution of the ARE.

First, we consider the special case $Q = 0$. Then the integral to be minimized equals:

$$J(x_0, u) = \int_0^\infty \|u\|^2 \,\mathrm{d}t.$$

Also assume that A has no eigenvalues on the imaginary axis. The solution of the free-endpoint problem is trivially $u = 0$ for all t. If in the zero-endpoint case, A is a stability matrix, $u = 0$ is still optimal, but otherwise we have to abandon this solution because the state does not tend to zero. Assume first that A has all its eigenvalues in $\mathbb{C}^+ := \{s \in \mathbb{C} \mid \Re e\, s > 0\}$, so that $-A$ is a stability matrix. Then the stabilizability of (A, B) implies that (A, B) is controllable. The Riccati equation in this case is

$$A^{\mathrm{T}} P + PA - PBB^{\mathrm{T}} P = 0. \tag{10.25}$$

We claim that this equation has a real symmetric positive definite solution P. This we show by giving a positive definite solution L of the (so-called) *Liapunov equation*

$$LA^{\mathrm{T}} + AL = BB^{\mathrm{T}}. \tag{10.26}$$

Then $P := L^{-1}$ is a positive definite solution of (10.25). The following integral is a solution of (10.26):

$$L := \int_0^\infty e^{-tA} BB^{\mathrm{T}} e^{-tA^{\mathrm{T}}} \,\mathrm{d}t. \tag{10.27}$$

In fact, because of our assumptions on A, there exist positive numbers M and γ such that $\|e^{-tA}\| \leqslant Me^{-\gamma t}$ for all $t \geqslant 0$. Hence the integral converges. Also it is clear that L is symmetric and $x^{\mathrm{T}} L x = \int_0^\infty \|x^{\mathrm{T}} e^{-tA} B\|^2 \,\mathrm{d}t > 0$ whenever $x \neq 0$, because of controllability. Hence L is positive definite. Finally, L satisfies (10.26):

$$\begin{aligned} AL + LA^{\mathrm{T}} &= \int_0^\infty Ae^{-tA} BB^{\mathrm{T}} e^{-tA^{\mathrm{T}}} + e^{-tA} BB^{\mathrm{T}} e^{-tA^{\mathrm{T}}} A^{\mathrm{T}} \,\mathrm{d}t \\ &= -\int_0^\infty \frac{\mathrm{d}}{\mathrm{d}t} (e^{-tA} BB^{\mathrm{T}} e^{-tA^{\mathrm{T}}}) \,\mathrm{d}t \\ &= BB^{\mathrm{T}} - \lim_{T \to \infty} e^{-TA} BB^{\mathrm{T}} e^{-TA^{\mathrm{T}}} \\ &= BB^{\mathrm{T}}. \end{aligned}$$

If we choose the feedback control $u(t) = -B^{\mathrm{T}}Px(t)$, the resulting state equation is $\dot{x}(t) = A_P x(t)$, where $A_P := A - BB^{\mathrm{T}}P$. The ARE can be written in terms of A_P instead of A, viz.

$$A_P^{\mathrm{T}}P + PA_P + PBB^{\mathrm{T}}P = 0. \tag{10.28}$$

We claim that A_P is a stability matrix. In fact, if $\lambda \in \sigma(A_P)$, say $A_P v = \lambda v$ and $v \neq 0$, then $0 = v^*(A_P^{\mathrm{T}}P + PA_P + PBB^{\mathrm{T}}P)v = 2(\Re\mathrm{e}\,\lambda)v^*Pv + \|v^*PB\|^2$. Here the asterisk denotes the conjugate transpose, $v^* = \bar{v}^{\mathrm{T}}$. Hence

$$\Re\mathrm{e}\,\lambda = -\|v^*PB\|^2/(2v^*Pv).$$

It follows that $\Re\mathrm{e}\,\lambda < 0$ unless $v^*PB = 0$, in which case $\Re\mathrm{e}\,\lambda = 0$. However, in that case, we have $\Re\mathrm{e}\,\lambda = 0$ and $Av = Av + BB^{\mathrm{T}}Pv = A_P v = \lambda v$. This is impossible since A has no eigenvalues on the imaginary axis. We have found a solution P of the ARE such that the corresponding feedback $F := -B^{\mathrm{T}}P$ stabilizes A. According to the remarks at the beginning of this section this implies that $u = Fx$ is optimal.

We have now solved the special cases $Q = 0$, A stable (with $u = 0$) and $Q = 0$, $-A$ stable (with $u = -B^{\mathrm{T}}Px$). Now assume that A has eigenvalues on both sides of the imaginary axis. Then we may perform a state space transformation such that in the resulting coordinate system, A and B have the form:

$$A = \begin{pmatrix} A_1 & 0 \\ 0 & A_2 \end{pmatrix}, \quad B = \begin{pmatrix} B_1 \\ B_2 \end{pmatrix},$$

where $\sigma(A_1) \subset \mathbb{C}^-$ and $\sigma(A_2) \subset \mathbb{C}^+$. We search for a solution P of the Riccati equation such that $A - BB^{\mathrm{T}}P$ is a stability matrix. Such a matrix can be found in the form of

$$P = \begin{pmatrix} 0 & 0 \\ 0 & P_2 \end{pmatrix},$$

where P_2 is the solution of the ARE corresponding to the (A_2, B_2) part of the problem, given in the equations above. It is easily seen that then P satisfies the ARE for (A, B) and that $A - BB^{\mathrm{T}}P$ is a stability matrix. We find that under the assumption that A has no eigenvalues on the imaginary axis, the problem has a unique solution, viz. $u(t) = -B^{\mathrm{T}}Px(t)$, where P is the unique real symmetric solution of the ARE such that $A - BB^{\mathrm{T}}P$ is stable. The minimal value of the integral is $x_0^{\mathrm{T}}Px_0$. It also follows that P is the maximal solution of the ARE.

If we drop the assumption $Q = 0$, we can reduce the problem to the special case $Q = 0$ by a preliminary feedback. In fact, we introduce the new control variable v by $u = -B^{\mathrm{T}}P^-x + v$, where P^- is the smallest real symmetric solution of the ARE, found in the previous section. Then the differential equation reads $\dot{x}(t) = A^-x(t) + Bv(t)$, where $A^- := A - BB^{\mathrm{T}}P^-$. Also, using (10.23), we find for the integral to be minimized:

$$J(x_0, u) = x_0^{\mathrm{T}}P^-x_0 + \int_0^\infty \|v(t)\|^2\,\mathrm{d}t,$$

which is exactly the situation $Q = 0$. If A^- has no eigenvalues on the imaginary axis, we find that there exists a solution $\Delta = \Delta^+$ of the ARE

$$(A^-)^{\mathrm{T}}\Delta + \Delta A^- - \Delta BB^{\mathrm{T}}\Delta = 0,$$

which is actually the maximal solution, for which $A^- - BB^{\mathrm{T}}\Delta^+$ is stable, so that the feedback $v(t) = -B^{\mathrm{T}}\Delta^+ x(t)$ is optimal. Consequently, the feedback control $u(t) = -B^{\mathrm{T}}Px(t)$, where $P := P^- + \Delta^+$, is the optimal control of the original problem. It is straightforward that P satisfies the original ARE, and is in fact the maximal and unique stabilizing solution. Hence, we denote P by P^+. We still have to express the condition that A^- has no purely imaginary eigenvalues in terms of the original data A, B, Q. According to the following lemma this corresponds to the condition that every eigenvalue of A on the imaginary axis is (Q, A)-observable (equivalently, after factorizing $Q = C^{\mathrm{T}}C$: (C, A)-observable).

Lemma 10.14 *Consider the system $\dot{x}(t) = Ax(t) + Bu(t)$ together with the cost functional*

$$J(x_0, u) := \int_0^\infty x(t)^T Q x(t) + u(t)^T u(t)\, dt,$$

with $Q \geqslant 0$. Assume that (A, B) is stabilizable. Assume that every eigenvalue of A on the imaginary axis is (Q, A)-observable. Then the following holds:

(*i*) *There exists exactly one real symmetric solution P^+ of the ARE such that $\sigma(A - BB^T P^+) \subset \mathbb{C}^-$.*

(*ii*) *P^+ is the largest real symmetric solution of the ARE, i.e., for every real symmetric solution P we have $P \leqslant P^+$. P^+ is positive semidefinite.*

(*iii*) *For every initial state x_0 we have*

$$J_0^*(x_0) = x_0^T P^+ x_0.$$

(*iv*) *For every initial state x_0 there is exactly one optimal input function, i.e., a function u^* such that $x(\infty) = 0$ and $J(x_0, u^*) = J_0^*(x_0)$. This optimal input function is generated by the time-invariant feedback law*

$$u(t) = -B^T P^+ x(t).$$

Proof : We only have to show that the observability condition implies that $A^- := A - BB^{\mathrm{T}}P^-$ has no eigenvalues on the imaginary axis. Suppose that $i\alpha \in \sigma(A^-)$ for some real α, and that v is the corresponding eigenvector. The ARE in terms of A^- reads:

$$(A^-)^{\mathrm{T}}P + PA^- + P^-BB^{\mathrm{T}}P^- + Q = 0.$$

Multiplying this equation from the left with v^* and from the right with v, we obtain $v^* Qv + \|B^{\mathrm{T}} P^- v\|^2 = 0$. This implies $Qv = 0$ and $B^{\mathrm{T}} P^- v = 0$. Hence $Av = i\alpha v$, $Qv = 0$, i.e. $i\alpha$ is an unobservable eigenvalue of (Q, A). ■

The above lemma is not yet completely satisfactory, because it only resolves the zero-endpoint problem under the assumption that A has no unobservable eigenvalues on the imaginary axis. However, the result can be used to tackle also the general situation. First we will show that also without the observability assumption, the ARE has a largest real symmetric solution. Furthermore, the optimal cost is still determined by this largest solution.

Lemma 10.15 *Assume that (A, B) is stabilizable. Then the ARE has a largest real symmetric solution P^+. P^+ is positive semidefinite. For every initial state x_0 we have $J_0^*(x_0) = x_0^T P^+ x_0$.*

Proof : Consider the 'perturbed' problem of minimizing

$$J_\varepsilon(x_0, u) = \int_0^\infty x(t)^{\mathrm{T}}(Q + \varepsilon^2 I)x(t) + u(t)^{\mathrm{T}} u(t)\, \mathrm{d}t$$

for $\varepsilon > 0$. Obviously, $J(x_0, u) \leqslant J_\varepsilon(x_0, u)$ for every ε. On the other hand, J_ε can be seen as the cost functional corresponding to the standard infinite horizon problem with $Q_\varepsilon = Q + \varepsilon^2 I$. The pair (Q_ε, A) is obviously observable. Therefore, because of the previous lemma, there exists a real symmetric solution P_ε^+ of the corresponding algebraic Riccati equation

$$A^{\mathrm{T}} P + PA - PBB^{\mathrm{T}} P + Q + \varepsilon^2 I = 0. \tag{10.29}$$

with the property that $A_\varepsilon := A - BB^{\mathrm{T}} P_\varepsilon^+$ is a stability matrix. Furthermore,

$$J_{0,\varepsilon}^*(x_0) := \inf\{J_\varepsilon(x_0, u) \mid u \in \mathbf{U},\ x(\infty) = 0\} = x_0^{\mathrm{T}} P_\varepsilon^+ x_0.$$

It follows from the definition of J_ε that $\varepsilon \mapsto J_\varepsilon(x_0, u)$ is non-decreasing for $\varepsilon > 0$, for all u and x_0. Consequently, P_ε^+ is a non-decreasing function of ε. Since obviously $P_\varepsilon^+ \geqslant 0$, it follows that $P_0 := \lim_{\varepsilon \to 0} P_\varepsilon^+$ exists. It is clear that $P_0 \geqslant 0$. By the continuity of the ARE, P_0 satisfies

$$A^{\mathrm{T}} P_0 + P_0 A - P_0 BB^{\mathrm{T}} P_0 + Q = 0, \tag{10.30}$$

We claim that, in fact, P_0 is the largest real symmetric solution of the ARE. Indeed, let P be any real symmetric solution of the ARE. Using (10.23), for all x_0 and all u such that $x(\infty) = 0$ we have $J_\varepsilon(x_0, u) \geqslant J(x_0, u) \geqslant x_0^{\mathrm{T}} P x_0$. Taking the infimum over u we obtain $x_0^{\mathrm{T}} P_\varepsilon^+ x_0 = J_{0,\varepsilon}^*(x_0) \geqslant x_0^{\mathrm{T}} P x_0$. Thus, $P_\varepsilon^+ \leqslant P$ for all $\varepsilon > 0$. By letting $\varepsilon \to 0$ we obtain $P_0 \geqslant P$, so $P^+ := P_0$ is the largest real symmetric solution of the ARE.

Next, we show that $J_0^*(x_0) = x_0^{\mathrm{T}} P^+ x_0$. First, it follows by applying (10.23) with $P = P^+$ that $J(x_0, u) \geqslant x_0^{\mathrm{T}} P^+ x_0$, for all x_0 and for all u such that $x(\infty) = 0$. By taking the infimum over all such u we obtain

$$J_0^*(x_0) \geqslant x_0^{\mathrm{T}} P^+ x_0.$$

On the other hand, for all x_0 and u and for all $\varepsilon > 0$ we have $J(x_0, u) \leqslant J_\varepsilon(x_0, u)$. Taking the infimum over all u such that $x(\infty) = 0$ on both sides in this inequality yields

$$J_0^*(x_0) \leqslant J_\varepsilon^*(x_0) = x_0^{\mathrm{T}} P_\varepsilon^+ x_0.$$

By taking the limit $\varepsilon \to 0$, we obtain

$$J_0^*(x_0) \leqslant x_0^{\mathrm{T}} P^+ x_0.$$

This completes the proof of the lemma. ∎

Next we show that the condition that every eigenvalue of A on the imaginary axis is (Q, A)-observable is in fact *necessary* for the existence of optimal controls for all initial states. This is illustrated by the following example:

Example 10.16 Let

$$A := \begin{pmatrix} 0 & 1 \\ -1 & 0 \end{pmatrix}, \quad B := \begin{pmatrix} 1 & 0 \\ 0 & 1 \end{pmatrix}, \quad Q := 0.$$

The eigenvalues of A (i and $-i$) are both unobservable. Take the input $u(t) := e^{t(A-\varepsilon I)} p$, with p still to be determined. The variation of constants formula yields

$$x(t) = e^{tA} x_0 + \int_0^t e^{(t-\tau)A} B u(\tau)\, \mathrm{d}\tau = e^{tA} \left[x_0 + \left(\int_0^t e^{-\varepsilon\tau}\, \mathrm{d}\tau \right) p \right].$$

Hence $e^{-tA} x(t) \to x_0 + p/\varepsilon$ $(t \to \infty)$. Since e^{tA} is an orthogonal matrix, it follows that $x(t) \to 0$ $(t \to \infty)$, if $p = -\varepsilon x_0$. Also, $\|u(t)\|^2 = e^{-2\varepsilon t} \varepsilon^2 \|x_0\|^2$, so that $\int_0^\infty \|u(t)\|^2\, \mathrm{d}t = (\varepsilon/2)\|x_0\|^2$. It follows that $J_0^*(x_0) = 0$. On the other hand, it is obvious that there is no u such that $x(\infty) = 0$ for which this infimum is attained.

Lemma 10.17 *Assume that (A, B) is stabilizable. If for every initial state x_0 there exists an optimal input, then every eigenvalue of A on the imaginary axis is (Q, A)-observable.*

Proof : Let P^+ be the largest real symmetric solution of the ARE:

$$A^{\mathrm{T}} P^+ + P^+ A - P^+ B B^{\mathrm{T}} P^+ + Q = 0. \tag{10.31}$$

Assume that there exists a real α and a nonzero v such that $Av = i\alpha v$, $Qv = 0$. Then multiplying (10.31) from the right by v and from the left by v^*, we obtain

$\|B^{\mathrm{T}}P^{+}v\|^2 = 0$ so $B^{\mathrm{T}}P^{+}v = 0$. By multiplying (10.31) from the right by v we then obtain $A^{\mathrm{T}}P^{+}v = -i\alpha P^{+}v$. Define $\eta := v^{*}P^{+}$. Then we find $\eta A = i\alpha\eta$ and $\eta B = 0$. If $\eta \neq 0$, then $i\alpha$ is an uncontrollable eigenvalue of A, which contradicts the stabilizability of the pair (A, B). Hence $\eta = 0$, so $P^{+}v = 0$. We write $v = v_1 + v_2 i$, where v_1 and v_2 are real vectors. Then $P_{+}v_j = 0$ $(j = 1, 2)$. At least one of the vectors v_j is nonzero. Let us call this vector v_0. We claim that v_0 is an eigenvector of A with eigenvalue $i\alpha$. Indeed, $A(v_1 + v_2 i) = i\alpha(v_1 + v_2 i)$ so both $Av_1 = i\alpha v_1$ and $Av_2 = i\alpha v_2$. Since $P^{+}v_0 = 0$, v_0 is also an eigenvector of $A - BB^{\mathrm{T}}P^{+}$ with eigenvalue $i\alpha$. By our assumption, there exists an optimal control for this initial state, say $\bar{u}$. Using (10.23) with $P = P^{+}$ and the fact that $J(x_0, \bar{u}) = x_0^{\mathrm{T}}P^{+}x_0$ we find that $\bar{u}$ is generated by the feedback law $u(t) = -B^{\mathrm{T}}P^{+}x(t)$. Consequently, the optimal closed loop system is given by $\dot{x}(t) = (A - BB^{\mathrm{T}}P^{+})x(t)$. With initial state $x(0) = v_0$ this yields $x(t) = e^{i\alpha t}v_0$, which clearly does not converge to zero. This contradicts the assumption that $\bar{u}$ is optimal. ■

Collecting the above lemmas we obtain the following:

Theorem 10.18 *Consider the system $\dot{x}(t) = Ax(t) + Bu(t)$ together with the cost functional*

$$J(x_0, u) := \int_0^\infty x(t)^T Q x(t) + u(t)^T u(t)\, dt,$$

with $Q \geqslant 0$. Assume that (A, B) is stabilizable. Then

(i) *there exists a largest real symmetric solution of the ARE, i.e., there exists a real symmetric solution P^{+} such that for every real symmetric solution P we have $P \leqslant P^{+}$. P^{+} is positive semidefinite.*

(ii) *For every initial state x_0 we have*

$$J_0^{*}(x_0) = x_0^T P^{+} x_0.$$

(iii) *For every initial state x_0 there exists an optimal input function, i.e., a function $u^{*} \in \mathbf{U}$ with $x(\infty) = 0$ such that $J(x_0, u^{*}) = J_0^{*}(x_0)$ if and only if every eigenvalue of A on the imaginary axis is (Q, A) observable.*

Under this assumption we have:

(iv) *For every initial state x_0 there is exactly one optimal input function u^{*}. This optimal input function is generated by the time-invariant feedback law*

$$u(t) = -B^T P^{+} x(t).$$

(v) *The optimal closed loop system $\dot{x}(t) = (A - BB^T P^{+})x(t)$ is stable. In fact, P^{+} is the unique real symmetric solution of the ARE for which $\sigma(A - BB^T P^{+}) \subset \mathbb{C}^{-}$.*

10.5 The nonstandard problems

In this section we will apply the results obtained for the standard case to solve the general linear quadratic regulator problem. As before, we make a distinction between the free endpoint problem and the zero endpoint problem.

As explained in section 10.1, the general regular linear quadratic regulator problem is the problem of minimizing for each initial state x_0 the cost functional

$$J(x_0, u) := \int_0^\infty \|z(t)\|^2 \, \mathrm{d}t$$

for the system Σ given by the equations

$$\dot{x}(t) = Ax(t) + Bu(t), \quad z(t) = Cx(t) + Du(t),$$

where D is injective. It was explained in section 10.1 that, in turn, this problem can be reduced to a standard problem by introducing a new control variable v and by applying the feedback transformation $u = Fx + Gv$ with

$$F := -(D^{\mathrm{T}}D)^{-1}D^{\mathrm{T}}C, \quad G := (D^{\mathrm{T}}D)^{-\frac{1}{2}} \tag{10.32}$$

The Riccati equation associated with the transformed system is obtained by replacing in (10.18) A by A_F, B by BG and Q by $C_F^{\mathrm{T}}C_F$. After reorganizing the equation thus obtained we get:

$$A^{\mathrm{T}}P + PA + C^{\mathrm{T}}C - (PB + C^{\mathrm{T}}D)(D^{\mathrm{T}}D)^{-1}(PB + C^{\mathrm{T}}D)^{\mathrm{T}} = 0. \tag{10.33}$$

This equation is called the Algebraic Riccati Equation (ARE) associated with Σ. It follows immediately from theorem 10.13 that the ARE (10.33) has a real symmetric positive semidefinite solution if and only if the system

$$(A_F, BG, \begin{pmatrix} C_F \\ 0 \end{pmatrix}, \begin{pmatrix} 0 \\ I \end{pmatrix})$$

is output stabilizable (with F and G given by (10.32)). It is easy to show that this system is output stabilizable if and only if the original system $\Sigma = (A, B, C, D)$ is output stabilizable.

In this way we obtain the following generalization of theorem 10.13:

Theorem 10.19 *Consider the system Σ given by $\dot{x}(t) = Ax(t) + Bu(t)$, $z(t) = Cx(t) + Du(t)$, together with the cost functional*

$$J(x_0, u) := \int_0^\infty \|z(t)\|^2 \, dt.$$

Assume that D is injective. Then the following statements are equivalent:

(i) For every $x_0 \in \mathcal{X}$ there exists $u \in \mathbf{U}$ such that $J(x_0, u) < \infty$.

(ii) *The ARE (10.33) has a real symmetric positive semidefinite solution P,*

(iii) *The system Σ is output stabilizable,*

(iv) $\mathcal{V}(\Sigma) + \mathcal{X}_{\text{stab}} = \mathcal{X}$.

Assume that one of the above conditions hold. Then there exists a smallest real symmetric positive semidefinite solution of the ARE, i.e., there exists a real symmetric solution $P^- \geqslant 0$ such that for every real symmetric solution $P \geqslant 0$ we have $P^- \leqslant P$. For every x_0 we have

$$J^*(x_0) := \inf\{J(x_0, u) \mid u \in \mathbf{U}\} = x_0^T P^- x_0.$$

Furthermore, for every x_0, there is exactly one optimal input function, i.e., a function $u^* \in \mathbf{U}$ such that $J(x_0, u^*) = J^*(x_0)$. This optimal input is generated by the time-invariant feedback law

$$u(t) = -(D^T D)^{-1}(B^T P^- + D^T C)x(t).$$

In order to generalize theorem 10.18 to the nonstandard case, we have to translate the condition that the transformed system has no unobservable eigenvalues on the imaginary axis into a condition on the original system Σ. Again let F and G be given by (10.32). We claim that the condition that A_F has no (C_F, A_F) unobservable eigenvalues on the imaginary axis is equivalent to the condition that the system Σ has no zeros on the imaginary axis. Indeed, the (C_F, A_F) unobservable eigenvalues of A_F are exactly the zeros of Σ (see exercise 10.1). This leads to the following theorem:

Theorem 10.20 *Consider the system Σ given by $\dot{x}(t) = Ax(t) + Bu(t)$, $z(t) = Cx(t) + Du(t)$, together with the cost functional*

$$J(x_0, u) := \int_0^\infty \|z(t)\|^2 \, dt.$$

Assume that D is injective and (A, B) is stabilizable. Then

(i) *there exists a largest real symmetric solution of the ARE, i.e., there exists a real symmetric solution P^+ such that for every real symmetric solution P we have $P \leqslant P^+$. P^+ is positive semidefinite.*

(ii) *For every initial state x_0 we have*

$$J_0^*(x_0) = x_0^T P^+ x_0.$$

(iii) *for every initial state x_0 there exists an optimal input function, i.e., a function $u^* \in \mathbf{U}$ with $x(\infty) = 0$ such that $J(x_0, u^*) = J_0^*(x_0)$ if and only if Σ has no zeros on the imaginary axis.*

Under this condition we have:

(iv) *For every initial state* x_0 *there is exactly one optimal input function* u^*. *This optimal input function is generated by the time-invariant feedback law*

$$u(t) = -(D^T D)^{-1}(B^T P^+ + D^T C)x(t).$$

(v) *The optimal closed loop system*

$$\dot{x}(t) = (A - B(D^T D)^{-1}(B^T P^+ + D^T C))x(t)$$

is stable. In fact, P^+ *is the unique real symmetric solution of the ARE for which*

$$\sigma(A - B(D^T D)^{-1}(B^T P^+ + D^T C)) \subset \mathbb{C}^-.$$

10.6 Exercises

10.1 Consider the system $\Sigma = (A, B, C, D)$, with input space $\mathcal{U} = \mathbb{R}^m$ and state space $\mathcal{X} = \mathbb{R}^n$. Assume that D is injective. Let $P_\Sigma(s)$ be the system matrix of Σ.

a. Show that the normal rank of P_Σ is equal to $n + m$.

b. Show that if $P_\Sigma(\lambda)\binom{x_0}{u_0} = \binom{0}{0}$ then $u_0 = -(DD^{\mathrm{T}})^{-1}D^{\mathrm{T}}Cx_0$.

c. Use (a) and (b) to show that λ is a zero of Σ if and only if λ is an unobservable eigenvalue of the system

$$(C - D(DD^{\mathrm{T}})^{-1}D^{\mathrm{T}}C, A - B(DD^{\mathrm{T}})^{-1}D^{\mathrm{T}}C).$$

10.2 Let $P : [0, \infty) \to \mathbb{R}^{n\times n}$ be such that $P(t) = P^{\mathrm{T}}(t)$ for all $t \geqslant 0$.

a. Prove that if $P(t)$ is monotonically non-decreasing (i.e., $P(t_1) \leqslant P(t_2)$ for $t_1 \leqslant t_2$) and if $x^{\mathrm{T}}P(t)x$ is bounded from above for all x (with upper bound depending on x), then $\lim_{t\to\infty} x^{\mathrm{T}}P(t)x$ exists for all x.

b. Prove that if $\lim_{t\to\infty} x^{\mathrm{T}}P(t)x$ exists for all x, then there exists a real symmetric matrix P_0 such that $\lim_{t\to\infty} P(t) = P_0$.

c. Prove that if $\lim_{t\to\infty} P(t)$ and $\lim_{t\to\infty} \dot{P}(t)$ exist, then $\lim_{t\to\infty} \dot{P}(t) = 0$.

10.3 Consider the system $\dot{x}(t) = Ax(t) + Bu(t)$, $x(0) = x_0$, $y(t) = Cx(t)$, with state space $\mathcal{X} = \mathbb{R}^n$. Define a cost functional by

$$H(x_0, u) := \int_0^\infty \| u(t) \|^2 \, dt + \int_1^\infty \| y_u(t, x_0) \|^2 \, dt$$

Let P^- be the smallest positive semi-definite real symmetric solution of the algebraic Riccati equation

$$A^{\mathrm{T}}P + PA + C^{\mathrm{T}}C - PBB^{\mathrm{T}}P = 0.$$

Define

$$H^*(x_0) := \inf_u H(x_0, u).$$

a. Show that $H^*(x_0) = x_0^{\mathrm{T}} P(1) x_0$, where $P(t)$ is the solution of the Riccati differential equation

$$\dot{P}(t) = A^{\mathrm{T}} P(t) + P(t)A - P(t)BB^{\mathrm{T}}P(t), \quad P(0) = P^-$$

b. Show that u is optimal, i.e., u satisfies $H(x_0, u) = H^*(x_0)$ if and only if

$$u(t) = \begin{cases} -B^{\mathrm{T}}P(1-t)x(t) & t \in [0,1] \\ -B^{\mathrm{T}}P^-x(t) & t \in [1,\infty). \end{cases}$$

10.4 Consider the algebraic Riccati equation $A^{\mathrm{T}}P + PA - PBB^{\mathrm{T}}P + Q = 0$ with Q positive semidefinite.

a. Assume P is a real symmetric positive semidefinite solution. Show that if (Q, A) is $\mathbb{C}^-$-detectable, then $\sigma(A - BB^{\mathrm{T}}P) \subset \mathbb{C}^-$.

b. Assume that P is a real symmetric positive semidefinite solution. Show that $\ker P$ is A-invariant and that $\ker P \subset \ker Q$.

c. Prove that if (Q, A) is observable, then every real symmetric positive semidefinite solution is positive definite.

d. Prove now that if (A, B) is $\mathbb{C}^-$-stabilizable and (Q, A) is $\mathbb{C}^-$-detectable, then the algebraic Riccati equation has exactly one real symmetric positive semidefinite solution. Also prove that if, in addition, (Q, A) is observable then this unique real symmetric positive semidefinite solution is positive definite.

10.5 Consider the algebraic Riccati equation $A^{\mathrm{T}}P + PA - PBB^{\mathrm{T}}P + Q = 0$. Assume that (A, B) is $\mathbb{C}^-$-stabilizable, and let P^+ be the largest real symmetric solution.

a. Show that $\sigma(A - BB^{\mathrm{T}}P^+) \subset \overline{\mathbb{C}^-} := \{s \in \mathbb{C} \mid \Re\mathrm{e}\, s \leqslant 0\}$, the *closed* left half plane.

b. Show that $\sigma(A - BB^{\mathrm{T}}P^+) \subset \mathbb{C}^-$, the open left half plane, if and only if every eigenvalue of A on the imaginary axis is (Q, A) observable.

10.6 (The linear matrix inequality.) As usual, let Σ be the system given by $\dot{x}(t) = Ax(t) + Bu(t)$, $z(t) = Cx(t) + Du(t)$, with (A, B) stabilizable and $D \in \mathbb{R}^{n \times m}$ injective. Introduce the following inequality in the unknown real symmetric matrix $P \in \mathbb{R}^{n \times n}$:

$$F(P) := \begin{pmatrix} A^{\mathrm{T}}P + PA + C^{\mathrm{T}}C & PB + C^{\mathrm{T}}D \\ B^{\mathrm{T}}P + D^{\mathrm{T}}C & D^{\mathrm{T}}D \end{pmatrix} \geqslant 0.$$

This inequality is called the *Linear Matrix Inequality* (LMI) associated with Σ.

a. Show that if P satisfies the ARE (10.33), then it satisfies the LMI.

b. Show that for every real symmetric matrix P we have

$$\operatorname{rank}(F(P)) = m + \operatorname{rank}\left[A^{\mathrm{T}}P + PA + C^{\mathrm{T}}C - (PB + C^{\mathrm{T}}D)(D^{\mathrm{T}}D)^{-1}(B^{\mathrm{T}}P + D^{\mathrm{T}}C)\right]$$

c. Show that $\min\{\operatorname{rank}(F(P)) \mid P \text{ real symmetric}\} = m$ and that a real symmetric matrix $\bar{P}$ satisfies the ARE if and only if $\operatorname{rank}(F(\bar{P})) = m$, i.e. $\bar{P}$ minimizes the rank of $F(P)$.

d. Show that if x, u and z satisfy the system equations, then we have

$$\begin{pmatrix} x(t) \\ u(t) \end{pmatrix}^{\mathrm{T}} F(P) \begin{pmatrix} x(t) \\ u(t) \end{pmatrix} = \frac{\mathrm{d}}{\mathrm{d}t}(x(t)^{\mathrm{T}}Px(t)) + \|z(t)\|^2.$$

10.7 (The singular linear quadratic regulator problem.) In this problem we consider the *singular* linear quadratic regulator problem. As before, consider the cost functional

$$J(x_0, u) = \int_0^\infty \|z(t)\|^2 \, \mathrm{d}t$$

for the system Σ given by the equations $\dot{x}(t) = Ax(t) + Bu(t)$, $z(t) = Cx(t) + Du(t)$. We call the problem singular if D is not injective. Note that in this case $D^{\mathrm{T}}D$ does not have an inverse, so the ARE (10.33) is no longer well-defined. One way to deal with the problem of minimizing $J(x_0, u)$ over all input functions u such that $x(\infty) = 0$ is to perturb the problem so that it becomes regular. For $\varepsilon > 0$ consider the system Σ_ε: $\dot{x}(t) = Ax(t) + Bu(t)$, $z_\varepsilon(t) = C_1x(t) + D_\varepsilon u(t)$, where C_1 and D_ε are defined by

$$C_1 := \begin{pmatrix} C \\ 0 \end{pmatrix}, \quad D_\varepsilon := \begin{pmatrix} D \\ \varepsilon I \end{pmatrix}.$$

Here, I is the $m \times m$ identity matrix. Let

$$J_\varepsilon(x_0, u) := \int_0^\infty \|z_\varepsilon(t)\|^2 \, \mathrm{d}t$$

a. Show that $J^*_{0,\varepsilon}(x_0) := \inf\{J_\varepsilon(x_0, u) \mid u \in \mathbf{U}, x(\infty) = 0\} = x_0^{\mathrm{T}} P^+_\varepsilon x_0$, where P^+_ε is the largest real symmetric solution of the ARE

$$A^{\mathrm{T}}P + PA + C^{\mathrm{T}}C - (PB + C^{\mathrm{T}}D)(D^{\mathrm{T}}D + \varepsilon^2 I)^{-1}(B^{\mathrm{T}}P + D^{\mathrm{T}}C)) = 0.$$

b. Show that $P_0 := \lim_{\varepsilon \to 0} P^+_\varepsilon$ exists

c. Show that $J^*_0(x_0) := \inf\{J(x_0, u) \mid u \in \mathbf{U}, x(\infty) = 0\}$ satisfies $J^*_0(x_0) \leqslant x_0^{\mathrm{T}} P_0 x_0$ for all x_0.

In order to proceed, consider the linear matrix inequality (LMI) associated with the system Σ, in the unknown real symmetric matrix $P \in \mathbb{R}^{n \times n}$:

$$\begin{pmatrix} A^{\mathrm{T}}P + PA + C^{\mathrm{T}}C & PB + C^{\mathrm{T}}D \\ B^{\mathrm{T}}P + D^{\mathrm{T}}C & D^{\mathrm{T}}D \end{pmatrix} \geqslant 0.$$

d. Show that P_0 satisfies the LMI.

e. Show that if P satisfies the LMI, then for all x, u and z that satisfy the equations of the system Σ we have

$$\frac{\mathrm{d}}{\mathrm{d}t}(x(t)^{\mathrm{T}}Px(t)) + \|z(t)\|^2 \geqslant 0 \text{ for all } t.$$

f. Show that for all $u \in \mathbf{U}$ such that $x(\infty) = 0$ we have $J(x_0, u) \geqslant x_0^{\mathrm{T}} P_0 x_0$, and conclude that $J_0^*(x_0) = x_0^{\mathrm{T}} P_0 x_0$ for all x_0.

g. Show that P_0 is the *largest* real symmetric solution P of the LMI, i.e., if the real symmetric matrix P satisfies the LMI, then we have $P \leqslant P_0$.

10.8 Consider the system $\dot{x}(t) = Ax(t) + Bu(t)$. Let $Q \geqslant 0$. In this exercise we investigate the existence of a *smallest* real symmetric solution of the associated ARE. Assume that (A, B) is $\mathbb{C}^+$- stabilizable.

a. Prove that the ARE $-A^{\mathrm{T}}P - PA - PBB^{\mathrm{T}}P + Q = 0$ has a largest real symmetric solution, say $\hat{P}$, with the property that $\hat{P} \geqslant 0$.

b. Show that the original ARE $A^{\mathrm{T}}P + PA - PBB^{\mathrm{T}}P + Q = 0$ has a smallest real symmetric solution P^- with the property that $P^- \leqslant 0$.

c. Show that if A has no (Q, A) unobservable eigenvalues on the imaginary axis, then $\sigma(A - BB^{\mathrm{T}}P^-) \subset \mathbb{C}^+$.

d. Consider the system on the interval $(-\infty, 0]$, with terminal state $x(0) = x_0$, and define a cost functional $J_-(x_0, u)$ by

$$J_-(x_0, u) := -\int_{-\infty}^{0} x(t)^{\mathrm{T}}Qx(t) + u(t)^{\mathrm{T}}u(t)\,\mathrm{d}t.$$

e. Prove that $J_-^*(x_0) := \sup\{J_-(x_0, u) \mid u \in \mathbf{U}, x(-\infty) = 0\} = x_0^{\mathrm{T}} P^- x_0$.

f. Show that if (A, B) is controllable, then there exist real symmetric solutions P^- and P^+ of the ARE $A^{\mathrm{T}}P + PA - PBB^{\mathrm{T}}P + Q = 0$ such that for any real symmetric solution P we have $P^- \leqslant P \leqslant P^+$.

10.7 Notes and references

The linear quadratic regulator problem and the Riccati equation were introduced by Kalman in [91]. Extensive treatments of the problem can be found in the textbooks [25] by Brockett, [105] by Kwakernaak and Sivan, and [5] by Anderson and Moore. For a detailed study of the Riccati differential equation and the algebraic Riccati equation we refer to the work of Wonham [221]. Treatments of the algebraic Riccati equation, including classifications of all real symmetric solutions in terms of the invariant subspaces of the Hamiltonian matrix associated with the problem, can also be found in Potter [150], Mårtensson [121], and Kucera [102]. The connection between output stabilizability and the existence of a real positive semi-definite solution of the algebraic Riccati equation (see theorems 10.13 and 10.19) is due to Geerts [58], see also Geerts and Hautus [56].

We briefly indicate how to solve Riccati equations in section 13.4. A standard reference for numerical algorithms for solving the algebraic Riccati equation is the work of Laub [108]. We also mention more recent work by Kenney, Laub and Wette [98], and Laub and Gahinet [109].

Extensions of the linear quadratic regulator problem to linear quadratic optimization problems where the integrand of the cost functional is a possibly indefinite quadratic function of the state and input variable, were studied in the classical paper [213] by Willems. This paper also provides a geometric characterization of all real symmetric solution of the algebraic Riccati equation, and establishes the intuitively appealing connection between real symmetric solutions of the algebraic Riccati equation, and storage functions of the underlying control system. For additional material on this connection, we refer to Molinari [125] and to Trentelman and Willems [200]. A further reference for the geometric classification of all real symmetric solutions is the work of Coppel [33]. In [198], Trentelman studied the free endpoint indefinite linear quadratic problem using this geometric classification.

The question what system performance can be obtained if, in the cost functional, the weighting matrix of the control input is singular or nearly singular, leads to singular and nearly singular linear quadratic regulator problems, and 'cheap control' problems. The asymptotic behavior of the optimal performance in case that the control weighting matrix tends to zero was studied in Kwakernaak and Sivan [106]. An early reference for a discussion on the singular problem is the work of Clements and Anderson [31]. Several approaches have been developed to study the singular linear quadratic regulator problem. One approach has been to approximate the (singular) control weighting matrix by a positive definite (regular) one, and subsequently study the behavior of the optimal cost and optimal control inputs and trajectories as the regular weighting matrix approaches the singular one. This method was worked out, for example, in Jameson and O'Malley [89], O'Malley [136], Francis [45], Fujii [52], and in Trentelman [197]. In this context, we also mention O'Malley and Jameson [138] and [139], where the singular problem was studied using the method of singular perturbations. In the singular problem, the role of the algebraic Riccati equation is taken over by a linear matrix inequality (see also exercise 10.6 and exercise 10.7). Details about this can be found in Willems [213], Clements, Anderson and Moylan [32], and in Schumacher [172]. A second method to approach the singular problem has been to allow for distributions as inputs. This method makes use of the full geometric machinery around the weakly unobservable subspace and strongly reachable subspace as developed in chapters 7 and 8 of this book, and was worked out in detail in Hautus and Silverman [74] and in Willems, Kitapçi and Silverman [219]. In this context we also mention the work of Geerts [57].

Chapter 11

The H_2 optimal control problem

In this chapter we consider the H_2 optimal control problem. Given a controlled linear system with a white noise disturbance input, we show that the size (in an appropriate sense) of the controlled output is equal to the H_2 norm of the closed loop transfer matrix. Motivated by this, we define the performance of the controlled system to be the H_2 norm of the closed loop transfer matrix. This gives rise to the H_2 optimal control problem: for a given control system, minimize the square of the H_2 norm of the closed loop transfer matrix over the class of all internally stabilizing feedback controllers.

The outline of this chapter is as follows. Section 11.1 contains an informal discussion on stochastic inputs to linear systems. In this section we also briefly recall the notion of white noise. It is explained how the H_2 norm of the transfer matrix is related to the size of the output process, if the input is taken to be a standard white noise process. In section 11.2 we treat the static state feedback version of the H_2 optimal control problem. Finally, in section 11.3 we discuss the dynamic measurement feedback version of the H_2 optimal control problem.

11.1 Stochastic inputs to linear systems

Consider the finite-dimensional, linear, time-invariant system Σ given by

$$\begin{aligned} \dot{x}(t) &= Ax(t) + Ev(t), \\ z(t) &= Cx(t). \end{aligned} \tag{11.1}$$

Assume that the system is internally stable, i.e. that $\sigma(A) \subset \mathbb{C}^-$. Let $T(t)$ be the impulse response matrix of the system. In this section we take the point of view that the input functions v are all samples of one and the same vector stochastic process v

on $\mathbb{R}$. We assume that every sample v has the property that the integral

$$x(t) = \int_{-\infty}^{t} e^{A(t-s)} Ev(s)\, ds \tag{11.2}$$

converges for every $t \in \mathbb{R}$ (a sufficient condition is, for example, that v is bounded on $(-\infty, c)$ for some $c \in \mathbb{R}$, or that v is square-integrable over $(-\infty, c)$ for some $c \in \mathbb{R}$). The function x defined by (11.2) is a solution of the differential equation $\dot{x}(t) = Ax(t) + Ev(t)$, and is in fact the unique solution x with the property that $\lim_{t\to-\infty} x(t) = 0$. For any sample v, the corresponding output is given by

$$z(t) = \int_{-\infty}^{t} T(t-s)v(s)\, ds. \tag{11.3}$$

In other words, $z(t), t \in \mathbb{R}$ is the output of the system corresponding to the initial state $x(-\infty) = 0$. In this way, every sample v gives rise to an output function z. The set of all output functions obtained in this way can in turn be interpreted as the set of all samples of a vector stochastic process, that is also denoted by z. In this way the expression (11.3) can be interpreted as giving the relation between the input stochastic process v and the output stochastic process z.

We assume that the input process v has *zero mean*. In other words, the expected value $m_v(t) := E\{v(t)\}$ satisfies $m_v(t) = 0$ for all $t \in \mathbb{R}$. We also assume that v is *wide-sense stationary*, i.e., the covariance $E\{v(t_1)v^{\mathrm{T}}(t_2)\}$ depends only on the difference $t_1 - t_2$. The covariance matrix of v is thus given by

$$R_v(\tau) := E\{v(t+\tau)v^{\mathrm{T}}(t)\} \ (t, \tau \in \mathbb{R}).$$

We are interested in the mean and covariance of the output process z. It can be shown that the output process z is again zero-mean, i.e., the mean m_z of z satisfies $m_z(t) = 0$ for all $t \in \mathbb{R}$. Also, z is wide-sense stationary and its covariance matrix is

$$R_z(\tau) = \int_0^\infty \int_0^\infty T(\tau_1)R_v(\tau + \tau_2 - \tau_1)T^{\mathrm{T}}(\tau_2)\, d\tau_2\, d\tau_1. \tag{11.4}$$

In control systems with *stochastic* disturbance inputs, one frequently encounters scalar stochastic processes w with the property that, even for values of $|t_1 - t_2|$ small, we have $R_w(t_2 - t_1) \approx 0$, that is $w(t_1)$ and $w(t_2)$ are *uncorrelated* even for t_1 and t_2 close. In order to model this property, the covariance function of such stochastic process can be put $R_v = \delta$, the Dirac distribution (see the Appendix.) A zero-mean, wide-sense stationary scalar stochastic process w on $\mathbb{R}$ with covariance $R_w = \delta$ is called a *standard (scalar) white noise process*. We can extend this notion to vector-valued stochastic processes: any zero-mean, wide-sense stationary vector stochastic process w on $\mathbb{R}$ with covariance matrix $R_w = I\delta$) is called *a standard white noise process*.

In a strict mathematical sense white noise processes do not exist. We ignore this fact here and proceed in a formal way, considering the following question. Given the system Σ, let the input process v be a standard white noise process, so $v = w$. What can we then say about the output process z? According to the discussion above, z is a zero-mean, wide-sense stationary process. If we insert the expression $R_w = I\delta$ into the formula (11.4) for the covariance matrix of the output process z, then (with a slight abuse of notation) we find that z has covariance matrix

$$R_z(\tau) = \int_0^\infty \int_0^\infty T(\tau_1) I \delta(\tau + \tau_2 - \tau_1) T^{\mathrm{T}}(\tau_2)\,\mathrm{d}\tau_2\,\mathrm{d}\tau_1 = \int_0^\infty T(\tau_1) T^{\mathrm{T}}(\tau_1 - \tau)\,\mathrm{d}\tau_1. \tag{11.5}$$

The conclusion we draw from this is the following: given the internally $\mathbb{C}^-$-stable system $\dot{x}(t) = Ax(t) + Ev(t)$, $z(t) = Cx(t)$, with input process v equal to a standard white noise process, the output process z has variance matrix $R_z(0)$ given by

$$R_z(0) = \int_0^\infty T(\tau) T^{\mathrm{T}}(\tau)\,\mathrm{d}\tau, \tag{11.6}$$

where $T(t) = Ce^{At}E$ is the impulse response matrix of the system.

We now use this conclusion in the issue of how to define in a natural way the concept of *performance* of the system Σ. We take the point of view that z is an output that we would like to have 'small' in the presence of a white noise disturbance. In the present context it is natural to measure the size of the output at time t by the expected value $E\{\|z(t)\|^2\}$. It turns out that, in fact, $E\{\|z(t)\|^2\}$ is independent of t. Indeed, for any $t \in \mathbb{R}$ we have

$$E\{\|z(t)\|^2\} = E\{z^{\mathrm{T}}(t)z(t)\} = E\{\mathrm{trace}\, z(t)z^{\mathrm{T}}(t)\} = \mathrm{trace}\, E\{z(t)z^{\mathrm{T}}(t)\} = \mathrm{trace}\, R_z(0).$$

Thus we see that for the system $\dot{x}(t) = Ax(t) + Ev(t)$, $z(t) = Cx(t)$, with input process v standard white noise, the size $E\{\|z(t)\|^2\}$ of the output process is the trace of the variance matrix $R_z(0)$. By the previous discussion, this is equal to

$$\begin{aligned} \mathrm{trace}\, R_z(0) = \mathrm{trace} \int_0^\infty T(\tau) T^{\mathrm{T}}(\tau)\,\mathrm{d}\tau &= \int_0^\infty \mathrm{trace}\, T(\tau) T^{\mathrm{T}}(\tau)\,\mathrm{d}\tau \\ &= \int_0^\infty \|T(\tau)\|^2\,\mathrm{d}\tau. \end{aligned}$$

We conclude that for all $t \in \mathbb{R}$

$$E\{\|z(t)\|^2\} = \int_0^\infty \|T(\tau)\|^2\,\mathrm{d}\tau,$$

the square of the L_2-norm of the impulse response matrix of the system Σ. Here, for a given matrix M, $\|M\|$ denotes the Frobenius-norm of M: $\|M\|^2 := \sum_{i,j} M_{ij}^2$. Thus, we define the performance of the system as the square of the L_2-norm of the impulse response matrix:

$$J_\Sigma := \int_0^\infty \|T(\tau)\|^2\,\mathrm{d}\tau. \tag{11.7}$$

This quantity is often called the *H_2-performance* of the system Σ. The latter terminology stems from the fact that J_Σ is equal to the square of the H_2-norm of the transfer matrix $G(s) = C(Is - A)^{-1}E$ of Σ:

$$J_\Sigma = \|G\|_2^2 = \frac{1}{2\pi}\int_{-\infty}^{\infty} \operatorname{trace}[G^{\mathrm{T}}(-i\omega)G(i\omega)]\, d\omega.$$

It is important to note that, once we have agreed to take (11.7) as the definition of performance of the system Σ, *we can completely abandon stochastics from the discussion.* From that moment on, the performance of the system $\Sigma = (A, E, C)$ equals the square of the L_2-norm of the impulse response matrix $T(t) = Ce^{At}E$, and the stochastic interpretation of this remains on the background. Starting with this definition of performance, we now formulate the corresponding synthesis problems of minimizing the H_2-performance over certain classes of feedback controllers. The next section discusses the case that this minimization takes place over the class of all static state feedback laws.

11.2 H_2 optimal control by state feedback

Consider the system Σ given by

$$\begin{aligned} \dot{x}(t) &= Ax(t) + Bu(t) + Ed(t), \\ z(t) &= Cx(t) + Du(t). \end{aligned} \tag{11.8}$$

The impulse response of the controlled system resulting from the state feedback control law $u(t) = Fx(t)$ is given by $T_F(t) = C_F e^{A_F t}E$, while the corresponding closed loop transfer matrix is equal to $G_F(s) = C_F(sI - A_F)^{-1}E$. As explained in the previous section, we measure the performance of the controlled system by the square of the L_2-norm of its impulse response matrix. Thus, for a state feedback map F such that $\sigma(A + BF) \subset \mathbb{C}^-$, we define the associated cost by

$$J_\Sigma(F) := \int_0^\infty \operatorname{trace}[T_F^{\mathrm{T}}(t)T_F(t)]\, dt. \tag{11.9}$$

As noted before, this is equal to the square of the H_2-norm of the closed loop transfer matrix:

$$J_\Sigma(F) = \int_{-\infty}^{\infty} \operatorname{trace}[G_F^{\mathrm{T}}(-i\omega)G_F(i\omega)]d\omega.$$

The *H_2 optimal control problem by state feedback* is to minimize the cost functional (11.9) over the set of all state feedback control laws such that the controlled system is internally stable:

Problem 11.1 *Consider the system (11.8), together with the cost functional (11.9). The H_2 optimal control problem by state feedback is to find*

$$J_\Sigma^* := \inf\left\{\, J_\Sigma(F) \mid F : \mathcal{X} \to \mathcal{U} \text{ such that } \sigma(A_F) \subset \mathbb{C}^- \,\right\}, \tag{11.10}$$

and to find, if it exists, an optimal state feedback control law, i.e., to find a map $F^* : \mathcal{X} \to \mathcal{U}$ *such that* $\sigma(A_{F^*}) \subset \mathbb{C}^-$ *and such that*

$$J_\Sigma(F^*) = J_\Sigma^*.$$

In this book, we restrict ourselves to the case that the direct feedthrough map D from the control input to the output to be controlled is *injective*. Under this assumption the H_2 optimal control problem is called *regular*. The H_2 problem by state feedback is said to be *in standard form* if $D^{\mathrm T}C = 0$ and $D^{\mathrm T}D = I$. Similar to the linear quadratic regulator problems studied in chapter 10, the general problem can be reduced to a problem in standard form by a preliminary state feedback transformation (see section 10.1). In this section we assume that the problem is in standard form. In order to assure the existence of a map F such that $\sigma(A_F) \subset \mathbb{C}^-$, it is also a standing assumption that the pair (A, B) is $\mathbb{C}^-$-stabilizable.

Like the optimal control problems studied in chapter 10, the solution of the H_2 optimal control problem uses completion of the squares. Let P be a real symmetric solution of the ARE

$$A^{\mathrm T}P + PA - PBB^{\mathrm T}P + C^{\mathrm T}C = 0 \tag{11.11}$$

Furthermore, let F be such that $\sigma(A_F) \subset \mathbb{C}^-$. Define

$$X(t) := e^{A_F t}E, \quad U(t) := Fe^{A_F t}E.$$

Then we have

$$\frac{\mathrm d}{\mathrm dt}X(t) = AX(t) + BU(t), \quad X(0) = E.$$

Now calculate (omitting the dependence on t)

$$\begin{aligned}\frac{\mathrm d}{\mathrm dt}X^{\mathrm T}PX &= X^{\mathrm T}(A^{\mathrm T}P + PA)X + U^{\mathrm T}B^{\mathrm T}PX + X^{\mathrm T}PBU \\ &= X^{\mathrm T}PBB^{\mathrm T}PX - X^{\mathrm T}C^{\mathrm T}CX + U^{\mathrm T}B^{\mathrm T}PX + X^{\mathrm T}PBU \\ &= (B^{\mathrm T}PX + U)^{\mathrm T}(B^{\mathrm T}PX + U) - (CX + DU)^{\mathrm T}(CX + DU).\end{aligned} \tag{11.12}$$

By integrating this from 0 to T we obtain

$$\begin{aligned}X(T)^{\mathrm T}PX(T) - E^{\mathrm T}PE = &\int_0^T (B^{\mathrm T}PX(t) + U(t))^{\mathrm T}(B^{\mathrm T}PX(t) + U(t))\,\mathrm dt \\ &- \int_0^T (CX(t) + DU(t))^{\mathrm T}(CX(t) + DU(t))\,\mathrm dt.\end{aligned}$$

Since $\sigma(A_F) \subset \mathbb{C}^-$, we have $X(T)^{\mathrm T}PX(T) \to 0$ $(T \to \infty)$, so we get

$$\begin{aligned}&\int_0^\infty (CX(t) + DU(t))^{\mathrm T}(CX(t) + DU(t))\,\mathrm dt \\ &\qquad = E^{\mathrm T}PE + \int_0^\infty (B^{\mathrm T}PX(t) + U(t))^{\mathrm T}(B^{\mathrm T}PX(t) + U(t))\,\mathrm dt.\end{aligned} \tag{11.13}$$

Now, obviously

$$CX(t) + DU(t) = C_F e^{A_F t} E$$

and

$$B^{\mathrm{T}} P X(t) + U(t) = (B^{\mathrm{T}} P + F) e^{A_F t} E.$$

Substituting these expressions in (11.13) and taking the trace on both sides of the resulting equation, we obtain

$$\int_0^\infty \operatorname{trace}[T_F^{\mathrm{T}}(t) T_F(t)]\, \mathrm{d}t = \operatorname{trace}(E^{\mathrm{T}} P E) + \int_0^\infty \operatorname{trace}[T_{P,F}^{\mathrm{T}}(t) T_{P,F}(t)]\, \mathrm{d}t, \quad (11.14)$$

where we define

$$T_{P,F}(t) := (B^{\mathrm{T}} P + F) e^{A_F t} E.$$

Observe that this is equal to the closed loop impulse response matrix obtained by applying the state feedback law $u = Fx$ to the auxiliary system Σ_P, represented by the equations

$$\begin{aligned} \dot{x}(t) &= Ax(t) + Bu(t) + Ed(t), \\ z(t) &= B^{\mathrm{T}} P x(t) + u(t). \end{aligned} \qquad (11.15)$$

Σ_P is obtained from Σ by replacing the output equation $z = Cx + Du$ by $z = B^{\mathrm{T}} P x + u$. Note that the second integral in (11.14) is equal to $J_{\Sigma_P}(F)$, the performance of the controlled system obtained by applying the state feedback control law $u(t) = Fx(t)$ to the system Σ_P. Thus we obtain the following lemma:

Lemma 11.2 *Consider the system (11.8). Assume that $D^T C = 0$, $D^T D = I$ and that (A, B) is stabilizable. Let P be a real symmetric solution of the ARE (11.11). A state feedback control law $u(t) = Fx(t)$ is internally stabilizing for Σ if and only if it is internally stabilizing for Σ_P. For any such F we have*

$$J_\Sigma(F) = \operatorname{trace}(E^T P E) + J_{\Sigma_P}(F). \qquad (11.16)$$

Since the quantity $\operatorname{trace}(E^{\mathrm{T}} P E)$ does not depend on F, this lemma shows that we can replace the minimization of the cost functional associated with the *original* system Σ by the minimization of the cost functional associated with the *new* system Σ_P. Of course, this is useful only if the minimization of the new cost functional is a simpler problem. A first attempt to minimize $J_{\Sigma_P}(F)$ is to choose $F := -B^{\mathrm{T}} P$. This choice yields $J_{\Sigma_P}(F) = 0$. Unfortunately, this choice of F does not necessarily give $\sigma(A_F) \subset \mathbb{C}^-$. This can however be repaired by taking for P *not just any* real symmetric solution of the ARE, but a real symmetric solution that does lead to a stable closed loop system. According to lemma 10.14, if every eigenvalue of A on the imaginary axis is (C, A) observable, then such solution, P^+, indeed exists (P^+ is the largest real symmetric solution of the ARE). Thus, by applying lemma 11.2 with $P = P^+$ we obtain

Lemma 11.3 *Consider the system (11.8). Assume that $D^TC = 0$, $D^TD = I$, that (A, B) is stabilizable, and that every eigenvalue of A on the imaginary axis is (C, A) observable. Let P^+ be the largest real symmetric solution of the ARE (11.11). Then we have*

(*i*) $J^*_\Sigma = \text{trace}(E^TP^+E)$.

(*ii*) *The feedback law $u(t) = -B^TP^+x(t)$ is optimal, i.e., $J_\Sigma(-B^TP^+) = J^*_\Sigma$ and $\sigma(A - BB^TP^+) \subset \mathbb{C}^-$.*

Similar to the linear quadratic regulator problems studied in chapter 10, the assumption on the eigenvalues of A on the imaginary axis can be removed. Again, the method is to consider a perturbed version of the H_2 optimal control problem. Given $\varepsilon > 0$, we introduce the perturbed system Σ_ε given by the equations

$$\begin{aligned} \dot{x}(t) &= Ax(t) + Bu(t) + Ed(t), \\ z(t) &= C_\varepsilon x(t) + D_0 u(t), \end{aligned}$$

with

$$C_\varepsilon := \begin{pmatrix} C \\ \varepsilon I \end{pmatrix}, \qquad D_0 := \begin{pmatrix} D \\ 0 \end{pmatrix}.$$

Clearly, for all F we have $J_\Sigma(F) \leqslant J_{\Sigma_\varepsilon}(F)$. The H_2 problem for Σ_ε is still in standard form. The crucial point however is that (C_ε, A) is observable so that we can apply lemma 11.3 to obtain $J^*_{\Sigma_\varepsilon} = \text{trace}(E^TP^+_\varepsilon E)$, where P^+_ε is the largest real symmetric solution of the perturbed algebraic Riccati equation

$$A^TP + PA - PBB^TP + C^TC + \varepsilon^2 I = 0 \tag{11.17}$$

It was shown in the proof of lemma 10.15 that P^+_ε converges to P^+, the largest real symmetric solution of the ARE (11.11). Since $J^*_\Sigma \leqslant J^*_{\Sigma_\varepsilon} = \text{trace}(E^TP^+_\varepsilon E)$, this implies

$$J^*_\Sigma \leqslant \text{trace}(E^TP^+E).$$

It follows from (11.16) with $P = P^+$ that also the converse inequality holds. This proves that, also without the assumption on the eigenvalues of A on the imaginary axis, the optimal cost is equal to $\text{trace}(E^TP^+E)$, with P^+ the largest real symmetric solution of the ARE.

If A has (C, A)-unobservable eigenvalues on the imaginary axis, then an optimal F will not always exist. This is illustrated by the following example:

Example 11.4 Consider the system Σ given by $\dot{x}(t) = u(t) + d(t)$, $z(t) = u(t)$. The optimal cost J^*_Σ is most easily computed by evaluating the corresponding ARE, which in this case takes the form $p^2 = 0$. Clearly, $p^+ = 0$, so $J^*_\Sigma = 0$. The closed loop transfer function resulting from $u(t) = fx(t)$ is equal to $f(s - f)^{-1}$, so f can only be optimal if $f = 0$. However, $f = 0$ does not internally stabilize the system, so an optimal f does not exist.

In order to study the question of existence of optimal control laws, we consider the system Σ_P defined by (11.15) for $P = P^+$. Apparently, for all F such that A_F is $\mathbb{C}^-$-stable, we have

$$J_\Sigma(F) = J^*_\Sigma + J_{\Sigma_{P^+}}(F),$$

so F is optimal if and only if A_F is $\mathbb{C}^-$-stable and $J_{\Sigma_{P^+}}(F) = 0$. Since $J_{\Sigma_{P^+}}(F) = 0$ if and only if the closed loop transfer matrix

$$G_{P^+,F}(s) := (B^{\mathrm{T}}P^+ + F)(sI - A_F)^{-1}E = 0,$$

this can be restated as: F is optimal if and only if it achieves disturbance decoupling with internal stability for the system Σ_{P^+}. Thus, we should consider the disturbance decoupling problem with internal stability, DDPS, for the system Σ_{P^+} given by (11.15) with $P = P^+$.

To be slightly more general, we shall consider DDPS for the system given by

$$\begin{aligned} \dot{x}(t) &= Ax(t) + Bu(t) + Ed(t), \\ z(t) &= Cx(t) + u(t). \end{aligned} \tag{11.18}$$

Assume there exists a state feedback map F such that $\sigma(A_F) \subset \mathbb{C}^-$ and $(C+F)(sI - A_F)^{-1}E = 0$. It is then immediate that (A, B) is stabilizable. If in the closed loop system we take $d = 0$ and $x_0 \in \operatorname{im} E$, then the corresponding state trajectory x is $\mathbb{C}^-$-stable and the corresponding output z is equal to zero. As a consequence, the closed loop input u must be given by $u(t) = -Cx(t)$ for all t. But then x satisfies the differential equation $\dot{x} = (A - BC)x$, $x(0) = x_0$. Since x is $\mathbb{C}^-$-stable, this immediately implies $x_0 \in \mathcal{X}^-(A - BC)$, the $\mathbb{C}^-$-stable subspace of $A - BC$ (see definition 2.13). Thus we have proven the 'only if' part of the following lemma:

Lemma 11.5 *Consider the system (11.18). There exists a map $F : \mathcal{X} \to \mathcal{U}$ such that $\sigma(A + BF) \subset \mathbb{C}^-$ and $(C + F)(sI - A_F)^{-1}E = 0$ if and only if (A, B) is stabilizable and* $\operatorname{im} E \subset \mathcal{X}^-(A - BC)$.

Proof : ($\Leftarrow$) Let $\mathcal{V} := \mathcal{X}^-(A - BC)$. We claim that there exists a feedback map F such that $\sigma(A_F) \subset \mathbb{C}^-$, $A_F\mathcal{V} \subset \mathcal{V}$ and $\mathcal{V} \subset \ker(C + F)$. In order to prove this, choose a basis of the state space $\mathcal{X}$ adapted to $\mathcal{V}$. With respect to this basis we have

$$A - BC = \begin{pmatrix} A_{11} & A_{12} \\ 0 & A_{22} \end{pmatrix}, \qquad B = \begin{pmatrix} B_1 \\ B_2 \end{pmatrix},$$

where $\sigma(A_{11}) \subset \mathbb{C}^-$ and (A_{22}, B_2) is stabilizable. Let F_2 be such that $\sigma(A_{22} + B_2F_2) \subset \mathbb{C}^-$ and define F_0 by $F_0 := (0 \;\; F_2)$. Then we have

$$A - BC + BF_0 = \begin{pmatrix} A_{11} & A_{12} \\ 0 & A_{22} + B_2F_2 \end{pmatrix}.$$

Finally, define $F := F_0 - C$. Then it is clear that F satisfies the required properties. The proof of the lemma is completed by noting that $\mathcal{V}$ is an A_F-invariant subspace such that $\operatorname{im} E \subset \mathcal{V} \subset \ker(C+F)$. According to theorem 4.6 this yields $(C+F)(sI-A_F)^{-1}E=0$. ■

By applying this lemma to the system Σ_{P^+}, we immediately obtain:

Theorem 11.6 *Consider the system Σ. Assume that $D^TC=0$, $D^TD=I$, and that (A,B) is stabilizable. Let P^+ be the largest real symmetric solution of the ARE (11.11). Then we have $J^*_\Sigma = \operatorname{trace}(E^TP^+E)$. Furthermore, the following conditions are equivalent:*

(*i*) *there exists an optimal feedback law, i.e., a map F such that $\sigma(A+BF)\subset\mathbb{C}^-$, and $J_\Sigma(F)=J^*_\Sigma$,*

(*ii*) *there exists F such that the feedback law $u(t)=Fx(t)$ achieves disturbance decoupling with internal stability for the system Σ_{P^+}, i.e., $\sigma(A+BF)\subset\mathbb{C}^-$, and $(B^TP^++F)(sI-A-BF)^{-1}E=0$,*

(*iii*) $\operatorname{im} E \subset \mathcal{X}^-(A-BB^TP^+)$.

Furthermore, $F:\mathcal{X}\to\mathcal{U}$ is optimal if and only if $\sigma(A+BF)\subset\mathbb{C}^-$ and $(B^TP^++F)(sI-A-BF)^{-1}E=0$.

It is an easy exercise to extend this theorem to the non-standard case (see exercise 11.1).

11.3 H_2 optimal control by measurement feedback

In this section we consider the H_2 optimal control problem by dynamic measurement feedback. Consider the system Σ

$$\begin{aligned}\dot{x}(t) &= Ax(t) + Bu(t) + Ed(t),\\ y(t) &= C_1x(t) + D_1d(t),\\ z(t) &= C_2x(t) + D_2u(t).\end{aligned} \tag{11.19}$$

In these equations, d represents the disturbance, u the control input, z the output to be controlled, and y the measured output. If we control the system by means of a dynamic feedback controller Γ, given by the equations

$$\begin{aligned}\dot{w}(t) &= Kw(t) + Ly(t),\\ u(t) &= Mw(t) + Ny(t),\end{aligned} \tag{11.20}$$

then the controlled system is given by the equations

$$\begin{aligned}\dot{x}_e(t) &= A_ex_e(t) + E_ed(t),\\ z(t) &= C_ex_e(t) + D_ed(t),\end{aligned} \tag{11.21}$$

with

$$A_e := \begin{pmatrix} A + BNC & BM \\ LC_1 & K \end{pmatrix}, \qquad E_e := \begin{pmatrix} E + BND_1 \\ LD_1 \end{pmatrix},$$
$$C_e := \begin{pmatrix} C_2 + D_2NC_1 & D_2M \end{pmatrix}, \qquad D_e := D_2ND_1.$$

The corresponding closed loop impulse response matrix between the disturbance input d and the output z is equal to

$$T_\Gamma(t) = C_e e^{A_e t} E_e + \delta D_e,$$

while the corresponding transfer matrix is given by

$$G_\Gamma(s) = C_e(sI - A_e)^{-1} E_e + D_e.$$

Like in the previous section, as a measure of performance of the controlled system we take the squared L_2-norm of the closed loop impulse response matrix. The L_2-norm of the impulse response matrix is defined only if it is $\mathbb{C}^-$-stable and the distributional part is not present (equivalently, the closed loop transfer matrix is strictly proper). For any feedback controller Γ such that the controlled system satisfies these properties, i.e., T_Γ is $\mathbb{C}^-$-stable and $D_e = 0$, we define the associated cost by

$$J_\Sigma(\Gamma) := \int_0^\infty \operatorname{trace}[T_\Gamma^{\mathrm{T}}(t) T_\Gamma(t)] \, \mathrm{d}t. \tag{11.22}$$

This is equal to the squared H_2-norm of the closed loop transfer matrix:

$$J_\Sigma(\Gamma) = \frac{1}{2\pi} \int_{-\infty}^{\infty} \operatorname{trace}[G_\Gamma^{\mathrm{T}}(-i\omega) G_\Gamma(i\omega)] \mathrm{d}\omega.$$

In the H_2 problem by dynamic measurement feedback we want to minimize the cost functional $J_\Sigma(\Gamma)$ over the set of all feedback controllers Γ such that the resulting closed loop transfer matrix is strictly proper and the resulting closed loop system is internally stable:

Definition 11.7 *Consider the system given by (11.19) together with the cost functional given by (11.22). The* H_2 optimal control problem by dynamic measurement feedback *is to find*

$$J_\Sigma^* := \inf \left\{ J_\Sigma(\Gamma) \mid \Gamma = (K, L, M, N) \text{ such that } D_e = 0 \text{ and } \sigma(A_e) \subset \mathbb{C}^- \right\},$$

and to find, if it exists, an optimal feedback controller, i.e., to find

$$\Gamma^* = (K^*, L^*, M^*, N^*)$$

such that $D_2 N^* D_1 = 0$, *the closed loop system is internally stable, i.e.,*

$$\sigma \begin{pmatrix} A + BN^*C_1 & BM^* \\ L^*C_1 & K^* \end{pmatrix} \subset \mathbb{C}^-,$$

and

$$J_\Sigma(\Gamma^*) = J_\Sigma^*.$$

A necessary and sufficient condition for the existence of a dynamic compensator such that the closed loop system is internally stable is that (A, B) is stabilizable and that (C_1, A) is detectable (both with respect to the stability domain $\mathbb{C}^-$). This will be a standing assumption is this section.

The closed loop transfer function $G_\Gamma(s)$ is strictly proper if and only if $D_e = D_2 N D_1 = 0$. We will call a controller *admissible* if it yields $D_e = 0$ and $\sigma(A_e) \subset \mathbb{C}^-$.

In this book, we restrict ourselves to the case that the direct feedthrough map D_1 from the disturbance input to the measured output is *surjective*, and the direct feedthrough map D_2 from the control input to the output to be controlled is *injective*. Under these two assumptions, the H_2 optimal control problem is called *regular*. In the regular case we have $D_2 N D_1 = 0$ if and only if $N = 0$. Hence, a controller $\Gamma = (K, L, M, N)$ is admissible if and only if $N = 0$ and $\sigma(A_e) \subset \mathbb{C}^-$. Note that if $N = 0$ then the closed loop system map A_e is equal to

$$A_e = \begin{pmatrix} A & BM \\ LC_1 & K \end{pmatrix}.$$

The H_2 optimal control problem by measurement feedback is said to be *in standard form* if $D_2^{\mathrm{T}} C_2 = 0$ and $D_2^{\mathrm{T}} D_2 = I$, and $D_1 E^{\mathrm{T}} = 0$ and $D_1 D_1^{\mathrm{T}} = I$. In this section we assume that the problem is in standard form.

Let P be any real symmetric solution of the algebraic Riccati equation ARE associated with the system (A, B, C_2, D_2) (the part of the system Σ that represents the open loop transfer from u to z):

$$A^{\mathrm{T}} P + PA - PBB^{\mathrm{T}} P + C_2^{\mathrm{T}} C_2 = 0. \tag{11.23}$$

For any admissible controller $\Gamma = (K, L, M, 0)$, define $X(t)$, $W(t)$ and $U(t)$ by

$$\begin{pmatrix} X(t) \\ W(t) \end{pmatrix} := e^{A_e t} \begin{pmatrix} E \\ LD_1 \end{pmatrix}, \quad U(t) := MW(t).$$

It is easily verified that

$$\frac{\mathrm{d}}{\mathrm{d}t} X(t) = AX(t) + BU(t), \quad X(0) = E,$$

and hence, according to (11.12), we have

$$\frac{\mathrm{d}}{\mathrm{d}t} X^{\mathrm{T}} PX = (B^{\mathrm{T}} PX + U)^{\mathrm{T}} (B^{\mathrm{T}} PX + U) - (C_2 X + D_2 U)^{\mathrm{T}} (C_2 X + D_2 U).$$

Again by integrating the above from 0 to T and letting $T \to \infty$, we find

$$\int_0^\infty (C_2 X(t) + D_2 U(t))^{\mathrm{T}} (C_2 X(t) + D_2 U(t))\, \mathrm{d}t$$
$$= E^{\mathrm{T}} PE + \int_0^\infty (B^{\mathrm{T}} PX(t) + U(t))^{\mathrm{T}} (B^{\mathrm{T}} PX(t) + U(t))\, \mathrm{d}t. \tag{11.24}$$

It is also easily verified that

$$C_2X(t) + D_2U(t) = C_e e^{A_e t} E_e = T_\Gamma(t)$$

and

$$B^{\mathrm T}PX(t) + U(t) = C_{P,e} e^{A_e t} E_e,$$

with $C_{P,e} := \begin{pmatrix} B^{\mathrm T}P & M \end{pmatrix}$. Substituting these expressions in (11.24) and taking the trace on both sides, we obtain

$$\int_0^\infty \mathrm{trace}[T_\Gamma^{\mathrm T}(t)T_\Gamma(t)]\,\mathrm{d}t = \mathrm{trace}[E^{\mathrm T}PE] + \int_0^\infty \mathrm{trace}[T_{P,\Gamma}^{\mathrm T}(t)T_{P,\Gamma}(t)]\,\mathrm{d}t, \quad (11.25)$$

where we have defined

$$T_{P,\Gamma}(t) := C_{P,e} e^{A_e t} E_e.$$

Observe that $T_{P,\Gamma}$ is equal to the closed loop impulse response obtained by applying the controller Γ to the auxiliary system Σ_P, defined by

$$\begin{aligned} \dot{x}(t) &= Ax(t) + Bu(t) + Ed(t), \\ y(t) &= C_1x(t) + D_1d(t), \\ z(t) &= B^{\mathrm T}Px(t) + u(t). \end{aligned} \quad (11.26)$$

Note that Σ_P is obtained from Σ by replacing the output equation $z = C_2x + D_2u$ by the new output equation $z = B^{\mathrm T}Px + u$. Also observe that the second integral in (11.25) is equal to $J_{\Sigma_P}(\Gamma)$, i.e., the performance of the closed loop system obtained by applying the controller Γ to the system Σ_P. Thus we have obtained the following lemma:

Lemma 11.8 *Consider the system Σ. Assume that $D_2^T C_2 = 0$ and $D_2^T D_2 = I$, and that (A, B) is stabilizable. Let P be a real symmetric solution of the ARE (11.23). A controller $\Gamma = (K, L, M, 0)$ internally stabilizes Σ if and only if it internally stabilizes Σ_P. For any such controller we have*

$$J_\Sigma(\Gamma) = \mathrm{trace}(E^T PE) + J_{\Sigma_P}(\Gamma). \quad (11.27)$$

Like in the state feedback case, the quantity trace$(E^{\mathrm T}PE)$ does no depend on Γ, so the lemma shows that we can replace the minimization of the cost functional J_Σ associated with the original system Σ by the minimization of the cost functional J_{Σ_P} associated with the transformed system Σ_P. However, in contrast with the state feedback case, it is not immediately clear which controller minimizes the functional J_{Σ_P}. It turns out that we need a *second* transformation step in order to arrive at a transformed system for which the H_2 optimal control problem does have an obvious solution.

Again consider the system Σ given by (11.19). Let the dual system $\Sigma^{\mathrm T}$ be given by

$$\begin{aligned} \dot{x}(t) &= A^{\mathrm T}x(t) + C_1^{\mathrm T}u(t) + C_2^{\mathrm T}d(t), \\ y(t) &= B^{\mathrm T}x(t) + D_2^{\mathrm T}d(t), \\ z(t) &= E^{\mathrm T}x(t) + D_1^{\mathrm T}u(t). \end{aligned}$$

For a given controller $\Gamma = (K, L, M, 0)$, let $\Gamma^{\mathrm{T}} := (K^{\mathrm{T}}, M^{\mathrm{T}}, L^{\mathrm{T}}, 0)$ be its dual. Clearly, Γ internally stabilizes Σ if and only if Γ^{T} internally stabilizes Σ^{T}. Furthermore, the closed loop impulse response matrix of the interconnection of Σ^{T} and Γ^{T} is equal to the transpose $T_\Gamma{}^{\mathrm{T}}$ of T_Γ. Since the L_2-norm of a matrix function does not change under transposition, we find

$$J_\Sigma(\Gamma) = J_{\Sigma^{\mathrm{T}}}(\Gamma^{\mathrm{T}}).$$

Note that if $D_1 E^{\mathrm{T}} = 0$, $D_1 D_1^{\mathrm{T}} = I$, and (C_1, A) is detectable, then the system Σ^{T} satisfies the assumptions of lemma 11.8. Thus, we can apply lemma 11.8 to Σ^{T}. For this, we write down the algebraic Riccati equation associated with the system $(A^{\mathrm{T}}, C_1^{\mathrm{T}}, E^{\mathrm{T}}, D_1^{\mathrm{T}})$:

$$AQ + QA^{\mathrm{T}} - QC_1^{\mathrm{T}}C_1Q + EE^{\mathrm{T}} = 0. \tag{11.28}$$

This Riccati equation will be called *the dual ARE*. For any real symmetric solution Q, we consider the auxiliary system $(\Sigma^{\mathrm{T}})_Q$ given by

$$\begin{aligned} \dot{x}(t) &= A^{\mathrm{T}}x(t) + C_1^{\mathrm{T}}u(t) + C_2^{\mathrm{T}}d(t), \\ y(t) &= B^{\mathrm{T}}x(t) + D_2^{\mathrm{T}}d(t), \\ z(t) &= C_1Qx(t) + u(t), \end{aligned}$$

By applying lemma 11.8 we find that for any controller $\Gamma = (K, L, M, 0)$ that internally stabilizes Σ we have

$$J_{\Sigma^{\mathrm{T}}}(\Gamma^{\mathrm{T}}) = \mathrm{trace}(C_2QC_2^{\mathrm{T}}) + J_{(\Sigma^{\mathrm{T}})_Q}(\Gamma^{\mathrm{T}}).$$

We would like to reformulate this in terms of the original system Σ. For this, define the system Σ_Q to be the dual of $(\Sigma^{\mathrm{T}})_Q$, i.e., the system given by the equations

$$\begin{aligned} \dot{x}(t) &= Ax(t) + Bu(t) + QC_1^{\mathrm{T}}d(t), \\ y(t) &= C_1x(t) + d(t), \\ z(t) &= C_2x(t) + D_2u(t). \end{aligned} \tag{11.29}$$

Σ_Q is obtained from Σ by replacing the differential equation $\dot{x} = Ax + Bu + Ed$ by $\dot{x} = Ax + Bu + QC^{\mathrm{T}}d$, and the output equation $y = C_1x + D_1d$ by $y = C_1x + d$. We then immediately obtain:

Lemma 11.9 *Consider the system Σ. Assume that $D_1E^T = 0$ and $D_1D_1^T = I$, and that (C_1, A) is detectable. Let Q be a real symmetric solution of the dual ARE (11.28). A controller $\Gamma = (K, L, M, 0)$ internally stabilizes Σ if and only if it internally stabilizes Σ_Q. For any such controller we have*

$$J_\Sigma(\Gamma) = \mathrm{trace}(C_2QC_2^T) + J_{\Sigma_Q}(\Gamma). \tag{11.30}$$

The idea is now to apply lemma 11.9 to the system Σ_P. As before, let P be a real symmetric solution of the algebraic Riccati equation (11.23) and let Σ_P be given by (11.26). Observe that the dual ARE associated with Σ_P coincides with the dual

ARE associated with Σ and is given by (11.28). Let Q be a real symmetric solution of the dual ARE and define $\Sigma_{PQ} := (\Sigma_P)_Q$. In other words, Σ_{PQ} is obtained from the original system Σ in two steps: first transform Σ to Σ_P, and subsequently transform the systems Σ_P to $(\Sigma_P)_Q$. The transformed system Σ_{PQ} is then given by the equations

$$\begin{aligned} \dot{x}(t) &= \quad Ax(t) + Bu(t) + QC_1^{\mathrm{T}}d(t), \\ y(t) &= \quad C_1x(t) + \qquad\qquad\quad d(t), \\ z(t) &= B^{\mathrm{T}}Px(t) + \quad u(t). \end{aligned} \tag{11.31}$$

By applying lemma 11.8 to Σ and subsequently lemma 11.9 to Σ_P we then obtain:

Lemma 11.10 *Consider the system Σ. Assume that $D_2^TC_2 = 0$, $D_2^TD_2 = I$, $D_1E^T = 0$, $D_1D_1^T = I$, (A, B) is stabilizable, and (C_1, A) is detectable. Let P be a real symmetric solution of the ARE (11.23) and let Q be a real symmetric solution of the dual ARE (11.28). A controller $\Gamma = (K, L, M, 0)$ internally stabilizes Σ if and only if it internally stabilizes Σ_{PQ}. For any such controller we have*

$$J_\Sigma(\Gamma) = \mathrm{trace}(E^TPE) + \mathrm{trace}(B^TPQPB) + J_{\Sigma_{PQ}}(\Gamma). \tag{11.32}$$

This lemma shows that the H_2 optimal control problem for the original system Σ can be replaced by the H_2 problem for the transformed system Σ_{PQ}. Let us investigate Σ_{PQ}. If we were allowed to use state feedback, then a first attempt to minimize $J_{\Sigma_{PQ}}$ would be to take the state feedback control law $u(t) = -B^{\mathrm{T}}Px(t)$. However, we are only allowed to use the measured output $y(t)$ for feedback. Note however that the measured output of Σ_{PQ} equals $y(t) = C_1x(t) + d(t)$. Thus, we know that the (unknown!) disturbance $d(t)$ is actually equal to $d(t) = y(t) - C_1x(t)$. In turn, this implies that the state trajectory of Σ_{PQ} satisfies

$$\begin{aligned} \dot{x}(t) &= Ax(t) + Bu(t) + QC_1^{\mathrm{T}}d(t) \\ &= Ax(t) + Bu(t) + QC_1^{\mathrm{T}}(y(t) - C_1x(t)) \\ &= (A - QC_1^{\mathrm{T}}C_1)x(t) + Bu(t) + QC_1^{\mathrm{T}}y(t) \end{aligned} \tag{11.33}$$

Since we are only dealing with closed loop transfer matrices, we have $x(0) = 0$. Together with the fact that $x(t)$ satisfies the differential equation (11.33), this implies that we can actually reconstruct $x(t)$ exactly, using the measurements $y(\tau)$, $\tau \leqslant t$. After having reconstructed $x(t)$, we can then apply the control $u(t) = -B^{\mathrm{T}}Px(t)$, which will yield zero output $z(t)$. To be more concrete, introduce the system Ω given by the equation

$$\dot{w}(t) = (A - QC_1^{\mathrm{T}}C_1)w(t) + Bu(t) + QC_1^{\mathrm{T}}y(t).$$

This system is a *state observer* for Σ_{PQ} (see section 3.11). Indeed, if we define the error by $e := w - x$, then e can be seen to satisfy $\dot{e}(t) = (A - QC_1^{\mathrm{T}}C_1)e(t)$. Thus, if $w(0) = x(0)$ then for any input function u we have $w(t) = x(t)$ for all t. Since we actually have $x(0) = w(0) = 0$, $w(t)$ is equal to $x(t)$ for all $t \geqslant 0$. Thus, if we apply

$u(t) = -B^{\mathrm{T}}Pw(t)$ to Σ_{PQ}, then we have $z(t) = 0$ for all $t \geqslant 0$. This control input $u(t)$ is exactly generated by the the controller Γ given by

$$\begin{aligned} \dot{w}(t) &= (A - BB^{\mathrm{T}}P - QC_1^{\mathrm{T}}C_1)w(t) + QC_1^{\mathrm{T}}y(t), \\ u(t) &= -B^{\mathrm{T}}Pw(t). \end{aligned} \tag{11.34}$$

Thus, if we interconnect the system Σ_{PQ} with the controller Γ, then the output z will be identically equal to zero for all disturbance inputs. Stated alternatively: the controller Γ achieves disturbance decoupling for the system Σ_{PQ}. Consequently, the closed loop transfer function between d and z will be equal to zero. We have now proven the following lemma:

Lemma 11.11 *Let P and Q be real symmetric solutions of the ARE (11.23) and the dual ARE (11.28), respectively. Let Γ be defined by (11.34). Then $J_{\Sigma_{PQ}}(\Gamma) = 0$.*

Up to now, we have not taken into account the requirement that an optimal controller should be internally stabilizing. Again, let P and Q be arbitrary real symmetric solutions of the ARE (11.23) and the dual ARE (11.28), respectively. Let $\Sigma_{PQ} := (\Sigma_P)_Q$ be the auxiliary system given by (11.31), obtained from Σ by the transformation $\Sigma \mapsto \Sigma_P \mapsto (\Sigma_P)_Q$. If we interconnect Σ_{PQ} and Γ, then the resulting closed loop system map is equal to

$$A_e = \begin{pmatrix} A & -BB^{\mathrm{T}}P \\ QC_1^{\mathrm{T}}C_1 & A - BB^{\mathrm{T}}P - QC_1^{\mathrm{T}}C_1 \end{pmatrix}$$

It is easily seen that

$$\sigma(A_e) = \sigma(A - BB^{\mathrm{T}}P) \cup \sigma(A - QC_1^{\mathrm{T}}C_1).$$

Consequently, Γ is an internally stabilizing controller if and only if $A - BB^{\mathrm{T}}P$ and $A - QC_1^{\mathrm{T}}C_1$ are $\mathbb{C}^-$-stable. By lemma 10.14 we know that if A has no (C_2, A)-unobservable eigenvalues on the imaginary axis, then there is exactly one real symmetric solution of the ARE (11.23), P^+, such that $\sigma(A - BB^{\mathrm{T}}P^+) \subset \mathbb{C}^-$ (P^+ is the largest real symmetric solution of the ARE). Also, if A^{T} has no $(E^{\mathrm{T}}, A^{\mathrm{T}})$-unobservable eigenvalues on the imaginary axis, then the dual ARE (11.28) has exactly one real symmetric solution, Q^+, such that $\sigma(A^{\mathrm{T}} - C_1^{\mathrm{T}}C_1Q) \subset \mathbb{C}^-$. Equivalently: if A has no (A, E)-uncontrollable eigenvalues on the imaginary axis then there is exactly one real symmetric solution, Q^+, of the dual ARE such that $\sigma(A - Q^+C_1^{\mathrm{T}}C_1) \subset \mathbb{C}^-$ (Q^+ is the largest real symmetric solution of the dual ARE).

We conclude that if all eigenvalues of A on the imaginary axis are (C_2, A)-observable and (A, E)-controllable, then the controller Γ given by (11.34), with the particular choices $P = P^+$ and $Q = Q^+$, achieves disturbance decoupling with internal stability for the system $\Sigma_{P^+Q^+}$. Thus, by applying lemma 11.10 with $P = P^+$ and $Q = Q^+$, we arrive at the following:

Theorem 11.12 *Consider the system Σ. Assume that (A, B) is stabilizable, that (C_1, A) is detectable, and that all eigenvalues of A on the imaginary axis are (C_2, A)-observable and (A, E)-controllable. Furthermore, assume that $D_2^T C_2 = 0$, $D_2^T D_2 =$*

I, and $D_1E^T = 0$, $D_1D_1^T = I$. Let P^+ be the largest real symmetric solution of the ARE (11.23) and let Q^+ be the largest real symmetric solution of the dual ARE (11.28). Then we have:

(*i*) $J^*_\Sigma = \text{trace}(E^T P^+ E + \text{trace}(B^T P^+ Q^+ P^+ B)$,

(*ii*) *the controller* Γ *given by*

$$\begin{aligned} \dot{w}(t) &= (A - BB^T P^+ - Q^+ C_1^T C_1)w(t) + Q^+ C_1^T y(t), \\ u(t) &= -B^T P^+ w(t) \end{aligned} \tag{11.35}$$

is optimal, i.e., $J_\Sigma(\Gamma) = J^*_\Sigma$ *and* Γ *is internally stabilizing.*

The above theorem gives a complete solution to the H_2 optimal control problem with measurement feedback for the case that the system does not have unobservable and uncontrollable eigenvalues on the imaginary axis.

Like in the state feedback version of the H_2 optimal control problem, we can remove the observability and controllability assumption on the eigenvalues of A on the imaginary axis by considering the problem for a perturbed system. Given Σ, define a perturbed system Σ_ε by

$$\begin{aligned} \dot{x}(t) &= Ax(t) + Bu(t) + E_\varepsilon \tilde{d}(t), \\ y(t) &= C_1 x(t) + D_{1,0}\tilde{d}(t), \\ z(t) &= C_{2,\varepsilon}x(t) + D_{2,0}u(t), \end{aligned} \tag{11.36}$$

with $\varepsilon > 0$ and

$$E_\varepsilon := \begin{pmatrix} E & \varepsilon I \end{pmatrix}, \quad D_{1,0} := \begin{pmatrix} D_1 & 0 \end{pmatrix}, \quad C_{2,\varepsilon} := \begin{pmatrix} C_2 \\ \varepsilon I \end{pmatrix}, \quad D_{2,0} := \begin{pmatrix} D_2 \\ 0 \end{pmatrix}.$$

Note that the H_2 optimal control problem associated with Σ_ε is in standard form. Also note that $(C_{2,\varepsilon}, A)$ is observable and that (A, E_ε) is controllable. The ARE and dual ARE associated with Σ_ε are given by

$$\begin{aligned} A^T P + PA - PBB^T P + C_2^T C_2 + \varepsilon^2 I &= 0, \\ AQ + QA^T - QC_1^T C_1 Q + EE^T + \varepsilon^2 I &= 0, \end{aligned}$$

respectively. Denote the largest real symmetric solutions of these equations by P^+_ε and Q^+_ε, respectively. It was shown in the proof of lemma 10.15 that $P^+_\varepsilon \to P^+$ ($\varepsilon \to 0$), where P^+ is the largest real symmetric solution of the ARE (11.23). Likewise, Q^+_ε converges to Q^+, the largest real symmetric solution of the dual ARE (11.28).

It is easily seen that for any admissible controller Γ we have

$$J_{\Sigma_\varepsilon}(\Gamma) \geqslant J_\Sigma(\Gamma),$$

which immediately implies that

$$J^*_{\Sigma_\varepsilon} \geqslant J^*_\Sigma.$$

By theorem 11.12, $J^*_{\Sigma_\varepsilon} = \text{trace}(E_\varepsilon^{\mathrm{T}} P_\varepsilon^+ E_\varepsilon + \text{trace}(B^{\mathrm{T}} P_\varepsilon^+ Q_\varepsilon^+ P_\varepsilon^+ B)$. Thus, by letting $\varepsilon \to 0$, we find

$$\text{trace}(E^{\mathrm{T}} P^+ E + B^{\mathrm{T}} P^+ Q^+ P^+ B) \geqslant J^*_\Sigma.$$

On the other hand, it follows from lemma 11.10 that also the converse inequality holds, so we may conclude that also without the observability and controllability assumption on the eigenvalues of A we have

$$J^*_\Sigma = \text{trace}(E^{\mathrm{T}} P^+ E + \text{trace}(B^{\mathrm{T}} P^+ Q^+ P^+ B).$$

By applying lemma 11.10 with $P = P^+$ and $Q = Q^+$ we then find: for any $\Gamma = (K, L, M, 0)$ that internally stabilizes Σ (equivalently $\Sigma_{P^+Q^+}$) we have

$$J_\Sigma(\Gamma) = J^*_\Sigma + J_{\Sigma_{P^+Q^+}}(\Gamma).$$

It follows immediately from this that a controller $\Gamma = (K, L, M, 0)$ is optimal *if and only if* it achieves disturbance decoupling with internal stability for $\Sigma_{P^+Q^+}$. Recall that if all eigenvalues of A on the imaginary axis are (C_2, A) observable and (A, E) controllable, then the controller Γ given by (11.35) achieves disturbance decoupling with internal stability for $\Sigma_{P^+Q^+}$. However, this observability and controllability condition is sufficient, but not necessary for the existence of such a compensator.

In order to obtain conditions that are both necessary *and* sufficient, we will now study the disturbance decoupling problem by measurement feedback and internal stability, DDPMS, for the system that is slightly more general than (11.31)

$$\begin{aligned} \dot{x}(t) &= Ax(t) + Bu(t) + Ed(t), \\ y(t) &= C_1 x(t) + d(t), \\ z(t) &= C_2 x(t) + u(t), \end{aligned} \tag{11.37}$$

For a given feedback controller $\Gamma = (K, L, M, 0)$, denote the closed loop transfer matrix by G_Γ and the closed loop system map by A_e. For a given map M, let $\mathcal{X}^-(M)$ and $\mathcal{X}^+(M)$ denote the $\mathbb{C}^-$-stable and $\mathbb{C}^-$-unstable subspace of M, respectively (see definition 2.13).

Lemma 11.13 *Consider the system (11.37). There exists a controller Γ with realization $(K, L, M, 0)$ such that $G_\Gamma(s) = 0$ and $\sigma(A_e) \subset \mathbb{C}^-$ if and only if (A, B) is stabilizable, (C_1, A) is detectable,*

$$\text{im}\, E \subset \mathcal{X}^-(A - BC_2), \tag{11.38}$$

$$\mathcal{X}^+(A - EC_1) \subset \ker C_2, \tag{11.39}$$

$$\mathcal{X}^+(A - EC_1) \subset \mathcal{X}^-(A - BC_2), \tag{11.40}$$

and

$$A\mathcal{X}^+(A - EC_1) \subset \mathcal{X}^-(A - BC_2). \tag{11.41}$$

Proof : ($\Rightarrow$) Assume a strictly proper controller Γ such that $G_\Gamma(s) = 0$ and $\sigma(A_e) \subset \mathbb{C}^-$ exists. The properties of stabilizability and detectability are immediate. The closed loop system is given by $\dot{x}_e = A_e x_e + E_e d$, $z = C_e x_e$, with

$$A_e := \begin{pmatrix} A & BM \\ LC_1 & K \end{pmatrix}, \quad E_e := \begin{pmatrix} E \\ L \end{pmatrix}, \quad C_e := \begin{pmatrix} C_2 & M \end{pmatrix}.$$

By theorem 4.6 there exists an A_e-invariant subspace $\mathcal{V}_e$ of the extended state space $\mathcal{X}_e$ such that $\operatorname{im} E_e \subset \mathcal{V}_e \subset \ker C_e$. Using the notation of section 6.1, let $\mathcal{S} := i(\mathcal{V}_e)$ be the intersection of $\mathcal{V}_e$ with the original state space $\mathcal{X}$ and let $\mathcal{V} := p(\mathcal{V}_e)$ be the projection of $\mathcal{V}_e$ onto $\mathcal{X}$. Clearly we have

$$\mathcal{S} \subset \mathcal{V}. \tag{11.42}$$

Furthermore, it is easily verified that

$$A\mathcal{S} \subset \mathcal{V}, \tag{11.43}$$

$$\operatorname{im} E \subset \mathcal{V}, \tag{11.44}$$

and

$$\mathcal{S} \subset \ker C_2. \tag{11.45}$$

Thus, in order to prove (11.38) to (11.41) it suffices to show that

$$\mathcal{V} \subset \mathcal{X}^-(A - BC_2) \tag{11.46}$$

and

$$\mathcal{X}^+(A - EC_1) \subset \mathcal{S}. \tag{11.47}$$

To show (11.46), let $x_0 \in \mathcal{V}$. Being the projection of $\mathcal{V}_e$ onto $\mathcal{X}$, there exists w_0 such that $(x_0^T, w_0^T)^T \in \mathcal{V}_e$. As initial state of the closed loop system, take (x_0^T, w_0^T). As disturbance input take $d = 0$. Then the corresponding state trajectory $\binom{x}{w}$ is $\mathbb{C}^-$-stable, while for the closed loop output we have $z = 0$. Since $z = C_2 x + u$, we find that the closed loop input signal u is equal to $-C_2 x$. This implies that x satisfies the differential equation

$$\dot{x} = (A - BC_2)x, \quad x(0) = x_0.$$

Since x is $\mathbb{C}^-$-stable this immediately implies that $x_0 \in \mathcal{X}^-(A - BC_2)$. In order to prove (11.47), note that the dual controller Γ^T achieves disturbance decoupling with internal stability for the dual of (11.37). Using the same argument as before, it follows that $p(\mathcal{V}_e^\perp) \subset \mathcal{X}^-(A^T - C_1^T E^T)$. Since $p(\mathcal{V}_e^\perp) = i(\mathcal{V}_e) = \mathcal{S}^\perp$ (see exercise 6.1), and since $\mathcal{X}^-(A^T - C_1^T E^T) = \mathcal{X}^+(A - EC_1)^\perp$, this yields (11.47).

($\Leftarrow$) Assume (A, B) is stabilizable, that (C_1, A) is detectable and that (11.38) to (11.41) hold. Define $\mathcal{V} := \mathcal{X}^-(A + BC_2)$ and $\mathcal{S} := \mathcal{X}^+(A - EC_1)$. It was shown in the proof of lemma 11.5 that there exists a feedback map F such that

$$(A + BF)\mathcal{V} \subset \mathcal{V}, \tag{11.48}$$

$$\mathcal{V} \subset \ker(C_2 + F), \tag{11.49}$$

and

$$\sigma(A + BF) \subset \mathbb{C}^-. \tag{11.50}$$

By dualization, the previous result can be applied to obtain the existence of an output injection map G such that the following properties hold:

$$(A + GC_1)\mathcal{S} \subset \mathcal{S}, \tag{11.51}$$
$$\text{im}(E + G) \subset \mathcal{S}, \tag{11.52}$$

and

$$\sigma(A + GC_1) \subset \mathbb{C}^-. \tag{11.53}$$

Now define a strictly proper controller Γ as follows. For the state space of Γ take $\mathcal{X}$, the state space of the original system. Define $K := A + BF + GC_1$, $L := -G$ and $M := F$. We claim that this controller achieves disturbance decoupling with internal stability. Indeed, with this compensator we have

$$A_e := \begin{pmatrix} A & BF \\ -GC_1 & A + BF + GC_1 \end{pmatrix}, \quad E_e := \begin{pmatrix} E \\ -G \end{pmatrix}, \quad C_e := \begin{pmatrix} C_2 & F \end{pmatrix}.$$

Obviously, $\sigma(A_e) = \sigma(A + BF) \cup \sigma(A + GC_1) \subset \mathbb{C}^-$, so Γ yields internal stability. In order to show that it achieves decoupling, define a subspace of the extended state space by

$$\mathcal{V}_e := \left\{ \begin{pmatrix} x_1 \\ 0 \end{pmatrix} + \begin{pmatrix} x_2 \\ x_2 \end{pmatrix} \middle| \, x_1 \in \mathcal{S}, x_2 \in \mathcal{V} \right\}.$$

It is straightforward to verify that $\mathcal{V}_e$ is A_e-invariant and that $\text{im}\, E_e \subset \mathcal{V}_e \subset \ker C_e$. The result then follows from theorem 4.6. ■

By applying the previous lemma to the system $\Sigma_{P^+Q^+}$, we arrive at:

Theorem 11.14 *Consider the system Σ. Assume that (A, B) is stabilizable and that (C_1, A) is detectable. Furthermore, assume that $D_2^T C_2 = 0$, $D_2^T D_2 = I$, and $D_1 E^T = 0$, $D_1 D_1^T = I$. Let P^+ be the largest real symmetric solution of the ARE (11.23) and let Q^+ be the largest real symmetric solution of the dual ARE (11.28). Then we have*

$$J_\Sigma^* = \text{trace}(E^T P^+ E + \text{trace}(B^T P^+ Q^+ P^+ B).$$

Furthermore, the following conditions are equivalent:

(i) There exists an optimal controller, i.e., a controller $\Gamma = (K, L, M, 0)$ such that $J_\Sigma(\Gamma) = J_\Sigma^*$ and Γ is internally stabilizing,

(ii) there exists a controller $\Gamma = (K, L, M, 0)$ that achieves disturbance decoupling with internal stability for the system $\Sigma_{P^+Q^+}$.

(*iii*) *The following four conditions are satisfied:*

$$\begin{aligned} &\operatorname{im} Q^+C_1^T \subset \mathcal{X}^-(A - BB^TP^+), \\ &\mathcal{X}^+(A - Q^+C_1^TC_1) \subset \ker B^TP^+, \\ &\mathcal{X}^+(A - Q^+C_1^TC_1) \subset \mathcal{X}^-(A - BB^TP^+) \end{aligned}$$

and

$$A\mathcal{X}^+(A - Q^+C_1^TC_1) \subset \mathcal{X}^-(A - BB^TP^+).$$

If any of the conditions (i), (ii) or (iii) hold, then any controller $\Gamma = (K, L, M, 0)$ *that achieves disturbance decoupling with internal stability for the system* $\Sigma_{P^+Q^+}$ *is optimal.*

11.4 Exercises

11.1 Formulate and prove the analogue of theorem 11.6 for a system Σ given by (11.8) for which the H_2 problem is not necessarily in standard form. Do assume that D is injective and (A, B) is stabilizable.

11.2 (The H_2-problem without internal stability.) In addition to the H_2 problem by state feedback treated in this chapter, we can also consider the version of this problem in which it is not required that the state feedback map F internally stabilizes the system. We will call this problem *the H_2 problem without stability*: given the system $\dot{x}(t) = Ax(t) + Bu(t) + Ed(t)$, $z(t) = Cx(t) + Du(t)$, find

$$J_\Sigma^* = \inf\{\, J_\Sigma(F) \mid F : \mathcal{X} \to \mathcal{U} \,\},$$

and find an optimal F, i.e. an F such that $J_\Sigma(F) = J_\Sigma^*$. Assuming that the problem is regular and in standard form, formulate and prove the analogue of theorem 11.6 for this problem. Try to be as general as possible, do not assume that (A, B) is stabilizable.

11.3 (The H_2-filtering problem.) In this exercise we study the *filtering problem*. Consider the following system Σ:

$$\begin{aligned} \dot{x}(t) &= Ax(t) + Ed(t), \\ y(t) &= C_1x(t) + D_1d(t), \\ z(t) &= C_2x(t), \end{aligned}$$

where it is assumed that $\sigma(A) \subset \mathbb{C}^-$. In these equations d represents a disturbance, the variable y represents an output that can be measured and the variable z is an output that we want to estimate on the basis of the output y. To this end, we want to construct a finite-dimensional linear time-invariant system Φ (a *filter*):

$$\begin{aligned} \dot{w}(t) &= Kw(t) + Ly(t), \\ \zeta(t) &= Mw(t) + Ny(t), \end{aligned}$$

such that ζ is an estimate of z, in the following sense. Let $e := z - \zeta$, the error between z and ζ. Let $\mathcal{W}$ be the state space of the system Φ. The interconnection of Σ and Φ is a system with state space $\mathcal{X} \times \mathcal{W}$, described by the equations

$$\begin{aligned} \dot{x}_e(t) &= A_e x_e(t) + E_e d(t), \\ e(t) &= C_e x_e(t) + D_e d(t), \end{aligned} \tag{11.54}$$

where we have introduced the following notation:

$$A_e := \begin{pmatrix} A & 0 \\ LC_1 & K \end{pmatrix}, \; E_e := \begin{pmatrix} E \\ LD_1 \end{pmatrix}, \; C_e := \begin{pmatrix} C_2 - NC_1 & -M \end{pmatrix},$$

$$D_e := -ND_1, \quad x_e(t) := \begin{pmatrix} x(t) \\ w(t) \end{pmatrix}.$$

Assume that $\sigma(A_e) \subset \mathbb{C}^-$, and that d is a standard white noise process. The size of the estimation error e is measured by the expected value $E\{\|e(t)\|^2\}$. It was explained in section 11.1 that this quantity is independent of t, and is equal to the square of the L_2 norm of the impulse response matrix $T_e(t) = C_e e^{A_e t} E_e + D_e \delta$ from d to e. In the filtering problem we want to minimize the cost functional

$$J_\Sigma(\Phi) := \int_0^\infty \|T_e(t)\|^2 \, \mathrm{d}t$$

over all filters Φ such that $\sigma(A_e) \subset \mathbb{C}^-$. We assume that (C_1, A) is $\mathbb{C}^-$-detectable. We assume in this exercise that the filtering problem is *regular*, i.e., that D_1 is surjective. Also, to simplify the problem, we assume that the problem is in *standard form*, i.e. that $D_1 E^{\mathrm{T}} = 0$ and that $D_1 D_1^{\mathrm{T}} = I$.

a. A filter is called admissible if $\sigma(A_e) \subset \mathbb{C}^-$ and $D_e = 0$. Show that a filter is admissible if and only if $\sigma(A_e) \subset \mathbb{C}^-$ and $N = 0$.

b. Consider the dual algebraic Riccati equation $AQ + QA^{\mathrm{T}} - QC_1^{\mathrm{T}} C_1 Q + EE^{\mathrm{T}} = 0$. For any real symmetric solution Q, define the system Σ_Q by

$$\begin{aligned} \dot{x}(t) &= Ax(t) + QC_1^{\mathrm{T}} d(t), \\ y(t) &= C_1 x(t) + d(t), \\ z(t) &= C_2 x(t). \end{aligned}$$

Show that Φ is an admissible filter for Σ if and only if it is an admissible filter for Σ_Q, and prove that for every admissible filter Φ we have

$$J_\Sigma(\Phi) = \mathrm{trace}(C_2 Q C_2^{\mathrm{T}}) + J_{\Sigma_Q}(\Phi).$$

c. Assume that every eigenvalue of A on the imaginary axis is (A, E)-controllable. Let Q^+ be the largest real symmetric solution of the dual ARE. Prove that the filter Φ^*:

$$\begin{aligned} \dot{w}(t) &= (A - Q^+ C_1^{\mathrm{T}} C_1) w(t) + Q^+ C_1^{\mathrm{T}} y(t), \\ \zeta(t) &= C_2 w(t), \end{aligned}$$

is admissible, and that $J_{\Sigma_{Q^+}}(\Phi^*) = 0$.

d. Prove that Φ^* is an optimal filter and that

$$J_\Sigma(\Phi^*) = J_\Sigma^* = \text{trace}(C_2 Q^+ C_2^{\mathrm{T}}).$$

11.4 (The singular H_2-problem.) In this exercise we study the singular H_2-problem. Consider the system Σ: $\dot{x}(t) = Ax(t) + Bu(t) + Ed(t)$, $z(t) = Cx(t) + Du(t)$. Let (A, B) be $\mathbb{C}^-$-stabilizable. The H_2-optimal control problem is called *singular* if D is not injective. One way to deal with the singular H_2-problem is via the perturbed system Σ_ε: $\dot{x}(t) = Ax(t) + Bu(t) + Ed(t)$, $z_\varepsilon(t) = C_1 x(t) + D_\varepsilon u(t)$, with

$$C_1 := \begin{pmatrix} C_1 \\ 0 \end{pmatrix}, \quad D_\varepsilon := \begin{pmatrix} D \\ \varepsilon I \end{pmatrix},$$

where $\varepsilon > 0$.

a. Show that for any F such that $\sigma(A_F) \subset \mathbb{C}^-$ we have $J_{\Sigma_\varepsilon}(F) \geqslant J_\Sigma(F)$.

b. Show that $J_{\Sigma_\varepsilon}^* = \text{trace}(E^{\mathrm{T}} P_\varepsilon^+ E)$, where P_ε^+ is the largest real symmetric solution of the ARE

$$A^{\mathrm{T}}P + PA + C^{\mathrm{T}}C - (PB + C^{\mathrm{T}}D)(D^{\mathrm{T}}D + \varepsilon^2)^{-1}(B^{\mathrm{T}}P + D^{\mathrm{T}}C) = 0.$$

In exercise 10.7 it was shown that $P_\varepsilon^+ \to P^+$ ($\varepsilon \to 0$), where P^+ is the largest real symmetric solution of the linear matrix inequality

$$F(P) := \begin{pmatrix} A^{\mathrm{T}}P + PA + C^{\mathrm{T}}C & PB + C^{\mathrm{T}}D \\ B^{\mathrm{T}}P + D^{\mathrm{T}}C & D^{\mathrm{T}}D \end{pmatrix} \geqslant 0.$$

c. Factorize $F(P^+) = (C_+ \;\; D_+)^{\mathrm{T}}(C_+ \;\; D_+)$, and introduce the system Σ_{P+} by $\dot{x}(t) = Ax(t) + Bu(t) + Ed(t)$, $z(t) = C_+ x(t) + D_+ u(t)$. Show that for all F such that $\sigma(A_F) \subset \mathbb{C}^-$ we have

$$J_\Sigma(F) = \text{trace}(E^{\mathrm{T}} P^+ E) + J_{\Sigma_{P+}}(F).$$

d. Prove that $J_\Sigma^* = \text{trace}(E^{\mathrm{T}} P^+ E)$.

e. Prove that there exists an optimal F^*, i.e., F^* such that $\sigma(A+BF^*) \subset \mathbb{C}^-$ and $J_\Sigma(F^*) = J_\Sigma^*$, if and only if DDPS is solvable for the system Σ_{P+}, and that in that case F^* is optimal if and only if $\sigma(A + BF^*) \subset \mathbb{C}^-$ and

$$(C_+ + D_+ F^*)(Is - A - BF^*)^{-1} E = 0.$$

We consider the special case that $D = 0$, i.e., Σ is given by $\dot{x}(t) = Ax(t) + Bu(t) + Ed(t)$, $z(t) = Cx(t)$. The H_2 problem is then called *totally singular*. For a given subspace $\mathcal{K}$ of $\mathcal{X}$, let $\mathcal{V}_g^*(\mathcal{K})$ be the largest $\mathbb{C}^-$-stabilizability subspace contained in $\mathcal{K}$.

f. Show that there exists an optimal F^* if and only if

$$\operatorname{im} E \subset \mathcal{V}_g^*(\ker(A^{\mathrm{T}}P^+ + P^+A + C^{\mathrm{T}}C)),$$

and that F^* is optimal if and only if $\sigma(A + BF^*) \subset \mathbb{C}^-$ and

$$(A^{\mathrm{T}}P^+ + P^+A + C^{\mathrm{T}}C)(Is - A - BF^*)^{-1}E = 0.$$

11.5 (The H_2-problem in the non-standard case.) In section 11.3, the H_2-problem was treated only for the standard case, i.e., under the assumptions that $D_2^{\mathrm{T}}C_2 = 0$, $D_2^{\mathrm{T}}D_2 = I$, $D_1E^{\mathrm{T}} = 0$, and $D_1D_1^{\mathrm{T}} = I$. The regular, non-standard case can be treated analogously, yielding only more complicated formulas.

a. By redoing the analysis of section 11.3, prove the following theorem:
Theorem 1: Consider the system Σ given by (11.19). Assume that D_1 is surjective, D_2 is injective, (A, B) is stabilizable, and (C_1, A) is detectable. Also assume that (A, B, C_2, D_2) and (A, E, C_1, D_1) have no zeros on the imaginary axis. Let P^+ and Q^+ be the largest real symmetric solutions of the ARE

$$A^{\mathrm{T}}P + PA + C_2^{\mathrm{T}}C_2 - (PB + C_2^{\mathrm{T}}D_2)(D_2^{\mathrm{T}}D_2)^{-1}(PB + C_2^{\mathrm{T}}D_2)^{\mathrm{T}} = 0,$$

and dual ARE

$$AQ + QA^{\mathrm{T}} + EE^{\mathrm{T}} - (QC_1^{\mathrm{T}} + ED_1^{\mathrm{T}})(D_1D_1^{\mathrm{T}})^{-1}(QC_1^{\mathrm{T}} + ED_1^{\mathrm{T}})^{\mathrm{T}} = 0,$$

respectively. Then we have:

1. $J_\Sigma^* = \operatorname{trace}(E^{\mathrm{T}}P^+E) + \operatorname{trace}(B^{\mathrm{T}}P^+Q^+P^+B)$,
2. the controller Γ given by

$$\begin{aligned} \dot{w} &= (A - (Q^+C_1^{\mathrm{T}} + ED_1^{\mathrm{T}})C_1)w + Bu \\ &\qquad + (Q^+C_1^{\mathrm{T}} + ED_1^{\mathrm{T}})(D_1D_1^{\mathrm{T}})^{-1}y, \\ u &= (D_2D_2^{\mathrm{T}})^{-1}(B^{\mathrm{T}}P^+ + D_2^{\mathrm{T}}C_2)w \end{aligned}$$

is optimal, i.e., $J_\Sigma(\Gamma) = J_\Sigma^*$ and Γ is internally stabilizing.

b. Also prove the following theorem:
Theorem 2: Consider the system Σ given by (11.19). Assume that D_1 is surjective, D_2 is injective, (A, B) is stabilizable, and (C_1, A) is detectable. Let P^+ and Q^+ be the largest real symmetric solutions of the ARE and dual ARE, respectively. Then we have

$$J_\Sigma^* = \operatorname{trace}(E^{\mathrm{T}}P^+E) + \operatorname{trace}(B^{\mathrm{T}}P^+Q^+P^+B).$$

Furthermore, the following conditions are equivalent:

1. There exists an optimal controller, i.e., a controller $\Gamma = (K, L, M, 0)$ such that $J_\Sigma(\Gamma) = J_\Sigma^*$ and Γ is internally stabilizing,

2. there exists a controller $\Gamma = (K, L, M, 0)$ that achieves disturbance decoupling with internal stability for the auxiliary system $\Sigma_{P^+Q^+}$ given by

$$\begin{aligned}\dot{x} &= Ax + Bu + (Q^+C_1^{\mathrm{T}} + ED_1^{\mathrm{T}})(D_1D_1^{\mathrm{T}})^{-1}d,\\ y &= C_1x + d,\\ z &= (D_2^{\mathrm{T}}D_2)^{-1}(B^{\mathrm{T}}P^+ + D_2^{\mathrm{T}}C_2)x + u.\end{aligned}$$

3. The following four conditions hold:

$$\operatorname{im}(Q^+C_1^{\mathrm{T}} + ED_1^{\mathrm{T}}) \subset \mathcal{X}^-,$$

$$\mathcal{X}^+ \subset \ker(B^{\mathrm{T}}P^+ + D_2^{\mathrm{T}}C_2),$$

$$\mathcal{X}^+ \subset \mathcal{X}^-,$$

and

$$A\mathcal{X}^+ \subset \mathcal{X}^-,$$

where we have denoted

$$\mathcal{X}^- = \mathcal{X}^-(A - B(D_2^{\mathrm{T}}D_2)^{-1}(B^{\mathrm{T}}P^+ + D_2^{\mathrm{T}}C_2)),$$

and

$$\mathcal{X}^+ = \mathcal{X}^+(A - (Q^+C_1^{\mathrm{T}} + ED_1^{\mathrm{T}})(D_1D_1^{\mathrm{T}})^{-1}C_1).$$

If any of the conditions 1, 2 or 3 holds, then a controller $\Gamma = (K, L, M, 0)$ that achieves disturbance decoupling with internal stability for the system $\Sigma_{P^+Q^+}$ is optimal.

11.5 Notes and references

The H_2 optimal control problem is the modern version of what is commonly known as the linear quadratic Gaussian (LQG) problem. As indicated in section 11.1, minimization of the H_2-norm of the closed loop transfer matrix can be given the stochastic interpretation of minimizing the expected value of the squared norm of the output, in case that the disturbance input is a standard white noise process. It is exactly the minimization of this expected value that the classical formulation of the LQG-problem deals with. Although, in the sixties, it was already generally known that the H_2-norm of the transfer matrix was the thing to be minimized, a direct approach to perform this minimization was not yet known. Hence, the general approach to find optimal LQG-controllers was to compute the optimal state feedback law from the deterministic linear quadratic regulator problem, and to apply this feedback law to the optimal estimate of the state, obtained from the Kalman-Bucy filter as developed by Kalman and Bucy [95] and Kalman [92]. The fact that this two-stage method indeed leads to an optimal LQG-controller is called *the separation principle*, that was proven rigorously for the first time by Wonham in [222]. The period 1960-1970 showed intensive research efforts on the LQG-problem. This led to a special issue of the IEEE Transactions on Automatic Control on the LQG-problem in 1971 [9], edited by M. Athans. Many classical references on the LQG problem can be found there. Since then, several textbooks have appeared dealing with the subject of LQG-control, among which we mention the work of Kwakernaak and Sivan [105], and Anderson and Moore [6].

The 'modern' way to treat the LQG-problem is to get rid of the stochastics in the problem formulation, and to perform a direct minimization of the H_2-norm of the closed loop transfer matrix. This approach, earlier studied in papers by Youla, Bongiorno and Jabr [226,227], was brought back to attention more recently in Doyle, Glover, Khargonekar and Francis [41]. For a treatment of the singular version, i.e., the H_2 problem without assumptions on the direct feedthrough matrices from disturbance input to measurement output, and from control input to output to-be-controlled, we refer to Stoorvogel [185]. In connection with exercise 11.3 on the H_2-filtering problem, we also refer to Schumacher [174].

Chapter 12

H_∞ control and robustness

In the following chapters, we will study the H_∞ control problem and the related problem of robust stability. A main motivation for this problem was robust stability. We would like to guarantee the stability of a plant even if the model on the basis of which the controller was designed is not a perfect representation of the behavior of the plant. On the other hand it also yields a clear link between several classical frequency-domain methods and more recent state space methods. This book has focused completely on state space methods. This is in contrast with some of the classical control books which focus almost entirely on frequency domain methods. H_∞ control is actually very useful as a link between these two approaches. In this chapter, we introduce the H_∞ control problem and show its connection with robust stability and some of its links to classical frequency domain methods.

12.1 Robustness analysis

Control theory is concerned with the control of processes with inputs and outputs. We would like to achieve desired specifications on the controlled system by choosing our inputs appropriately.

Example 12.1 Assume that we have a paper-making machine. This machine has certain inputs: wood-pulp, water, pressure and steam. The wood-pulp is diluted with water. Then the fibers are separated from the water and a web is formed. Water is pressed out of the mixture and the paper is then dried on steam-heated cylinders (this is of course a very simplified view of the process). The product of the plant is the paper. More precisely we have two outputs: the thickness of the paper and the mass of fibers per unit area (expressing the quality of the paper). We would like both outputs to be equal to some desired value. Thus, we have a process with a number of inputs and two goals: we would like to make the deviation from the desired values of the thickness and of the mass of fibers per unit area of the paper produced as small as possible.

The first step is to find a mathematical model describing the behavior of our plant. The second step is to use mathematical tools to find suitable input signals for our plant based on measurements we make of all, or of a subset, of our outputs; in other words to design a feedback controller. However, ultimately we apply this controller to our physical plant and not to our model. Since our model should be sufficiently simple for the mathematical tools of the second step (for instance, in this book we require the model to be linear) and since we never have complete information regarding our physical plant, the model will not describe the plant exactly. Because we do not know how sensitive our objectives are with respect to the differences between model and plant, the behavior obtained might differ significantly from the mathematically predicted behavior. Hence our controller will in general not be suitable for our plant and the behavior we actually obtain can be differ from the desired behavior in a critical way.

Therefore, it is extremely important that, when we search for a control law for our model, we keep in mind that our model is far from perfect. This leads to the so-called robustness analysis of our plant and suggested controllers. Robustness of a system says nothing more than that the stability of the system (or another design specification we want to achieve for the system) is preserved if the model is replaced by a more complex system (such as the real plant), as long as this new system is close to our model.

An approach used for multivariable controller design which stems from the 1960s is the Linear Quadratic Gaussian (LQG) or H_2 theory discussed in chapter 11. In that approach the uncertainty is modeled as a white noise Gaussian process added as an extra (vector) input to the system. The major problem of this approach is that our uncertainty cannot always be modeled as white noise. While measurement noise can be quite well described by a random process, this is not the case with parameter uncertainty. If we model $a = 0.9$ instead of $a = 1$, then the error is not random but deterministic. In particular, the noise is biased. The only problem is that the deterministic error is unknown. Another problem of main importance with parameter uncertainty is that uncertainty in the transfer from inputs to outputs cannot be modeled as state or output disturbances, i.e. as extra inputs. This is due to the fact that the size of the errors is relative to the size of the inputs and can hence only be modeled as an extra input in a *non-linear* framework. We will show in exercise 12.5 that state feedback H_2 optimal controllers still exhibit a certain robustness. However dynamic measurement feedback H_2 optimal controllers can have arbitrarily bad robustness margins.

In the last few years several approaches to robustness have been studied, mainly with one goal: to obtain internal stability, where instead of trying to obtain this for one system, we seek one single controller that stabilizes any element from a certain class of systems. It is then hoped that the controller that stabilizes all elements of this class of systems also stabilizes the plant itself. Either because the plant is in this class of systems or because the controller has become insensitive to plant variations.

In this chapter, several approaches to this problem will be briefly discussed. Each of these approaches will result in an H_∞ control problem. In the next two chapters the solution of the H_∞ control problem will be discussed.

12.2 The H_∞ control problem

We now state the H_∞ control problem. Assume that we have a system Σ: As usual,

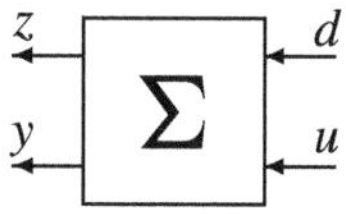

we assume Σ to be a finite-dimensional linear time-invariant system. We note that Σ is a system with two kinds of inputs and two kinds of outputs. The input d is an exogenous input representing the disturbance acting on the system. The output z is an output of the system, whose dependence on the exogenous input d we want to minimize. The output y is a measurement we make on the system, which we shall use to choose our input u, which in turn is the tool we have to minimize the effect of d on z. A constraint we impose is that this mapping from y to u should be such that the closed-loop system is internally stable. This is quite natural since we do not want the states to become too large while we try to regulate our performance. The effect of w on z after closing the loop is measured in terms of the energy and the worst disturbance d. Our performance measure, which will turn out to be equal to the closed-loop H_∞ norm, is the supremum over all disturbances unequal to zero of the quotient of the energy flowing out of the system and the energy flowing into the system.

In the previous chapter the H_2 norm of a transfer matrix was introduced by:

$$\|G\|_2^2 := \frac{1}{2\pi} \int_{-\infty}^{+\infty} \operatorname{trace} G(i\omega)G^*(i\omega)\, \mathrm{d}\omega. \tag{12.1}$$

The H_∞ norm of a transfer matrix is defined by:

$$\|G\|_\infty := \sup_{\omega \in \mathbb{R}} \|G(i\omega)\| < \infty. \tag{12.2}$$

where $\|M\|$ denotes the largest singular value of the complex matrix M. The H_2 and H_∞ norms are only defined for stable transfer matrices. However, for the H_2 norm to be finite we need in addition that the transfer matrix is strictly proper. We denote by H_2 and H_∞ the space of all stable transfer matrices for which, respectively, (12.1) and (12.2) are well-defined and finite. Of course (12.1) and (12.2) are also defined for unstable transfer matrices as long as the transfer matrix has no poles on the imaginary axis (and in the case of (12.1), is strictly proper). In that case we refer to (12.2) as the L_∞ norm. By L_∞, we denote the space of all transfer matrices for which (12.2) is well-defined and finite.

Both H_2 and H_∞ are special cases of so-called Hardy spaces named after G.H. Hardy (1877–1947). They are formally defined as spaces of functions defined on the open right half plane. On the other hand L_2 and L_∞ are spaces of functions defined on the imaginary axis. The above definitions are therefore not very precise but sufficient for us since we only consider rational transfer matrices in this book. However it is useful to realize that for stable transfer matrices we have:

$$\|G\|_\infty := \sup_{s \in \mathbb{C}^+} \|G(s)\| < \infty. \tag{12.3}$$

where $\mathbb{C}^+$ denotes the open right half complex plane.

Define L_2^n as the set of functions f from $[0, \infty)$ to $\mathbb{R}^n$ for which

$$\|f\|_2 := \left(\int_0^\infty \|f(t)\|^2 \, dt \right)^{1/2} \tag{12.4}$$

is well-defined and finite, where $\|g\|^2 := g^\mathrm{T} g$ denotes the Euclidian norm. This norm is induced by the following inner product

$$\langle f, g \rangle_2 := \int_0^\infty g(t)^\mathrm{T} f(t) \, dt,$$

In other words $\|f\|_2^2 = \langle f, f \rangle_2$. By L_2 we denote the set of all f for which an n exists such that $f \in L_2^n$.

Note that, since we call the extension of the H_∞ norm to unstable transfer matrices the L_∞ norm, it would be natural to call the extension of the H_2 norm to unstable transfer matrices the L_2 norm. We do not do this and, instead, we use the name L_2 norm for the expression (12.4). This is slightly inconsistent. However, the expressions (12.1) and (12.4) are closely related. If we have a function in L_2 then its Laplace transform will be in H_2 and (12.4) for the function will be equal to (12.1) for its Laplace-transform (a consequence of the so-called Parseval's theorem). It is good to note that the H_2 norm will only be applied to transfer matrices (in the frequency domain) while the L_2 norm will only be used for input-output signals (in the time-domain).

In this book, instead of the above definition, we shall frequently use a time-domain characterization of the H_∞ norm. Let $\Sigma \times \Gamma$ be the closed-loop system obtained by applying a linear controller Γ to a given system Σ. If the closed-loop system is stable, then the closed-loop transfer matrix G_{cl} is in H_∞. Given the closed loop system, there exists a map which associates to every input d an output z, given zero initial state. We will denote this map, sometimes called the *input-output operator* by $\mathcal{G}_{cl}$. It is easy to check that $\mathcal{G}_{cl}$ is a well-defined linear mapping. The H_∞ norm of G_{cl} is equal to the L_2-induced operator norm of $\mathcal{G}_{cl}$, i.e.

$$\|G_{cl}\|_\infty = \|\mathcal{G}_{cl}\|_\infty := \sup_d \left\{ \frac{\|\mathcal{G}_{cl} d\|_2}{\|d\|_2} \;\middle|\; d \in L_2^\ell,\ d \neq 0 \right\}. \tag{12.5}$$

Note that we will sometimes use the terminology H_∞ norm for an input-output operator $\mathcal{G}$ even though it is formally better to call this the L_2-induced operator norm.

In order to see this equality let the closed loop system have impulse response g. Then we have:

$$z(t) = (\mathcal{G}_{cl}d)(t) = \int_0^t g(t-\tau)d(\tau)\,\mathrm{d}\tau$$

If we apply the Laplace transform to z we obtain:

$$\hat{z}(s) = G_{cl}(s)\hat{d}(s)$$

where we used the fact that the transfer matrix is equal to the Laplace transform of the impulse response. On the other hand, Parseval's theorem, that we already mentioned before, tells us:

$$\|z\|_2^2 = \frac{1}{2\pi}\int_0^\infty z^*(i\omega)z(i\omega)\,\mathrm{d}\omega.$$

Combining these properties yields:

$$\begin{aligned}\|\mathcal{G}_{cl}d\|_2^2 &= \int_0^\infty \|G_{cl}(i\omega)\hat{d}(i\omega)\|^2\,\mathrm{d}\omega \leqslant \left(\sup_\omega \|G_{cl}(i\omega)\|^2\right)\|\hat{d}\|_2^2 \\ &= \|G_{cl}\|_\infty^2\|d\|_2^2.\end{aligned}$$

Finally, it is not very hard to see that this inequality becomes an equality for those $\hat{d}$ that contain only sharp peaks in the direction and at the frequency where G_{cl} is maximal. Since we take the supremum over all possible d in (12.5), it is then not difficult to see that we do get the equality (12.5).

Because of this equality we often refer to the L_2-induced operator norm of the closed-loop operator $\mathcal{G}_{cl}$ as the H_∞ norm of $\mathcal{G}_{cl}$. The H_∞ norm of a stable system is defined as the H_∞ norm of the corresponding transfer matrix of that system.

We note that the above alternative interpretation also has the advantage that it gives a natural way to extend the H_∞ norm to nonlinear or time-varying systems. We will use a finite-horizon version in the next chapter, which is defined as:

$$\|\mathcal{G}_{cl}\|_{\infty,\mathrm{T}} := \sup_d \left\{ \frac{\|\mathcal{G}_{cl}d\|_{2,\mathrm{T}}}{\|d\|_{2,\mathrm{T}}} \;\middle|\; d \in L_2^\ell[0,T],\ d \neq 0 \right\}. \tag{12.6}$$

where

$$\|f\|_{2,\mathrm{T}} := \left(\int_0^T \|f(t)\|^2\,\mathrm{d}t\right)^{1/2} < \infty,$$

This is clearly connected to the time-domain interpretation of the H_∞ norm given above. It is not possible to connect this finite-horizon version to the frequency domain interpretation (12.2) or (12.1).

The H_∞ norm is motivated as a tool to achieve robustness, i.e. a method to design controllers that are insensitive to variations in the model. However, it is very important to realize that minimizing the closed loop H_∞ norm in itself does not have any connection with robustness. One can only guarantee robustness in connection with the small-gain theorem, which will be discussed in the next section.

Example 12.2 Assume that we have the following system:

$$\Sigma: \begin{aligned} \dot{x} &= \quad -u + d, \\ y &= x, \\ z &= \quad\quad u. \end{aligned}$$

It can be checked that a feedback controller that minimizes the closed loop H_∞ norm from d to z is given by:

$$u = \varepsilon x$$

for any $\varepsilon > 0$. Compared to the H_2 control problem (where in most cases the optimal controller is unique), it is interesting to see that any stabilizing state feedback is optimal. On the other hand it is easily seen that a small perturbation of the system parameters might yield an unstable closed-loop system when we have a small ε. For robust stability it is better to choose ε large. This is however not predicted by the H_∞ norm from d to z which gives no preference for one ε over another.

Hence for the controller with small ε, internal stability of the closed-loop system is certainly not robust with respect to perturbations of the state matrix, even though the controller is optimal in the sense of the H_∞ control problem.

12.3 The small-gain theorem

The connection between the H_∞ norm and the problem of robust stabilization is made via the small-gain theorem. In this section we present and prove this theorem. Consider the following interconnection:

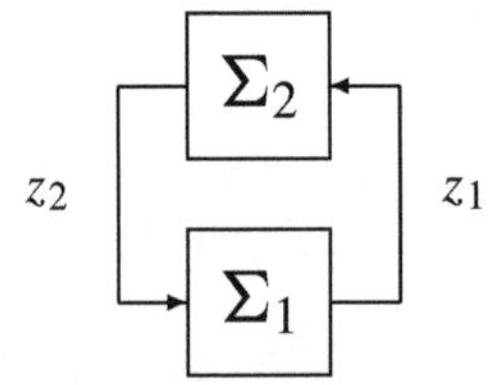

Figure 12.1

Before we can discuss the stability of the interconnection in Figure 12.1 we need a precise definition of stability of such an interconnection. The definition consistent with chapter 3 is to consider the state space model of the interconnection based on state space models for the two systems Σ_1 and Σ_2 and require the state space model of the interconnection to be internally stable. However, in the context of the small gain theorem this definition is not very easy to work with. We first give an alternative but equivalent characterization of internal stability of such an interconnection. We consider the following figure:

Lemma 12.3 *The interconnection of Σ_1 and Σ_2 as in Figure 12.1 is internally stable if and only if:*

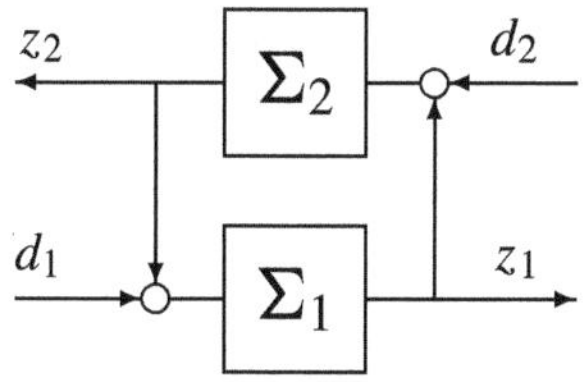

Figure 12.2

(*i*) Σ_1 *and* Σ_2 *are stabilizable and detectable.*

(*ii*) *Given the interconnection in Figure 12.2, we have* $z_1, z_2 \in L_2$ *for all signals* $d_1, d_2 \in L_2$ *and zero initial conditions.*

We could have combined the two parts by allowing arbitrary initial conditions in part (ii). In that case, part (i) is not needed. However, the current formulation will be easier to work with. We can now present the small gain theorem:

Theorem 12.4 *Let* Σ_1 *be a given, stable system. The interconnection (12.1) is internally stable for all internally stable systems* Σ_2 *with* H_∞ *norm less than or equal to* 1 *if and only if the* H_∞ *norm of* Σ_1 *is strictly less than* 1.

Proof : Clearly part (i) of lemma 12.3 is always automatically satisfied since both systems are internally stable.

Suppose Σ_1 has H_∞ norm strictly less than 1. Let Σ_2 be any internally stable system with H_∞ norm less than or equal to 1. Denote the transfer matrices of Σ_1 and Σ_2 by G_1 and G_2, respectively. The transfer matrix from d_1, d_2 to z_1, z_2 in the interconnection (12.2) is equal to:

$$\begin{pmatrix} G_1(I - G_2G_1)^{-1} & G_1(I - G_2G_1)^{-1}G_2 \\ G_2G_1(I - G_2G_1)^{-1} & (I - G_2G_1)^{-1}G_2 \end{pmatrix}$$

To prove internal stability of the closed loop system it is therefore sufficient to guarantee that $I - G_2G_1$ has a stable inverse (remember that G_1 and G_2 are stable). We know that $\|G_2G_1\|_\infty < 1$. But then for any point s in the closed right half plane (using the alternative characterization of the H_∞ norm in (12.3)) we have $\|G_1(s)G_2(s)\| < 1$, which guarantees that $I - G_1(s)G_2(s)$ is invertible. This implies that $I - G_1G_2$ has the required stable inverse.

Conversely, suppose that Σ_1 has H_∞ norm larger than or equal to 1. We will construct a stable Σ_2 with H_∞ norm less than or equal to 1 which yields an unstable interconnection (12.1). Since Σ_1 has H_∞ norm larger than or equal to 1, there exists a point s in the closed right half plane (possibly $s = \infty$) for which the transfer matrix G_1 of Σ_1 satisfies $\|G_1(s)\| \geqslant 1$. In other words there exist $u, v \neq 0$ such that

$G_1(s)u = v$ and $\|u\| \leqslant \|v\|$. In that case we could choose Σ_2 equal to the constant gain D, where

$$D = \frac{uv^*}{\|v\|^2}$$

Then Σ_2 will have H_∞ norm less than or equal to 1 and it will destabilize the interconnection (12.1) since the closed loop system has a pole at s. However, Σ_2 would in general be a constant, *complex* gain. We can also find a real-valued, destabilizing system Σ_2 but in that case Σ_2 must in general be dynamic. Since it is rather technical, an explicit construction will not be given. ■

Remark 12.5 The small gain theorem as defined above yields as an easy corollary that the interconnection of two stable systems with transfer matrices G_1 and G_2 is stable if $\|G_1 G_2\|_\infty < 1$. This implies that the interconnection is stable if

$$\sup_\omega \|G_1(i\omega)\| \|G_2(i\omega)\| < 1 \tag{12.7}$$

This basically implies that if for certain values of ω, i.e. for certain frequencies $G_1(i\omega)$ is quite large then for these frequencies $G_2(\omega)$ must be very small.

Note that the above theorem in fact holds for any induced operator norm. The H_∞ norm, which is equal to the L_2-induced operator norm (as we saw in the previous section), is just a particular example. Note that the H_2 norm is not an induced operator norm (see exercise 12.7) and we can therefore not find a comparable result for the H_2 norm.

12.4 Stabilization of uncertain systems

As already mentioned, a method for handling the problem of robustness is to treat the uncertainty as additional input(s) to the system. The LQG design method treats these inputs as white noise, and we noted in the beginning of this chapter that parameter and dynamic uncertainty are not suited for treatment as white noise. Also the idea of treating the error as extra inputs is not suitable because the size of the error depends in general on the size of the state. For instance, if we have a system $\dot{x} = ax + bu$ then variations in a have an effect which depends on the size of the state x.

This motivates an approach where parameter or dynamic uncertainty is modeled as a disturbance *system* taking values in some range and modeled in a feedback setting (which allows us to incorporate the "relative" character of the error). We would like to know the effect with respect to stability of the "worst" disturbance in the prescribed range (we want guaranteed performance so, even if the worst happens, it should still be acceptable). If this disturbance does not destabilize the system, then we are certain (under the assumption that the physical plant is exactly described by a system associated with a particular choice of parameter values within the prescribed range) that the plant is stabilized by our control law.

In a linear setting, parameter uncertainty can very often be modeled as in Figure 12.3. Here the system Δ represents the uncertainty and if the transfer matrix of

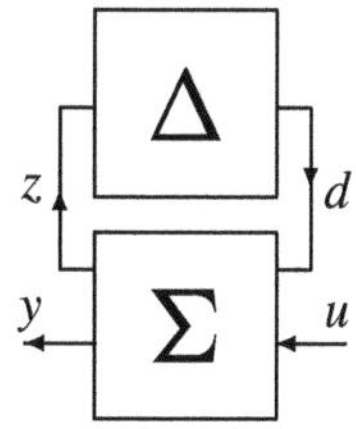

Figure 12.3

Δ is zero, then we obtain our nominal model from u to y. The system Δ might contain uncertain parameters, ignored dynamics after model reduction or discarded non-linearities. The goal is to find a feedback controller that stabilizes the model for a large range of systems Δ. In this chapter we shall give some examples of different kinds of uncertainties that can be modeled in the above sense. We shall also show what the results of this book, applied to these problems, look like. At this point, we only show by means of an example that a large class of parameter uncertainties can be considered as an interconnection of the form depicted in Figure 12.3.

Example 12.6 Assume that we have a single-input, single-output system with two unknown parameters:

$$\Sigma_n : \begin{aligned} \dot{x} &= -ax + bu, \\ y &= \quad x. \end{aligned}$$

where a and b are parameters with values in the ranges $[a_0 - \varepsilon, a_0 + \varepsilon]$ and $[b_0 - \delta, b_0 + \delta]$ respectively. We can consider this system as an interconnection of the form (12.3) by choosing the system Σ to be equal to:

$$\Sigma : \begin{aligned} \dot{x} &= -a_0 x + b_0 u + d, \\ y &= x, \\ z &= \begin{pmatrix} -1 \\ 0 \end{pmatrix} x + \begin{pmatrix} 0 \\ 1 \end{pmatrix} u. \end{aligned}$$

and the system Δ to be the following static system:

$$d = \begin{pmatrix} a - a_0 & b - b_0 \end{pmatrix} z.$$

It is easily seen that by scaling we may assume that $\varepsilon = \delta = 1$.

On the other hand, if we want to find a controller from y to u such that the closed-loop system is internally stable for all *internally stable* systems Δ with H_∞ norm less than or equal to γ, then the problem is equivalent to the following problem: find a controller which is such that the closed-loop system (if the transfer matrix of Δ is zero) is internally stable and the H_∞ norm from d to z is strictly less than γ^{-1}. The latter result is a direct consequence of the small gain theorem treated in the previous section. Suppose Γ is some arbitrary controller from y to u. After interconnection, Figure 12.3 then has the additional structure depicted in Figure 12.4.

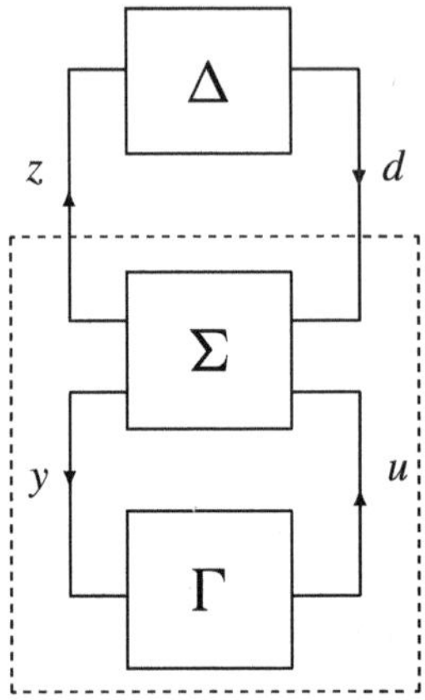

Figure 12.4

Denote the interconnection of Γ and Σ in the dotted box by Σ_1. The interconnection should be stable for all Δ with norm less than 1. Clearly this implies that the interconnection must be stable for $\Delta = 0$ which yields the requirement that Σ_1 must be stable. The rest then follows from the small gain theorem. The interconnection of Δ and Σ_1 is stable for all stable systems Δ with H_∞ norm less than or equal to 1 if and only if Σ_1 has H_∞ norm less than 1. To scale 1 to γ is an easy exercise.

In this section we will apply the results of this book to three specific types of uncertainty:

- Additive perturbations
- Multiplicative perturbations
- Coprime-factor perturbations.

Each time we find a problem which can be reduced to an H_∞ control problem. The first two problems can be found in [123,207]. The last problem is discussed in [123].

12.4.1 Additive perturbations

Assume that we have a system Σ

$$\Sigma : \begin{aligned} \dot{x} &= Ax + Bu, \\ y &= Cx + Du. \end{aligned} \tag{12.8}$$

being an imperfect model of a certain plant. We assume that the error is additive, i.e. we assume that the actual plant can be exactly described by the interconnection in Figure 12.5. Here Δ is some arbitrary system such that the transfer matrices of Σ

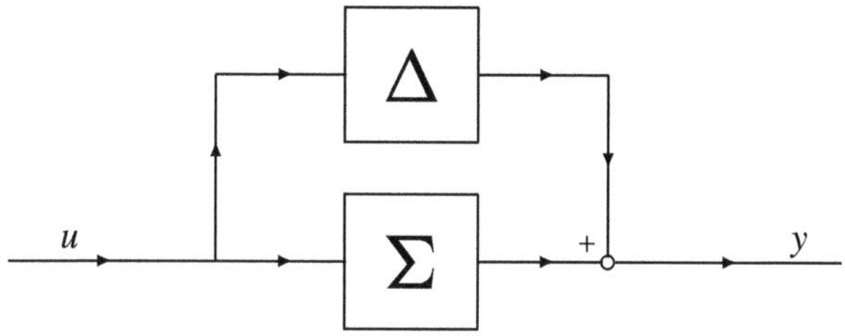

Figure 12.5

and $\Sigma + \Delta$ have the same number of unstable poles. Thus we assume that the plant is described by the system Σ interconnected as in Figure 12.5 with another system Δ. The system Δ represents the uncertainty and is hence, by definition, unknown. In this subsection we derive conditions under which a controller Γ of the form

$$\Sigma_F : \begin{array}{l} \dot{w} = Kw + Ly, \\ u = Mw + Ny. \end{array} \tag{12.9}$$

exists such that the interconnection (12.5) is stabilized by this controller for all systems Δ which do not change the number of unstable poles and which have L_∞ norm less than or equal to some, a priori given, positive number γ. Note that since the error system is assumed to have a finite L_∞ norm, it can not have poles on the imaginary axis. It should be noted that an assumption like fixing the number of unstable poles is needed since otherwise there are always arbitrary small perturbations that destabilize the closed-loop system. We have the following result:

Lemma 12.7 *Assume that Σ given by (12.8) is stabilizable and detectable. Let Γ be a controller of the form (12.9). Let $\gamma > 0$. The following conditions are equivalent:*

(*i*) *The controller Γ from y to u applied to the interconnection in Figure 12.5, yields a well-posed and internally stable closed-loop system for every system Δ with L_∞ norm less than or equal to γ and such that Σ and $\Sigma + \Delta$ have the same number of unstable poles.*

(*ii*) *The controller Γ from y to u internally stabilizes the following system*

$$\Sigma_{na} : \begin{array}{l} \dot{x} = Ax + Bu, \\ y = Cx \qquad + d, \\ z = \qquad u. \end{array} \tag{12.10}$$

and is such that the H_∞ norm of the closed-loop transfer matrix from d to z is strictly less than γ^{-1}.

Proof : A proof for the case that Σ is not internally stable is a bit technical. We will only give a proof for the case when Σ is internally stable.

We want to use the technique described in the beginning of this section. We can get an interconnection of the form (12.4) quite easily since applying the controller Γ from y to u in Figure 12.5 then this corresponds to the interconnection in Figure 12.6. The small gain theorem then gives the equivalence between (i) and (ii). ■

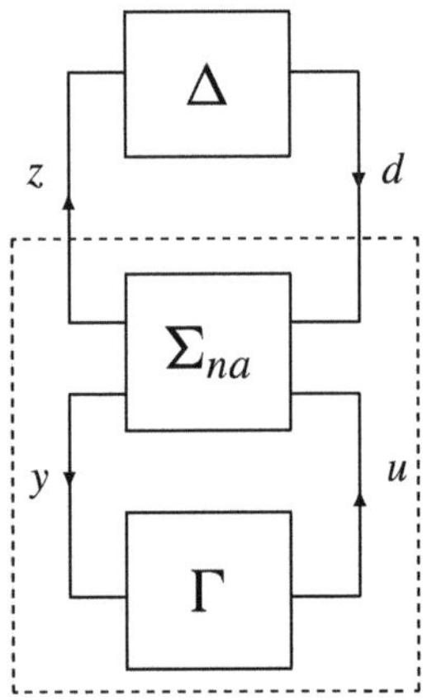

Figure 12.6

Remark 12.8 It is quite easy to find a counterexample for the above theorem if we do not impose the condition that Σ and $\Sigma + \Delta$ have the same number of unstable poles. Actually, without that condition for any any given controller, we can always find perturbations Δ with arbitrarily small L_∞ norm which yield an unstable closed loop system.

Remark 12.9 We can actually also use the arguments presented in (12.5) instead of the small gain theorem. This will enable us to get a stronger result which looks more at the model uncertainty at each frequency, i.e. at each value of ω. It tells us, for instance, that if for high frequencies we have more model uncertainty i.e. if for large values of ω we have that $\Delta(i\omega)$ will be large, then we have to make sure that the closed-loop transfer matrix from d to z is extra small for these large frequencies.

12.4.2 Multiplicative perturbations

We assume that again we have system Σ of the form (12.8) being an imperfect model of a certain plant. In addition to the additive error used in the previous subsection, it is often useful to describe uncertainty via a relative instead of an absolute error. This implies that this time we assume that the error is multiplicative, i.e. we assume that the plant is exactly described by the interconnection in Figure 12.7. Here Δ is some arbitrary system such that the interconnection in Figure 12.7 has the same number

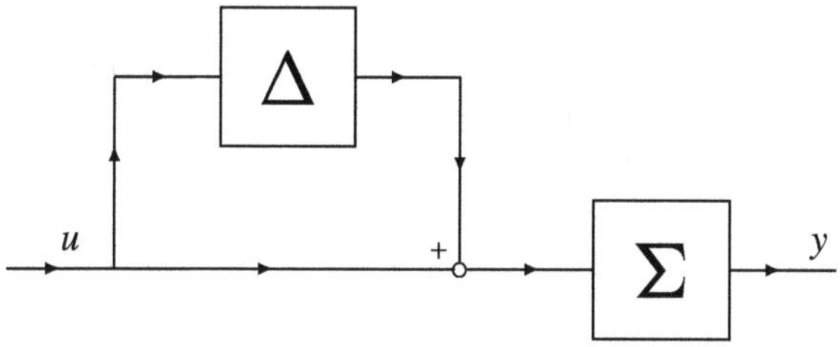

Figure 12.7

of unstable poles as Σ. In other words, we assume that the plant is described by the system Σ interconnected as in Figure 12.7 with another system Δ. The system Δ represents the uncertainty. As in the case of additive perturbations, our goal is to find conditions under which a controller Γ of the form (12.9) from y to u exists such that the interconnection in Figure 12.7 is stabilized by this controller for all systems Δ which do not change the number of unstable poles of the interconnection in Figure 12.7 and which have L_∞ norm less than *or equal to* some, a priori given, positive number γ. It is easy to adapt the proof of lemma 12.7 to obtain a result for multiplicative perturbations:

Lemma 12.10 *Assume that Σ given by (12.8) is stabilizable and detectable. Let Γ be a controller of the form (12.9). Let $\gamma > 0$. The following conditions are equivalent:*

(*i*) *The controller Γ from y to u in the interconnection of Figure 12.7, yields a well-posed and internally stable closed-loop system for every system Δ with L_∞ norm less than or equal to γ and such that Σ and the interconnection in Figure 12.7 have the same number of unstable poles.*

(*ii*) *The controller Γ from y to u stabilizes the following system*

$$\Sigma_{nm}: \begin{array}{l} \dot{x} = Ax + Bu + Bd, \\ y = Cx + Du + Dd, \\ z = \qquad\quad u. \end{array} \tag{12.11}$$

and is such that the closed-loop H_∞ norm from d to z is strictly less than γ^{-1}.

Remark 12.11 One can again find counterexamples for the above lemma if we do not impose the condition that Σ and the interconnection in Figure 12.7 have the same number of unstable poles. Actually, without that condition for any given controller, we can always find perturbations Δ with arbitrarily small L_∞ norm which yield an unstable closed loop system.

Remark 12.12 Note that in the interconnection of Figure 12.7 we have the uncertainty at the input of the system. We can also formulate multiplicative uncertainty with the uncertainty at the output of the system. These two descriptions are obviously identical in case y and u are scalar signals but are identical in general. We can trivially formulate a similar result as lemma 12.10 for the case of multiplicative uncertainty at the output of the plant. See exercise 12.2.

Remark 12.13 As already stated in remark 12.9 we can use the arguments presented in (12.5) instead of the small gain theorem. This will enable us to get a stronger result which looks more at the model uncertainty at each frequency, i.e. at each value of ω.

12.4.3 Coprime factor perturbations

In the previous two subsections we treated the two most common versions of unstructured uncertainty: the absolute or the relative error.

However, we have only considered perturbations of the system which do not change the number of unstable poles and as noted in remarks 12.8 and 12.11 the results are no longer true without this condition. Hence we need a different method for the case that we do not have sufficient information on the unstable poles.

Moreover, when using an additive or multiplicative model uncertainty structure, systems with poles close to the imaginary axis are very sensitive to variations in these poles in the sense that very small variations in the pole location still lead to very large Δ-blocks. Therefore, it is hard to guarantee stability of the closed loop system.

In this section we will present an alternative approach which tries to handle these problems. Suppose we have a (possibly unstable) system with transfer matrix G. We can always factorize $G = M^{-1}N$ where N and M are both stable, proper transfer matrices. Moreover we can guarantee that M is biproper (i.e. it has a proper but not necessarily stable inverse). We can now perturb the factors N and M, i.e. we assume our real system P can be described by $(M - \Delta_M)^{-1}(N + \Delta_N)$. We get the interconnection in Figure 12.8. Although the structure of the picture looks quite

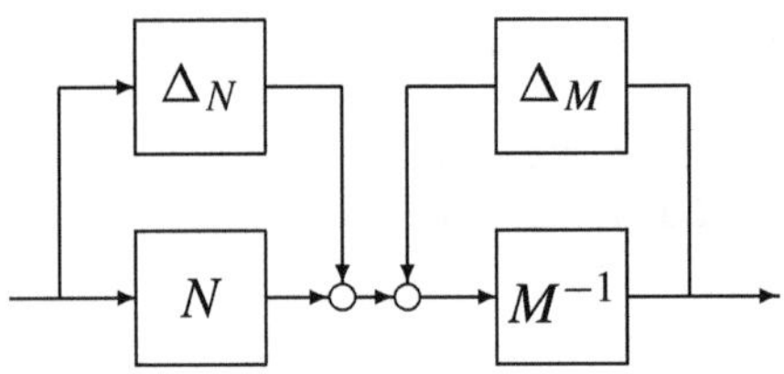

Figure 12.8

different from the additive and multiplicative uncertainty case since there are two uncertainty blocks, this is only a visual difference. By defining $\Delta := (\Delta_N \quad \Delta_M)$, we have one matrix-valued uncertainty block.

A more intricate question in this approach is the factorization of G into N and M. This factorization is not unique and the results can strongly depend on the precise choice of N and M. A first obvious requirement is that N and M are left-coprime, i.e. there exist stable matrices $X, Y \in H_\infty$ such that $NX + MY = I$. This is natural because it is a direct extension to the matrix-valued case of the requirement of no unstable pole-zero cancellations between N and M. After all if $M(s_0) = 0$ and $N(s_0) = 0$ with $s_0 \in \mathbb{C}^+$ then the stability of X and H implies that $N(s_0)X(s_0) + M(s_0)Y(s_0)$ which is in contradiction with $NX + MY = I$.

We can choose N and M such that we obtain a normalized coprime factorization, i.e. $NN^* + MM^* = I$, where $N^*(s) = N^{\mathrm{T}}(-s)$. It is obvious that multiplying N and M by a factor α will result in an error Δ which is also multiplied by the factor α. Therefore some kind of normalization is quite natural. The following lemma tells us how to obtain such a normalized coprime factorization:

Lemma 12.14 *Assume a system Σ given by (12.8) is stabilizable and detectable and has transfer matrix G. There exists a solution of the following algebraic Riccati equation:*

$$0 = AX + XA^T - (BD^T + XC^T)(I + DD^T)^{-1}(DB^T + CX) + BB^T$$

such that

$$A - (BD^T + XC^T)(I + DD^T)^{-1}C$$

is a stability matrix. Then $G = NM^{-1}$ is a normalized left-coprime factorization were N and M are the transfer matrices of systems Σ_N and Σ_M respectively which are given by:

$$\Sigma_N = \left(A + HC, B + HD, (I + DD^T)^{-1/2}C, (I + DD^T)^{-1/2}D\right), \quad (12.12a)$$

$$\Sigma_M = \left(A + HC, H, (I + DD^T)^{-1/2}C, (I + DD^T)^{-1/2}\right). \quad (12.12b)$$

where $H = -(BD^T + XC^T)(I + DD^T)^{-1}$.

Proof : The existence of a stabilizing solution to the algebraic Riccati equation is a direct consequence of theorem 10.20 with (A, B, C, D) replaced by

$$(A^{\mathrm{T}}, C^{\mathrm{T}}, (B \quad 0)^{\mathrm{T}}, (D \quad I)^{\mathrm{T}})$$

To prove that N and M are left-coprime it suffices to check that $NX + MY = I$ where X and Y are the transfer matrices of the systems Σ_X and Σ_Y respectively which are given by

$$\Sigma_X = \left(A + BF, -H(I + DD^{\mathrm{T}})^{1/2}, -F, 0\right),$$

$$\Sigma_Y = \left(A + BF, -H(I + DD^{\mathrm{T}})^{1/2}, C + DF, (I + DD^{\mathrm{T}})^{1/2}\right).$$

and F is an arbitrary matrix such that $A + BF$ is stable. Moreover, we have to check that $NN^* + MM^* = I$ which shows that it is a normalized factorization. Both of these tasks require only straightforward algebraic manipulations. ■

Finally, we conclude this section by showing that the problem of stabilizing (12.8) for all possible Δ with $\|\Delta\|_\infty < \gamma$ reduces once again to a standard H_∞ control problem. The proof of this result is again a simple adaptation of the proof of lemma 12.7.

Lemma 12.15 *Assume that Σ given by (12.8) is stabilizable and detectable. Let G be the transfer matrix of Σ and let $G = M^{-1}N$ be a coprime factorization as constructed in lemma 12.14. Also let a controller Γ of the form (12.9) be given. The following conditions are equivalent:*

(i) *The controller Γ from y to u applied to the interconnection in Figure 12.8, yields a closed-loop system which is well posed and internally stable for every stable system $\Delta := (\Delta_N \quad \Delta_M)$ such that $\|\Delta\|_\infty < \gamma$.*

(ii) *The controller Γ from y to u stabilizes the following system and is such that the closed-loop H_∞ norm from d to z is strictly less than γ^{-1}:*

$$\Sigma_{cf}: \quad \begin{aligned} \dot{x} &= Ax + Bu - H(I + DD^T)^{1/2} d, \\ y &= Cx + Du + (I + DD^T)^{1/2} d, \\ z &= \begin{pmatrix} C \\ 0 \end{pmatrix} x + \begin{pmatrix} D \\ I \end{pmatrix} u + \begin{pmatrix} (I + DD^T)^{1/2} \\ 0 \end{pmatrix} d. \end{aligned} \tag{12.13}$$

Remark 12.16 As already stated in remark 12.9 we can use the arguments presented in (12.5) instead of the small gain theorem. This will enable us to get a stronger result which looks more at the model uncertainty at each frequency, i.e. at each value of ω.

To conclude this section, we should note the following. In the case of additive and multiplicative perturbations we had to impose a restriction that Δ should not change the number of unstable poles of the open loop system. This restriction need not be made for perturbations of coprime factors as described in this subsection. Assume the real system has transfer matrix G_r with coprime factorization N_r, M_r. One possible choice is $\Delta_N = N_r - N$ and $\Delta_M = M_r - M$. Since a coprime factorization is not unique, this might not yield the smallest possible Δ. But this particular choice does yield stable perturbations of N and M.

The unstable poles of the model are the zeros of M. On the other hand the unstable poles of the original system are the zeros of $M + \Delta_M$. Clearly by appropriate choices of Δ_M the number of unstable poles of the system can vary. Of course, the freedom in the number of unstable poles is still implicitly limited by the size of the allowable perturbations. But this time it enters much more naturally. On the other hand, coprime factorizations appear to have a rather weak system theoretic interpreation and therefore the whole structure presented in this subsection seems to be less natural compared to the other two cases.

12.5 The mixed-sensitivity problem

The mixed-sensitivity problem is a special kind of H_∞ control problem. In the mixed-sensitivity problem it is assumed that the system under consideration can be written as the following interconnection where Γ is the controller which has to satisfy certain prerequisites.

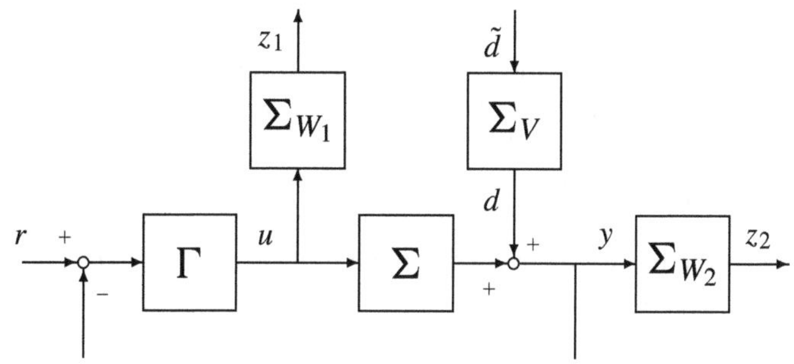

Figure 12.9

Many H_∞ control problems can be formulated in terms of an interconnection depicted in Figure 12.9. As an example, we show how the tracking problem can be formulated in the setting described by the diagram in Figure 12.9. First look at the following interconnection:

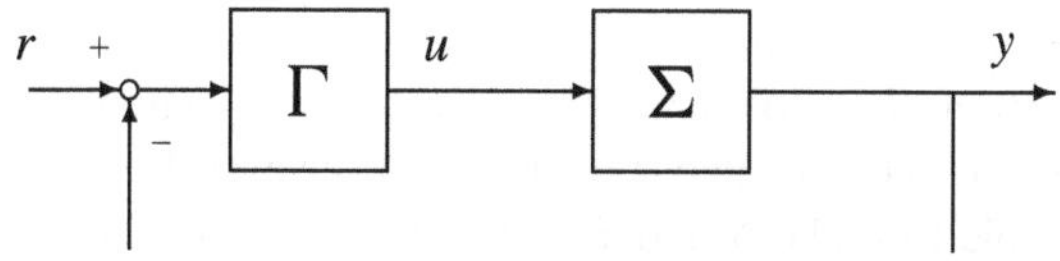

Figure 12.10

The problem is to regulate the output y of the system Σ to look like some given reference signal r by designing a precompensator Γ which has as its input the error signal, i.e. the input of the controller is the difference between the output y of Σ and the reference signal r. To prevent undesirable surprises we require internal stability. We could formulate the problem as "minimizing" the transfer function from r to $r - y$. As one might expect we shall minimize the H_∞ norm of this transfer function under the constraint of internal stability. The transfer matrix from r to u should also be under consideration. In practice, the process inputs will often be restricted by physical constraints. This yields a bound on the transfer matrix from r to u. These transfer matrices from r to $r - y$ and from r to u are given by:

$$S := (I + GH)^{-1},$$
$$T := H\,(I + GH)^{-1},$$

respectively, where G and H denote the transfer matrices of Σ and Γ. Here S is called the *sensitivity function* and T is called the *control sensitivity function*. A small function S expresses good tracking properties while a small function T expresses small inputs u. Note that $S + GT = I$ and therefore there is a trade-off: making S smaller will in general make T larger. We add a signal d to the output y as in Figure 12.9 on the preceding page. Then the transfer matrix from d to y is equal to the sensitivity matrix S and the transfer matrix from d to u is equal to the control sensitivity matrix T.

Although we assume the tracking signal to be, a priori, unknown, in most cases we know that our tracking signal will have a limited frequency spectrum, i.e. the Fourier transform of the reference signal $\hat{r}$ will be small for many frequencies, i.e. $\|\hat{r}(i\omega)\|$ is small for many values of ω. In most cases, we only track slowly varying signals which implies that $\|\hat{r}(i\omega)\|$ is small for large frequencies ω. By minimizing $\|W_2 S\|_\infty$, where W_2 is large for small values of ω and small for large values of ω, we actually put more effort in achieving good tracking for small values of ω. In other words, we put most effort in tracking slowly time-varying signals r.

On the other hand it is in general very difficult to implement an input u which varies very quickly. Therefore we want to ensure in particular that T is small for large frequencies. By minimizing $\|W_1 T\|_\infty$, where W_1 is small for small values of ω and large for large values of ω, we actually put more emphasis on avoiding fast changes over time in the input u.

The transfer matrices W_1 and W_2 are referred to as weighting functions to express the above information. It helps us to incorporate our requirements that we want S to be small at low frequencies and T to be small at high frequencies. Note that it is general impossible to make S and T both small for the same frequency because $S + GT = I$. It is obvious that if we want track signals up to higher frequencies, i.e with faster time-variations, yet have severe bandwidth limitations on the controller, i.e. the controller can not vary very fast over time then we will not be able to obtain satisfactory results.

Therefore, in Figure 12.9 on the preceding page, the systems Σ_{W_1}, Σ_{W_2} and Σ_V are weights which are chosen in such a way that we put more effort in regulating frequencies of interest instead of one uniform bound. To handle the above conflicting requirements the choice of these weights is the crucial component in the design.

In the above, we have given a sketch of the ideas resulting in the structure of the interconnection depicted in the interconnection (12.9). Note that the transfer matrix from the disturbance $\tilde{d}$ to z_1 and z_2 is

$$\begin{pmatrix} W_1 T V \\ W_2 S V \end{pmatrix} \tag{12.14}$$

where W_1, W_2 and V are the transfer matrices of Σ_{W_1}, Σ_{W_2} and Σ_V, respectively.

Note that we can also use these weights to stress the relative importance of minimizing the sensitivity matrix S with respect to the importance of minimizing the control sensitivity matrix T by multiplying W_1 by a scalar.

We want to find a controller which minimizes the H_∞ norm of the transfer matrix (12.14) and which yields internal stability. This problem can be solved using the techniques we present in this book.

12.6 The bounded real lemma

As we have shown in the previous sections, the H_∞ norm can be used as a tool in the study of multivariable control systems. In the next two chapters we will study the minimization of the H_∞ norm using stabilizing controllers. In the present section, we will establish methods to determine the H_∞ norm of a system Σ:

$$\begin{aligned} \dot{x} &= Ax + Ed, \\ z &= Cx + Dd. \end{aligned} \tag{12.15}$$

We note that the time-domain interpretation of the H_∞ norm as the L_2-induced operator norm suggests to study the following problem:

$$\sup_{d \in L_2} \left\{ \|z\|_2^2 - \gamma^2 \|d\|_2^2 \mid x(0) = \xi \right\}. \tag{12.16}$$

If the system has initial state 0, i.e. $\xi = 0$ then this expression is finite only if the H_∞ norm of the system is less than or equal to γ. After all if we use the expression (12.5) for the H_∞ norm then we see that if the H_∞ norm is larger than γ then there exists a signal $d \in L_2$ such that

$$\frac{\|z\|_2}{\|d\|_2} > \gamma$$

but taking the square and then multiplying with $\|d\|_2^2$ we obtain:

$$\|z\|_2^2 - \gamma^2 \|d\|_2^2 =: r > 0. \tag{12.17}$$

If we replace d by αd then the resulting output is equal to αz because the initial state is zero. But then the resulting cost is equal to $\alpha^2 r$. Since $r > 0$ we see that by choosing large enough α we can make the cost function arbitrary large and the supremum will hence not be finite.

This problem is closely related to the linear quadratic regulator problem and we will solve this problem using similar techniques. As a matter of fact, it is an example of a so-called indefinite linear quadratic regulator problem. If we rewrite our objective as maximizing:

$$\int_0^\infty x(t)^{\mathrm{T}} C^{\mathrm{T}} C x(t) + x(t)^{\mathrm{T}} C^{\mathrm{T}} D d(t) + d(t)^{\mathrm{T}} D^{\mathrm{T}} C x(t) + d(t)^{\mathrm{T}} \left(D^{\mathrm{T}} D - \gamma^2 I \right) d(t) \, \mathrm{d}t$$

or equivalently of minimizing:

$$\int_0^\infty -x(t)^{\mathrm{T}}C^{\mathrm{T}}Cx(t) - x(t)^{\mathrm{T}}C^{\mathrm{T}}Dd(t) - d(t)^{\mathrm{T}}D^{\mathrm{T}}Cx(t) + d(t)^{\mathrm{T}}\left(\gamma^2 I - D^{\mathrm{T}}D\right)d(t)\,\mathrm{d}t$$

then the connection to the theory of chapter 10 should be obvious. However, the quadratic term in $x(t)$ is negative semi-definite and hence the resulting cost could actually be negative and this does yield some crucial differences compared to the results presented in chapter 10.

We first study the finite horizon version and then take the limit as time tends to infinity. This will be done in the following two subsections.

12.6.1 A finite horizon

Consider the system (12.15). We first note that if the $L_2[0, T]$-induced operator norm of this system is smaller than some bound $\gamma > 0$ then we must have $\|D\| < \gamma$.

Lemma 12.17 *Consider the system (12.15). If the $L_2[0, T]$-induced operator norm is smaller than some bound $\gamma > 0$ then $\|D\| < \gamma$.*

Proof : We prove this by contradiction. Assume $\|D\| \geqslant \gamma$. Then there exists a vector d_0 such that $\|Dd_0\| \geqslant \gamma\|d_0\|$. We then apply a disturbance $d$of the form:

$$d(t) = \begin{cases} \varepsilon^{-1/2}d_0 & 0 < t < \varepsilon, \\ 0 & \text{elsewhere,} \end{cases}$$

to the system (12.15), and it can be easily seen that:

$$\lim_{\varepsilon\to 0} \|z_\varepsilon\|_{2,\mathrm{T}} = \|Dd_0\| \geqslant \gamma\|d_0\| = \gamma\|d_\varepsilon\|_{2,\mathrm{T}}$$

(the last equality holds for all ε). Using (12.5) we then immediately find that the H_∞ norm of the system (12.15) is larger than γ. This yields the required contradiction. ∎

If the $L_2[0, T]$-induced operator norm is smaller than some bound $\gamma > 0$ then

$$\sup_d \left\{\|z\|_{2,\mathrm{T}}^2 - \gamma^2\|d\|_{2,\mathrm{T}}^2\right\} \leqslant 0 \tag{12.18}$$

for initial condition $x(0) = 0$. Assume there exists a $d \in L_2[0, T]$ such that (12.18) is positive. Then there exists $d_e \in L_2$ defined by:

$$d_e(t) = \begin{cases} d(t) & t \leqslant T \\ 0 & t > T \end{cases}$$

such that

$$\|z\|_2^2 - \gamma^2\|d\|_2^2 \geqslant \|z\|_{2,\mathrm{T}}^2 - \gamma^2\|d\|_2^2 = \|z\|_{2,\mathrm{T}}^2 - \gamma^2\|d\|_{2,\mathrm{T}}^2 > 0.$$

But this input d and the resulting output z then yield a contradiction that the H_∞ norm is less than γ by using the characterization (12.5) for the H_∞ norm.

It will turn later that the use of non-zero initial state will help us in the proof of our main result. We have the following lemma:

Lemma 12.18 *Let system (12.15) be given. Then*

$$\sup_{d\in L_2} \left\{ \|z\|_{2,T}^2 - \gamma^2\|d\|_{2,T}^2 \;\middle|\; x(0) = \xi \right\} \tag{12.19}$$

is finite for all $\xi \in \mathbb{R}^n$ if the $L_2[0,T]$-induced operator norm is strictly less than γ.

Proof : Suppose the $L_2[0,T]$-induced operator norm is strictly less than γ, say δ. Then we find:

$$\begin{aligned}
\|z\|_{2,\mathrm{T}}^2 - \gamma^2\|d\|_{2,\mathrm{T}}^2 &= \|z_{0,d} + z_{\xi,0}\|_{2,\mathrm{T}}^2 - \gamma^2\|d\|_{2,\mathrm{T}}^2 \\
&\leqslant \|z_{0,d}\|_{2,\mathrm{T}}^2 + \|z_{\xi,0}\|_{2,\mathrm{T}}^2 + 2\|z_{0,d}\|_{2,\mathrm{T}} \cdot \|z_{\xi,0}\|_{2,\mathrm{T}} - \gamma^2\|d\|_{2,\mathrm{T}}^2 \\
&\leqslant (\delta^2 - \gamma^2)\|d\|_{2,\mathrm{T}}^2 + 2\delta\|z_{\xi,0}\|_{2,\mathrm{T}} \cdot \|d\|_{2,\mathrm{T}} + \|z_{\xi,0}\|_{2,\mathrm{T}}^2 \\
&\leqslant \sup_{\alpha\in[0,\infty)} (\delta^2 - \gamma^2)\alpha^2 + 2\delta\|z_{\xi,0}\|_{2,\mathrm{T}} \cdot \alpha + \|z_{\xi,0}\|_{2,\mathrm{T}}^2, \\
&< \infty
\end{aligned}$$

where $z_{0,d}$ denotes the output of the system with disturbance d and zero initial state while $z_{\xi,0}$ denotes the output of the system with disturbance 0 and initial state $x(0) = \xi$. The last inequality is a consequence of the fact that $\delta < \gamma$. This proves that (12.19) is finite for all $\xi \in \mathbb{R}^n$. ■

We define

$$\mathcal{C}(d, \xi, \tau) := \int_\tau^T z^{\mathrm{T}}(t)z(t) - \gamma^2 d^{\mathrm{T}}(t)d(t)\,\mathrm{d}t$$

where z is the output of the system with input d and initial state $x(\tau) = \xi$. We will investigate in detail the following criterion:

$$\mathcal{C}^*(\xi, \tau) := \sup_d \mathcal{C}(d, \xi, \tau) \tag{12.20}$$

for arbitrary initial state $x(\tau) = \xi$, and where we maximize over $d \in L_2[\tau, T]$. Lemma 12.18 guarantees that $\mathcal{C}^*(\xi, 0)$ is bounded from above. Moreover, for $d = 0$, we have $\mathcal{C}(0, \xi, 0) \geqslant 0$ and hence $\mathcal{C}^*(\xi, 0)$ is also bounded from below. We get that if the $L_2[0,T]$-induced operator norm is less than γ then $\mathcal{C}^*(\xi, 0)$ is bounded.

Next we claim that $\mathcal{C}^*(\xi, \tau)$ is an non-increasing function of τ. This is a consequence of the following arguments, where $\tau_1 \leqslant \tau_2$.

$$
\begin{aligned}
\mathcal{C}^*(\xi, \tau_1) &:= \sup_d \ \mathcal{C}(d, \xi, \tau_1) \\
&= \sup_d \int_{\tau_1}^{T} \|z(t)\|^2 - \gamma^2 \|d(t)\|^2 \, \mathrm{d}t \\
&\geqslant \sup_d \left\{ \int_{\tau_1}^{\tau} \|z(t)\|^2 - \gamma^2 \|d(t)\|^2 \, \mathrm{d}t \right. \\
&\qquad\qquad \left. + \int_{\tau}^{T} \|z(t)\|^2 \, \mathrm{d}t \;\middle|\; d(t) = 0, \ t > T \right\} \\
&\geqslant \sup_d \ \inf_u \int_{\tau_1}^{\tau} \|z(t)\|^2 - \gamma^2 \|d(t)\|^2 \, \mathrm{d}t \\
&= \mathcal{C}^*(\xi, \tau_2)
\end{aligned}
$$

where $\tau = T - \tau_2 + \tau_1$ and where, in the last step, we use that the system is time-invariant.

Hence we note that, if $\mathcal{C}^*(\xi, 0)$ is bounded, then $\mathcal{C}^*(\xi, \tau)$ is bounded for all $\tau \in [0, T]$. We will use theorem 10.7. We apply this theorem, with time reversed, to the following Riccati differential equation:

$$
-\dot{P} = A^{\mathrm{T}}P + PA + C^{\mathrm{T}}C + \left(PE + C^{\mathrm{T}}D\right)\left(\gamma^2 I - D^{\mathrm{T}}D\right)^{-1}\left(E^{\mathrm{T}}P + D^{\mathrm{T}}C\right) \quad (12.21)
$$

Note that we already concluded that $\|D\|$ must be less than γ, and therefore the inverse is well-defined. We know there exists $T_1 \geqslant 0$ such that this Riccati equation has a solution P on the interval $[T_1, T]$ with $P(T) = 0$. If we show that this solution is uniformly bounded on the interval $[T_1, T]$ with a bound independent of T_1 then theorem 10.7 guarantees that a solution P exists on the whole interval $[0, T]$. The bound on P is a consequence of the following lemma.

Lemma 12.19 *Assume the Riccati equation (12.21) has a solution P with $P(T) = 0$ on the interval $[T_1, T]$. Then we have:*

$$
\mathcal{C}^*(\xi, \tau) = \xi^T P(\tau) \xi
$$

for all $\xi \in \mathbb{R}^n$ and all $\tau \in [T_1, T]$.

Proof : We will use a *completion of the squares* argument similar to the one used in chapter 10. We get

$$
\begin{aligned}
\mathcal{C}^*(\xi, \tau) &:= \sup_d \ \mathcal{C}(d, \xi, \tau) \\
&:= \xi^{\mathrm{T}} P(\tau) \xi + \sup_d \int_{\tau}^{T} \|z(t)\|^2 - \gamma^2 \|d(t)\|^2 \\
&\qquad + \left(\frac{\mathrm{d}}{\mathrm{d}t} x^{\mathrm{T}}(t) P(t) x(t) \right) \mathrm{d}t
\end{aligned}
$$

$$= \xi^{\mathrm{T}} P(\tau)\xi + \sup_{d} -\|(\gamma^2 I - D^{\mathrm{T}} D)^{1/2}(d - Fx)\|_{2,\mathrm{T}}^2$$

where $F(t) = (\gamma^2 I - D^{\mathrm{T}} D)^{-1}(E^{\mathrm{T}} P(t) + D^{\mathrm{T}} C)$. It is obvious that the maximizing d is given by $d = Fx$. ■

We have shown that, if the $L_2[0, T]$-induced operator norm is less than γ, then $\|D\| < \gamma$ and $\mathcal{C}^*(\xi, 0)$ is bounded for all $\xi \in \mathbb{R}^n$ and, moreover, we have shown that $\mathcal{C}^*(\xi, 0) \geqslant \mathcal{C}^*(\xi, \tau)$ for all $\tau \in [0, T]$. Therefore, the above lemma shows that P is uniformly bounded. But we already discussed before lemma 12.19 that this implies that a solution P of the Riccati equation (12.21) exists on the complete interval $[0, T]$.

We will prove the converse. Assume that $\|D\| < \gamma$ and there exists a solution P of the Riccati equation (12.21) with $P(T) = 0$. We show that this implies that the $L_2[0, T]$-induced operator norm is strictly less than γ. If we use the completion of the squares argument from the proof of lemma (12.19) then, for zero initial conditions, we obtain:

$$\|z\|_{2,\mathrm{T}}^2 - \gamma^2 \|d\|_{2,\mathrm{T}}^2 = -\|(\gamma^2 I - D^{\mathrm{T}} D)^{1/2}(d - Fx)\|_{2,\mathrm{T}}^2 \leqslant 0$$

for all $d \in L_2[0, T]$. This immediately implies that the $L_2[0, T]$-induced operator norm is less than or equal to γ. We still have to prove that the norm is strictly less than γ. Introduce a new variable $v = (\gamma^2 I - D^{\mathrm{T}} D)^{1/2}(d - Fx)$ and consider the following system

$$\Sigma_{vd} : \begin{array}{l} \dot{x} = (A + EF)x + E(\gamma^2 I - D^{\mathrm{T}} D)^{-1/2} v, \\ d = \qquad\quad Fx + \;\; (\gamma^2 I - D^{\mathrm{T}} D)^{-1/2} v. \end{array}$$

This system defines an operator $\mathcal{G}_{vd}$ mapping v to d, which has a finite $L_2[0, T]$-induced operator norm. Then we get

$$\|z\|_{2,\mathrm{T}}^2 - \gamma^2 \|d\|_{2,\mathrm{T}}^2 = -\gamma^2 \|v\|_{2,\mathrm{T}}^2 \leqslant -\gamma^2 \|\mathcal{G}_{vd}\|_{\infty,\mathrm{T}}^{-2} \|d\|_{2,\mathrm{T}}^2$$

which guarantees that the $L_2[0, T]$-induced operator norm is less than or equal to

$$\gamma \sqrt{1 - \|\mathcal{G}_{vd}\|_{\infty,\mathrm{T}}^{-2}} < \gamma.$$

Thus, we have proven the following lemma:

Lemma 12.20 *Let the system Σ be given by (12.15). Then the $L_2[0, T]$-induced operator norm of Σ is less than γ if and only if $\|D\| < \gamma$ and there exists a solution P of (12.21) with $P(T) = 0$.*

Note that from the proof it can be seen that the above lemma is also valid for time-varying systems (where we need to require that $\|D(t)\| < \gamma$ for all $t \in [0, T]$). However, in the next subsection we study the infinite horizon case where our approach will depend on the fact that the system is time-invariant.

12.6.2 Infinite horizon

We again consider the system (12.15). We will basically apply the limit as $T \to \infty$ to lemma 12.20. Assume the system is internally stable, and has H_∞ norm less than γ. Then, similarly as in lemma 12.18, we can show that:

$$\mathcal{C}^*(\xi) := \sup_d \int_0^\infty z^{\mathrm{T}}(t)z(t) - \gamma^2 d^{\mathrm{T}}(t)d(t)\,\mathrm{d}t \tag{12.22}$$

is bounded for all initial conditions $x(0) = \xi$. We find:

$$\begin{aligned}
\mathcal{C}^*(0) :=& \sup_{d \in L_2[0,\infty)} \int_0^\infty \|z(t)\|^2 - \gamma^2 \|d(t)\|^2\,\mathrm{d}t \\
\geqslant& \sup_{d \in L_2[0,\infty)} \left\{ \int_0^\infty \|z(t)\|^2\,\mathrm{d}t \right.\\
& \left. \qquad -\gamma^2 \int_0^T \|d(t)\|^2\,\mathrm{d}t \;\middle|\; d(t) = 0,\ t > T \right\} \\
\geqslant& \sup_{d \in L_2[0,T]} \int_0^T \|z(t)\|^2 - \gamma^2 \|d(t)\|^2\,\mathrm{d}t.
\end{aligned} \tag{12.23}$$

This shows that the $L_2[0,T]$-induced operator norm is less than or equal to the H_∞ norm for all T. Consider the following Riccati differential equation:

$$\dot{Q} = A^{\mathrm{T}}Q + QA + C^{\mathrm{T}}C + \left(QE + C^{\mathrm{T}}D\right)\left(\gamma^2 I - D^{\mathrm{T}}D\right)^{-1}\left(E^{\mathrm{T}}Q + D^{\mathrm{T}}C\right),$$

with $Q(0) = 0$. We have $Q(t) = P(T-t)$ when P satisfies (12.21) with $P(T) = 0$. We also know that the $L_2[0,T]$-induced operator norm is less than the H_∞ norm and hence less than γ for all T. Therefore, the results of the previous subsection imply that a solution Q exists for all t. Moreover Q is increasing and bounded. This obviously implies that a limit $\bar{Q}$ exists, which must satisfy the algebraic Riccati equation:

$$0 = A^{\mathrm{T}}\bar{Q} + \bar{Q}A + C^{\mathrm{T}}C + \left(\bar{Q}E + C^{\mathrm{T}}D\right)\left(\gamma^2 I - D^{\mathrm{T}}D\right)^{-1}\left(E^{\mathrm{T}}\bar{Q} + D^{\mathrm{T}}C\right). \tag{12.24}$$

We need an additional condition on $\bar{Q}$ for which we need to do some work. We know that the H_∞ norm from d to z is less than γ, say δ. But then, for all $T > 0$, the $L_2[0,T]$-induced operator norm is less than or equal to δ. We obtain:

$$\begin{aligned}
\|z\|_{2,\mathrm{T}}^2 - \gamma^2 \|d\|_{2,\mathrm{T}}^2 &= \|z_{0,d} + z_{\xi,0}\|_{2,\mathrm{T}}^2 - \gamma^2 \|d\|_{2,\mathrm{T}}^2 \\
&\leqslant \|z_{0,d}\|_{2,\mathrm{T}}^2 + \|z_{\xi,0}\|_{2,\mathrm{T}}^2 + 2\|z_{0,d}\|_{2,\mathrm{T}} \cdot \|z_{\xi,0}\|_{2,\mathrm{T}} - \gamma^2 \|d\|_{2,\mathrm{T}}^2 \\
&\leqslant \|z_{0,d}\|_{2,\mathrm{T}}^2 + \|z_{\xi,0}\|_2^2 + 2\|z_{0,d}\|_{2,\mathrm{T}} \cdot \|z_{\xi,0}\|_2 - \gamma^2 \|d\|_{2,\mathrm{T}}^2 \\
&\leqslant (\delta^2 - \gamma^2)\|d\|_{2,\mathrm{T}}^2 + \|z_{\xi,0}\|_2^2 + 2\delta \|d\|_{2,\mathrm{T}}^2 \cdot \|z_{\xi,0}\|_2 \\
&\leqslant (\gamma^2 - \delta^2)\left(c_2 - (\|d\|_{2,\mathrm{T}} - c_1)^2\right)
\end{aligned}$$

for certain constants c_1 and c_2, independent of T. Choose an arbitrary T. We note that for $d_T = F_T x$ with $F_T(t) = (\gamma^2 I - D^T D)^{-1}(E^T Q(T-t) + D^T C)$ we have

$$\|z\|_{2,T}^2 - \gamma^2 \|d_T\|_{2,T}^2 = \xi^T Q(T-t)\xi \geqslant 0$$

for all initial conditions $x(0) = \xi$ and hence

$$(\|d_T\|_{2,T} - c_1)^2 \leqslant c_2.$$

This implies that $\|d_T\|_{2,T} < \alpha$, with α independent of T. It is trivial to see that $\|d_T\|_{2,T_1} < \|d_T\|_{2,T} < \alpha$ for all $T \geqslant T_1$. Finally, $F_T(t) \to F$ $(T \to \infty)$ for all t, where $F = (\gamma^2 I - D^T D)^{-1}(E^T \bar{Q} + D^T C)$, and therefore $d_T(t) \to d(t)$ as $T \to \infty$ for all t, where $d = Fx$. But then

$$\|d\|_{2,T_1} = \lim_{T\to\infty} \|d_T\|_{2,T_1} < \alpha$$

for all T_1, which guarantees that $d \in L_2$. On the other hand, d is determined by

$$\dot{x} = Ax + Ed = (A + EF)x.$$

Since A is a stability matrix and $Ed \in L_2$ we have $x \in L_2$ for all initial states (see exercise 13.1). This implies that

$$A + EF = A + E(\gamma^2 I - D^T D)^{-1}(E^T \bar{Q} + D^T C) \tag{12.25}$$

is asymptotically stable. Finally, it is obvious that $\bar{Q} \geqslant 0$.

We will now show the converse. Assume there exists a matrix $\bar{Q} \geqslant 0$ which satisfies (12.24) and is such that (12.25) is stable. First, we will check whether the system is stable. Assume $x \neq 0$ is an eigenvector of A with eigenvalue λ. From (12.24) we obtain:

$$2(\Re e\, \lambda)x^* \bar{Q} x = -\|(\gamma^2 I - D^T D)^{-1/2}(E^T \bar{Q} + D^T C)x\|^2 - \|Cx\|^2.$$

Suppose $\Re e\, \lambda \geqslant 0$. Then the above equation implies that we must have $(E^T \bar{Q} + D^T C)x = 0$. This implies $(A + EF)x = \lambda x$ which yields a contradiction since $A + EF$ is a stability matrix. Therefore, A must be a stability matrix.

It is also easy to check that we have:

$$\|z\|_{2,T}^2 - \gamma^2 \|d\|_{2,T}^2 = -\|(\gamma^2 I - D^T D)^{1/2}(d - Fx)\|_{2,T}^2 - x^T(T)\bar{Q}x(T).$$

Since the system is stable, and $d \in L_2$, we have $z \in L_2$ and $x(T) \to 0$ as $T \to \infty$. Thus we get:

$$\|z\|_2^2 - \gamma^2 \|d\|_2^2 = -\int_0^\infty [d(t) - Fx(t)]^T \left(\gamma^2 I - D^T D\right) [d(t) - Fx(t)]\, dt$$

for initial state $x(0) = 0$. Define a new variable v by putting $v = (\gamma^2 I - D^T D)^{1/2}(d - Fx)$ and consider the system:

$$\Sigma_{vd} : \begin{array}{rl} \dot{x} = & (A + EF)x - E(\gamma^2 I - D^T D)^{-1/2} v \\ d = & Fx - (\gamma^2 I - D^T D)^{-1/2} v \end{array}$$

Since $A + EF$ is a stability matrix, this system has a finite H_∞ norm, say δ. We obtain:

$$\|z\|_2^2 - \gamma^2\|d\|_2^2 = -\|v\|_2^2 \leqslant -\delta^{-1}\|d\|_2^2$$

for all $d \in L_2$. This implies that the H_∞ norm of Σ is strictly less than γ. In the above, we have proven the *bounded real lemma*:

Lemma 12.21 *Consider the system Σ given by (12.15). Assume (C, A) is detectable. Then A is stable and the H_∞ norm of Σ is less than γ if and only if $\|D\| < \gamma$ and there exists a solution P of the algebraic Riccati equation (12.24) such that the matrix (12.25) is asymptotically stable.*

12.7 Exercises

12.1 Give a proof of lemma 12.3.

12.2 As noted in remark 12.12, we can also have multiplicative uncertainty at the output of the system. Derive the equivalent of lemma 12.10 for the case of multiplicative uncertainty at the output of the system. Impose the additional assumption that the system Σ is internally stable and prove this result with the help of the small gain theorem.

12.3 Consider lemma 12.14. Verify that the systems described by (12.12) indeed yield a normalized coprime factorization of the transfer matrix G.

12.4 Consider a system Σ and an internally stabilizing controller Γ. Look at the following setup:

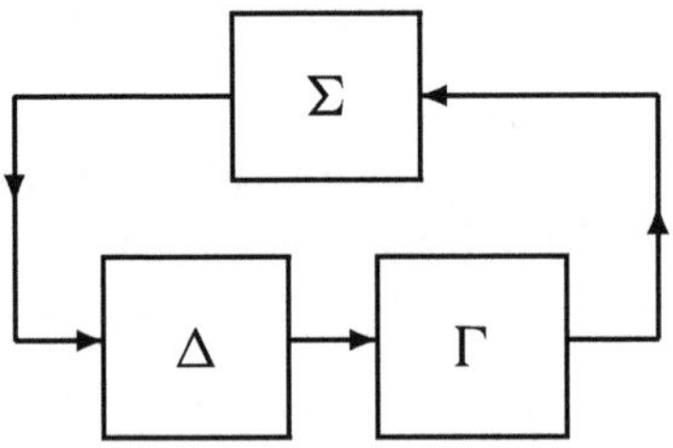

(12.26)

The gain margin $[m, M]$ and phase margin Θ of the interconnection of Σ and Γ are defined as:

$$m = \max\{20\log_{10}\delta \mid (12.26) \text{ is unstable for } \Delta = \delta I \text{ where } \delta \in [0, 1]\},$$
$$M = \min\{20\log_{10}\delta \mid (12.26) \text{ is unstable for } \Delta = \delta I \text{ where } \delta \in [1, \infty)\},$$
$$\Theta = \min\{\frac{180^\circ|\theta|}{\Pi} \mid (12.26) \text{ is unstable for } \Delta = e^{i\theta} I \text{ where } \theta \in [-\pi, \pi)\},$$

where $m = -\infty$ ($M = \infty$, $\Theta = 180^\circ$) if there does not exists a m (M, Θ) which destabilizes the system.

Let G and H denote the transfer matrices of Σ and Γ respectively. Assume that $\|(I - GH)^{-1}\|_\infty = \gamma$.

a. Derive bounds on the gain and phase margin in terms of γ.

b. Is it possible to derive bounds on the H_∞ norm γ in terms of the gain and phase margin?

12.5 Consider the infinite horizon linear quadratic regulator problem which we studied in chapter 10. We have the system $\Sigma : \dot{x}(t) = Ax(t) + Bu(t), x(0) = x_0$ where we try to find an input u which minimizes the following criterion:

$$J(x_0, u) := \int_0^\infty x(t)^{\mathrm{T}} Q x(t) + u(t)^{\mathrm{T}} u(t) \mathrm{d}t$$

subject to $x(t) \to 0$ as $t \to \infty$. Assume that $Q \geqslant 0$ is such that (Q, A) is detectable. We know that the solution of this problem is related to the largest real symmetric solution P of the Riccati equation:

$$A^{\mathrm{T}} P + PA - PBB^{\mathrm{T}} P + Q = 0,$$

The optimal feedback for the above minimization problem is given by $u = -B^{\mathrm{T}} P x$. We will derive some properties of the resulting closed loop system.

a. Using the Riccati equation, show that:

$$\begin{aligned} &I + B^{\mathrm{T}}(-i\omega I - A^{\mathrm{T}})^{-1} Q (i\omega I - A)^{-1} B \\ &\quad = \left[I + B^{\mathrm{T}}(-i\omega I - A^{\mathrm{T}})^{-1} PB\right]\left[I + B^{\mathrm{T}} P (i\omega I - A)^{-1} B\right]. \end{aligned}$$

b. Show that

$$\left\| \left(I + B^{\mathrm{T}} P (i\omega I - A)^{-1} B\right)^{-1} \right\|_\infty = 1.$$

c. In order to investigate the gain and phase margin of the closed loop system, we study the following interconnection:

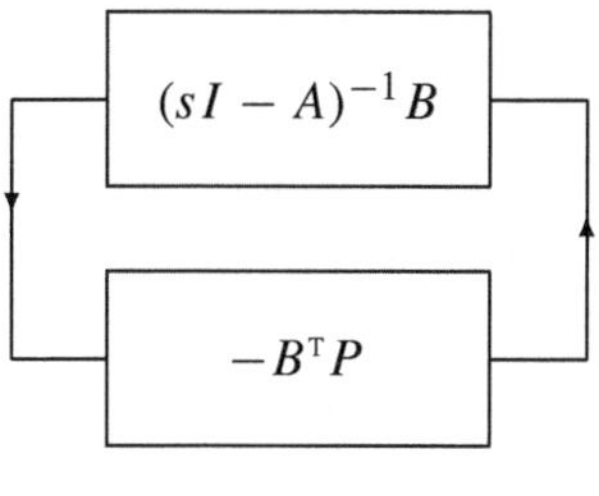

Show that the gain margin of this interconnection is at least equal to $[-20\log_{10}2, \infty)$ and the phase margin at least equal to $60°$.

12.6 Consider the following system:

$$\Sigma: \quad \begin{aligned} \begin{pmatrix} \dot{x}_1 \\ \dot{x}_2 \end{pmatrix} &= \begin{pmatrix} 1 & 1 \\ 0 & 1 \end{pmatrix} \begin{pmatrix} x_1 \\ x_2 \end{pmatrix} + \begin{pmatrix} 0 \\ 1 \end{pmatrix} u + \begin{pmatrix} 1 & 0 \\ 1 & 0 \end{pmatrix} d \\ y &= \begin{pmatrix} 1 & 0 \end{pmatrix} \begin{pmatrix} x_1 \\ x_2 \end{pmatrix} + \begin{pmatrix} 0 & \alpha \end{pmatrix} d \\ z &= \begin{pmatrix} \beta & \beta \\ 0 & 0 \end{pmatrix} \begin{pmatrix} x_1 \\ x_2 \end{pmatrix} + \begin{pmatrix} 0 \\ 1 \end{pmatrix} u \end{aligned}$$

a. Derive the H_2 optimal dynamic measurement controller Γ for Σ as a function of α and β using the theory of chapter 11.

b. Determine analytically or numerically (by trying several values for α and β) that the gain margin of the interconnection of Σ_{ci} and Γ becomes arbitrarily small, i.e. $m, M \to 0$dB, as α and β increase. Here Σ_{ci} is the subsystem of Σ from u to y.

12.7 Let Σ be the single-input, single output linear system given by

$$\begin{aligned} \dot{x} &= Ax + bu \\ y &= cx \end{aligned}$$

Let G be its transfer matrix.

a. Show that for zero initial condition:

$$|y(t)| \leqslant \left(\int_0^t c e^{A\tau} b b^{\mathrm{T}} e^{A^{\mathrm{T}}\tau} c^{\mathrm{T}} \mathrm{d}\tau \right)^{1/2} \left(\int_0^t u^2(\tau) \mathrm{d}\tau \right)^{1/2}.$$

b. Define $|y|_\infty := \sup_t |y(t)|$. Show that the H_2 norm of Σ is equal to the L_2 to L_∞ induced operator norm, i.e.

$$\|G\|_2 = \sup_{u \neq 0} \frac{|y|_\infty}{\|u\|_2} \tag{12.27}$$

for zero initial conditions.

c. For a vector valued function $y : [0, \infty) \to \mathbb{R}^n$ we define

$$|y|_\infty := \sup_{i,t} |y_i(t)|$$

where $y_i(t)$ denotes the ith component of $y(t)$. Show, e.g. via an example, that for a multi-input, multi-output system Σ the H_2 norm is not necessarily equal to the L_2 to L_∞ induced operator norm defined by (12.27).

12.8 Notes and references

The H_∞ control problem was originally formulated in the work of Zames, [230]. Early solutions, based on frequency domain techniques, are described in Zames, [231] and Francis, [46]. The first more or less complete solution in a state space setting was given in the work by Doyle, Glover Khargonekar and Francis, [41]. For a more precise and detailed exposition of the spaces H_∞, H_2, L_∞, L_2 we refer to Rudin, [156] and Young, [228]. The small gain theorem described in section 12.4 is described in detail in the book [212] by Willems. Our proof of theorem 12.4 worked for simplicity with a static and complex uncertainty system. The effect of allowing for dynamic, time-varying or nonlinear and the difference between real and complex uncertainty systems are described in [79] by Hinrichsen and Pritchard.

If we want to use some special structure of the system Δ (e.g. that Δ is static and not dynamic), then we have to resort to the so-called μ-synthesis. μ-synthesis has the disadvantage that this method is so general that no reasonably efficient algorithms are available to improve robustness via μ-synthesis. The available technique for μ-synthesis is to use an H_∞ upper bound (depending on scales) for μ which is consequently minimized by choosing an appropriate controller. The result is then minimized over all possible choices for the scales. One of the first papers on μ-analysis and -synthesis is by Doyle, [39]. A book which describes this technique in quite some detail and contains a lot of references is the book [232] by Zhou, Doyle and Glover.

Additive and multiplicative model uncertainty has been studied by McFarlane and Glover in [123], and by Vidyasagar in [207]. The use of coprime factorizations in control is studied extensively by Vidyasagar in the book [207]. More details regarding coprime-factor uncertainty can be found in the work of McFarlane and Glover, [123]. Note that the two coprime factors actually yield a kernel representation within the framework of behavioral systems. See for instance the book [146] by Polderman and Willems and the paper [217] by Willems. Therefore this might yield a very natural motivation for the use of coprime factorizations.

For single-input, single-output systems expressing performance criteria into requirements on the desired shape of the magnitude Bode diagram is well established. This can be translated into appropriate choices for the weights in the mixed-sensitivity problem. See any book on classical control such as Van de Vegte [202], Phillips and Harbor [145], Franklin, Power and Emami-Naeini [49], and Kuo [101]. On the other hand, for multi-input, multi-output systems it is in general very hard to translate practical performance criteria into an appropriate choice for the weights. For more details we refer to the books by Freudenberg and Looze [50] and by Horowitz, [81].

Chapter 13

The state feedback H_∞ control problem

13.1 Introduction

In the next two chapters we will present a solution to the H_∞ control problem as formulated in the previous chapter. In this book we will present a time-domain oriented solution of the H_∞ control problem. A main advantage is the great similarity with the techniques used for the solution of the H_2 control problem.

We will first study the state feedback H_∞ control problem. There is a quite a bit of similarity with the linear quadratic regulator problem studied in chapter 10 and the H_2 control problem of chapter 11. However, there is a fundamental difference between H_2 and H_∞. In the H_2 control problem we have found an expression for the *optimal* state feedback controller. The latter was expressed in terms of the largest solution of a certain Riccati equation. In the H_∞ control problem we will ask ourselves whether a controller exists that makes the H_∞ norm smaller than an, a priori given, bound γ. It will turn out that if a certain Riccati equation has a solution, then such controller exists. An expression for one suitable controller is then obtained in terms of a particular solution of this Riccati equation. In other words, while in the H_2 context a solution of the Riccati equation always exists, in the H_∞ context a solution of the Riccati equation only exists if we have chosen the parameter γ sufficiently large. Moreover, we will not find an optimal controller, but a *suboptimal* controller, i.e. a controller that makes the H_∞ norm less than γ. We therefore need to search for the smallest γ for which the Riccati equation has a solution. Except for some special cases, there is no analytic expression for this smallest value of γ and therefore, in general, we have to do a numerical search.

A Riccati equation is a quadratic equation for the unknown matrix P of the form $F^{\mathrm{T}}P + PF + PRP + Q$. In linear quadratic and H_2 optimal control we always have $R \leqslant 0$ and $Q \geqslant 0$ as we have seen in the previous chapters. The Riccati equation appearing in H_∞ has an indefinite quadratic term R. Riccati equations of this type first appeared in the theory of differential games, and have different properties compared to the Riccati equation used in linear quadratic and H_2 optimal control. In section 13.4 we will discuss a method to find solutions of the algebraic Riccati equation.

As in chapter 10, we will first study a finite horizon version of the state feedback H_∞ control problem, and then use these results to obtain a solution of the standard infinite horizon H_∞ control problem. The finite horizon version of the H_∞ control problem is based on the time-domain interpretation of the H_∞ norm as the L_2-induced operator norm. The latter has been discussed in section 12.2.

13.2 The finite horizon H_∞ control

Consider the linear, time-invariant, finite-dimensional system:

$$\Sigma: \begin{array}{l} \dot{x} = Ax + Bu \ + Ed, \\ z = Cx + Du, \end{array} \tag{13.1}$$

where, as before, $x \in \mathbb{R}^n$ is the state, $u \in \mathbb{R}^m$ the control input, $d \in \mathbb{R}^\ell$ the disturbance and $z \in \mathbb{R}^q$ the output to be controlled. A, B, E, C and D are matrices of appropriate dimensions. Note that we could have added a direct feedthrough matrix from d to z. This would not make the problem more difficult to solve, but it makes the formulas a lot more messy. Hence, for ease of exposition, we have set this matrix equal to 0.

We want to minimize the effect of the disturbance d on the output z by finding an appropriate control input u. More precisely, we seek a controller Γ described by a time-varying state feedback law $u(t) = F(t)x(t)$ such that after applying this feedback law to the system (13.1), the resulting closed-loop system $\Sigma \times \Gamma$ has minimal $L_2[0, T]$-induced operator norm defined by:

$$\|\mathcal{G}_{cl}\|_{\infty,T} := \sup_d \left\{ \frac{\|\mathcal{G}_{cl} d\|_{2,T}}{\|d\|_{2,T}} \;\middle|\; d \in L_2^\ell[0, T],\ d \neq 0 \right\}. \tag{13.2}$$

where $\mathcal{G}_{cl}$ is the closed loop operator mapping d to z for zero initial state in (13.1) and

$$\|f\|_{2,T} := \left(\int_0^T \|f(t)\|^2 \,dt \right)^{1/2} < \infty.$$

Although minimizing the $L_2[0, T]$-induced operator norm is our ultimate goal, we shall only derive necessary and sufficient conditions under which we can find a controller that makes the resulting $L_2[0, T]$-induced operator norm of the closed-loop system strictly less than some a priori given bound γ. We are now in the position to formulate our result.

Theorem 13.1 *Consider the system (13.1). Let $T > 0$ and let $\gamma > 0$. Assume that the matrix D is injective. Then the following statements are equivalent:*

(i) *There exists a time-varying state feedback law $u(t) = F(t)x(t)$ such that after applying this controller to the system (13.1) the resulting closed-loop operator $\mathcal{G}_{cl}$ has $L_2[0,T]$-induced operator norm less than γ, i.e. $\|\mathcal{G}_{cl}\|_{\infty,T} < \gamma$.*

(ii) *There exists a differentiable function $P : [0,T] \to \mathbb{R}^{n\times n}$ such that $P(t) \geqslant 0$ for all t and P satisfies the following Riccati differential equation*

$$-\dot{P} = A^T P + PA + C^T C + \gamma^{-2} P E E^T P - \left(PB + C^T D\right)\left(D^T D\right)^{-1}\left(B^T P + D^T C\right) \quad (13.3)$$

with P(T)=0.

If P satisfies the conditions in part (ii), then a controller satisfying the conditions in part (i) is given by:

$$F(t) := -\left(D^T D\right)^{-1}\left[D^T C + B^T P(t)\right]. \quad (13.4)$$

We will now start with our proof of theorem 13.1. Let $0 \leqslant \tau \leqslant T$ and define

$$\mathcal{C}(u,d,\xi,\tau) := \int_\tau^T z^{\mathrm{T}}(t)z(t) - \gamma^2 d^{\mathrm{T}}(t)d(t)\,\mathrm{d}t$$

where z is the output of the system with inputs u, d and initial condition $x(\tau) = \xi$. We will investigate in detail the following criterion:

$$\mathcal{C}^*(\xi,\tau) := \sup_d \inf_u \{\ \mathcal{C}(u,d,\xi,\tau)\ \} \quad (13.5)$$

for arbitrary initial state $x(\tau) = \xi$. Here $d \in L_2[0,T]$ and $u \in L_2[0,T]$. Since we optimize over u first, i.e. for a fixed d, it does not matter whether we choose u in open or closed loop. Here open loop refers to the fact that we choose u to be an element of $L_2[0,T]$ while closed loop refers to the fact that u is chosen as a function of x and, possibly, d. So a state feedback is a typical example of a u chosen in closed loop. However, especially in the next section, it is essential that d is chosen in open loop, i.e. we optimize over $d \in L_2[0,T]$ and not over some feedback strategy.

We will first show the implication (i) $\Rightarrow$ (ii) of theorem (13.1). We can apply lemma 12.18 to the closed loop system obtained after applying a feedback $u = F(t)x(t)$ which yields an $L_2[0,T]$-induced operator norm strictly less than γ. Thus we obtain that:

$$\sup_d \left\{\ \|z\|_{2,\mathrm{T}}^2 - \gamma^2\|d\|_{2,\mathrm{T}}^2\ \middle|\ u(t) = F(t)d(t),\ x(0) = \xi\ \right\} \quad (13.6)$$

is finite for all $\xi \in \mathbb{R}^n$.

It is easy to see that (13.6) is larger than $\mathcal{C}^*(\xi,0)$. Therefore $\mathcal{C}^*(\xi,0)$ is bounded from above. Moreover, by making the suboptimal choice $d = 0$ in the optimization (13.5) we see that $\mathcal{C}^*(\xi,0) \geqslant 0$. Hence $\mathcal{C}^*(\xi,0)$ is bounded for all $\xi \in \mathbb{R}^n$.

Next we claim that $\mathcal{C}^*(\xi, \tau)$ is a decreasing function of τ. This is a consequence of the following implications,

$$
\begin{aligned}
\mathcal{C}^*(\xi, \tau_1) &:= \sup_d \inf_u \mathcal{C}(u, d, \xi, \tau_1) \\
&= \sup_d \inf_u \int_{\tau_1}^T \|z(t)\|^2 - \gamma^2 \|d(t)\|^2 \, dt \\
&\geqslant \sup_d \inf_u \left\{ \int_{\tau_1}^T \|z(t)\|^2 - \gamma^2 \|d(t)\|^2 \, dt \;\middle|\; d(t) = 0, \ \forall t > \tau \right\} \\
&\geqslant \sup_d \inf_u \int_{\tau_1}^{\tau} \|z(t)\|^2 - \gamma^2 \|d(t)\|^2 \, dt \\
&= \mathcal{C}^*(\xi, \tau_2)
\end{aligned}
$$

where $\tau_1 \leqslant \tau_2$ and $\tau = T - \tau_2 + \tau_1$. Moreover, in the last step, we use that the (open-loop) system is time-invariant.

Since $\mathcal{C}^*(\xi, 0)$ is bounded, we note that the above implies that the cost function $\mathcal{C}^*(\xi, \tau)$ is bounded for all $\tau \in [0, T]$. We will now use theorem 10.7. This will all be very similar to subsection 12.6.1. We apply this theorem, with the time reversed, to the Riccati equation (13.3). We therefore note there exists $T_1 \geqslant 0$ such that the Riccati equation (13.3) has a solution P on the interval $[T_1, T]$ with $P(T) = 0$. If we show that this solution is uniformly bounded on the interval $[T_1, T]$ with a bound independent of T_1 then theorem 10.7 guarantees that a solution P exists on the whole interval $[0, T]$ and we have proven part (ii) of theorem 13.1. The bound on P is a consequence of the following lemma.

Lemma 13.2 *Assume the Riccati equation (13.3) has a solution P on the interval $[T_1, T]$ with $P(T) = 0$. Then we have:*

$$
\mathcal{C}^*(\xi, \tau) = \xi^T P(\tau) \xi \tag{13.7}
$$

for all $\xi \in \mathbb{R}^n$ and all $\tau \in [T_1, T]$.

Proof : We will use a completion of the squares argument similar to the one used in subsection 12.6.1. We get

$$
\begin{aligned}
\mathcal{C}^*(\xi, \tau) &:= \sup_d \inf_u \mathcal{C}(u, d, \xi, \tau) \\
&:= \xi^{\mathrm{T}} P(\tau) \xi + \sup_d \inf_u \int_{\tau}^T \|z(t)\|^2 - \gamma^2 \|d(t)\|^2 \\
&\qquad\qquad + \left(\frac{\mathrm{d}}{\mathrm{d}t} x^{\mathrm{T}}(t) P(t) x(t) \right) \mathrm{d}t \\
&= \xi^{\mathrm{T}} P(\tau) \xi + \sup_d \inf_u \int_{\tau}^T \|D[u(t) - F(t) x(t)]\|^2 \\
&\qquad\qquad - \gamma^2 \|d(t) - F_2(t) x(t)\|^2 \, \mathrm{d}t
\end{aligned}
$$

where $F = -(D^{\mathrm{T}}D)^{-1}(B^{\mathrm{T}}P + D^{\mathrm{T}}C)$ and $F_2 = \gamma^{-2}E^{\mathrm{T}}P$. It is obvious that if $u = Fx$ then, no matter what d is, $\mathcal{C}(u, d, \xi, \tau) \geqslant \xi^{\mathrm{T}}P(\tau)\xi$.

The problem is the converse. It seems obvious to choose $d = F_2x$. However, we cannot choose d in feedback; we must choose d only depending on the initial condition ξ. It is natural for d to expect that u will be chosen equal to Fx. Hence, we choose d equal to $\bar{d} = F_2p$ where $\dot{p} = (A + BF + EF_2)p$ and $p(0) = \xi$ (this is the unique open-loop policy such that $d = F_2x$ whenever $u = Fx$). Moreover, define v by $v = D(u - Fx)$, i.e. $u = Fx + (D^{\mathrm{T}}D)^{-1}D^{\mathrm{T}}v$. If we can prove that the optimal choice for v equals 0 then we have $d = F_2x$ and the cost equal to $\xi^{\mathrm{T}}P(\tau)\xi$. This would complete the proof.

We rewrite the cost criterion for the specific disturbance $\bar{d}$ and in terms of v. We get:

$$\inf_u \mathcal{C}(u, \bar{d}, \xi, \tau) = \xi^{\mathrm{T}}P(\tau)\xi + \inf_v \int_\tau^T \|v(t)\|^2 - \gamma^2\|\bar{z}(t)\|^2 \, \mathrm{d}t \tag{13.8}$$

where $\bar{z}$ is generated by the following system:

$$\bar{\Sigma}: \begin{array}{rl} \dot{\bar{x}} = & (A + BF)\bar{x} + B(D^{\mathrm{T}}D)^{-1}D^{\mathrm{T}}v, \qquad \bar{x}(0) = 0 \\ \bar{z} = & F_2\bar{x}. \end{array}$$

Equation (13.8) can be checked by using the variations of constant formula to determine $\bar{d}(t) - F_2(t)x(t)$ in terms of v and ξ. Note the surprising fact that the integral in (13.8) is independent of the initial condition ξ. The input $v = 0$ is clearly optimal as soon as we have shown that the $L_2\,[0, T]$-induced operator norm from v to $\bar{z}$ is less than γ^{-1}. Because the equations become easier, we prove instead that the following system has $L_2\,[0, T]$-induced operator norm less than 1:

$$\bar{\Sigma}_e: \begin{array}{rl} \dot{\bar{x}} = & (A + BF)\bar{x} + B(D^{\mathrm{T}}D)^{-1}D^{\mathrm{T}}v, \qquad \bar{x}(0) = 0 \\ \bar{z}_e = & \begin{pmatrix} \gamma F_2 \\ \left[I - D(D^{\mathrm{T}}D)^{-1}D^{\mathrm{T}}\right]^{1/2} C \end{pmatrix} \bar{x}. \end{array}$$

It is easy to see that if $\bar{\Sigma}_e$ has norm less than 1 then clearly $\bar{\Sigma}$ has norm less than γ^{-1}. We now apply lemma 12.20, and we see that we must check the existence of a solution to the following Riccati differential equation:

$$\begin{aligned} -\dot{X} = (A + BF)^{\mathrm{T}}X + X(A + BF) + \gamma^2F_2^{\mathrm{T}}F_2 + XB(D^{\mathrm{T}}D)^{-1}B^{\mathrm{T}}X \\ + C^{\mathrm{T}}[I - D(D^{\mathrm{T}}D)^{-1}D^{\mathrm{T}}]C \end{aligned}$$

where $X(T) = 0$. We see that $X(t) = Q(T - t)$ satisfies the above Riccati equation and therefore we have shown that $v = 0$ is optimal. This guarantees that the $L_2\,[0, T]$-induced operator norm of $\bar{\Sigma}$ is less than γ^{-1}. As mentioned above, this guarantees that the optimal choice for u equals Fx when $d = \bar{d}$. This proves (13.7). ■

We have shown that if part (i) of theorem 13.1 is satisfied then $\mathcal{C}^*(\xi, 0)$ is bounded for all $\xi \in \mathbb{R}^n$ and, moreover, we have shown that $\mathcal{C}^*(\xi, 0) \geqslant \mathcal{C}^*(\xi, \tau)$ for all $\tau \in$

$[0, T]$. Therefore, the above lemma shows that P is uniformly bounded. But we already discussed before lemma 13.2 that this implies that a solution P of the Riccati equation (13.3) exists on the complete interval $[0, T]$, i.e. part (ii) of theorem 13.1 is satisfied.

It remains to show that part (ii) of theorem 13.1 implies part (i). We will prove that the feedback $u = Fx$ defined by (13.4) satisfies part (i). If we use the completion of the squares argument from the proof of lemma (13.2) then, for zero initial state and $u = Fx$, we obtain:

$$\int_0^T \|z(t)\|^2 - \gamma^2\|d(t)\|^2 \,\mathrm{d}t = -\int_0^T \gamma^2\|d(t) - F_2(t)x(t)\|^2 \,\mathrm{d}t \leqslant 0$$

for all $d \in L_2[0, T]$. This immediately implies that the $L_2[0, T]$-induced operator norm is less than or equal to γ. We still have to prove that the norm is *strictly* less than γ. We use the same technique as in subsection 12.6.1. Introduce the variable v by $v = d - F_2x$, and consider the system

$$\Sigma_{vd} : \begin{array}{rl} \dot{x} = & (A + BF + EF_2)x + Ev \quad x(0) = 0 \\ d = & \qquad\qquad -F_2x + \quad v. \end{array}$$

This system defines an operator $\mathcal{G}_{vd}$ mapping v to d, with a finite $L_2[0, T]$-induced operator norm. Thus we get

$$\|z\|_{2,T}^2 - \gamma^2\|d\|_{2,T}^2 = -\gamma^2\|v\|_{2,T}^2 \leqslant -\gamma^2\|\mathcal{G}_{vd}\|_{\infty,T}^{-2}\|d\|_{2,T}^2,$$

which guarantees that the $L_2[0, T]$-induced operator norm is strictly less than γ. This implies that $u = Fx$ satisfies part (i) of theorem 13.1 and hence we have completed the proof of theorem 13.1.

13.3 Infinite horizon H_∞ control problem

We will show in this section that the results for the finite horizon case can be used to derive a similar result for the infinite horizon H_∞ control problem. The approach used is very similar to the techniques used in section 10.3 for the linear quadratic regulator problem.

Theorem 13.3 *Consider the system (13.1) and let $\gamma > 0$. Assume that the system (A, B, C, D) has no zeros on the imaginary axis and assume that the matrix D is injective. Then the following statements are equivalent:*

(i) There exists a static state feedback law $u(t) = Fx(t)$ such that, after applying this controller to the system (13.1), the resulting closed loop transfer matrix G_F has H_∞ norm less than γ, i.e. $\|G_F\|_\infty < \gamma$.

(ii) *There exists a matrix $P \in \mathbb{R}^{n \times n}$ such that $P \geqslant 0$ and such that P satisfies the following algebraic Riccati equation*

$$0 = A^T P + PA + C^T C + \gamma^{-2} P E E^T P - \left(PB + C^T D\right)\left(D^T D\right)^{-1}\left(B^T P + D^T C\right) \quad (13.9)$$

with A_{cl} a stability matrix where

$$A_{cl} := A + \gamma^{-2} E E^T P - B\left(D^T D\right)^{-1}\left(B^T P + D^T C\right). \quad (13.10)$$

If P satisfies the conditions in part (ii), then a controller satisfying the conditions in part (i) is given by:

$$F := -\left(D^T D\right)^{-1}\left(D^T C + B^T P\right). \quad (13.11)$$

Remark 13.4 Note that we require explicitly that the solution P must be positive semi-definite. This is a very crucial requirement. In the linear quadratic regulator problem, the stabilizing solution of the algebraic Riccati equation is automatically positive semi-definite. On the other hand, in H_∞ the stabilizing solution might be indefinite or even negative definite. It can be shown that this would imply that the feedback $u = Fx$ is no longer stabilizing. In the linear quadratic regulator problem the stability requirement for the algebraic Riccati equation is imposed on a matrix which is equal to the resulting closed loop state matrix $A + BF$ after applying the optimal feedback. Hence the stability of the closed loop system is automatically guaranteed. In the H_∞ control problem, the stability requirement for the algebraic Riccati equation is imposed on a matrix which is different from the closed loop state matrix $A + BF$ resulting from the suggested feedback. Therefore we still have to prove stability of the closed loop system and for this we need the requirement that $P \geqslant 0$.

We will first show the implication (i) $\Rightarrow$ (ii) of theorem (13.3). As we did in subsection 12.6, we would like to obtain these results from the finite horizon results. In general, however, the solution P does not converge to the correct limit as $T \to \infty$. We saw the same kind of problem occurring in section 10.4. There we could use an argument which, unfortunately, does not work for the H_∞ control problem. This time we will derive the results from a finite horizon problem with a very specific endpoint penalty. Unfortunately, we have to repeat parts of the previous section to obtain similar results for a non-zero endpoint penalty.

In section 10.5, we have shown that there exists a solution L to the algebraic Riccati equation

$$A^T L + LA + C^T C - (LB + C^T D)(D^T D)^{-1}(LB + C^T D)^T = 0 \quad (13.12)$$

such that

$$A - B(D^T D)^{-1}(B^T L + D^T C) \quad (13.13)$$

is a stability matrix. Define

$$\mathcal{C}_L(u,d,\xi,T) := \int_0^T z^{\mathrm{T}}(t)z(t) - \gamma^2 d^{\mathrm{T}}(t)d(t)\,\mathrm{d}t + x^{\mathrm{T}}(T)Lx(T),$$

where z is the output of the system with inputs u, d and initial condition $x(0) = \xi$. We will investigate in detail the following criterion:

$$\mathcal{C}_L^*(\xi,T) := \sup_d \inf_u \; \{\; \mathcal{C}_L(u,d,\xi,T) \;\} \tag{13.14}$$

for arbitrary initial state $x(0) = \xi$. Here $d \in L_2[0,T]$ and $u \in L_2[0,T]$. This is the same criterion we studied in the previous section but we have included a specific endpoint penalty. Note that we now vary the final time of the optimization problem while we varied the initial time in the previous section. Clearly one wonders why we have chosen this particular end point penalty. For the infinite-horizon problem we have to optimize over $d \in L_2[0,\infty)$. Since d is square integrable, it is not a bad approximation to assume $d(t) = 0$ for all $t > T$ where T is chosen very large. Then, after time T, we basically have a linear quadratic control problem since we do not impose that $u(t) = 0$ after time T and hence we still have to minimize over u. This is a linear quadratic regulator problem with initial condition $x(T)$ and, because of stability, we impose a zero endpoint. From chapter 10, we know that the cost of this zero-endpoint linear quadratic regulator problem is $x^{\mathrm{T}}(T)Lx(T)$. But we still have to optimize over d and u on the interval $[0,T]$. However, we know the cost after time T will be $x^{\mathrm{T}}(T)Lx(T)$ and hence we include this as an endpoint penalty. In this way we obtain the criterion (13.14). When we let $T \to \infty$, the above showes intuitively why this will converge to the solution of the infinite horizon problem. Of course, we will give a precise proof.

We first show that, if there exists a state feedback $u = Fx$ which makes the H_∞ norm less than γ, then (13.14) is bounded for all $T > 0$ and for all initial states $x(0) = \xi$. The argument is a direct consequence of the above intuitive reasoning. We have:

$$\begin{aligned}
&\mathcal{C}_L^*(\xi,T) = \\
&\quad = \sup_{d\in L_2[0,T]} \inf_{u\in L_2[0,T]} \left\{ \int_0^T \|z(t)\|^2 - \gamma^2\|d(t)\|^2\,\mathrm{d}t + x^{\mathrm{T}}(T)Lx(T) \right\} \\
&\quad = \sup_{d\in L_2[0,T]} \inf_{u\in L_2[0,T]} \left(\int_0^T \|z(t)\|^2 - \gamma^2\|d(t)\|^2\,\mathrm{d}t + \inf_{u\in L_2[T,\infty)} \right. \\
&\qquad\qquad \left. \left\{ \int_T^\infty \|z(t)\|^2\,\mathrm{d}t \;\middle|\; d(t) = 0 \text{ and } x(t) \to 0 \text{ as } t \to \infty \right\} \right) \\
&\quad = \sup_{d\in L_2[0,\infty)} \inf_{u\in L_2[0,\infty)} \left\{ \int_0^\infty \|z(t)\|^2 - \gamma^2\|d(t)\|^2\,\mathrm{d}t \;\right| \\
&\qquad\qquad\qquad \left. d(t) = 0, \forall t > T \text{ and } x(t) \to 0 \text{ as } t \to \infty \right\}
\end{aligned}$$

$$\leqslant \sup_{d\in L_2[0,\infty)} \inf_{u\in L_2[0,\infty)} \left\{ \int_0^\infty \|z(t)\|^2 - \gamma^2\|d(t)\|^2 \, \mathrm{d}t \;\middle|\; x(t) \to 0 \text{ as } t \to \infty \right\} \tag{13.15}$$

It is easy to see that $\mathcal{C}_L^*(\xi, T)$ is bounded from below by 0 by choosing $d = 0$. The above sequence of inequalities proves that it is bounded from above as soon as we show that the last expression is bounded from above. It was already mentioned in subsection 12.6.2 that:

$$\sup_d \left\{ \|z\|_2^2 - \gamma^2\|d\|_2^2 \;\middle|\; u(t) = Fd(t),\ x(0) = \xi,\ x(t) \to 0 \text{ as } t \to \infty \right\} \tag{13.16}$$

is finite for all $\xi \in \mathbb{R}^n$, if the closed loop H_∞ norm after applying the stabilizing feedback $u(t) = Fx(t)$ to the system (13.1) is strictly less than γ. Clearly (13.16) is larger than (13.15) and hence we find that $\mathcal{C}_L^*(\xi, T)$ is bounded, uniformly with respect to T.

We will study the following Riccati differential equation with $P(0) = L$:

$$\dot{P} = A^{\mathrm{T}}P + PA + C^{\mathrm{T}}C + \gamma^{-2}PEE^{\mathrm{T}}P - \left(PB + C^{\mathrm{T}}D\right)\left(D^{\mathrm{T}}D\right)^{-1}\left(B^{\mathrm{T}}P + D^{\mathrm{T}}C\right) \tag{13.17}$$

Once again, we will use theorem 10.7. We apply this theorem to the above Riccati equation. We note there exists $T_1 \geqslant 0$ such that the Riccati equation (13.3) has a solution P on the interval $[0, T_1]$ with $P(T) = 0$. If we show that this solution is uniformly bounded on the interval $[0, T_1]$ with a bound independent of T_1 then theorem 10.7 guarantees that a solution P exists on the whole interval $[0, \infty)$. The bound on P is a consequence of the following lemma.

Lemma 13.5 *Assume that the Riccati equation (13.17) has a solution P on the interval $[0, T_1]$ with $P(0) = L$. Then we have $\mathcal{C}_L^*(\xi, T) = \xi^T P(T)\xi$ for all $\xi \in \mathbb{R}^n$ and all $T \in [0, T_1]$.*

Proof : We will use a completion of the squares argument identical to the one used in the proof of lemma 13.2. We get

$$\begin{aligned}
\mathcal{C}_L^*(\xi, T) &:= \sup_d \inf_u \ \mathcal{C}_L(u, d, \xi, T) \\
&= \xi^{\mathrm{T}}P(T)\xi + \sup_d \inf_u \int_0^T \|z(t)\|^2 - \gamma^2\|d(t)\|^2 \\
&\qquad\qquad + \left(\frac{\mathrm{d}}{\mathrm{d}t}x^{\mathrm{T}}(t)P(T-t)x(t)\right) \mathrm{d}t \\
&= \xi^{\mathrm{T}}P(T)\xi + \sup_d \inf_u \int_0^T \|D[u(t) - F(t)x(t)]\|^2 \\
&\qquad\qquad - \gamma^2\|d(t) - F_2(t)x(t)\|^2 \, \mathrm{d}t,
\end{aligned}$$

where $F(t) = (D^{\mathrm{T}}D)^{-1}(B^{\mathrm{T}}P(T-t)+D^{\mathrm{T}}C)$ and $F_2(t) = E^{\mathrm{T}}P(T-t)$. It is obvious that if $u = Fx$ then, no matter what d is, $\mathcal{C}_L(u, d, \xi, T) \geqslant \xi^{\mathrm{T}}P(T)\xi$.

The converse is again the most difficult part. We would like to choose $d = F_2x$, but we are constrained to choose d in open loop, i.e. independent of u. We follow the same course of action as in the proof of lemma 13.2. We choose d equal to $\bar{d} = F_2 p$ where $\dot{p} = (A+BF-EF_2)p$ and $p(0) = \xi$. Moreover, define $v = D(u - Fx)$. We get:

$$\inf_u \mathcal{C}_L(u, \bar{d}, \xi, T) = \xi^{\mathrm{T}}P(T)\xi + \inf_v \int_0^T \|v(t)\|^2 - \gamma^2\|\bar{z}(t)\|^2 \,\mathrm{d}t, \tag{13.18}$$

where $\bar{z}$ is generated by the following system:

$$\bar{\Sigma} : \begin{array}{rl} \dot{\bar{x}} = & (A + BF)\bar{x} + B(D^{\mathrm{T}}D)^{-1}D^{\mathrm{T}}v, \qquad \bar{x}(0) = 0 \\ \bar{z} = & F_2\bar{x}. \end{array}$$

The input $v = 0$ is clearly optimal as soon as we have shown that the $L_2[0, T]$-induced operator norm from v to $\bar{z}$ is less than γ^{-1}. Because the equations become easier, we prove instead that the following system has $L_2[0, T]$-induced operator norm less than 1:

$$\bar{\Sigma}_e : \begin{array}{rl} \dot{\bar{x}} = & (A + BF)\bar{x} + B(D^{\mathrm{T}}D)^{-1}D^{\mathrm{T}}v, \qquad \bar{x}(0) = 0 \\ \bar{z}_e = & \begin{pmatrix} \gamma F_2 \\ \left[I - D(D^{\mathrm{T}}D)^{-1}D^{\mathrm{T}}\right]^{1/2} C \end{pmatrix} \bar{x}. \end{array}$$

It is easy to see that if $\bar{\Sigma}_e$ has norm less than 1 then clearly $\bar{\Sigma}$ has norm less than γ^{-1}. Clearly, $X(t) := Q(T-t)$ satisfies the following Riccati differential equation:

$$\begin{aligned} -\dot{X} = (A + BF)^{\mathrm{T}}X + X(A + BF) + \gamma^2 F_2^{\mathrm{T}}F_2 + XB(D^{\mathrm{T}}D)^{-1}B^{\mathrm{T}}X \\ + C^{\mathrm{T}}[I - D(D^{\mathrm{T}}D)^{-1}D^{\mathrm{T}}]C \end{aligned}$$

where $X(T) = L$. To apply lemma 12.20 we need a solution of this equation for zero endpoint penalty. However the existence of a solution to the above Riccati equation with a positive semi-definite endpoint (see exercise 13.2) also guarantees that the $L_2[0, T]$-induced operator norm from v to $\bar{z}_e$ is less than 1. Hence we get $\mathcal{C}_L(u, d, \xi, T) \geqslant \xi^{\mathrm{T}}P(T)\xi$. ■

We have shown that if part (i) of theorem 13.3 is satisfied then $\mathcal{C}_L^*(\xi, T)$ is bounded by (13.16) for all T. Therefore, the above lemma shows that P is uniformly bounded. But we already discussed before lemma 13.5 that this implies that a solution P with $P(0) = L$ of the Riccati equation (13.17) exists on the complete interval $[0, \infty)$. Moreover, using the argument from (13.15) it is not hard to see that P is an increasing, bounded function. Therefore $P(t) \to \bar{P}$ as $t \to \infty$. Moreover $\bar{P} \geqslant L$. It is not difficult to see that $\bar{P}$ satisfies the algebraic Riccati equation (13.9) for $P = \bar{P}$.

However, we still have to prove that $\bar{P}$ is such that the matrix in (13.10) for $P = \bar{P}$ is stable. First we show that $A + BF$ is a stability matrix where F is defined by (13.11) for $P = \bar{P}$. We can rewrite the algebraic Riccati equation as:

$$(A + BF)^{\mathrm{T}}\bar{P} + \bar{P}(A + BF) + (C + DF)^{\mathrm{T}}(C + DF) + \gamma^{-2}\bar{P}EE^{\mathrm{T}}\bar{P} = 0$$

Therefore, if x is an eigenvector of $A + BF$, with eigenvalue λ we get:

$$2(\Re\mathrm{e}\,\lambda)x^*\bar{P}x = -\gamma^{-2}\|E^{\mathrm{T}}\bar{P}x\|^2 - \|(C + DF)x\|^2.$$

We see that $\Re\mathrm{e}\,\lambda \leqslant 0$ or $\bar{P}x = 0$. If $\bar{P}x = 0$ then we also have $Lx = 0$ (remember that $\bar{P} \geqslant L \geqslant 0$) so x is also an eigenvector with eigenvalue λ of (13.13). The latter matrix is stable and hence $\Re\mathrm{e}\,\lambda < 0$. On the other hand, assume $\Re\mathrm{e}\,\lambda = 0$. Then we have $(A + BF)x = \lambda x$ and $(C + DF)x = 0$, which implies that (A, B, C, D) has a zero λ on the imaginary axis (see exercise 13.3), which contradicts a basic assumption of theorem 13.3. In conclusion, we find that $A + BF$ is asymptotically stable.

Next, we apply a similar technique as in subsection 12.6.2. We know there exists a feedback $u = F_1x$ such that the closed loop H_∞ norm from d to z is less than γ, say δ. But then, for all $T > 0$, the $L_2[0, T]$-induced operator norm is less than or equal to δ. We obtain:

$$\begin{aligned}
\mathcal{C}_L(F_1x, d, \xi, T) &:= \int_0^T \|z(t)\|^2 - \gamma^2\|d(t)\|^2 \,\mathrm{d}t \; + x^{\mathrm{T}}(T)Lx(T) \\
&\leqslant \int_0^\infty \|z(t)\|^2 - \gamma^2\|d(t)\|^2 \,\mathrm{d}t \qquad\qquad (d(t) = 0, \forall t > T) \\
&\leqslant (\delta^2 - \gamma^2)\|d\|_{2,\mathrm{T}}^2 + 2\delta\|z_{\xi,0}\|_2 \cdot \|d\|_{2,\mathrm{T}} + \|z_{\xi,0}\|_2^2 \\
&\leqslant (\delta^2 - \gamma^2)(\|d\|_{2,\mathrm{T}} - c_1)^2 + c_2,
\end{aligned}$$

for certain constants c_1 and c_2, independent of T. Choose an arbitrary T. We note that for $d_{\mathrm{T}} = F_{2,\mathrm{T}}x$ with $F_{2,\mathrm{T}}(t) = \gamma^{-2}E^{\mathrm{T}}P(T - t)$ and for all initial state $x(0) = \xi$ we have $\mathcal{C}_L(u, d, \xi, T) \geqslant \xi^{\mathrm{T}}P(T - t)\xi \geqslant 0$ and hence

$$(\delta^2 - \gamma^2)(\|d_{\mathrm{T}}\|_{2,\mathrm{T}} - c_1)^2 + c_2 \geqslant 0.$$

This implies (since $\delta < \gamma$) that $\|d_{\mathrm{T}}\|_{2,\mathrm{T}} < \alpha$ with α independent of T. Moreover, $\|d_{\mathrm{T}}\|_{2,T_1} < \|d_{\mathrm{T}}\|_{2,\mathrm{T}} < \alpha$ for all $T \geqslant T_1$. Finally, since $F_{2,\mathrm{T}}(t) \to F_2$ $(T \to \infty)$ for all t where $F_2 = \gamma^{-2}E^{\mathrm{T}}\bar{P}$, we have $d_{\mathrm{T}}(t) \to d(t)$ as $T \to \infty$ for all t with $d = F_2x$. But then

$$\|d\|_{2,T_1} = \lim_{T\to\infty} \|d_{\mathrm{T}}\|_{2,T_1} < \alpha$$

for all T_1, which guarantees that $d \in L_2$. On the other hand, d is determined by

$$\dot{x} = (A + BF)x + Ed = (A + BF + EF_2)x.$$

Since $A + BF$ is stable and $Ed \in L_2$, we have $x \in L_2$ for all initial states (see exercise 13.1). This implies that A_{cl} defined by (13.10) with $P = \bar{P}$ is a stability matrix. Finally, it is obvious that $\bar{P} \geqslant 0$. This completes the proof of the implication (ii) $\Rightarrow$ (i) of theorem 13.1

It remains to show that part (ii) of theorem 13.1 implies part (i). We will prove that the feedback $u = Fx$ defined by (13.4) satisfies part (i). We first have to show that this feedback is stabilizing. We have

$$(A + BF)^{\mathrm{T}}P + P(A + BF) + (C + DF)^{\mathrm{T}}(C + DF) + \gamma^{-2}PEE^{\mathrm{T}}P = 0.$$

Therefore if $x \neq 0$ is an eigenvector of $A + BF$ with eigenvalue λ we get:

$$2(\Re e\,\lambda)x^*Px = -\gamma^{-1}\|E^{\mathrm{T}}Px\|^2 - \|(C + DF)x\|^2.$$

We see that if $\Re e\,\lambda \geqslant 0$ then $E^{\mathrm{T}}Px = 0$. But then $(A+BF+\gamma^{-2}EE^{\mathrm{T}}P)x = \lambda x$. The latter matrix is a stability matrix and hence $\Re e\,\lambda < 0$ which yields a contradiction. In conclusion we find that $A + BF$ is a stability matrix.

If we use a completion of the squares argument similar to the one used in the proof of lemma 13.5, then for zero initial state and $u = Fx$ we obtain:

$$\int_0^\infty \|z(t)\|^2 - \gamma^2\|d(t)\|^2\,\mathrm{d}t = -\int_0^\infty \gamma^2\|d(t) - F_2(t)x(t)\|^2\,\mathrm{d}t \leqslant 0 \quad (13.19)$$

for all $d \in L_2[0, T]$. Here we used that, since $A + BF$ is stable and $d \in L_2$, we have that $x(t) \to 0$ as $t \to \infty$ (see exercise 13.1). (13.19) immediately implies that the H_∞ norm is less than or equal to γ. We still have to prove that the norm is strictly less than γ. Define $v = d - F_2x$ and consider the system

$$\Sigma_{vd} : \begin{array}{rl} \dot{x} = & (A + BF + EF_2)x + Ev, \\ d = & F_2x + v. \end{array}$$

Since the state matrix of Σ_{vd} is equal to the matrix in (13.10), and therefore stable we find that this system with transfer matrix G_{vd} has a finite H_∞ norm. Thus we get

$$\|z\|_2^2 - \gamma^2\|d\|_2^2 = -\gamma^2\|v\|_2^2 \leqslant -\gamma^2\|G_{vd}\|_\infty^{-2}\|d\|_2^2,$$

which guarantees that the H_∞ norm is less than γ. Therefore $u = Fx$ satisfies part (i) of theorem 13.1 and we have completed the proof of theorem 13.1.

13.4 Solving the algebraic Riccati equation

We have expressed conditions for our finite horizon problems in terms of the Riccati differential equation. To find a solution we can use standard numerical tools to solve the differential equation.

For the infinite horizon problems however, we obtain an algebraic Riccati equation. Solving such a nonlinear equation is not easy. Moreover, it might have several or even an infinite number of solutions. How do we solve this equation and obtain the specific solution which satisfies the required stability requirement? The algebraic Riccati equation in this chapter, the algebraic Riccati equation from the bounded real lemma in section 12.6 as well as the algebraic Riccati equation in the zero-endpoint, linear quadratic regulator problem from section 10.4 have a similar structure. We seek a matrix X such that:

$$0 = F^{\mathrm{T}}X + XF + XRX + Q \tag{13.20}$$

and such that $A + RP$ is a stability matrix, where Q and R are symmetric. Such a matrix is then called the stabilizing solution of the algebraic Riccati equation. For instance the Riccati equation (13.9) we get by choosing:

$$\begin{aligned} F &:= A - B(D^{\mathrm{T}}D)^{-1}D^{\mathrm{T}}C, \\ R &:= \gamma^{-2}EE^{\mathrm{T}} - B(D^{\mathrm{T}}D)^{-1}B^{\mathrm{T}}, \\ Q &:= C^{\mathrm{T}}\left[I - D(D^{\mathrm{T}}D)^{-1}D^{\mathrm{T}}\right]C. \end{aligned}$$

To solve this equation we note that we can rewrite (13.20) as:

$$\begin{pmatrix} X & -I \end{pmatrix} \begin{pmatrix} F & R \\ -Q & -F^{\mathrm{T}} \end{pmatrix} \begin{pmatrix} I \\ X \end{pmatrix}.$$

This in turn implies that

$$\mathcal{X} = \operatorname{im} \begin{pmatrix} I \\ X \end{pmatrix}$$

is an invariant subspace of the matrix:

$$H = \begin{pmatrix} F & R \\ -Q & -F^{\mathrm{T}} \end{pmatrix}.$$

H is called a *Hamiltonian matrix* because it has a special structure:

$$\begin{pmatrix} 0 & -I \\ I & 0 \end{pmatrix} H \begin{pmatrix} 0 & I \\ -I & 0 \end{pmatrix} = -H^{\mathrm{T}} \tag{13.21}$$

This implies that H and $-H^{\mathrm{T}}$ are similar, which guarantees, for instance, that if λ is an eigenvalue of H then $-\lambda$ is also an eigenvalue of H. Let $\mathcal{X}_-(H)$, $\mathcal{X}_0(H)$ and $\mathcal{X}_+(H)$ denote the $\mathbb{C}^-$, $\mathbb{C}^0$ and $\mathbb{C}^+$ stable subspaces of H respectively (these spaces were defined in definition 2.13). Then (13.21) guarantees that $\dim \mathcal{X}_-(H) = \dim \mathcal{X}_+(H)$.

We have:

$$H \begin{pmatrix} I \\ X \end{pmatrix} = \begin{pmatrix} I \\ X \end{pmatrix} (F + RX)$$

We see that X satisfies the Riccati equation and the connected stability condition if and only if $\mathcal{X} \subset \mathcal{X}_-(H)$. The condition (13.21) guarantees that $\dim \mathcal{X}_-(H) =$

$\dim \mathcal{X}_+(H)$. Let $X \in \mathbb{R}^{n\times n}$. Then $\mathcal{X}$ is an n-dimensional subspace. On the other hand the dimension of the modal subspaces $\mathcal{X}_-(H)$ and $\mathcal{X}_+(H)$ must both be at least n. Since H has in total $2n$ eigenvalues this guarantees that if a stabilizing solution of the Riccati equation exists then H can not have eigenvalues on the imaginary axis. Moreover:

$$\mathcal{X}_-(H) = \operatorname{im} \begin{pmatrix} I \\ X \end{pmatrix}$$

Since there are efficient ways to determine the $\mathbb{C}^-$-stable subspace this yields a method to determine the stabilizing solution of the algebraic Riccati equation. It also shows the stabilizing solution of the algebraic Riccati equation must be unique.

Note that if (Q, F) has an unobservable eigenvalue on the imaginary axis, i.e. there exists a vector $x \neq 0$ such that $Fx = \lambda x$ and $Qx = 0$ with $\lambda \in \mathbb{C}^0$ then:

$$H \begin{pmatrix} x \\ 0 \end{pmatrix} = \lambda \begin{pmatrix} x \\ 0 \end{pmatrix}$$

Hence $\dim \mathcal{X}_0(H) > 0$ and this guarantees that $\dim \mathcal{X}_-(H) < n$. The latter implies that the algebraic Riccati equation can not have a stabilizing solution. The assumptions made about zeros on the imaginary axis made in several of our main theorems (such as theorem 13.3) are equivalent to the requirement that (Q, F) has no unobservable eigenvalues on the imaginary axis and are hence necessary conditions for the existence of a stabilizing solution of the algebraic Riccati equation.

13.5 Exercises

13.1 Consider the following system:

$$\Sigma : \begin{array}{l} \dot{x} = Ax + Ed, \\ z = Cx + Dd. \end{array} \tag{13.22}$$

a. Show that $d \in L_2$ and A a stability matrix imply that $x \in L_2$ for any initial state.

b. Show that if $d \in L_2$ and A a stability matrix then $x(t) \to 0$ as $t \to 0$ for arbitrary initial statres.

c. Prove that Σ is $\mathbb{C}^-$ stable if and only if

- (C, A) is detectable,
- (A, B) stabilizable, and
- for all $d \in L_2$, and zero initial state, we have $z \in L_2$.

13.2 Consider the following two Riccati differential equation:

$$\begin{aligned} \dot{P} &= A^{\mathrm{T}}P + PA + PRP + Q, \quad P(0) = P_0, \\ \dot{X} &= A^{\mathrm{T}}X + XA + XRX + Q, \quad X(0) = X_0, \end{aligned}$$

where $Q \geqslant 0$ and $R \geqslant 0$. We know that there exist an interval $[0, T]$ with $T > 0$ on which both equations have a solution.

a. Show that both $P(\tau)$ and $X(\tau)$, for any $\tau \in [0, T]$, represent the optimal cost of an optimization problem, where the two optimization problems differ only in their endpoint penalty.

b. Prove that $P_0 \geqslant X_0$ implies $P(\tau) \geqslant X(\tau)$ for all $\tau \in [0, T]$.

c. Show that $P_0 \geqslant X_0 \geqslant 0$ implies that the solution P to the first Riccati equation exists on a time-interval which is not longer than the time-interval on which a solution X exists to the second Riccati equation.

d. Consider the system $\Sigma = (A, B, C, D)$. Prove that the $L_2[0, T]$-induced operator norm of Σ is less than γ if and only if there exists a solution P on $[0, T]$ of

$$-\dot{P} = A^{\mathrm{T}}P + PA + C^{\mathrm{T}}C + \left(PE + C^{\mathrm{T}}D\right)\left(\gamma^2 I - D^{\mathrm{T}}D\right)^{-1}\left(B^{\mathrm{T}}P + D^{\mathrm{T}}C\right)$$

with $P(T) \geqslant 0$.

13.3 Consider the system $\Sigma = (A, B, C, D)$, with D injective. Show that λ is a zero of Σ if and only if there exists a matrix $F \in \mathbb{R}^{n \times m}$ and an eigenvector $x \neq 0$ with eigenvalue λ of $A + BF$ such that $(C + DF)x = 0$. Does the same result hold if D is not injective?

13.6 Notes and references

The first solutions of the H_∞ control problem (see e.g. Francis [46]) were all in the frequency domain and based on different types of factorizations of rational matrices (inner-outer factorization, spectral factorization, etc.). These methods had difficulty with the order of the controller which could be much higher than the order of the plant. In the last few years this frequency domain approach has been refined via the introduction of the so-called J-spectral factorization (see e.g. Green, Glover, Limebeer and Doyle [65]). At the moment this is quite an elegant theory. On the other hand, in this book we will present a time-domain oriented solution of the H_∞ control problem (see e.g. Doyle, Glover, Khargonekar and Francis [41], Petersen and Hollot [144] and Stoorvogel [184]). The state feedback problem studied in this chapter plays a crucial role in this time-domain approach.

Note that in theorem 13.1 our assumption that D is injective is clearly needed since otherwise the inverse in the Riccati differential equation (13.3) will never exist. Conditions for the general case where D is not injective can be found in Stoorvogel and Trentelman [190]. Similarly, in theorem 13.3 our assumption that D is injective is clearly needed since otherwise the inverse in the algebraic Riccati equation (13.9) will never exist. Conditions for the general case where D is not injective can be found in Stoorvogel and Trentelman [189] or in Stoorvogel [184]. If the subsystem (A, B, C, D) has zeros on the imaginary axis then it can be shown that the Riccati equation (13.9) can never have a solution such that A_{cl} in (13.10) is a stability matrix. The extension to the general case is studied by Scherer [164].

To find the minimal achievable norm γ^*, we have to try several values of γ. If, for a certain γ the conditions of theorem 13.1 or, equivalently, theorem 13.3, then we have $\gamma > \gamma^*$. Otherwise $\gamma \leqslant \gamma^*$. In this way, we can find γ^* via a binary search. Note that for $\gamma = \gamma^*$ there does not exist a solution P. Therefore the above theorem can not be used to find an optimal controller achieving a closed loop norm γ^*. There still is no good method to check for the existence of an optimal controller directly. Under some strong assumptions, direct characterization of γ^* is studied in Chen [28]. Optimal controllers are studied in Scherer [165] and in Glover, Limebeer, Doyle, Kasenally and Safonov [63]. Improving the numerical search has been studied by Scherer [162].

Note that for $\gamma = \infty$ we get the same Riccati equation as in the linear quadratic control theory from chapter 10. Hence for γ large, the controller presented in theorem 13.1 and, equivalently theorem 13.3 is very close to the controller of the related linear quadratic control problem. To understand this connection more deeply it is worthwhile to study the minimum entropy interpretation of H_∞ controllers (see section 15.4).

We noted after theorem (13.3), the importance of guaranteeing that the solution of the algebraic Riccati equation is positive semi-definite. To see the consequence of dropping this requirement one should have a look at the work of Stoorvogel [187]

In solving the Riccati equation, a crucial step was the determination of the $\mathbb{C}^-$-stable subspace. It should be noted that for this we do not need to compute the eigenvectors or more general the Jordan form of the Hamiltonian matrix as done in Potter [150]. We get much better numerical properties if we use the ordered Schur decomposition (see Stewart [182]). For details we refer to Laub [108]. An alternative, more recent, approach to solve the Riccati equation is the so-called sign method. This method is especially better suited for parallel algorithms. We refer to Roberts [154] and Byers [26]. For solving Riccati equations we also mention more recent work by Kenney, Laub and Wette [98] and Laub and Gahinet [109].

Chapter 14

The H_∞ control problem with measurement feedback

14.1 Introduction

In the previous chapter we have studied the H_∞ control problem under the assumption that the entire state vector is available for feedback. In this chapter we will study the H_∞ control problem with measurement feedback. Instead of the entire state vector we assume that only a (noisy) measurement is available. The technique we will use is very similar to that of section 11.3. That is, we will apply two system transformations to obtain a system for which the disturbance decoupling problem is solvable. Moreover, a controller internally stabilizes this new system and yields an H_∞ norm less than γ if and only if this controller is internally stabilizing and yields an H_∞ norm less than γ when applied to the original system. Note that the transformation does not preserve the H_∞ norm but does preserve the property that the norm is smaller than γ. One specific suitable controller is then determined by solving this disturbance decoupling problem. Subsequently, we will parameterize all suboptimal solutions.

Under some assumptions, the necessary and sufficient conditions for the existence of internally stabilizing controllers which make the H_∞ norm of the closed-loop system less than some, a priori given, bound γ are the following: two Riccati equations should have certain positive semi-definite solutions and the product of these two matrices should have spectral radius less than γ^2. One of these equations is equal to the Riccati equation from chapter 13. Hence this Riccati equation is related to the state feedback H_∞ control problem. The second Riccati equation is dual to the first and is related to the problem of state estimation. The final condition that the product of the solutions of these Riccati equations should have spectral radius less than γ^2 is at this moment hard to explain intuitively. It is a kind of coupling condition which expresses whether state estimation and state feedback combined in some suitable manner yield the desired result: an internally stabilizing feedback which makes the H_∞ norm less than γ. The appearance of the coupling condition is related to an intrinsic difference with the H_2 optimal control problem. The optimal state feedback and the optimal estimator are coupled: if we use a different cost function, then we will obtain a different estimator. In the H_2 optimal control problem the estimator does not depend on the cost functional.

14.2 Problem formulation and main results

We consider the linear, time-invariant, finite-dimensional system:

$$\Sigma : \begin{array}{rl} \dot{x} = & Ax + Bu + Ed, \\ y = & C_1 x + D_1 d, \\ z = & C_2 x + D_2 u, \end{array} \tag{14.1}$$

where $x \in \mathbb{R}^n$ is the state, $u \in \mathbb{R}^m$ is the control input, $d \in \mathbb{R}^\ell$ is the unknown disturbance, $y \in \mathbb{R}^p$ is the measured output and $z \in \mathbb{R}^q$ is the output to be controlled. A, B, E, C_1, C_2, D_1, and D_2 are matrices of appropriate dimensions. We assume that two direct-feedthrough matrices are zero. The direct feedthrough matrix from d to z is no problem and can be handled using the techniques presented in this chapter. It does, however, yield rather messy formulas. Including a direct feedthrough matrix from u to y does not change much. After all u is known and by adding something known to the measurement, the information supplied by the measurement does not change. However, a subtle problem of well-posedness can occur (see section 3.13).

As in the previous chapter we would like to minimize the effect of the disturbance d on the output z by finding an appropriate control input u. This time, however, the measured output y is not necessarily x but a more general linear function of the state and the disturbance. The controller has less information and hence the necessary conditions for the existence of internally stabilizing controllers which make the H_∞ norm of the closed-loop system less than some given bound will be stronger. It turns out that we need an extra algebraic Riccati equation which tests how well we can observe the state.

More precisely, we seek a *dynamic* controller Γ described by:

$$\Gamma : \begin{array}{rl} \dot{w} = & Kw + Ly, \\ u = & Mw + Ny. \end{array} \tag{14.2}$$

such that, after applying the feedback Γ to the system (14.1), the resulting closed-loop system is internally stable and has H_∞ norm strictly less than some a priori given bound γ. We shall derive necessary and sufficient conditions under which such a controller exists.

Recall that $\rho(M)$ denote the spectral radius of a matrix M (see section 2.5). We willl now formulate the main result of this chapter.

Theorem 14.1 *Consider the system (14.1). Assume that (A, B, C_2, D_2) and (A, E, C_1, D_1) have no zeros on the imaginary axis. Moreover, assume that D_1 is surjective and D_2 is injective. Then the following two statements are equivalent:*

(*i*) *For the system (14.1) a time-invariant, dynamic controller* Γ *of the form (14.2) exists such that the resulting closed-loop system, with transfer matrix* G_F*, is internally stable and has* H_∞ *norm less than* γ*, i.e.* $\|G_F\|_\infty < \gamma$.

(*ii*) *There exist positive semi-definite real symmetric solutions* P *and* Q *of the algebraic Riccati equations*

$$A^T P + PA + C_2^T C_2 + \gamma^{-2} PEE^T P - \left(PB + C_2^T D_2\right)\left(D_2^T D_2\right)^{-1}\left(B^T P + D_2^T C_2\right) = 0, \quad (14.3)$$

and

$$AQ + QA^T + EE^T + \gamma^{-2} QC_2^T C_2 Q - \left(QC_1^T + ED_1^T\right)\left(D_1 D_1^T\right)^{-1}\left(C_1 Q + D_1 E^T\right) = 0, \quad (14.4)$$

respectively, such that $\rho(PQ) < \gamma^2$ *amd such that the following two matrices are stability matrices:*

$$A + \gamma^{-2} EE^T P - B\left(D_2^T D_2\right)^{-1}\left(B^T P + D_2^T C_2\right), \quad (14.5)$$

$$A + \gamma^{-2} QC_2^T C_2 - \left(QC_1^T + ED_1^T\right)\left(D_1 D_1^T\right)^{-1} C_1. \quad (14.6)$$

If P *and* Q *satisfying the conditions in part (ii) exist, then a controller satisfying part (i) is given by:*

$$\Gamma : \begin{array}{l} \dot{\hat{x}} = A\hat{x} + Bu + E\hat{d}_{worst} + L(y - C_1\hat{x} - D_1\hat{d}_{worst}), \\ u = F\hat{x}, \end{array}$$

where $\hat{d}_{worst} = \gamma^{-2} E^T P\hat{x}$ *and*

$$\begin{array}{lcl} F & := & -\left(D_2^T D_2\right)^{-1}\left(D_2^T C_2 + B^T P\right), \\ L & := & (I - \gamma^{-2} QP)^{-1}(ED_1^T + QC_1^T)(D_1 D_1^T)^{-1}. \end{array}$$

Remarks :

- Note that the conditions on P in part (ii) of theorem 13.3 are exactly the same as the conditions on P in part (ii) of the above theorem. Hence the conditions on P are related to the state feedback H_∞ control problem. The conditions on Q are exactly dual to the conditions on P. It can be shown that the existence of Q is related to the question how well we are able to estimate the state x on the basis of our observations y. The test whether we are able to estimate and control simultaneously with the desired effect, is expressed by the coupling condition $\rho(PQ) < \gamma^2$.

- We will prove this theorem only for the case $\gamma = 1$. The general result can then be easily obtained by scaling. This scaling implies that we define a modified system by replacing E and D_1 by E/γ and D_1/γ respectively.

  Suppose that we have a controller which make the H_∞ norm less than γ for the original system, then it is easy to check that the same controller makes the H_∞ norm for this modified system less than 1 and conversely. If for this modified system there exists solutions P and Q of the algebraic Riccati equations defined in theorem 14.1 for $\gamma = 1$ then we see that for the original system the conditions of theorem 14.1 are satisfied by P and $\gamma^2 Q$.

- In the state feedback problem we can consider the following optimization problem:

$$\sup_{w \in L_2} \inf_{u \in L_2} \{ \|z\|_2^2 - \gamma^2 \|w\|_2^2 \mid x \in L_2 ,\ x(0) = \xi \} \tag{14.7}$$

  and we would find that the solution of this optimization problem is given by $u = Fx$ and $d = \gamma^{-2} E^{\mathrm{T}} P x$. This d makes the above criterion as large as possible while we, as controller, want to minimize this criterion. Therefore this d can be considered the worst possible disturbance and hence sometimes denotes by d_{worst}.

  Note that in the last chapter we did not really look at the optimization problem (14.7) but finite horizon versions. But the above interpretation of d_{worst} is useful in understanding the structure of the controller we describe in theorem 14.1.

- Note the special structure for the controller as a state observer (see section 3.11) interconnected with a state feedback. The state feedback is equal to the one given in theorem 13.3 for the state feedback case. The differences with a standard state observer interconnected with a state feedback are the terms with $\hat{d}_{\text{worst}}$. In an observer, if we have known inputs, then we would add them in the way we have added $\hat{d}_{\text{worst}}$ above. But we do not know whether d equals $\hat{d}_{\text{worst}}$. The measurements do not give us information what d is actually going to be. But, from the above comment, we know that in a certain sense the worst d that can occur is $d = \gamma^{-2} E^{\mathrm{T}} P x$. H_∞ control is a kind of worst case analysis and hence we expect d to be equal to $\gamma^{-2} E^{\mathrm{T}} P x$. However, since we do not know x we replace x by our estimate $\hat{x}$. We will of course prove formally that this controller has the required properties.

- Note that the controller converges to the H_2 optimal controller derived in section 11.3 as $\gamma \to \infty$. However, there is a major difference between H_2 and H_∞. In H_∞ control, the observer depends explicitly on the control problem. For instance, the observer depends on the matrices C_2 and D_2 which determine the cost criterion. In H_2 control the observer, often called the Kalman filter (see exercise 11.3), is completely independent of the control problem so, in particular, independent of the matrices C_2 and D_2. A classical way to solve the H_2 control problem is to use the separation principle, which states that we can design a controller and an observer independently. This no longer holds for the H_∞ control problem. The technique we used in this book to solve the H_2 control problem with measurement feedback was not based on the separation principle, and can be adapted to yield a solution of the H_∞ control problem with measurement feedback.

14.3 Inner systems and Redheffer's lemma

In this section we present some preliminary results and definitions needed in the proof of theorem 14.1.

Consider a system $\Sigma = (A, B, C, D)$ with input–output operator $\mathcal{G}$ (i.e. the map which associates to every input d an output z, given zero initial state). Σ is called *inner* if the system is internally stable and the input-output operator $\mathcal{G}$ is *unitary*, i.e. $\mathcal{G}$ maps L_2^m into itself and $\mathcal{G}$ has the property that for all $f \in L_2^m$ we have

$$\|\mathcal{G}f\|_2 = \|f\|_2.$$

Often, inner is defined as a property of the transfer matrix, but in our setting the above is a more natural definition. It can be shown that $\mathcal{G}$ is unitary if and only if the transfer matrix of the system, denoted by G, satisfies:

$$G^{\mathrm{T}}(-s)G(s) = I. \tag{14.8}$$

A transfer matrix G satisfying (14.8) is called *unitary*. Note that if G is unitary then G need not be $\mathbb{C}^-$ stable. If the transfer matrix is unitary and stable we call it inner (if the transfer matrix is unitary but not necessarily stable then we call it *all-pass*). In general, for an operator from L_2^m to L_2^p, in the literature two concepts, inner and co-inner are defined. A system $\Sigma = (A, B, C, D)$ is called *co-inner* if the dual system $\Sigma^{\mathrm{T}} = (A^{\mathrm{T}}, C^{\mathrm{T}}, B^{\mathrm{T}}, D^{\mathrm{T}})$ is inner. In other words, the system is co-inner if it is stable and its transfer matrix satisfies

$$G(s)G^{\mathrm{T}}(-s) = I.$$

Note that for square systems the concepts of inner and co-inner coincide. We now formulate a lemma which yields a test to check whether a system is inner:

Lemma 14.2 *Consider the system Σ described by:*

$$\Sigma : \begin{array}{l} \dot{x} = Ax + Bu, \\ z = Cx + Du, \end{array} \tag{14.9}$$

with A a stability matrix. The system Σ is inner if there exists a matrix X satisfying:

(*i*) $A^T X + XA + C^T C = 0,$

(*ii*) $D^T C + B^T X = 0,$

(*iii*) $D^T D = I$.

Remarks :

- If (A, B) is controllable the reverse of the above implication is also true. However, in general, the reverse does not hold. A simple counter example is given by $\Sigma := (-1, 0, 1, 1)$, which is inner but for which (ii) does not hold for any choice of X.
- Note that since A is a stability matrix, the (unique) matrix X satisfying part (i) of lemma 14.2 is equal to the observability gramian of (C, A) (see section 3.8). We know that $X > 0$ if and only if (C, A) is observable. In general we only have $X \geqslant 0$.

Inner systems are very important in H_∞ control. We will present a lemma which is a main ingredient in our proof of theorem 14.1, and which makes use of inner systems. However, we first need the following preliminary lemma.

Lemma 14.3 *Let $\mathcal{G}$ be the input-output operator of a linear time-invariant system Σ. We define:*

$$\mathcal{M} = \left\{ r \in L_2 \mid \mathcal{G}r \in L_2 \right\} \tag{14.10}$$

If for all $r \in \mathcal{M}$ we have $\|Qr\|_2 \leqslant \|r\|_2$, then $\mathcal{M}$ is a closed subset of L_2.

Proof : Assume we have $r_i \in \mathcal{M}$ such that $r_i \to \bar{r} \in L_2$ as $i \to \infty$. We have to prove that $\bar{r} \in \mathcal{M}$. First note that since $\{r_i\}$ is a convergent sequence it must be bounded in norm, i.e. $\|r_i\|_2 < \alpha$ for some $\alpha > 0$. Moreover, note that $\mathcal{G}$ can also be viewed as a system from $L_2[0, T]$ to $L_2[0, T]$ with a finite norm $\|\mathcal{G}\|_{\infty,\mathrm{T}}$ (if the system is unstable we will have $\|\mathcal{G}\|_{\infty,\mathrm{T}} \to \infty$ as $T \to \infty$). We obtain:

$$\begin{aligned}
\int_0^T \|(\mathcal{G}\bar{r})(t)\|^2 \,\mathrm{d}t &= \int_0^T \|(\mathcal{G}r_i)(t) + (\mathcal{G}[\bar{r} - r_i])(t)\|^2 \,\mathrm{d}t \\
&\leqslant \int_0^T 2\|(\mathcal{G}r_i)(t)\|^2 + 2\|(\mathcal{G}[\bar{r} - r_i])(t)\|^2 \,\mathrm{d}t \\
&\leqslant 2\|\mathcal{G}r_i\|_2^2 + 2\|\mathcal{G}\|_{\infty,\mathrm{T}}^2 \int_0^T \|[\bar{r} - r_i](t)\|^2 \,\mathrm{d}t \\
&\leqslant 2\alpha^2 + 2\|\mathcal{G}\|_{\infty,\mathrm{T}}^2 \|\bar{r} - r_i\|_2^2 \\
&\to 2\alpha^2
\end{aligned}$$

as $i \to \infty$. Since α does not depend on T, this clearly implies that $\mathcal{G}\bar{r} \in L_2$ and hence $\bar{r} \in \mathcal{M}$. ■

We now give a result which shows the importance of inner systems in H_∞ control. It is often referred to as "Redheffer's lemma".

Lemma 14.4 *Consider the linear time-invariant systems Σ and Ψ. Suppose Σ has inputs w and u and outputs z and y, while Ψ has input y and output u. Consider the interconnection depicted in the diagram in Figure 14.1. Assume that Σ is inner and*

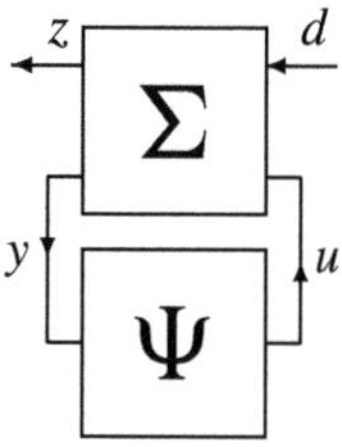

Figure 14.1

its input-output operator $\mathcal{G}$ has the following decomposition:

$$\mathcal{G}\begin{pmatrix} d \\ u \end{pmatrix} =: \begin{pmatrix} \mathcal{G}_{11} & \mathcal{G}_{12} \\ \mathcal{G}_{21} & \mathcal{G}_{22} \end{pmatrix}\begin{pmatrix} d \\ u \end{pmatrix} = \begin{pmatrix} z \\ y \end{pmatrix}, \tag{14.11}$$

which is compatible with the sizes of d, u, z and y, such that the $\mathcal{G}_{21}$ is invertible and $\mathcal{G}_{21}^{-1}$ has a finite L_2-induced operator norm.

Under the above assumptions the following two statements are equivalent:

(i) *The interconnection in Figure 14.1 is internally stable and its closed-loop transfer matrix has H_∞ norm less than* 1.

(ii) *The system Ψ is internally stable and its transfer matrix has H_∞ norm less than* 1.

Proof : Suppose part (ii) is satisfied. As $\|\Psi\|_\infty < 1$ and $\|\mathcal{G}_{22}\|_\infty \leqslant \|\mathcal{G}\|_\infty = 1$ an application of the small gain theorem implies that $(I - \mathcal{G}_{22}\Psi)^{-1}$ exists and has a finite L_2 induced operator norm. This implies internal stability of the interconnection (14.1) since both $\mathcal{G}$ and Ψ are stable (see section 12.3). Since $\mathcal{G}$ is inner, we have

$$\|d\|_2^2 + \|u\|_2^2 = \|z\|_2^2 + \|y\|_2^2$$

Combined with $\|\Psi\|_\infty < 1$, this yields:

$$\|z\|_2^2 - \|d\|_2^2 = \|u\|_2^2 - \|y\|_2^2 \leqslant -\varepsilon\|y\|_2^2 \leqslant -\frac{\varepsilon}{\beta}\|d\|_2^2 \tag{14.12}$$

for some $\varepsilon > 0$, where β is the L_2-induced operator norm of $\mathcal{G}_{21}^{-1}(I - \mathcal{G}_{22}\Psi)$ which is finite since by assumption $\mathcal{G}_{21}^{-1}$ has a finite L_2-induced operator norm. Formula (14.12) guarantees that the closed loop system has H_∞ norm strictly less than 1.

Conversely, assume part (i) is satisfied. Define the set $\mathcal{M}$ by (14.10). For any $y \in \mathcal{M}$ we have $u = \Psi y \in L_2$ and:

$$d = \mathcal{G}_{21}^{-1}y - \mathcal{G}_{21}^{-1}\mathcal{G}_{22}u \in L_2$$

since $\mathcal{G}_{21}^{-1}$ has a finite L_2-induced operator norm. But then:

$$\|u\|_2^2 - \|y\|_2^2 = \|z\|_2^2 - \|d\|^2 \leqslant -\varepsilon\|d\|_2^2 \leqslant -\frac{\varepsilon}{\beta}\|y\|_2^2$$

for some $\varepsilon > 0$, where β is the L_2-induced operator norm of the closed loop system mapping d to y which must be finite since the interconnected system is internally stable. Hence we get that

$$\sup_{y \in \mathcal{M}, y \neq 0} \frac{\|\Psi y\|_2}{\|y\|_2} < 1. \tag{14.13}$$

Then, by lemma 14.3, we know that $\mathcal{M}$ is a closed subspace of L_2. Suppose $\mathcal{M}$ is not equal to the whole of L_2. Then there must be a signal $0 \neq w \in \mathcal{M}^\perp \subset L_2$. Since the closed loop system is stable we know that $\Psi(I - \mathcal{G}_{22}\Psi)^{-1}$ and $(I - \mathcal{G}_{22}\Psi)^{-1}$ must both be stable. But then $r = (I - \mathcal{G}_{22}\Psi)^{-1}w \in \mathcal{M}$. We get:

$$\|\mathcal{G}_{22}\Psi r\|_2^2 = \|r - w\|_2^2 = \|r\|_2^2 + \|w\|_2^2 \geqslant \|r\|_2^2 \tag{14.14}$$

The second equality follows from the fact that $r \in \mathcal{M}$ and $w \in \mathcal{M}^\perp$. But we know that $r \in \mathcal{M}$ guarantees that $\|\Psi r\|_2 < \|r\|$ and we also know that $\|\mathcal{G}_{22}\|_\infty \leqslant 1$ which together contradict (14.14). Hence $\mathcal{M} = L_2$, which guarantees that Ψ is stable, while (14.13) guarantees that $\|\Psi\|_\infty < 1$. ■

14.4 Proof of theorem 14.1

In this section theorem 14.1 will be proven. We will show that the problem of finding a suitable controller Γ for the system (14.1) is equivalent to finding a suitable controller Γ for a *new system* which has some very nice structural properties. We can show that for this new system the disturbance decoupling problem with measurement feedback and stability is solvable (see lemma 11.13). Clearly, this implies that we can find for this new system, and hence also for our original system, a suitable controller. We recall that in the remainder of this chapter it is assumed that $\gamma = 1$.

Lemma 14.5 *Assume that the systems (A, B, C_2, D_2) and (A, E, C_1, D_1) have no zeros on the imaginary axis and assume that D_1 and D_2 are surjective and injective, respectively. If a controller Γ exists such that the resulting closed-loop system is internally stable and has H_∞ norm less than 1, then the following two conditions are satisfied:*

(i) *There exists a real symmetric solution $P \geqslant 0$ of the algebraic Riccati equation (14.3) such that (14.5) is a stability matrix (with $\gamma = 1$).*

(ii) *There exists a symmetric solution $Q \geqslant 0$ of the algebraic Riccati equation (14.4) such that (14.6) is a stability matrix (with $\gamma = 1$).*

Proof : Since there exists an internally stabilizing feedback controller that makes the H_∞ norm of the closed-loop system less than 1 for the problem with measurement feedback, it is easy to check that part (i) of theorem 13.3 is satisfied. Note that we need to do some work because part (i) requires the existence of a static state feedback and we have a dynamic function of both the state and the disturbance. But checking the proof of theorem 13.3, the existence of a matrix P satisfying the conditions in part (i) of the above lemma is guaranteed when the expression (13.15) is bounded and if we make the suboptimal choice of determining u on the basis of our dynamic measurement feedback it is not hard to see that (13.15) is indeed bounded.

For a system $\Gamma = (F, G, H, J)$, as before we define the dual system Γ^{T} as the system with realization $(F^{\mathrm{T}}, H^{\mathrm{T}}, G^{\mathrm{T}}, J^{\mathrm{T}})$. It is not hard to see (using the frequency domain interpretation of the H_∞ norm) that the H_∞ norm of a system is equal to the H_∞ norm of its dual system. Define the dual system of Σ:

$$\Sigma^{\mathrm{T}}: \begin{aligned} \dot{x}_d &= A^{\mathrm{T}}x_d + C_1^{\mathrm{T}}u_d + C_2^{\mathrm{T}}d_d, \\ y_d &= B^{\mathrm{T}}x_d + \phantom{C_1^{\mathrm{T}}u_d +{}} D_2^{\mathrm{T}}d_d, \\ z_d &= E^{\mathrm{T}}x_d + D_1^{\mathrm{T}}u_d, \end{aligned}$$

and the dual of our controller Γ:

$$\Sigma_F^{\mathrm{T}}: \begin{aligned} \dot{w}_d &= K^{\mathrm{T}}w_d + M^{\mathrm{T}}y_d, \\ u_d &= L^{\mathrm{T}}w_d + N^{\mathrm{T}}y_d. \end{aligned}$$

It is easy to see that the dual system of $\Sigma \times \Gamma$ is equal to $\Sigma^{\mathrm{T}} \times \Gamma^{\mathrm{T}}$. Therefore, since Γ stabilizes Σ and yields a closed loop H_∞ norm less than 1, we have that Γ^{T} stabilizes Σ^{T} and yields the same closed loop H_∞ norm, less than 1. But then the state feedback H_∞ control problem for Σ^{T} is also solvable and by applying theorem 13.3 (the required assumptions are satisfied since $(A^{\mathrm{T}}, C_1^{\mathrm{T}}, E^{\mathrm{T}}, D_1^{\mathrm{T}})$ has no zeros on the imaginary axis and D_1^{T} is injective) we find that the Riccati equation related to Σ^{T} has a stabilizing real symmetric solution, say Q, with $Q \geqslant 0$. This solution turns out to satisfy the conditions in part (ii) of our lemma. ■

Note that exercise 14.1 yields an alternative proof of part (ii) without resorting to the concept of duality.

Assume that there exist matrices P and Q satisfying conditions (i) and (ii) in lemma 14.5. In the previous chapter we have seen that:

$$\|z\|_2^2 - \|d\|_2^2 = \|D_2(u - Fx)\|_2^2 - \|d - E^{\mathrm{T}}Px\|_2^2$$

where $F = -(D_2^{\mathrm{T}}D_2)^{-1}(B^{\mathrm{T}}P + D_2^{\mathrm{T}}C_2)$. We define $z_P = D_2(u - Fx)$ and $d_P = d - E^{\mathrm{T}}Px$. Then from the above it is intuitively clear that if we find a controller from y to u which makes the H_∞ norm from d to z less than 1 then the same controller will also make the H_∞ norm from d_P to z_P equal to 1. This new system with disturbance d_P, input u, measurement y and to be controlled output z_P has the following form:

$$\Sigma_P: \begin{aligned} \dot{x}_P &= A_Px_P + Bu + Ed_P, \\ y &= C_{1,P}x_P + \phantom{Bu +{}} D_1d_P, \\ z_P &= C_{2,P}x_P + D_2u, \end{aligned} \tag{14.15}$$

where $A_P := A + EE^{\mathrm{T}}P$, $C_{1,P} := C_1 + D_1 E^{\mathrm{T}}P$ and $C_{2,P} := -D_2 F$.

We will prove that a controller Γ stabilizes Σ and yields a closed loop system $\Sigma \times \Gamma$ with H_∞ norm less than 1 if and only if the same controller stabilizes Σ_P and also yields a closed loop system $\Sigma_P \times \Gamma$ with H_∞ norm less than 1. Note that Σ_P has the property that the stabilizing state feedback $u = Fx$ yields a closed loop system with H_∞ norm equal to 0, i.e. for Σ_P the disturbance decoupling problem with state feedback and internal stability is solvable. We first derive the following lemma:

Lemma 14.6 *Assume that there exists a solution of the Riccati equation (14.3) such that the matrix in (14.5) is a stability matrix. Moreover, assume that the system* (A, B, C_2, D_2) *has no zeros on the imaginary axis. In that case, the systems* $(A_P, B, C_{2,P}, D_2)$ *and* $(A_P, E, C_{1,P}, D_1)$ *have no zeros on the imaginary axis either.*

Proof : The system $(A_P, E, C_{1,P}, D_1)$ can be obtained from the system (A, E, C_1, D_1) by applying the preliminary feedback $u = E^{\mathrm{T}}Px + v$. Therefore, the zeros of the two systems coincide. Hence the system $(A_P, E, C_{1,P}, D_1)$ has no zeros on the imaginary axis.

Similarly, $(A_P, B, C_{2,P}, D_2)$ and $(A_P + BF, B, 0, D_2)$ have the same zeros since the second system can be obtained from the first via the preliminary feedback $u = Fx + v$. Since $A_P + BF$ is equal to the matrix in (14.5) and hence asymptotically stable, these systems are both minimum-phase and in particular have no zeros on the imaginary axis. ■

We will now formally prove that a controller is suitable for Σ if and only if it is suitable for Σ_P. This result will be a consequence of "Redheffer's lemma" (lemma 14.4).

Lemma 14.7 *Let* P *satisfy the conditions (i) of lemma 14.5. Moreover, let an arbitrary dynamic controller* Γ *be given, described by (14.2). Consider the two systems in Figure 14.2, where the system on the left is the interconnection of (14.1) and (14.2) and the system on the right is the interconnection of (14.15) and (14.2). Then the*

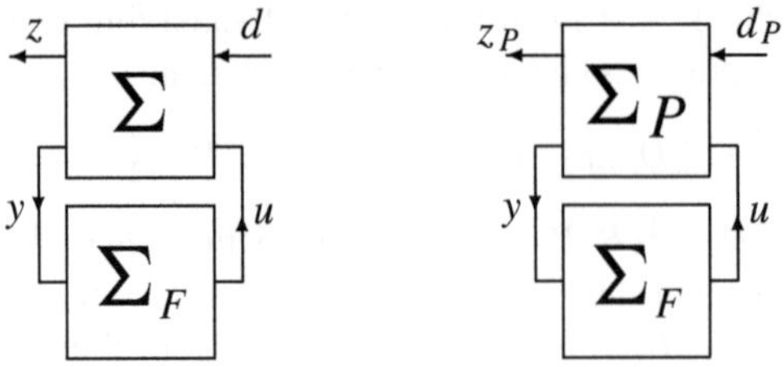

Figure 14.2

following statements are equivalent:

(*i*) *The system on the left is internally stable and its transfer matrix from* d *to* z *has* H_∞ *norm less than* 1.

(*ii*) *The system on the right is internally stable and its transfer matrix from* d_P *to* z_P *has* H_∞ *norm less than* 1.

Proof : Define the system Σ_U by

$$\Sigma_U: \begin{aligned} \dot{x}_U &= \tilde{A}x_U + B(D_2^{\mathrm{T}}D_2)^{-1}D_2^{\mathrm{T}}z_P + Ed, \\ d_P &= -E^{\mathrm{T}}Px_U + d, \\ z_U &= \tilde{C}_2x_U + z_P, \end{aligned} \tag{14.16}$$

where

$$\tilde{A} := A - B(D_2^{\mathrm{T}}D_2)^{-1}(B^{\mathrm{T}}P + D_2^{\mathrm{T}}C_2),$$
$$\tilde{C}_2 := C_2 - D_2(D_2^{\mathrm{T}}D_2)^{-1}(B^{\mathrm{T}}P + D_2^{\mathrm{T}}C_2).$$

The system Σ_U is inner. This is seen by noting that P satisfies the conditions of lemma 14.2 for the system Σ_U.

The input-output operator $\mathcal{U}$ of the system Σ_U has the following decomposition:

$$\mathcal{U}\begin{pmatrix} d \\ z_P \end{pmatrix} =: \begin{pmatrix} \mathcal{U}_{11} & \mathcal{U}_{12} \\ \mathcal{U}_{21} & \mathcal{U}_{22} \end{pmatrix}\begin{pmatrix} d \\ z_P \end{pmatrix} = \begin{pmatrix} z_U \\ d_P \end{pmatrix}, \tag{14.17}$$

which is compatible with the sizes of d, z_P, z_U and d_P. The input-output operator $\mathcal{U}_{21}$ is associated to the system $\Sigma_{U,21} = (\tilde{A}, E, -E^{\mathrm{T}}P, I)$ and, since $\tilde{A} + EE^{\mathrm{T}}P$ is equal to (14.5) and hence a stability matrix, the input-output operator $\mathcal{U}_{21}$ is invertible and $\mathcal{U}_{21}^{-1}$ has a finite L_2-induced operator norm.

Now compare the two systems in Figure 14.3. The system on the left is the

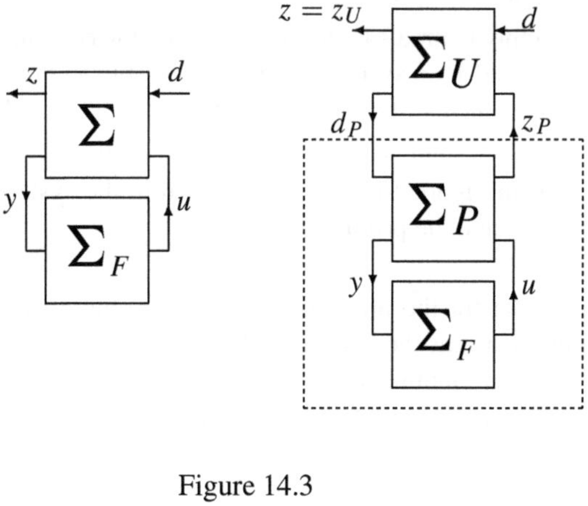

Figure 14.3

same as the system on the left in Figure 14.2 on the facing page and the system on the right is described by the system (14.16) interconnected with the system on the right in Figure 14.2. The system on the right in Figure 14.3 on the preceding page is described by:

$$
\begin{aligned}
\begin{pmatrix} \dot{x}_U - \dot{x}_P \\ \dot{x}_P \\ \dot{w} \end{pmatrix} &= \begin{pmatrix} \tilde{A} + EE^{\mathrm{T}}P & 0 & 0 \\ -(E + BND_1)E^{\mathrm{T}}P & A + BNC_1 & BM \\ -LD_1E^{\mathrm{T}}P & LC_1 & K \end{pmatrix} \begin{pmatrix} x_U - x_P \\ x_P \\ w \end{pmatrix} \\
&\quad + \begin{pmatrix} 0 \\ E + BND_1 \\ LD_1 \end{pmatrix} d, \\
z_U &= \begin{pmatrix} \tilde{C}_2 - D_2ND_1E^{\mathrm{T}}P & C_2 + D_2NC_1 & D_2M \end{pmatrix} \begin{pmatrix} x_U - x_P \\ x_P \\ w \end{pmatrix} \\
&\quad + D_2ND_1d.
\end{aligned}
$$

If we also derive the system equations for the system on the left in Figure 14.3 on the preceding page we get:

$$
\begin{aligned}
\begin{pmatrix} \dot{x}_P \\ \dot{w} \end{pmatrix} &= \begin{pmatrix} A + BNC_1 & BM \\ LC_1 & K \end{pmatrix} \begin{pmatrix} x_P \\ w \end{pmatrix} + \begin{pmatrix} E + BND_1 \\ LD_1 \end{pmatrix} d, \\
z &= \begin{pmatrix} C_2 + D_2NC_1 & D_2M \end{pmatrix} \begin{pmatrix} x_P \\ w \end{pmatrix} + D_2ND_1d.
\end{aligned}
$$

We immediately see that, since $\tilde{A} + EE^{\mathrm{T}}P$ is asymptotically stable, the two systems in Figure 14.3 on the preceding page have the same transfer matrix and one is internally stable if and only if the other one is internally stable. Hence the system on the left is stable and has H_∞ norm less than 1 if, and only if, the system on the right is stable and has H_∞ norm less than 1.

We can now apply lemma 14.4 to the system on the right in Figure 14.3 and hence we find that the closed-loop system is internally stable and has H_∞ norm less than 1 if, and only if, the dashed system is internally stable and has H_∞ norm less than 1.

Since the dashed system is exactly the system on the right in Figure 14.2 and the system on the left in Figure 14.3 is exactly equal to the system on the left in Figure 14.2 we have completed the proof. ∎

We assumed that for the original system (14.1) there exists an internally stabilizing controller such that the resulting closed-loop matrix has H_∞ norm less than 1. Hence, by applying lemma 14.7, we know that the *same* controller is internally stabilizing for the new system (14.15) and yields a closed-loop transfer matrix with H_∞ norm less than 1. Moreover we know by lemma 14.6 that Γ satisfies the assumptions on the invariant zeros needed to apply lemma 14.5. Therefore, if we consider for this new system the two Riccati equations we know that there are positive semi-definite, stabilizing solutions. We shall now formalize this in the following lemma.

Consider the following Riccati equations:

$$A_P^{\mathrm{T}} X + X A_P + C_{2,P}^{\mathrm{T}} C_{2,P} + X E E^{\mathrm{T}} X \\ - \left(X B + C_{2,P}^{\mathrm{T}} D_2\right) \left(D_2^{\mathrm{T}} D_2\right)^{-1} \left(B^{\mathrm{T}} X + D_2^{\mathrm{T}} C_{2,P}\right) = 0, \quad (14.18)$$

$$A_P Y + Y A_P^{\mathrm{T}} + E E^{\mathrm{T}} + Y C_{2,P}^{\mathrm{T}} C_{2,P} Y \\ - \left(Y C_{1,P}^{\mathrm{T}} + E D_1^{\mathrm{T}}\right) \left(D_1 D_1^{\mathrm{T}}\right)^{-1} \left(C_{1,P} Y + D_1 E^{\mathrm{T}}\right) = 0, \quad (14.19)$$

in the unknowns X and Y in $\mathbb{R}^{n \times n}$. In addition, for given X, Y consider the following two matrices:

$$A_P + E E^{\mathrm{T}} X - B(D_2^{\mathrm{T}} D_2)^{-1}(B^{\mathrm{T}} X + D_2^{\mathrm{T}} C_{2,P}), \quad (14.20)$$

$$A_P + Y C_{2,P}^{\mathrm{T}} C_{2,P} - \left(Y C_{1,P}^{\mathrm{T}} + E D_1^{\mathrm{T}}\right) \left(D_1 D_1^{\mathrm{T}}\right)^{-1} C_{1,P}. \quad (14.21)$$

Then we have:

Lemma 14.8 *Let P and Q satisfy conditions (i) and (ii) in lemma 14.5. Then we have the following two results:*

(*i*) *$X = 0$ is a solution of the algebraic Riccati equation (14.18) such that (14.20) is stable.*

(*ii*) *The algebraic Riccati equation (14.19) has a real symmetric matrix Y such that (14.21) is a stability matrix if and only if $I - QP$ is invertible. Moreover, in this case there is a unique solution $Y := (I - QP)^{-1} Q$. This matrix Y is positive semi-definite if and only if*

$$\rho\,(PQ) < 1. \quad (14.22)$$

Proof : Part (i) can be checked straightforwardly.

We know that Q is the stabilizing solution of the algebraic Riccati equation (14.4). Using the results from section 13.4 we know therefore

$$\mathcal{X}_{old} = \operatorname{im} \begin{pmatrix} I \\ Q \end{pmatrix}$$

is the $\mathbb{C}^-$-stable subspace of the Hamiltonian matrix:

$$H_{old} = \begin{pmatrix} A^{\mathrm{T}} - C_1^{\mathrm{T}}(D_1 D_1^{\mathrm{T}})^{-1} D_1 E^{\mathrm{T}} & C_2^{\mathrm{T}} C_2 - C_1^{\mathrm{T}}(D_1 D_1^{\mathrm{T}})^{-1} C_1 \\ -E\left[I - D_1^{\mathrm{T}}(D_1 D_1^{\mathrm{T}})^{-1} D_1\right] E^{\mathrm{T}} & -A + E D_1^{\mathrm{T}}(D_1 D_1^{\mathrm{T}})^{-1} C_1 \end{pmatrix}.$$

Define a new Hamiltonian matrix by

$$H_{new} = \begin{pmatrix} I & -P \\ 0 & I \end{pmatrix} H_{old} \begin{pmatrix} I & P \\ 0 & I \end{pmatrix}. \quad (14.23)$$

It is then easy to check that for Y to be a stabilizing solution of the algebraic Riccati equation (14.19) requires that

$$\mathcal{X}_{new} = \operatorname{im} \begin{pmatrix} I \\ Y \end{pmatrix}$$

is the $\mathbb{C}^-$-stable subspace of the Hamiltonian matrix H_{new}. However, (14.23) implies that the $\mathbb{C}^-$-stable subspace of H_{new} is equal to:

$$\operatorname{im} \begin{pmatrix} I & -P \\ 0 & I \end{pmatrix} \begin{pmatrix} I \\ Q \end{pmatrix} = \operatorname{im} \begin{pmatrix} I - PQ \\ Q \end{pmatrix}.$$

Therefore we know that a stabilizing solution to the algebraic Riccati equation (14.21) exists if and only if $I - PQ$ is invertible and in this case the solution is given by $Y = Q(I - PQ)^{-1}$. The requirement $Y \geqslant 0$ is satisfied if and only if $\rho(PQ) < 1$, which can be checked by noting that:

$$Y = Q^{1/2} \left(I - Q^{1/2} P Q^{1/2} \right)^{-1} Q^{1/2}$$

and $\rho(PQ) = \rho(Q^{1/2} P Q^{1/2})$. This completes the proof. ■

This completes the proof of the implication (i) ⇒ (ii) in theorem 14.1. The existence of $P \geqslant 0$ and $Q \geqslant 0$ satisfying the algebraic Riccati equations (14.3) and (14.4) such that the matrices (14.5) and (14.6) are stability matrices can be obtained directly from lemma 14.5. We know by lemma 14.6 that the two subsystems $(A_P, B, C_{2,P}, D_2)$ and $(A_P, E, C_{1,P}, D_1)$ have no zeros on the imaginary axis. We also know by lemma 14.7 that a controller exists for the transformed system Σ_P which internally stabilizes the system and makes the H_∞ norm of the closed-loop system less than 1. By applying lemma 14.5 to this new system we find that a matrix $Y \geqslant 0$ exists that satisfies the algebraic Riccati equation (14.19), such that (14.21) is a stability matrix. Hence by lemma 14.8 we have (14.22) and therefore all the conditions in theorem 14.1, part (ii) are satisfied.

We will now prove the reverse implication (ii) ⇒ (i) in theorem 14.1. Hence, assume that matrices P and Q exist satisfying the conditions of part (ii) of theorem 14.1.

In order to prove the implication (ii) ⇒ (i) we transform the system (14.15) once again, this time however using the dualized version of the original transformation. In other words, we define the dual system Σ_P^{T}, apply the same transformation from Σ to Σ_P, but this time we transform Σ_P^{T} into a new system which we call Σ_{PQ}^{T} and when we dualize Σ_{PQ}^{T} we obtain our new system Σ_{PQ}. The transformation from Σ to Σ_P depended on the solution of a Riccati equation which is determined by the realization of Σ. This time we have to solve a similar Riccati equation for Σ_P^{T} It turns out this Riccati equation is equal to (14.19) and has, by lemma 14.8, a solution $Y = (I - QP)^{-1} Q \geqslant 0$.

In the way described above, we obtain the following system:

$$\Sigma_{PQ}: \begin{array}{rcl} \dot{x}_{PQ} &=& A_{PQ}x_{PQ} + B_{PQ}u + E_{PQ}d_{PQ}, \\ y &=& C_{1,P}x_{PQ} \qquad\qquad + \quad D_1 d_{PQ}, \\ z_{PQ} &=& C_{2,P}x_{PQ} + \ D_2 u, \end{array} \tag{14.24}$$

where

$$\begin{aligned} A_{PQ} &:= A_P + YC_{2,P}^{\mathrm{T}}C_{2,P}, \\ B_{PQ} &:= B + YC_{2,P}^{\mathrm{T}}D_2, \\ E_{PQ} &:= (YC_{1,P}^{\mathrm{T}} + ED_1^{\mathrm{T}})(D_1D_1^{\mathrm{T}})^{-1}D_1. \end{aligned}$$

By applying lemma 14.7 and its dualized version, the following corollary can be derived:

Corollary 14.9 *Let Γ be a controller of the form (14.2). The following two statements are equivalent:*

(*i*) *The controller Γ applied to the system Σ described by (14.1) is internally stabilizing, and the resulting closed-loop transfer matrix has H_∞ norm less than* 1.

(*ii*) *The controller Γ applied to the system Σ_{PQ} described by (14.24) is internally stabilizing, and the resulting closed-loop transfer matrix has H_∞ norm less than* 1.

Remark : We note that, even if for this new system we can make the H_∞ norm arbitrarily small, for the original system we are only sure that the H_∞ norm will be less than 1. It is possible that a controller for the new system yields an H_∞ norm of say 0.0001 while the same controller makes the H_∞ norm of the original plant only 0.9999.

We can apply the following controller to Σ_{PQ}:

$$\Gamma: \begin{array}{rcl} \dot{\hat{x}}_{PQ} &=& A_{PQ}\hat{x}_{PQ} + B_{PQ}u + K(y - C_{1,P}\hat{x}_{PQ}), \\ u &=& F\hat{x}_{PQ}, \end{array} \tag{14.25}$$

where

$$\begin{aligned} K &= (YC_{1,P}^{\mathrm{T}} + ED_1^{\mathrm{T}})(D_1D_1^{\mathrm{T}})^{-1}, \\ F &= -(D_2^{\mathrm{T}}D_2)^{-1}(D_2^{\mathrm{T}}C_2 + B^{\mathrm{T}}P). \end{aligned}$$

Note that

$$\begin{aligned} A_{PQ} + B_{PQ}F &= (A_P + BF) + YC_{2,P}^{\mathrm{T}}(C_{2,P} + D_2F) = A + EE^{\mathrm{T}}P + BF, \\ A_{PQ} - KC_{1,P} &= A_P + YC_{2,P}^{\mathrm{T}}C_{2,P} - KC_{1,P}, \end{aligned}$$

which are equal to the matrices in (14.5) and (14.21) respectively and hence stability matrices. It is then easy to check that the resulting closed loop system $\Sigma_{PQ} \times \Gamma$ is stable and achieves disturbance decoupling with measurement feedback and internal stability (see section 11.3). Hence by applying corollary 14.9 we see that this controller satisfies part (i) of theorem 14.1. This completes the implication (ii) $\Rightarrow$ (i) in theorem 14.1. Note that the controller (14.25) is equal to the controller given in theorem 14.1.

14.5 Characterization of all suitable controllers

In this section we will parameterize all stabilizing controllers which achieve the required H_∞ norm bound. We assume throughout this section that there exist matrices P and Q satisfying the conditions of part (ii) of theorem 14.1.

Let Σ and Σ_{PQ} be defined by (14.1) and (14.24) respectively. We define the following system:

$$\bar{\Sigma}_{PQ}: \begin{array}{rl} \dot{x}_{PQ} &= A_{PQ}x_{PQ} + B_{PQ}u + \bar{E}_{PQ}\bar{d}_{PQ}, \\ y &= C_{1,P}x_{PQ} + (D_1D_1^{\mathrm{T}})^{1/2}\bar{d}_{PQ}, \\ \bar{z}_{PQ} &= \bar{C}_{2,P}x_{PQ} + (D_2^{\mathrm{T}}D_2)^{1/2}u, \end{array} \tag{14.26}$$

where

$$\bar{C}_{2,P} := (D_2^{\mathrm{T}}D_2)^{-1/2}(B^{\mathrm{T}}P + D_2^{\mathrm{T}}C_2),$$
$$\bar{E}_{PQ} := (YC_{1,P}^{\mathrm{T}} + ED_1^{\mathrm{T}})(D_1D_1^{\mathrm{T}})^{-1/2}.$$

We have $\Sigma_{PQ} = \Pi_2\bar{\Sigma}_{PQ}\Pi_1$ where

$$\Pi_1 = (D_1D_1^{\mathrm{T}})^{-1/2}D_1 \text{ and } \Pi_2 = D_2(D_2^{\mathrm{T}}D_2)^{-1/2}.$$

Hence, it is straightforward that the class of stabilizing controllers for Σ_{PQ} is equal to the class of stabilizing controllers for $\bar{\Sigma}_{PQ}$. Moreover, for any interconnection $\Sigma_{PQ} \times \Gamma$ where Γ stabilizes Σ, we get a closed loop transfer matrix $Q = \Pi_2 X \Pi_1$ where X is the stable closed loop transfer matrix of $\bar{\Sigma}_{PQ} \times \Gamma$. The system $\bar{\Sigma}_{PQ}$ turns out to have a very special property:

Lemma 14.10 *Let $\bar{\Sigma}_{PQ}$ be given. For any internally stable system Σ_x with transfer matrix X there exists a controller which stabilizes $\bar{\Sigma}_{PQ}$ and which yields a closed loop transfer matrix equal to X.*

Proof : Consider the following system.

$$\Sigma_e: \begin{array}{rl} \dot{x}_c &= A_{PQ}x_c + B_{PQ}u + K(y - C_{1,P}x_c), \\ d_x &= (D_1D_1^{\mathrm{T}})^{-1/2}[y - C_{1,P}x_c], \\ u &= (D_2^{\mathrm{T}}D_2)^{-1/2}[z_x - \bar{C}_{2,P}x_c]. \end{array}$$

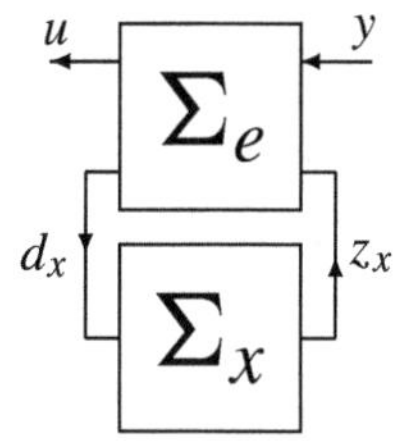

Figure 14.4

We choose a controller Γ defined by the interconnection in Figure 14.4.

The interconnection of Σ_e and Σ_{PQ} is given by:

$$\begin{aligned}
\dot{x}_c - \dot{x}_{PQ} &= (A_{PQ} - KC_{1,P})(x_c - x_{PQ}) \\
\dot{x}_{PQ} &= (A_P + BF)x_{PQ} + B_{PQ}(D_2^{\mathrm{T}}D_2)^{-1/2}[z_x - \bar{C}_{2,P}(x_c - x_{PQ})] \\
&\qquad + E_{PQ}\bar{d}_{PQ} \\
d_x &= -(D_1^{\mathrm{T}}D_1)^{-1/2}C_{1,P}(x_c - x_{PQ}) + \bar{d}_{PQ} \\
\bar{z}_{PQ} &= \bar{C}_{2,P}(x_{PQ} - x_c) + z_x
\end{aligned}$$

Since $A_{PQ} - KC_{1,P}$ and $A_P + BF$ are stability matrices, it is obvious that Γ stabilizes $\bar{\Sigma}_{PQ}$ and, for zero initial conditions, we have $x_c = X_{PQ}$, $d_x = \bar{d}_{PQ}$ and $\bar{z}_{PQ} = z_x$ and hence the closed loop transfer matrix is equal to X. ■

This yields the following parameterization of all stabilizing controllers which yield an H_∞ norm less than 1:

Lemma 14.11 *A controller Γ stabilizes Σ and yields a closed loop system $\Gamma \times \Sigma$ with H_∞ norm strictly less than* 1 *if and only if Γ is equal to the interconnection in Figure 14.4 for some stable system Σ_x with H_∞ norm strictly less than* 1.

Proof : Let a controller Γ be given which stabilizes Σ such that the closed loop system has H_∞ norm strictly less than 1. Then according to corollary 14.9, Γ stabilizes Σ_{PQ} and yields a closed loop system with transfer matrix Q where $\|Q\|_\infty < 1$. We then define a stable transfer matrix X by $\Pi_2^{\mathrm{T}} Q \Pi_1^{\mathrm{T}}$ which will have $\|X\|_\infty < 1$ and we have $Q = \Pi_2 X \Pi_1$ since $\operatorname{im} Q \subset \operatorname{im} D_2$ and $\ker Q \supset \ker D_1$. We then know from the proof of lemma 14.10 that the interconnection (14.4), where Σ_x is any internally stable system with transfer matrix X, yields a controller $\bar{\Gamma}$ which stabilizes $\bar{\Sigma}_{PQ}$ and yields a closed loop system X. But then it will also stabilize Σ_{PQ} and attain a closed loop system $Q = \Pi_2 X \Pi_1$ with norm less than 1. Then corollary 14.9 guarantees that $\bar{\Gamma}$ stabilizes Σ.

Since we have seen that the two interconnections in Figure 14.3 on page 319 achieve the same closed loop transfer matrix, it should be noted that if two controllers achieve the same closed loop transfer matrix when applied to Σ_P then they also achieve the same closed loop transfer matrix when applied to Σ. Since the connection between Σ_P and Σ_{PQ} is dual to the connection between Σ and Σ_P, we obtain that two controllers achieve the same closed loop transfer matrix when applied to Σ_{PQ} then they also achieve the same closed loop transfer matrix when applied to Σ_P. In conclusion, since we know that Γ and $\bar{\Gamma}$ yield the same closed loop transfer matrix when applied to Σ_{PQ}, we have that $\bar{\Gamma}$ yields the same closed loop transfer matrix as Γ when applied to Σ. Partition the transfer matrix G of Σ into Σ:

$$G = \begin{pmatrix} G_{11} & G_{12} \\ G_{21} & G_{22} \end{pmatrix} \tag{14.27}$$

compatible with the partitioning of (d, u) and $)(z, y)$. Let C and $\bar{C}$ be the transfer matrices associated to Γ and $\bar{\Gamma}$ respectively. We have:

$$G_{11} + G_{12}C(I - G_{22}C)^{-1}G_{21} = G_{11} + G_{12}\bar{C}(I - G_{22}\bar{C})^{-1}G_{21}.$$

Since $G_{21}(\infty)$ is surjective and $G_{12}(\infty)$ is injective we get that the associated transfer matrices are surjective and injective respectively. Hence:

$$C(I - G_{22}C)^{-1} = \bar{C}(I - G_{22}\bar{C})^{-1},$$

which yields

$$(I - \bar{C}G_{22})C = \bar{C}(I - G_{22}C).$$

Hence $C = \bar{C}$, which proves that Γ and $\bar{\Gamma}$ are the same or, in other words, our controller Γ is equal to the interconnection (14.4) for a stable system Σ_x with H_∞ norm strictly less than 1.

Conversely, suppose a stable Σ_x is given whose transfer matrix X has H_∞ norm less than 1. Define by Γ the controller given by the interconnection (14.4). Then lemma 14.10 guarantees that Γ stabilizes $\bar{\Sigma}_{PQ}$ and hence also stabilizes Σ_{PQ}. Moreover the closed loop transfer matrix of the interconnection of Σ_{PQ} and Γ is equal to $\Pi_2 X \Pi_1$ and has H_∞ norm strictly less then 1. Corollary 14.9 then guarantees that Γ stabilizes Σ and yields a closed loop system with H_∞ norm strictly less than 1. ■

Note that the above theorem parameterizes the class of all suitable controllers Γ by all stable systems with H_∞ norm strictly less than 1. The controller we obtain for Σ_x equal to 0 is in a certain sense the center of the parameterization and is hence often called the *central controller*. The controller given in theorem 14.1, turns out to be equal to this central controller.

14.6 Exercises

14.1 The dualization argument used in the proof of lemma 14.5 is of course valid but it is sometimes hard to get a good feeling for it. There is another derivation without using a dualization argument. We know d will be a worst-case disturbance. From the point of view of observing the state, intuitively the worst d could do is to guarantee that:

$$0 = C_1 x + D_1 d.$$

In other words:

$$d = -D_1^{\mathrm{T}}(D_1 D_1^{\mathrm{T}})^{-1} C_1 x + [I - D_1^{\mathrm{T}}(D_1 D_1^{\mathrm{T}})^{-1} D_1]^{1/2} v.$$

For that particular class of disturbances the closed loop system looks like:

$$\begin{aligned} \dot{x} &= \left(A - E D_1^{\mathrm{T}}(D_1 D_1^{\mathrm{T}})^{-1} C_1\right) x + E[I - D_1^{\mathrm{T}}(D_1 D_1^{\mathrm{T}})^{-1} D_1]^{1/2} v, \\ z &= C_2 x, \end{aligned}$$

and is independent of the particular controller since $y = 0$. Since we assume there exists a controller which makes the H_∞ norm less than 1, there exists $\delta < 1$ such that:

$$\int_0^\infty \|z(t)\|^2 \,\mathrm{d}t \leqslant \delta \int_0^\infty \|d(t)\|^2 \,\mathrm{d}t$$

for all $d \in L_2$ and zero initial state.

a. Show that the zeros of (A, E, C_1, D_1) are the eigenvalues of the matrix $A - E D_1^{\mathrm{T}}(D_1 D_1^{\mathrm{T}})^{-1} C_1$.

b. Show that we have:

$$\|d\|_2^2 = \|D_1^{\mathrm{T}}(D_1 D_1^{\mathrm{T}})^{-1} C_1 x\|_2^2 + \|[I - D_1^{\mathrm{T}}(D_1 D_1^{\mathrm{T}})^{-1} D_1]^{1/2} v\|_2^2$$

c. Solve the following optimization problem (using techniques from section 12.6)

$$\begin{aligned} \sup_v \int_{-T}^0 & \|z(t)\|^2 - \|D_1^{\mathrm{T}}(D_1 D_1^{\mathrm{T}})^{-1} C_1 x(t)\|^2 \\ & - \|[I - D_1^{\mathrm{T}}(D_1 D_1^{\mathrm{T}})^{-1} D_1]^{1/2} v(t)\|^2 \,\mathrm{d}t. \end{aligned}$$

where $x(0) = \xi$. In other words show that the Riccati differential equation:

$$\begin{aligned} \dot{Y} = {} & YA + A^{\mathrm{T}} Y + Y E E^{\mathrm{T}} Y + C_2^{\mathrm{T}} C_2 \\ & - \left(C_1^{\mathrm{T}} + Y E D_1^{\mathrm{T}}\right)\left(D_1 D_1^{\mathrm{T}}\right)^{-1}\left(C_1 + D_1 E^{\mathrm{T}} Y\right), \end{aligned}$$

with $Y(0) = 0$ has a solution Y on the interval $[0, T]$. Show that the optimal cost is equal to $\xi^{\mathrm{T}} Y(T) \xi$ and show that an optimal v is given by:

$$v(t) = [I - D_1^{\mathrm{T}}(D_1 D_1^{\mathrm{T}})^{-1} D_1]^{1/2} E^{\mathrm{T}} Y(t + T) x(t).$$

d. Assume (A, E, C_1, D_1) has no zeros in the closed left half plane. Show that

$$\sup_{v \in L_2(-\infty,0]} \int_{-\infty}^{0} \|z(t)\|^2 - \|D_1^{\mathrm{T}}(D_1 D_1^{\mathrm{T}})^{-1} C_1 x(t)\|^2 - \|[I - D_1^{\mathrm{T}}(D_1 D_1^{\mathrm{T}})^{-1} D_1]^{1/2} v(t)\|^2 \, \mathrm{d}t$$

is bounded for all initial conditions $x(0) = \xi$ with optimal cost $\xi^{\mathrm{T}} \bar{Y} \xi$ and optimal controller

$$v = [I - D_1^{\mathrm{T}}(D_1 D_1^{\mathrm{T}})^{-1} D_1]^{1/2} E^{\mathrm{T}} \bar{Y} x$$

where $\bar{Y} = \lim_{t \to \infty} Y(t)$.

e. Assume (A, E, C_1, D_1) has no zeros in the closed left half plane. Show that $\bar{Y}$ satisfies the algebraic Riccati equation

$$0 = \bar{Y} A + A^{\mathrm{T}} \bar{Y} + \bar{Y} E E^{\mathrm{T}} \bar{Y} + C_2^{\mathrm{T}} C_2 - \left(C_1^{\mathrm{T}} + \bar{Y} E D_1^{\mathrm{T}}\right) \left(D_1 D_1^{\mathrm{T}}\right)^{-1} \left(C_1 + D_1 E^{\mathrm{T}} \bar{Y}\right),$$

and is such that the matrix:

$$A - E D_1^{\mathrm{T}}(D_1 D_1^{\mathrm{T}})^{-1} C_1 + E[I - D_1^{\mathrm{T}}(D_1 D_1^{\mathrm{T}})^{-1} D_1]^{1/2} E^{\mathrm{T}} \bar{Y}$$

is antistable, i.e. all its eigenvalues are in the open right half plane $\mathbb{C}^+$.

f. Assume the antistabilizing solution of the algebraic Riccati equation in part (e) is invertible. Show that the inverse satisfies the algebraic Riccati equation (14.4) and is such that the matrix (14.6) is stable. Show that if (A, E, C_1, D_1) has no zeros in the closed *right* half plane then this invertibility assumption is satisfied. *Note:* to really get the existence of a stabilizing solution of the algebraic Riccati equation (14.4) we need to remove the assumption that (A, E, C_1, D_1) has no zeros in the close left half plane from part (d). But then the finite horizon problem does not converge to the infinite horizon problem unless we use endpoint penalties as we did in chapter 13.

14.7 Notes and references

The first solution to the measurement feedback H_∞ control problem was based on frequency domain techniques. see e.g. Francis [46]. These methods are based on different types of factorizations of rational matrices (inner-outer factorization, spectral factorization, etc.). These methods had difficulty with the order of the controller which could be much higher than the order of the plant.

The state space theory first presented in Doyle, Glover, Khargonekar and Francis [41] and independently in Tadmor [193] was the first solution that yielded solutions of the same dynamic order as the plant. Our presentation was strongly influenced by the results from Stoorvogel [184]. We made some basic assumptions in this chapter: no invariant zeros on the imaginary axis and direct feedthough matrices that should have full rank. The case when these assumptions are not satisfied is referred to as the singular case and will be discussed in the next chapter. The addition of direct feedthrough matrices from u to y or from w to z, can be approached using the techniques presented in Safonov, Limebeer and Chiang [160]. The general results in this case (regular but all direct feedthrough matrices present) can be found in Glover and Doyle [62].

In the last few years this frequency domain approach has been refined via the introduction of the so-called J-spectral factorization (see e.g. Green, Glover, Limebeer and Doyle [65]). At the moment this is quite an elegant theory and does not suffer any longer from the drawbacks of high-order controllers that were present in the original frequency domain methods. Another recent development is to solve the H_∞ control problem via linear matrix inequalities (LMI). See, for instance, Gahinet [53] and Iwasaki and Skelton [88]. Other techniques to solve H_∞ are based on interpolation methods, Limebeer and Anderson [110] and Zames and Francis [231] and polynomial methods, see Kwakernaak [103, 104].

In this chapter we prove the existence of solutions of the second Riccati equation and the existence of a suitable observer via duality. A beautiful interpretation of this second (observer) Riccati equation is presented in Khargonekar [100] by connecting it to the behavior of the zero dynamics, i.e. the behavior of the system under the constraint $y = 0$.

The theory of H_∞ has been extended to infinite dimensional systems. See, for instance, van Keulen [205] and the special issue [34]. Extensions to time-varying systems also exist. See for instance Limebeer, Anderson, Khargonekar, and Green [111] and Ravi, Nagpal and Khargonekar [152].

The discrete-time version of H_∞ has also been studied in detail. See for instance Iglesias and Glover [84], Stoorvogel [184] and Stoorvogel, Saberi and Chen [188].

Chapter 15

Some applications of the H_∞ control problem

15.1 Introduction

In chapter 12, we formulated a number of problems related to robust stabilization with additive, multiplicative or coprime factor model uncertainty. In this chapter, we would like to revisit these problems and see how the theory of the previous two chapter can give us additional insight.

We also like to consider in some detail what to do if some of the basic assumptions in H_∞ are not satisfied. These so-called singular problems are quite difficult but by presenting some ad-hoc techniques we hope to provide the reader with some additional insight for the difficulty of these problems.

Minimum entropy is an interpretation of the so-called central controller. It is useful to study this interpretation because it yields additional insight in the relation between H_∞ and H_2.

Finally, for those people who are familiar with classical control, we present a design example that might help to clarify the relations between classical control and H_∞.

15.2 Robustness problems and the H_∞ control problem

In section 12.4 we discussed stabilization of uncertain systems. In the present section we will apply the results of the previous two chapters to the three different types of uncertainty described in that section:

- Additive perturbations
- Multiplicative perturbations

- Coprime factor perturbations.

From section 12.4 we know how to reduce each of these problems to an H_∞ control problem. In this section we will apply the results from the previous chapters and show the additional insight that can be obtained from these results.

15.2.1 Additive perturbations

In this section we study the problem as posed in subsection 12.4.1. Assume that we have a system Σ, being an imperfect model of a certain plant. We assume that the error is additive, i.e. we assume that the plant is exactly described by the interconnection in Figure 15.1. Here, Δ is some arbitrary system such that Σ and $\Sigma + \Delta$ have

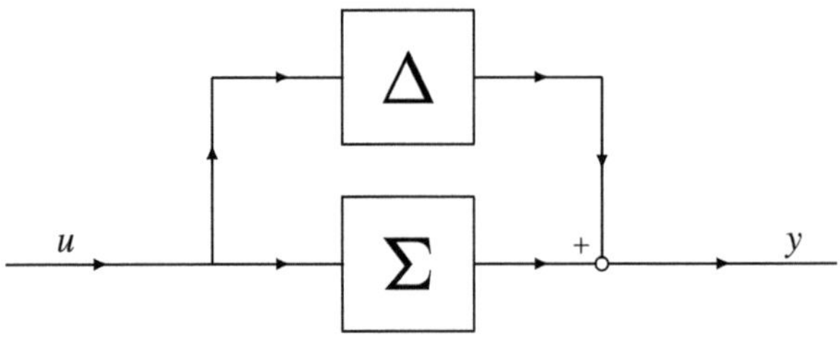

Figure 15.1

the same number of unstable poles. Thus we assume that the plant is described by the system Σ interconnected as in Figure 12.5 with another system Δ. The system Δ represents the uncertainty and is hence, by definition, unknown. In this subsection we derive conditions under which a controller Γ of the form

$$\Gamma: \begin{array}{l} \dot{w} = Kw + Ly, \\ u = Mw + Ny. \end{array} \tag{15.1}$$

exists such that the interconnection in figure 15.1 is stabilized by this controller for all systems Δ that do not change the number of unstable poles and that have L_∞ norm less than or equal to some, a priori given, positive number γ. We quote the following result from subsection 12.4.1:

Lemma 15.1 *Consider the system $\Sigma = (A, B, C, D)$. Assume Σ is stabilizable and detectable. Let $\gamma > 0$. Define a new system*

$$\Sigma_{na}: \begin{array}{l} \dot{x} = Ax + Bu, \\ y = Cx + Du + d, \\ z = \qquad\quad u. \end{array} \tag{15.2}$$

Let a controller Γ of the form (15.1) be given. The following conditions are equivalent:

(i) *The controller Γ from y to u applied to the interconnection in figure 12.5, yields a well-posed and internally stable closed loop system for every system Δ such that*

(a) *Δ has L_∞ norm less than or equal to γ.*

(b) *Σ and $\Sigma + \Delta$ have the same number of unstable poles, counting multiplicities.*

(ii) *Γ internally stabilizes Σ_{na}, and the closed loop system has H_∞ norm less than γ^{-1}.*

We would like to apply the results from the previous chapter but the system Σ_{na} is not of the form (14.1). It has an additional direct feedthrough matrix from d to y. That matrix however, does not play an essential role in the design of a suitable controller. Define the following system:

$$\bar{\Sigma}_{na}: \begin{array}{rl} \dot{x} = & Ax + Bu, \\ y = & Cx \qquad\quad + d, \\ z = & \qquad\quad u. \end{array} \tag{15.3}$$

Suppose a controller

$$\bar{\Gamma}: \begin{array}{rl} \dot{w} = & \bar{K}w + \bar{L}y, \\ u = & \bar{M}w + \bar{N}y, \end{array} \tag{15.4}$$

is given which internally stabilizes $\bar{\Sigma}_{na}$. Let G_{cl} be the closed loop transfer matrix G_{cl} of $\bar{\Sigma}_{na} \times \bar{\Gamma}$. Then the controller

$$\Gamma: \begin{array}{rl} \dot{w} = & \bar{K}w + \bar{L}(y - Du), \\ u = & \bar{M}w + \bar{N}(y - Du). \end{array} \tag{15.5}$$

internally stabilizes Σ_{na} and yields the same closed loop transfer matrix G_{cl} under the condition that $I + \bar{N}D$ is nonsingular since otherwise we cannot solve the second equation of (15.5) to find u. Conversely, suppose Γ given by (15.1) internally stabilizes Σ_{na}, and achieves a closed loop transfer matrix G_{cl}. Then the controller

$$\bar{\Gamma}: \begin{array}{rl} \dot{w} = & Kw + L(y + Du), \\ u = & Mw + N(y + Du). \end{array} \tag{15.6}$$

stabilizes $\bar{\Sigma}_{na}$ and yields the same closed loop transfer matrix G_{cl} again assuming that $I - ND$ is nonsingular, since otherwise we can not solve the second equation of (15.6) to find u.

The problem arising in the above is the problem of well-posedness of a feedback interconnection as discussed in section 3.13. In our case, we have a strictly proper controller such as the one presented in theorem 14.1 and therefore the interconnection is always is always well posed.

Given the above arguments, we can apply the results from the previous chapter to the system $\bar{\Sigma}_{na}$. In order to satisfy the assumptions of theorem 14.1 we impose that

A no eigenvalues on the imaginary axis. Then theorem 14.1 states that an internally stabilizing controller for the system Σ_{na} exists which yields an H_∞ norm less than γ if and only if there exist positive semi-definite real symmetric solutions of the Riccati equations

$$A^{\mathrm{T}}P + PA - PBB^{\mathrm{T}}P = 0, \tag{15.7}$$

$$AQ + QA^{\mathrm{T}} - QC^{\mathrm{T}}CQ = 0, \tag{15.8}$$

such that $\rho(PQ) < \gamma^2$ and such that the matrices $A - BB^{\mathrm{T}}P$ and $A + QC^{\mathrm{T}}C$ are stability matrices. Note that P and Q do not depend on γ. Moreover, the results from chapter 10 guarantee the existence real symmetric solutions P and Q of (15.7) and (15.8) such that $A - BB^{\mathrm{T}}P$ and $A + QC^{\mathrm{T}}C$ are stability matrices. We see that a stabilizing controller for Σ_{na} exists with a closed loop norm strictly less than γ if and only if γ is larger than $\rho(PQ)^{1/2}$.

We thus find the following theorem:

Theorem 15.2 *Assume that $\Sigma = (A, B, C, D)$ is stabilizable and detectable. Assume that A has no eigenvalues on the imaginary axis. We define the related system Σ_{na} by (15.2) and let $\gamma > 0$. The following conditions are equivalent:*

(i) There exists a controller Γ from y to u of the form (15.1) such that, when applied to the interconnection (15.1), the closed-loop system is well posed and internally stable for all systems Δ for which

(a) Δ has L_∞ norm less than or equal to γ,

(b) Σ and $\Sigma + \Delta$ have the same number of unstable poles.

(ii) We have $\rho(PQ) < \gamma^{-2}$, where P and Q are the solutions of the algebraic Riccati equations (15.7) and (15.8) for which $A - BB^{T}P$ and $A + QC^{T}C$ are stability matrices..

Moreover, if P and Q satisfy part (ii), a controller satisfying part (i) is described by:

$$\begin{aligned} \dot{\hat{x}} &= A\hat{x} + Bu + L(y - C\hat{x} - Du), \\ u &= -B^{T}P\hat{x}, \end{aligned}$$

where $L := (I - \gamma^2 QP)^{-1}QC^{T}$.

Remarks:

- Naturally the class of perturbations we have chosen is rather artificial. However, it is easy to show that, if we allow for perturbations which additional unstable poles, then there are arbitrarily small perturbations which destabilize the closed-loop system. On the other hand, our class of perturbations does include all stable systems Δ with H_∞ norm less than or equal to γ.

- We want to find a controller satisfying part (i) for a γ which is as large as possible. Note that part (ii) shows that, in fact, for every γ smaller than the bound $[\rho(PQ)]^{-1/2}$ we can find a controller satisfying part (i). Actually there exists a controller which makes the H_∞ norm equal to $[\rho(PQ)]^{-1/2}$ which is clearly the best we can do.
- It can be shown that the bound $[\rho(PQ)]^{-1/2}$ depends only on the unstable dynamics of Σ. Hence we could assume a priori that A has only eigenvalues in the open right half complex plane (we still have to exclude eigenvalues on the imaginary axis). In that case it can be shown that P and Q are the inverses of X and Y, respectively, where X and Y are the unique positive definite solutions of the following two Liapunov equations:

$$\begin{aligned} AX + XA^{\mathrm{T}} &= BB^{\mathrm{T}}, \\ A^{\mathrm{T}}Y + YA &= C^{\mathrm{T}}C. \end{aligned}$$

Recall that X and Y are referred to as the controllability and observability gramian respectively (see section 3.8).

15.2.2 Multiplicative perturbations

In this subsection we study the problem as posed in subsection 12.4.2.

We assume that, once again, we have a system Σ being an imperfect model of a certain plant. This time, we study multiplicative uncertainty, i.e. we assume that the plant is exactly described by the interconnection in Figure 15.2.

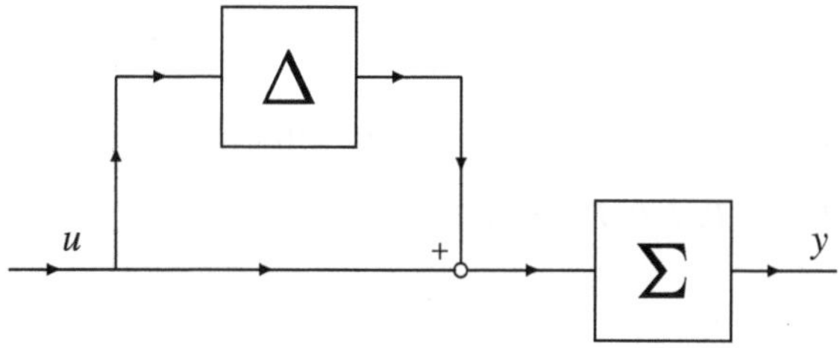

Figure 15.2

Here, Δ is some arbitrary system describing the uncertainty such that the interconnection in figure 15.2 has the same number of unstable poles (counting multiplicities) as Σ. In subsection 12.4.2, we have derived the following result

Lemma 15.3 *Assume that $\Sigma = (A, B, C, D)$ is stabilizable and detectable. Define a new system:*

$$\Sigma_{nm}: \quad \begin{aligned} \dot{x} &= Ax + Bu + Bd, \\ y &= Cx + Du + Dd, \\ z &= \qquad\quad u. \end{aligned} \tag{15.9}$$

Let a controller Γ of the form (15.1) be given. The following conditions are equivalent:

(i) *The controller Γ from y to u applied to the interconnection in figure 15.2, yields a well-posed and internally stable closed loop system for every system Δ such that*

   (a) *Δ has L_∞ norm less than or equal to γ,*

   (b) *The interconnection (15.2) has the same number of unstable poles as Σ.*

(ii) *Γ internally stabilizes Σ_{nm} and the closed loop system has H_∞ norm less than γ^{-1}.*

Thus, we find that our original problem formulation is equivalent to the problem of finding an internally stabilizing controller for Σ_{nm} which makes the H_∞ norm of the closed-loop system less than γ^{-1}. We would like to apply the results of the previous chapter to this problem. Note that, as in the previous subsection, we have a direct feedthrough matrix from u to y. The latter was not treated in chapter 14 but we have seen in the previous subsection that we can design a controller for the auxiliary system:

$$\bar{\Sigma}_{nm}: \begin{aligned} \dot{x} &= Ax + Bu + Bd, \\ y &= Cx \qquad\quad + Dd, \\ z &= \qquad\quad u, \end{aligned} \tag{15.10}$$

and via a simple transformation, find a controller for the original system Σ_{nm}. In order to satisfy the assumptions of theorem 14.1 we require that (A, B, C, D) has no zeros on the imaginary axis, that A has no poles on the imaginary axis and that D is surjective.

The results of chapter 10 tell us that algebraic Riccati equations

$$A^{\mathrm{T}} P_m + P_m A - P_m B B^{\mathrm{T}} P_m = 0, \tag{15.11}$$

and

$$A Q_m + Q_m A^{\mathrm{T}} + B B^{\mathrm{T}} - (Q_m C^{\mathrm{T}} + B D^{\mathrm{T}})(D D^{\mathrm{T}})^{-1}(C Q_m + D B^{\mathrm{T}}) = 0, \tag{15.12}$$

have a real symmetric solution P_m and Q_m for which the matrices $A - BB^{\mathrm{T}} P_m$ and $A - (Q_m C^{\mathrm{T}} + BD^{\mathrm{T}})(DD^{\mathrm{T}})^{-1} C$ are stability matrices. By applying theorem 14.1 to the system $\bar{\Sigma}_{nm}$, we find the following theorem:

Theorem 15.4 *Let $\Sigma = (A, B, C, D)$ be stabilizable and detectable. Let $\gamma > 0$. Assume that A has no eigenvalues on the imaginary axis, D is surjective, and (A, B, C, D) has no zeros on the imaginary axis. Under these assumptions the following conditions are equivalent:*

(i) *A controller Γ from y to u of the form (15.1) exists which, when applied to the interconnection (15.2), yields a closed-loop system that is well posed and internally stable for all systems Δ such that:*

(a) Δ *has* L_∞ *norm less than or equal to* γ.

(b) *The interconnection (15.2) and* Σ *have the same number of unstable poles.*

(ii) *Either A is a stability matrix or* $1 + \rho(P_m Q_m) < \gamma^{-2}$.

If part (ii) is satisfied then a controller satisfying the conditions of part (i) is given by $u = 0$ *if A is stable, and otherwise by:*

$$\begin{aligned}\dot{\hat{x}} &= (A - BB^T P_m)\hat{x} + L(y - C\hat{x} + DB^T P_m \hat{x})\\ u &= (\gamma^2 - 1)^{-1} B^T P_m \hat{x}\end{aligned}$$

where

$$L := \left(I - \frac{\gamma^2 Q_m P_m}{1-\gamma^2}\right)^{-1} (BD^T + Q_m C^T)(DD^T)^{-1}.$$

Proof : We apply theorem 14.1 to the system $\bar{\Sigma}_{nm}$ and we have to investigate four cases:

- *A is stable*: the matrices $P := 0$ and $Q := Q_m$ satisfy condition (ii) of theorem 14.1.

- *A is not stable and* $\gamma < 1$: we define

$$P := \frac{P_m}{1-\gamma^2}, \qquad Q := Q_m \tag{15.13}$$

  Then it is straightforward to check that P and Q are the unique matrices satisfying all requirements of condition (ii) of theorem 14.1 for Σ_{nm} except for possibly the condition on the spectral radius of PQ. Moreover, P_m and Q_m satisfy $1 + \rho(P_m Q_m) < \gamma^{-2}$ if, and only if, P and Q satisfy the requirement $\rho(PQ) < \gamma^{-2}$ (note that we have to replace γ by γ^{-1}).

- *A is not stable and* $\gamma = 1$: the stability requirement for P reduces to the requirement that A is stable which, by assumption, is not true.

- *A is not stable and* $\gamma > 1$: the matrices P and Q given by (15.13) are the stabilizing solutions of the two Riccati equations of part (iii) of theorem 14.1. We have seen in section 13.4 that the stabilizing solution is unique. However, the matrix P is not positive semi-definite. Therefore the conditions of part (ii) of theorem 14.1 are not satisfied.

Therefore theorem 14.1 guarantees that a suitable controller exists for $\bar{\Sigma}_{nm}$ if and only if the conditions of part (ii) of theorem 15.4 are satisfied. But the existence of a suitable controller for Σ_{nm} is equivalent to the existence of a suitable controller for $\bar{\Sigma}_{nm}$ as argued in the previous subsection.

Finally we need to show that the given controller has the required properties. This is done by showing that the controller given by theorem 14.1 for $\bar{\Sigma}_{nm}$ yields the same closed loop system as the controller of theorem 15.4 when applied to Σ_{nm}. ■

Remarks:

- Our class of perturbations is rather artificial but it includes all stable systems Δ with H_∞ norm less than or equal to γ.
- As for additive perturbations, we have an explicit bound for the allowable size of perturbations: part (ii) shows that for every γ smaller than the bound $[1 + \rho(P_m Q_m)]^{-1/2}$ we can find a controller satisfying part (i).
- For additive perturbations it was shown that the upper bound given by

  $$[\rho(P_a Q_a)]^{-1/2}$$

  depends only on the antistable part of Σ. It should be noted that this is *not* true for the bound $[1 + \rho(P_m Q_m)]^{-1/2}$ which we obtained for multiplicative perturbations.

15.2.3 Coprime factor perturbations

In this subsection we study the problem as posed in subsection 12.4.3.

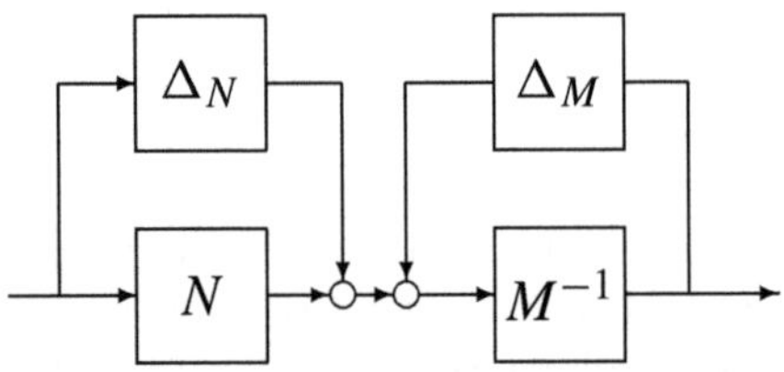

Figure 15.3

We assume that we have a system Σ being an imperfect model of a certain plant. This time we first construct a normalized coprime factorization for Σ using lemma 12.14. We then assume that the plant is exactly described by the interconnection in figure 15.3. Here $\Delta := (cc\Delta_N \quad \Delta_M)$ is some arbitrary system describing the uncertainty. Since the essential feature of coprime factors is that they are chosen stable it is natural to require Δ to be stable. In this subsection we will require that the system Σ has a strictly proper transfer matrix. This is only to avoid complicated formulas. In subsection 12.4.3, we derived the following result:

Lemma 15.5 *Assume that $\Sigma = (A, B, C, 0)$ is stabilizable and detectable. We define* *a new system:*

$$\Sigma_{cf}: \begin{array}{rl} \dot{x} = & Ax + \quad Bu + XC^T d, \\ y = & Cx \qquad\qquad + \quad d, \\ z = & \begin{pmatrix} C \\ 0 \end{pmatrix} x + \begin{pmatrix} 0 \\ I \end{pmatrix} u + \begin{pmatrix} I \\ 0 \end{pmatrix} d, \end{array} \tag{15.14}$$

where X is the unique real symmetric solution of

$$0 = AX + XA^T - XC^TCX + BB^T$$

such that $A - XC^TC$ is a stability matrix. Let a controller Γ of the form (15.1) be given. The following conditions are equivalent:

(*i*) *The controller Γ from y to u applied to the interconnection in Figure 15.3, yields a well-posed and internally stable closed-loop system for every stable system Δ with H_∞ norm less than or equal to γ,*

(*ii*) *Γ internally stabilizes Σ_{cf}, and the closed loop system has H_∞ norm less than γ^{-1}.*

We want to apply the results of the previous chapter to this problem. Note that we do not have a direct feedthrough matrix from u to y. However, the system Σ_{cf} does have a direct feedthrough matrix from d to z, which was not covered by the results of the previous chapter, and which we have not seen before. First we look at the closed loop transfer matrix for $s = \infty$. The closed loop transfer matrix G_{cl} satisfies

$$G_{cl}(\infty) = \begin{pmatrix} I \\ 0 \end{pmatrix} + \begin{pmatrix} 0 \\ I \end{pmatrix} N \tag{15.15}$$

where N is the direct feedthrough matrix of the controller. We minimize $\|G_{cl}(\infty)\|$ over N. We obtain that $N = 0$ is optimal and yields $\|G_{cl}(\infty)\| = 1$.

This first stage basically minimizes the direct feedthrough matrix from d to z. If the optimal N is not 0 then we would have applied a preliminary feedback $u = Ny + \bar{u}$. This first step did not make the direct feedthrough matrix equal to 0 but it is easy to check that the existence of a stabilizing controller which yields an H_∞ norm strictly less than γ^{-1} must imply that $\gamma^{-1} > \|G_{cl}(\infty)\| \geqslant 1$.

The second step makes use of a special case of Redheffer's lemma (lemma 14.4). Define the static system Σ_U by:

$$\Sigma_U: \begin{pmatrix} z_U \\ d \end{pmatrix} = \begin{pmatrix} \begin{pmatrix} \gamma I \\ 0 \\ -(1-\gamma^2)^{1/2} I \end{pmatrix} & \begin{pmatrix} (1-\gamma^2)^{1/2} I & 0 \\ 0 & I \\ (\gamma I & 0) \end{pmatrix} \end{pmatrix} \begin{pmatrix} d_U \\ \bar{z} \end{pmatrix}$$

We can now apply Redheffer's lemma on the interconnection in figure 15.4 on the next page. It is easy to check that Σ_U satisfies the requirements of lemma 14.4. Hence a controller stabilizes Σ_{cf} and yields an H_∞ norm strictly less than γ^{-1} if and only if the same controller stabilizes the dashed system in the above interconnection and

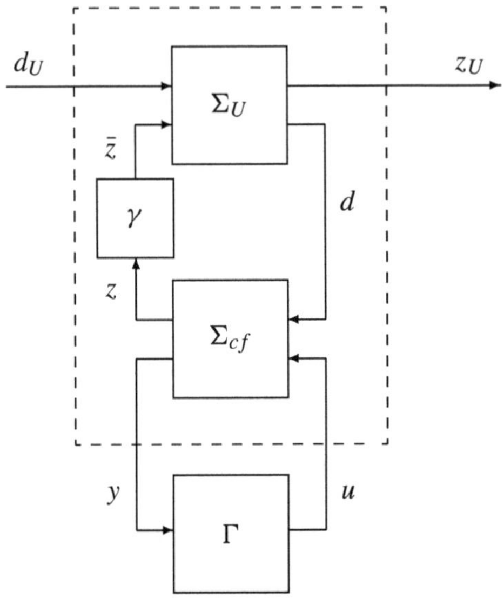

Figure 15.4

yields an H_∞ norm strictly less than 1. We denote the dashed system by $\bar{\Sigma}_{cf}$. This new system has the following representation

$$\bar{\Sigma}_{cf}: \begin{cases} \dot{x} = (A + \gamma^2(1-\gamma^2)^{-1} X C^{\mathrm{T}} C)x + Bu - (1-\gamma^2)^{-1/2} X C^{\mathrm{T}} d_U, \\ y = (1-\gamma^2)^{-1} C x - (1-\gamma^2)^{-1/2} d_U, \\ z_U = \begin{pmatrix} \gamma(1-\gamma^2)^{-1/2} C \\ 0 \end{pmatrix} x + \begin{pmatrix} 0 \\ \gamma I \end{pmatrix} u, \end{cases}$$

and we note that this new system has a direct feedthrough matrix from d_U to z_U equal to 0. In this way, we reduced the H_∞ control problem for a system with a direct feedthrough matrix from d to z to a system without such a direct feedthrough matrix. Basically, it amounts to minimizing the norm of this feedthrough matrix via a preliminary static feedback and then applying Redheffer's theorem with a static transformation. This method will always work. Note that we can now apply theorem 14.1 to the system $\bar{\Sigma}_{cf}$ to find the following theorem:

Theorem 15.6 *Let the system $\Sigma = (A, B, C, 0)$ be stabilizable and detectable. Let $\gamma > 0$ be given. There exist real symmetric matrices $X \geqslant 0$ and $Y \geqslant 0$ such that*

$$0 = AX + XA^{\mathrm{T}} - XC^{\mathrm{T}}CX + BB^{\mathrm{T}},$$
$$0 = YA + A^{\mathrm{T}}Y - YBB^{\mathrm{T}}Y + C^{\mathrm{T}}C,$$

and such that $A + XC^{\mathrm{T}}C$ and $A - BB^{\mathrm{T}}Y$ are stability matrices. Then the following conditions are equivalent:

(i) *A controller Γ from y to u of the form (15.1) exists which when applied to the interconnection in figure 15.3, yields a closed-loop system which is internally stable for every stable system Δ with H_∞ norm less than or equal to γ.*

(*ii*) $1 + \rho(XY) < \gamma^{-2}$.

Proof : The existence of X and Y follows from the results of chapter 10.

Next, we apply theorem 14.1 to the system $\bar{\Sigma}_{cf}$. The second algebraic Riccati equation gets a very simple form. In fact, the question becomes whether there exists a real symmetric solution $Q \geqslant 0$ to

$$(A - XC^{\mathrm{T}}C)Q + Q(A - XC^{\mathrm{T}}C)^{\mathrm{T}} - QC^{\mathrm{T}}CQQ = 0$$

such that $A - XC^{\mathrm{T}}C - QC^{\mathrm{T}}C$ is a stability matrix. Since $A - XC^{\mathrm{T}}C$ is stability matrix, it is easy to check that $Q = 0$ solves this equation. The first Riccati equation gets the following special form:

$$0 = PA + A^{\mathrm{T}}P - \gamma^{-2}PBB^{\mathrm{T}}P + \gamma^2C^{\mathrm{T}}C + (1-\gamma^2)^{-1}(\gamma^2 I + PX)C^{\mathrm{T}}C(\gamma^2 I + XP)$$

Via some calculations we obtain that $P = [(\gamma^{-2} - 1)I - YX]^{-1}Y$ is a stabilizing solution of this Riccati equation. Thus, theorem 14.1 guarantees that a suitable controller exists for $\bar{\Sigma}_{cf}$ if and only if $P \geqslant 0$, i.e. if and only if $1 + \rho(YX) < \gamma^{-2}$. ■

Remarks:

- The existence of a suitable controller for Σ_{cf} is equivalent to the existence of a suitable controller for $\bar{\Sigma}_{cf}$. Hence a preliminary static output feedback combined with theorem 14.1 yields a suitable controller satisfying part (i)

- Our class of perturbations is rather artificial but it does not suffer from the drawback regarding the fixed number of unstable zeros of the plant that we needed to require for additive and multiplicative perturbations.

  However, for any $\lambda \in \mathbb{C}^+$, we can find a perturbation Δ_M with $\|\Delta_M\|_\infty < \gamma$ such that $M + \Delta_M$ has an unstable zero $\lambda \in \mathbb{C}^+$ if and only if the smallest singular value of $M(\lambda)$ is less than γ. Note that unstable zeros of $M + \Delta_M$ are equal to unstable poles of the perturbed system. The number and the location of the unstable zeros is hence implicitly restricted by the allowed size for the perturbation Δ_M.

- Again, we have an explicit bound for the allowable size of perturbations: part (ii) shows that for every γ smaller than the bound $[1 + \rho(XY)]^{-1/2}$ we can find a suitable controller satisfying part (i).

15.3 Singular problems

In this section we shall discuss two methods which can be used to solve the H_∞ problem when the assumptions of theorem 14.1 are not satisfied.

15.3.1 Frequency-domain loop shifting

The basic method described in this section is based on applying a transformations acting on transfer matrices. Consider the following transformation:

$$z = \frac{\varepsilon + s}{1 + \varepsilon s}$$

with $\varepsilon > 0$ then the set of all $s \in \mathbb{C}$ located outside the circle, which we denote by $\mathbb{C}_\varepsilon^{\otimes}$ in figure 15.5 is mapped to the set of all $z \in \mathbb{C}$ in the open right half complex plane $\mathbb{C}^+$. We denote the inside of the circle in figure 15.5 by $\mathbb{C}^{\oplus}$.

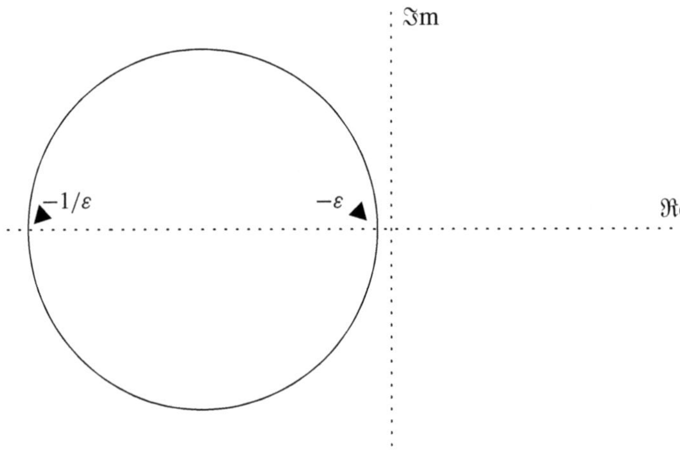

Figure 15.5

Let G be a given transfer matrix and define a new transfer matrix:

$$H_\varepsilon(\lambda) = G\left(\tfrac{\varepsilon+\lambda}{1+\varepsilon\lambda}\right) \tag{15.16}$$

Then G has all its poles in $\mathbb{C}_\varepsilon^{\oplus}$ if and only if H has all its poles in the open left half complex plane. Assume G has all its poles in $\mathbb{C}_\varepsilon^{\oplus}$. Then

$$\begin{aligned} \|H_\varepsilon\|_\infty &= \sup_{\lambda\in\mathbb{C}^+} \|H_\varepsilon(\lambda)\| = \sup_{\lambda\in\mathbb{C}_\varepsilon^{\oplus}} \|G(\lambda)\| \\ &\geqslant \sup_{\lambda\in\mathbb{C}^+} \|G(\lambda)\| = \|G\|_\infty. \end{aligned}$$

Moreover, it can be shown that:

$$\lim_{\varepsilon\downarrow 0} \|H_\varepsilon\|_\infty = \|G\|_\infty$$

Moreover if G has all its poles in the open left half complex plane then there exists ε^* such that H_ε has all its poles in the left half complex plane for all $0 < \varepsilon < \varepsilon^*$.

What makes the transformation from G to H_ε interesting? It actually helps us relax some of the assumptions made in our solution of the H_∞ control problem:

- We needed to make assumptions that two subsystems did not have zeros on the imaginary axis. But this transformation effectively replaces the imaginary axis by the circle in figure 15.5 and by suitable choosing ε we can always guarantee that we have no zeros on the circle and therefore we can avoid this assumption.
- We needed to make assumptions that two subsystems have a direct feedthrough matrix which is injective or surjective respectively. The point $s = \infty$ is by our transformation replaces by the point $s = -1/\varepsilon$. Now note that if a system $\Sigma = (A, B, C, D)$ has a direct feedthrough matrix which is not injective this might be caused by two issues:
  - The system Σ has a transfer matrix G which is not injective as a rational matrix, i.e. $\operatorname{rank} G(s)$ is less than the number of columns of G for all $s \in \mathbb{C}$.
  - The system Σ has an infinite zero.

  Our transformation can avoid the assumption that Σ has no infinite zeros because it replaces $s = \infty$ by the point $s = -1/\varepsilon$. However if the transfer matrix is not injective as a rational matrix then this transformation does not help us in removing this assumption.

There are many other similar transformations based on transfer matrices that we can use. For instance $z = s + \varepsilon$ which moves the imaginary axis to the line $\Re\mathrm{e}\, s = -\varepsilon$. The reason why the transformation described by (15.16) is popular is that it avoids that the closed loop system has poles close to the imaginary axis with a large imaginary part. These yield very fast oscillations in the closed loop system which are very badly damped and are highly undesirable in applications.

Let us next describe how we will use this transformation. Suppose we have a system of the form:

$$\Sigma : \begin{aligned} \dot{x} &= Ax + Bu + Ed, \\ y &= C_1x + D_{11}u + D_{12}d, \\ z &= C_2x + D_{21}u + D_{22}d, \end{aligned} \tag{15.17}$$

with transfer matrix G from (u, d) to (y, z). If $I - \varepsilon A$ is invertible we can define the following transformed system:

$$\tilde{\Sigma}_\varepsilon : \begin{aligned} \dot{x} &= \tilde{A}x + \tilde{B}u + \tilde{E}d, \\ y &= \tilde{C}_1x + \tilde{D}_{11}u + \tilde{D}_{12}d, \\ z &= \tilde{C}_2x + \tilde{D}_{21}u + \tilde{D}_{22}d, \end{aligned} \tag{15.18}$$

where

$$
\begin{aligned}
\tilde{A} &:= (A - \varepsilon I)(I - \varepsilon A)^{-1}, \\
\tilde{B} &:= (1 - \varepsilon^2)(I - \varepsilon A)^{-1} B, \\
\tilde{E} &:= (1 - \varepsilon^2)(I - \varepsilon A)^{-1} E, \\
\tilde{C}_1 &:= C_1 (I - \varepsilon A)^{-1}, \\
\tilde{C}_2 &:= C_2 (I - \varepsilon A)^{-1}, \\
\tilde{D}_{11} &:= D_{11} + \varepsilon C_1 (I - \varepsilon A)^{-1} B, \\
\tilde{D}_{12} &:= D_{12} + \varepsilon C_1 (I - \varepsilon A)^{-1} E, \\
\tilde{D}_{21} &:= D_{21} + \varepsilon C_2 (I - \varepsilon A)^{-1} B, \\
\tilde{D}_{22} &:= D_{22} + \varepsilon C_2 (I - \varepsilon A)^{-1} E.
\end{aligned}
$$

Note that the transfer matrix of this transformed system is given by:

$$\tilde{G}_\varepsilon(\lambda) = G\left(\tfrac{\varepsilon+\lambda}{1+\varepsilon\lambda}\right)$$

In other words, we basically applied exactly the transformation as described before.

Assume that a controller $\tilde{\Gamma}$ described by the transfer matrix $\tilde{\Gamma}$ exists for $\tilde{\Sigma}_\varepsilon$ such that the closed-loop system is internally stable and the closed-loop transfer matrix $\tilde{G}_{cl}$ has H_∞ norm less than 1. Then if we apply the feedback Γ described by the transfer matrix H where

$$H(\lambda) = \tilde{H}\left(\frac{\varepsilon - \lambda}{\varepsilon\lambda - 1}\right)$$

to our original system, then the closed-loop system $\Gamma \times \Sigma$, with transfer matrix G_{cl}, is related to the closed-loop system $\tilde{\Sigma}_\varepsilon \times \tilde{\Gamma}$, with transfer matrix $\tilde{G}_{cl}$, via the above transformation, i.e.

$$\tilde{G}_{cl}(\lambda) = G_{cl}\left(\frac{\varepsilon + \lambda}{1 + \varepsilon\lambda}\right).$$

Moreover, it can be shown that the state matrix of the closed-loop system $\Sigma \times \Gamma$ has all its eigenvalues inside $\mathbb{C}_\varepsilon^\oplus$. Hence the closed-loop system is certainly internally stable. Using the same arguments as before, we have

$$\|\tilde{G}_{cl}\|_\infty \geqslant \|G_{cl}\|_\infty.$$

Hence if $\tilde{\Gamma}$ applied to $\tilde{\Sigma}_\varepsilon$ makes the H_∞ norm less than some bound γ, then Γ makes the H_∞ norm of the closed-loop transfer matrix G_{cl} also less than γ.

On the other hand, if, for the system Σ, there is a stabilizing static controller Γ which makes the H_∞ norm of the closed-loop system strictly less than γ then it can be shown that there exists $\varepsilon_1 > 0$ such that for all $0 < \varepsilon \leqslant \varepsilon_1$ the transformed system $\tilde{\Sigma}_\varepsilon$ with the transformed controller $\tilde{\Gamma}$ is internally stable and the H_∞ norm of the closed-loop system is strictly less than γ.

Now for all but finitely many $\varepsilon > 0$ the system $\tilde{\Sigma}_\varepsilon$, described by (15.18) is such that the systems $(\tilde{A}, \tilde{B}, \tilde{C}_2, \tilde{D}_{21})$ and $(\tilde{A}, \tilde{E}, \tilde{C}_1, \tilde{D}_{12})$ have no zeros on the imaginary axis. However, $\tilde{D}_{21}$ and $\tilde{D}_{12}$ will not always be injective and surjective respectively. Either they are injective and surjective for all but finitely many ε, or they are not for any value of ε. This is in line with what we noted before as the advantages of this transformation. The matrices $\tilde{D}_{11}$ and $\tilde{D}_{22}$ will be there and therefore we can not directly apply theorem 14.1. But we have described methods to get rid of these matrices in the previous section.

Therefore, to check if for $\tilde{\Sigma}$ there is an internally stabilizing feedback controller which makes the H_∞ norm of the closed-loop system less than γ we can, in most cases, use theorem 14.1. Note that we have to check the conditions of theorem 14.1 for some $\varepsilon > 0$. The problem is that we do not know how small we must make ε. If for some $\varepsilon > 0$ the conditions of theorem 14.1 are satisfied, then we find a suitable controller. However, if the conditions of theorem 14.1 are not satisfied then either no suitable controller exists, or one does exist in which case the conditions of theorem 14.1 are satisfied for some smaller ε. Therefore we are never sure. Moreover, for small ε we very easily run into numerical difficulties.

15.3.2 Cheap control

In the previous section we described an approach to solve the H_∞ control problem if we have zeros on the imaginary axis or assumptions on the direct feedthrough matrices are not satisfied. This extends theorem 14.1 which can only be used if these assumptions are satisfied. In this subsection we shall briefly describe an alternative approach.

Assume that we have a system Σ of the form

$$\Sigma : \begin{array}{l} \dot{x} = Ax + Bu + Ed, \\ y = C_1 x + D_1 d, \\ z = C_2 x + D_2 u. \end{array} \tag{15.19}$$

For each $\varepsilon > 0$ we define the following perturbed system:

$$\tilde{\Sigma}_\varepsilon : \begin{array}{l} \dot{x} = Ax + Bu + \tilde{E}d, \\ y = C_1 x + \tilde{D}_1 d, \\ z = \tilde{C}_2 x + \tilde{D}_2 u, \end{array} \tag{15.20}$$

where

$$\tilde{C}_2 = \begin{pmatrix} C_2 \\ \varepsilon I \\ 0 \end{pmatrix}, \quad \tilde{D}_2 = \begin{pmatrix} D_2 \\ 0 \\ \varepsilon I \end{pmatrix}, \quad \tilde{E} = \begin{pmatrix} E & \varepsilon I & 0 \end{pmatrix}, \quad \tilde{D}_1 = \begin{pmatrix} D_1 & 0 & \varepsilon I \end{pmatrix}.$$

The structure of the perturbations on the matrices is such that it is easy to show that any controller Γ of the form (15.1) has the property that Γ is internally stabilizing when applied to Σ if and only if the same controller Γ is internally stabilizing when applied to $\tilde{\Sigma}_\varepsilon$. Actually the closed loop poles of these two systems coincide.

Let Γ be internally stabilizing when applied to Σ. Denote the closed-loop operator mapping w to z with zero initial state by $\mathcal{G}_{cl}$. Moreover, denote the closed-loop operator mapping w to z with zero initial state when the controller is applied to $\tilde{\Sigma}_\varepsilon$ by $\tilde{\mathcal{G}}_{cl,\varepsilon}$. Then we have

$$\|\mathcal{G}_{cl}\|_\infty \leqslant \|\tilde{\mathcal{G}}_{cl,\varepsilon_1}\|_\infty \leqslant \|\tilde{\mathcal{G}}_{cl,\varepsilon_2}\|_\infty$$

for all $0 \leqslant \varepsilon_1 \leqslant \varepsilon_2$. Hence, if we have an internally stabilizing controller which makes the H_∞ norm of the closed-loop system less than γ when applied to the system $\tilde{\Sigma}_\varepsilon$, then the same controller is internally stabilizing and makes the H_∞ norm of the closed-loop system less than γ when applied to the original system. Thus we obtain the following result:

Theorem 15.7 *Let a system Σ be given by (15.19). For all $\varepsilon > 0$ define $\tilde{\Sigma}_\varepsilon$ by (15.20). The following two statements are equivalent:*

(i) For the system Σ there is a feedback Γ of the form (15.1) which is internally stabilizing and which makes the H_∞ norm of the closed-loop system less than 1.

(ii) There exists $\varepsilon_1 > 0$ such that for all $0 < \varepsilon < \varepsilon_1$ there is a feedback Γ of the form (15.1) for $\tilde{\Sigma}_\varepsilon$ which is internally stabilizing and which makes the H_∞ norm of the closed-loop system less than 1.

Any controller satisfying part (ii) for some $\varepsilon > 0$ also satisfies part (i).

On the other hand, for the system $\tilde{\Sigma}_\varepsilon$ the subsystems $(A, B, \tilde{C}_2, \tilde{D}_2)$ and $(A, \tilde{E}, C_1, \tilde{D}_1)$ do not have zeros and hence certainly no zeros on the imaginary axis. Moreover, $\tilde{D}_1$ and $\tilde{D}_2$ are surjective and injective, respectively. Hence, we may apply the results of theorem 14.1 to the system $\tilde{\Sigma}(\varepsilon)$ to obtain necessary and sufficient conditions for the existence of internally stabilizing controllers which make the H_∞ norm of the closed-loop system less than 1.

If we compare the method of the previous subsection with the method of the current subsection, then we see that the method of this subsection is much easier. First of all, we do not have to apply transformations on the controller. Secondly, for all $\varepsilon > 0$ the system (15.20) is proper and satisfies the assumptions of chapter 13. In contrast, for some values of $\varepsilon > 0$, the system (15.18) is either not proper or the conditions of theorem 14.1 are not satisfied. Finally, the method of the previous subsection might yield systems which do not satisfy the assumptions of theorem 14.1 for any value of ε.

Both methods have the disadvantage that the conditions cannot actually be checked since the conditions in the previous subsection as well as the conditions in this subsection have to be checked for an infinite number of $\varepsilon > 0$.

One of the main reasons for using the method of the previous subsection is because all closed-loop poles will be placed inside a circle in the open left half complex plane and therefore the closed-loop system will not have ill-damped high-frequency poles.

15.4 The minimum entropy H_∞ control problem

We have been studying the H_∞ control problem. In H_∞ control we design controllers to minimize the peak value of the largest singular value of the transfer matrix on the imaginary axis. It can be shown that (near)optimal controllers have a flat magnitude Bode diagram. On the other hand for the classical H_2 or Linear Quadratic Gaussian (LQG) control problem (see chapter 11) the average value of the transfer matrix over the imaginary axis is minimized. The latter problem is not concerned with peaks as long as they have small width.

The above reasoning implies that H_∞ control is well-suited for robustness synthesis via the small-gain theorem. On the other hand, robust controllers might lead to a large closed loop H_2 norm. This motivates a new synthesis problem: minimize the H_2 norm under an H_∞ norm bound. It is hoped that the H_∞ norm bound yields the desired level of robustness while performance is optimized simultaneously via the minimization of the H_2 norm.

The minimum entropy H_∞ control problem is defined as the problem of minimizing an entropy function under the constraint of internal stability of the closed-loop system and under the constraint of an upper bound on the H_∞ norm of the closed-loop transfer matrix. It is a mixture between optimizing the H_∞ norm and the H_2 norm. It is also yields a better understanding of why the controllers we obtained in theorems 13.3 and 14.1 converge to the H_2 controllers from chapter 10. Therefore, it is hoped that by minimizing this entropy function, one can obtain a good trade-off between the H_∞ norm for robustness and the H_2 norm for performance.

15.4.1 Problem formulation and results

Consider the linear time-invariant system:

$$\Sigma: \begin{aligned} \dot{x} &= Ax + Bu + Ed, \\ y &= C_1 x + D_1 d, \\ z &= C_2 x + D_1 u. \end{aligned} \tag{15.21}$$

Here, A, B, E, C_1, C_2, D_1 and D_2 are real matrices of suitable dimension. Let G be a strictly proper real rational matrix which has no poles on the imaginary axis and which is such that $\|G\|_\infty < \gamma$. For such a transfer matrix G, we define the following entropy function:

$$\mathcal{J}_\gamma(G) := -\frac{\gamma^2}{2\pi} \int_{-\infty}^{\infty} \ln\det\left(I - \gamma^{-2} G^\sim(i\omega) G(i\omega)\right) \mathrm{d}\omega \tag{15.22}$$

where $G^\sim(s) := G^{\mathrm{T}}(-s)$. The minimum entropy H_∞ control problem is then defined as:

> Minimize $\mathcal{J}_\gamma(G_{cl})$ over all controllers which yield a strictly proper, internally stable closed-loop transfer matrix G_{cl} with H_∞ norm strictly less than γ.

We shall investigate controllers of the form (15.1). We can now formulate the main result of this chapter:

Theorem 15.8 *Consider the system (15.21). Assume that the systems* (A, B, C_2, D_2) *and* (A, E, C_1, D_1) *have no zeros on the imaginary axis, with* D_1 *and* D_2 *surjective and injective, respectively. Assume that a controller exists which is such that the closed-loop system is internally stable and has* H_∞ *norm strictly less than* γ. *The infimum of* $\mathcal{J}(G_{cl})$, *over all internally stabilizing controllers of the form (15.1) which are such that the closed system has* H_∞ *norm strictly less than* γ, *is equal to:*

$$\operatorname{trace}\left[E^T P E\right] + \operatorname{trace}\left[(A^T P + PA + C_2^T C_2 + \gamma^{-2} P E E^T P)(I - \gamma^{-2} Q P)^{-1} Q\right]$$

where P *and* Q *are such that part (ii) of theorem 14.1 is satisfied, and* G_{cl} *denotes the closed loop transfer matrix. The infimum is attained by the controller given in theorem 14.1.*

15.4.2 Properties of the entropy function

In this section we recall some basic properties of the entropy function as defined in (15.22). Note that in order to prove some of the results in this section some a priori knowledge of function theory is required. Those students not familiar with this area can skip the proofs in this section.

Lemma 15.9 *Let* G *be a strictly proper, stable rational matrix such that* $\|G\|_\infty \leqslant \gamma$. *Then we have*

- $\mathcal{J}_\gamma(G) \geqslant 0$ *and* $\mathcal{J}_\gamma(G) = 0$ *implies* $G = 0$,
- $\mathcal{J}_\gamma(G) = \mathcal{J}_\gamma(G^T)$.

Next, we relate our entropy function to the H_2 norm studied in chapter 10. We find:

Lemma 15.10 *Let a system* $\Sigma = (A, B, C, 0)$ *be such that* A *is stable. Let* G *be the transfer matrix of* Σ. *We have* $\mathcal{J}_\gamma(G) \geqslant \|G\|_2^2$. *Moreover,* $\mathcal{J}_\gamma(G) \to \|G\|_2^2$ *as* $\gamma \to \infty$.

Proof : For an arbitrary symmetric matrix A we have

$$-\ln\det(I - A) = -\ln \Pi(1 - \lambda_i) = -\Sigma \ln(1 - \lambda_i) \geqslant \Sigma - \lambda_i = -\operatorname{trace}[A]$$

where λ_i denote the eigenvalues of A. This immediately yields the first inequality:

$$-\frac{\gamma^2}{2\pi}\int_{-\infty}^{\infty} \ln\det\left(I - \gamma^{-2} G^\sim(i\omega) G(i\omega)\right) d\omega \geqslant \frac{1}{2\pi}\int_{-\infty}^{\infty} \operatorname{trace} G^\sim(i\omega) G(i\omega)\, d\omega$$

To prove the convergence as $\gamma \to \infty$ we will use $|\ln(1-\lambda) - 1 + \lambda| < \varepsilon\lambda^2$ for all λ such that $|\lambda| < \delta$. Note that for all $\gamma > \|G\|_\infty \delta^{-1}$ we obtain that all eigenvalues of $\gamma^{-2} G^{\sim}(i\omega)G(i\omega)$ will be less than δ in magnitude. We find:

$$\mathcal{J}_\gamma(G) - \|G\|^2 < \frac{\gamma^{-2}}{2\pi} \int_{-\infty}^{\infty} \operatorname{trace}[G^{\sim}(i\omega)G(i\omega)]^2 \, d\omega$$

which yields the desired convergence. ■

Next, we give two key lemmas. Of the first lemma, the first part is equal to lemma 14.4 while the second part originates from [133]. We give a separate proof of the second part.

Lemma 15.11 *Consider the linear time-invariant systems Σ and Ψ. Suppose Σ has inputs w and u and outputs z and y, while Ψ has input y and output u. Consider the interconnection depicted in the diagram in Figure 15.6. Assume that Σ is inner and*

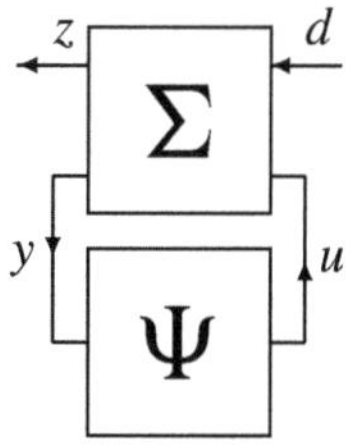

Figure 15.6

its input-output operator $\mathcal{G}$ has the following decomposition:

$$\mathcal{G}\begin{pmatrix} d \\ u \end{pmatrix} =: \begin{pmatrix} \mathcal{G}_{11} & \mathcal{G}_{12} \\ \mathcal{G}_{21} & \mathcal{G}_{22} \end{pmatrix} \begin{pmatrix} d \\ u \end{pmatrix} = \begin{pmatrix} z \\ y \end{pmatrix}, \tag{15.23}$$

which is compatible with the sizes of d, u, z and y. Let G_{11}, G_{12}, G_{21} and G_{22} be the transfer matrices associated with the operators $\mathcal{G}_{11}$, $\mathcal{G}_{12}$, $\mathcal{G}_{21}$ and $\mathcal{G}_{22}$ respectively. Assume G_{21} is invertible and G_{21}^{-1} has a finite H_∞ norm. Moreover the transfer matrices G_{11} and G_{22} are strictly proper.

Under the above assumptions the following two statements are equivalent:

(i) The interconnection in Figure 15.6 is internally stable and its closed-loop transfer matrix G_{cl} has H_∞ norm less than 1.

(ii) The system Ψ is internally stable and its transfer matrix has H_∞ norm less than 1.

Moreover, G_{cl} is strictly proper if, and only if, the transfer matrix of the system Ψ is strictly proper. Finally, if (i) holds and the transfer matrix of the system Ψ is *strictly proper then the following relation between the entropy functions for the different transfer matrices is satisfied:*

$$\mathcal{J}_1(G_{cl}) = \mathcal{J}_1(G_{11}) + \mathcal{J}_1(Q). \tag{15.24}$$

Proof : The first claim that the statements (i) and (ii) are equivalent, has been shown in lemma 14.4. We know that G_{11} and G_{22} are strictly proper. Combined with the fact that Σ is inner, this implies that G_{12} and G_{21} are bicausal, i.e. they are invertible and both the transfer matrices and their inverses are proper. Using this, it is easy to check that the transfer matrix H of Ψ is strictly proper if and only if G_{cl} is strictly proper.

We still have to prove equation (15.24). The following equality is easily derived using the property that Σ is inner (in particular that its transfer matrix satisfies $G^{\mathrm{T}}(s)G(s) = I$).

$$I - G_{cl}^{\sim}G_{cl} = G_{21}^{\sim}\left(I - H^{\sim}G_{22}^{\sim}\right)^{-1}\left(I - H^{\sim}H\right)\left(I - G_{22}H\right)^{-1}G_{21}.$$

Therefore, we find that

$$\begin{aligned}&\ln\det\left(I - G_{cl}^{\sim}G_{cl}\right)\\&\qquad = \ln\det\left(I - G_{11}^{\sim}G_{11}\right) + \ln\det\left(I - H^{\sim}H\right) - 2\ln\det\left(I - G_{22}H\right).\end{aligned}$$

Moreover, if statement (i) is satisfied and if H is strictly proper then we have

$$\mathcal{J}_1(H) = -\frac{1}{2\pi}\int_{-\infty}^{\infty} \ln\det\left(I - H^{\sim}(i\omega)H(i\omega)\right)\, d\omega, \tag{15.25}$$

$$\mathcal{J}_1(G_{11}) = -\frac{1}{2\pi}\int_{-\infty}^{\infty} \ln\det\left(I - G_{11}^{\sim}(i\omega)G_{11}(i\omega)\right)\, d\omega. \tag{15.26}$$

Using the fact that H is strictly proper, stable and has H_∞ norm strictly less than 1 and the fact that G_{22} is also stable, strictly proper and has H_∞ norm less than or equal to 1, we know that $\ln\det\left(I - G_{22}(\lambda)H(\lambda)\right)$ is an analytic function in the open right half plane and that a constant M exists such that

$$|\ln\det\left(I - G_{22}(\lambda)Q(\lambda)\right)| < \frac{M}{|\lambda|^2}, \qquad \forall\lambda \in \mathbb{C}^0 \cup \mathbb{C}^+$$

This implies, using Cauchy's theorem, that

$$\int_{-\infty}^{\infty} \ln\det\left(I - G_{22}(i\omega)Q(i\omega)\right)\, d\omega = 0 \tag{15.27}$$

Combining the above, we find (15.24). ■

The following lemma is an essential tool for actually calculating the entropy function for some specific system:

Lemma 15.12 *Let $\Sigma = (A, B, C, D)$ be stabilizable and detectable with transfer matrix G ans with* $\det D = 1$. *Also, assume that $G, G^{-1} \in H_\infty$ and G has H_∞ norm equal to* 1. *Then we have:*

$$\int_{-\infty}^{\infty} \ln|\det G(i\omega)|\, d\omega = -\pi \operatorname{trace}[BD^{-1}C]. \tag{15.28}$$

Proof : Denote the integral in (15.28) by $\mathcal{K}$ and define

$$a := -\operatorname{trace} BD^{-1}C.$$

We have (remember that $\ln|z| = \Re\mathrm{e} \ln z$):

$$\mathcal{K} = \Re\mathrm{e}\left(\int_{-\infty}^{\infty} \ln\det G(i\omega) - \frac{a}{1+i\omega}\,\mathrm{d}\omega\right) + a\int_{-\infty}^{\infty} \frac{\mathrm{d}\omega}{1+\omega^2} \tag{15.29}$$

Next, it is easily checked that $p(s) := \ln\det G(s) - \frac{a}{1+s}$ is a bounded analytic function in $\mathbb{C}^+$ such that $p(s) = O(1/s^2)$ ($|s| \to \infty$, $\operatorname{Re} s \geqslant 0$). Hence, using Cauchy's theorem we find

$$\int_{-\infty}^{\infty} p(i\omega)\,\mathrm{d}\omega = 0. \tag{15.30}$$

Combining (15.29) and (15.30) yields (15.28). ■

Corollary 15.13 *Let $\Sigma = (A, B, C, 0)$ be internally stable with strictly proper transfer matrix G. Assume G has H_∞ norm strictly less than γ. Then we have:*

$$\mathcal{J}_\gamma(G) = \operatorname{trace} B^T X B, \tag{15.31}$$

where X is the unique reak symmetric solution of the algebraic Riccati equation:

$$XA + A^T X + \gamma^{-2} X B B^T X + C^T C = 0,$$

such that $A + \gamma^{-2} B B^T X$ is asymptotically stable.

Proof : The existence and uniqueness of X is a consequence of the bounded real lemma (see section 12.6). It is easy to check that the transfer matrix M with realization $(A, B, -\gamma^{-2}B^T X, I)$ satisfies:

$$I - \gamma^{-2} G^\sim G = M^\sim M.$$

Moreover, $M, M^{-1} \in H_\infty$, i.e. M is a spectral factor of $I - \gamma^{-2} G^\sim G$. We have

$$\mathcal{J}_\gamma(G) = \frac{-\gamma^2}{\pi} \int_{-\infty}^{\infty} \ln|\det M(i\omega)|\,\mathrm{d}\omega,$$

and therefore (15.31) is a direct consequence of applying lemma 15.12 to the above equation. ■

15.4.3 A system transformation

We will only proof theorem 15.8 for $\gamma = 1$. The general result can then be obtained by scaling (as we also did in proving theorem 14.1). Throughout this section we assume that there are matrices P and Q satisfying the conditions in theorem 14.1. Note that this is no restriction when proving theorem 15.8. The existence of such P and Q is implied by our assumption that an internally stabilizing controller exists which makes the H_∞ norm strictly less than 1. We use the technique from chapter 14 of transforming the system twice such that the problem of minimizing the entropy function for the original system is equivalent to minimizing the entropy function for the new system we thus obtain. In the next section we shall show that this new system satisfies some desirable properties which enables us to solve the minimum entropy H_∞ control problem for this new system and hence also for the original system.

We define the following system:

$$\Sigma_P : \begin{array}{rl} \dot{x}_P = & A_P x_P + Bu + Ed_P, \\ y = & C_{1,P} x_P + D_1 d_P, \\ z_P = & C_{2,P} x_P + D_2 u, \end{array} \tag{15.32}$$

where $A_P := A + EE^{\mathrm{T}}P$, $C_{1,P} := C_1 + D_1 E^{\mathrm{T}} P$ and $C_{2,P} := D_2(D_2^{\mathrm{T}} D_2)^{-1}(B^{\mathrm{T}}P + D_2^{\mathrm{T}} C_2)$.

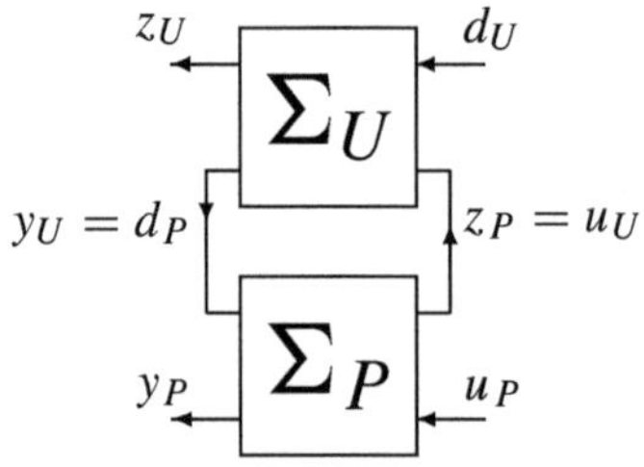

Figure 15.7

From chapter 14 we know that when we define Σ_U by 14.16 then Σ has the same transfer matrix as the interconnection in figure 15.7. Moreover, Σ is stabilizable and detectable if and only if the interconnection in figure 15.7 is stabilizable and detectable. We know from chapter 14 that Σ_U is inner. Moreover, the input-output operator $\mathcal{U}$ of the system Σ_U has the following decomposition:

$$\mathcal{U} \begin{pmatrix} d_U \\ z_P \end{pmatrix} =: \begin{pmatrix} \mathcal{U}_{11} & \mathcal{U}_{12} \\ \mathcal{U}_{21} & \mathcal{U}_{22} \end{pmatrix} \begin{pmatrix} d_U \\ z_P \end{pmatrix} = \begin{pmatrix} z_U \\ d_P \end{pmatrix}, \tag{15.33}$$

which is compatible with the sizes of d_U, z_P, z_U and d_P. The input-output operator $\mathcal{U}_{21}$ is associated to the system $\Sigma_{U,21} = (\tilde{A}, E, -E^{\mathrm{T}}P, I)$ and, since $\tilde{A} + EE^{\mathrm{T}}P$ is equal to (14.5) and hence a stability matrix, the input-output operator $\mathcal{U}_{21}$ is invertible and $\mathcal{U}_{21}^{-1}$ has a finite L_2-induced operator norm.

Moreover, by corollary 15.13, we have $\mathcal{J}_1(U_{11}) = \operatorname{trace} E^{\mathrm{T}}PE$. Combining the above with lemma 15.11, we find the following theorem:

Theorem 15.14 *Let the systems (15.21) and (15.32) be given. Moreover, let a controller Γ of the form (15.1) be given. The following two conditions are equivalent:*

- *Γ is internally stabilizing for Σ such that the closed-loop transfer matrix G_{cl} is strictly proper and has H_∞ norm strictly less than* 1.
- *Γ is internally stabilizing for Σ_P such that the closed-loop transfer matrix $G_{cl,P}$ is strictly proper and has H_∞ norm strictly less than* 1.

Moreover, if Γ satisfies the above conditions then we have

$$\mathcal{J}_1(G_{cl}) = \mathcal{J}_1(G_{cl,P}) + \operatorname{trace} E^{T}PE.$$

Next, we make another transformation from Σ_P to $\Sigma_{P,Q}$. This transformation is exactly dual to the transformation from Σ to Σ_P. We know there exists a controller which is internally stabilizing for Σ_P which makes the H_∞ norm of the closed-loop system strictly less than 1. Therefore if we apply lemma 14.8 together with theorem 14.1 to Σ_P we find that the matrix $Y = (I - QP)^{-1}Q$ is a real symmetric solution of (14.19) such that (14.21) is a stability matrix. We define the following system:

$$\Sigma_{P,Q}: \quad \begin{aligned} \dot{x}_{P,Q} &= A_{P,Q}x_{P,Q} + B_{P,Q}u + E_{P,Q}d, \\ y &= C_{1,P}x_{P,Q} + D_1 d_{P,Q}, \\ z_{P,Q} &= C_{2,P}x_{P,Q} + D_2 u, \end{aligned} \tag{15.34}$$

where

$$\begin{aligned} A_{P,Q} &:= A_P + YC_{2,P}^{\mathrm{T}}C_{2,P}, \\ B_{P,Q} &:= B + YC_{2,P}^{\mathrm{T}}D_2, \\ E_{P,Q} &:= (YC_{1,P}^{\mathrm{T}} + ED_1^{\mathrm{T}})(D_1D_1^{\mathrm{T}})^{-1}D_1. \end{aligned}$$

Using theorem 15.14 and a dualized version for the transformation from Σ_P to $\Sigma_{P,Q}$ we can derive the following corollary:

Corollary 15.15 *Let the systems (15.21) and (15.34) be given. Moreover, let a controller Γ of the form (15.1) be given. The following two conditions are equivalent:*

- *Γ is internally stabilizing for Σ such that the closed-loop transfer matrix G_{cl} is strictly proper and has H_∞ norm strictly less than* 1.
- *Γ is internally stabilizing for $\Sigma_{P,Q}$ such that the closed-loop transfer matrix $G_{cl,P,Q}$ is strictly proper and has H_∞ norm strictly less than* 1.

Moreover, if Γ satisfies the above conditions then we have

$$\mathcal{J}_1(G_{cl}) = \mathcal{J}_1(G_{cl,P,Q}) + \operatorname{trace} E^{T}PE + \operatorname{trace} C_{1,P}YC_{1,P}.$$

From this corollary it is immediate that it is sufficient to investigate $\Sigma_{P,Q}$ to prove the results in our main theorem 15.8. On the other hand, we know that the central controller from theorem 14.1 is such that, when applied to $\Sigma_{P,Q}$, it yields a closed loop transfer matrix equal to 0 and hence an entropy equal to 0. Since the entropy function is nonnegative, this implies that the central controller is optimal for $\Sigma_{P,Q}$ and therefore also for Σ. This completes the proof of theorem 15.8.

15.5 A design example: an inverted pendulum on a cart

In order to connect the theory presented in this book for the H_∞ control problem to classical control theory, we present in this section a small design example. Note that for those people who have not had a course on classical control theory some of the terminology in this chapter might be unknown and in that case it might be better to skip this section.

We shall consider the following physical example of an inverted pendulum on a cart depicted in figure 15.8.

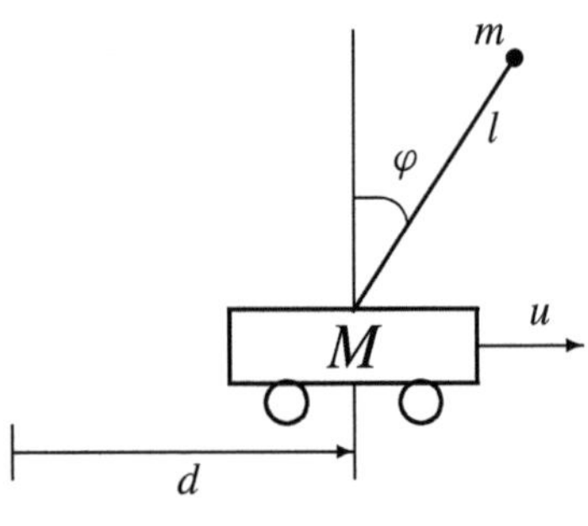

Figure 15.8

We assume the mass of the pendulum to be concentrated in the top with mass m. l is the length of the pendulum and M is the mass of the cart. To describe the position, d and φ express the distance of the cart from some reference point and the angle of the pendulum with respect to the vertical axis. The input u is the horizontal force applied to the cart. All motions are assumed to be in the plane. We assume that the system is completely stiff and the friction between the cart and the ground has friction coefficient F. Finally let g denote the acceleration of gravity. We then have the following non-linear model for this system:

$$
\begin{aligned}
(M+m)\ddot{d} + ml\ddot{\varphi}\cos\varphi - ml(\dot{\varphi})^2\sin\varphi + F\dot{d} &= u,\\
l\ddot{\varphi} - g\sin\varphi + \ddot{d}cos\varphi &= 0.
\end{aligned}
$$

If $l \neq 0$ and $M \neq 0$, then linearization around $\varphi = 0$ yields the following linear model:

$$\begin{pmatrix} \dot{d} \\ \ddot{d} \\ \dot{\varphi} \\ \ddot{\varphi} \end{pmatrix} = \begin{pmatrix} 0 & 1 & 0 & 0 \\ 0 & -\frac{F}{M} & -\frac{mg}{M} & 0 \\ 0 & 0 & 0 & 1 \\ 0 & \frac{F}{lM} & \frac{g(m+M)}{lM} & 0 \end{pmatrix} \begin{pmatrix} d \\ \dot{d} \\ \varphi \\ \dot{\varphi} \end{pmatrix} + \begin{pmatrix} 0 \\ \frac{1}{M} \\ 0 \\ -\frac{1}{lM} \end{pmatrix} u.$$

Denote these matrices by A and B respectively. Since $g \neq 0$ our linearized system is always controllable. Assuming our measurements are d and $\dot{d}$ then the system is observable if $m \neq 0$. If $m = 0$ the system is not even detectable (which has a clear physical interpretation). However, it turned out that although the system is observable, for many choices of the parameters an unstable pole and zero almost cancel out. This makes the system impossible to control. Therefore we have to add the angle φ as a measurement, i.e. our measurement vector equals $y = (d, \dot{d}, \varphi)$.

We would like to track a reference signal for the position. Moreover, we require our controller to yield a robustly stable system with respect to the several uncertainties that affect our system:

(*i*) Discarded non-linear dynamics

(*ii*) Uncertainties in the parameters F, m, M, l

(*iii*) Flexibility in the pendulum.

Finally we have to take into account the limit on the bandwidth and gain of our controller. This is essential due to limitations on the sampling rate for the digital implementation as well as limitation in the speed of the actuators. We look at the following setup in figure 15.9.

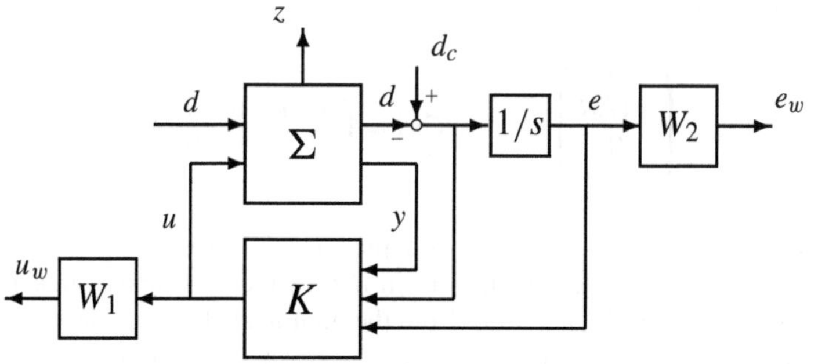

Figure 15.9

- d_c is the command signal for the position d. We minimize the weighted integrated tracking error where W_2 is a first order weight of the form

$$W_2(s) := \varepsilon \frac{1+\alpha s}{1+\beta s}$$

By choosing $\alpha << \beta$ we obtain a low pass filter. This expresses that we are only interested in tracking low-frequency signals. The integrator also expresses our interest in low frequencies: it is one way to guarantee zero steady state tracking error. Finally ε in the weight is used to express the relative importance of tracking over the other goals we have for the system.

- We also minimize u_w which is the weighted control input. W_1 has the same structure as W_2. However, this time we choose $\alpha >> \beta$ to obtain a high pass filter. We would like to constrain the open-loop bandwidth and gain of the controller. This facilitates digital implementation and it also prevents us pushing our actuators beyond their capabilities. Finally, it prevents the controller stimulating high-frequency uncertainty in the system dynamics, such as bending modes of the pendulum. Since we cannot incorporate open-loop limitations directly in an H_∞ design, we push the bandwidth and gain down indirectly via this weight on u. However, it turned out to be very hard to make the bandwidth small.

- z and d are new inputs and outputs we add to the system Σ to express robustness requirements, i.e. the system Σ is of the form:

$$\Sigma: \begin{array}{rcl} \dot{x} &=& Ax + Bu + Ed, \\ y &=& \begin{pmatrix} 1 & 0 & 0 & 0 \\ 0 & 1 & 0 & 0 \\ 0 & 0 & 1 & 0 \end{pmatrix} x, \\ z &=& C_2 x. \end{array} \tag{15.35}$$

The matrices A and B are as defined before. On the other hand, E and C_2 still have to be chosen. For instance if we want to guard against fluctuations in the parameters F and m, the friction and the mass of the pendulum, then we choose E and C_2 as:

$$E := \begin{pmatrix} 0 \\ -1/M \\ 0 \\ 1/lM \end{pmatrix}, \qquad C_2 := \begin{pmatrix} 0 & 1 & 0 & 0 \\ 0 & 0 & g & 0 \end{pmatrix}. \tag{15.36}$$

This amounts to a Δ-block such as depicted in (12.3) where $\Delta = (F_r - F \;\; m_r - m)$. An H_∞ norm less than γ from d to z then guarantees that if the true parameter values F_r, m_r differ from the nominal values m, r less than γ^{-1} (to be precise such that $\|\Delta\| \leqslant \gamma^{-1}$), the system will still be stable. A similar definition can be made to guard against fluctuations in all parameters of the 2 differential equations due to discarded non-linearities or flexibility of the beam. In that case we choose, instead of (15.36), E and C_2 as:

$$E := \begin{pmatrix} 0 & 0 \\ 1 & 0 \\ 0 & 0 \\ 0 & 1 \end{pmatrix}, \qquad C_2 := \begin{pmatrix} 1 & 0 & 0 & 0 \\ 0 & 1 & 0 & 0 \\ 0 & 0 & 1 & 0 \\ 0 & 0 & 0 & 1 \end{pmatrix}. \tag{15.37}$$

On the basis of the above we start designing a controller K for the interconnection in figure 15.9 to minimize the H_∞ norm from (d, d_c) to (z, u_w, e_w) where we manipulated, by hand, the parameters of the weights W_1 and W_2 on the basis of the properties of the controllers. It turned out that the system Σ is in general much more sensitive to perturbations in all parameters of the differential equation then to perturbations in the parameters F and m. Therefore we incorporate in our design robustness against all parameters of the differential equation. In other words Σ is given by (15.35) where E and C_2 are given by (15.37).

We would like to stress that optimal controllers have a tendency of ruining every nice property of the system which is not explicitly taken into account. Our design incorporated a γ iteration. However, we implemented the central controller for a γ approximately 10% larger than the infimum over all stabilizing controllers of the closed-loop H_∞ norm. It is our experience that this reduces the bandwidth of the resulting controllers.

We shall give some facts illustrated by frequency and time-responses of our controller and the resulting closed-loop system. We choose $M = 1, m = .1, l = 1$ and $F = 0.1$.

(*i*) The open-loop transfer matrix from u to x and from u to φ are given by

$$G_{xu}(s) := \frac{(s+3.13)(s-3.13)}{(s+3.29)(s-3.28)(s+0.09)},$$

$$G_{\varphi u}(s) := \frac{-s^2}{(s+3.29)(s-3.28)(s+0.09)}.$$

Hence, it is immediate that because of the near pole-zero cancellation in the right half plane that we really need a measurement of the angle φ.

(*ii*) Although not for every choice of the parameters, our final controller, whose Bode magnitude diagram is depicted in figure 15.10, is stable. At this moment this desirable property cannot, however, be incorporated in our design criteria. We obtained a stable controller by playing around with the parameters. The eigenvalues of the controller are $-45.7, -22.8, -4.04, -2.37, -1.3 \pm 0.51i$ and -0.667. The fact that the controller has both very fast and relatively slow modes suggests that via singular perturbation theory (see, e.g. [137]) we can reduce the order of the controller. There is, however, no theory available for such an approach.

(*iii*) Our final controller can, according to the theory developed in subsection 15.2.3, stand fluctuations in the parameters F and m of 400%. Moreover, fluctuations in the parameters of the differential equations of size less than 0.025 are allowed. The latter is not very much but since most parameters are 0 they are in general quite sensitive to fluctuations. Without taking robustness into account the final controller could only stand fluctuations of size less than 0.000025. Some simulations suggested good robustness properties.

(*iv*) The time-responses as given in figures 15.14, 15.15 and 15.16 show little overshoot and therefore the controller is expected to work well on the non-linear model.

(*v*) The final controller still has quite a large gain and bandwidth as can be seen from figure 15.10. The gain is for a large part due to the fact that the controller is translating angles in radians into forces in Newtons. In general the angles are much smaller than the forces, which is to be expected. The large bandwidth is needed to be able to stand sudden fluctuations in the angles. The bandwidth and gain of the transfer matrix from d_c to u is much smaller as shown in figure 15.12. This transfer matrix is weighted more strongly in our cost criterion than the effect of fluctuations in φ (since the angle is not steered directly via a command signal).

(*vi*) Figure 15.11 shows that we have good tracking properties of low-frequency command signals as required. Tracking can be improved to 1 rad/sec. However, this results in a controller of much larger bandwidth and gain.

(*vii*) The loop gain if we break the loop at the control input is given in figure 15.13. The surprising part here is the cross-over angle which is very small. The angle of the Bode diagram at the cross-over frequency is related to the phase margin. The fact that this angle is very small suggests a good phase margin.

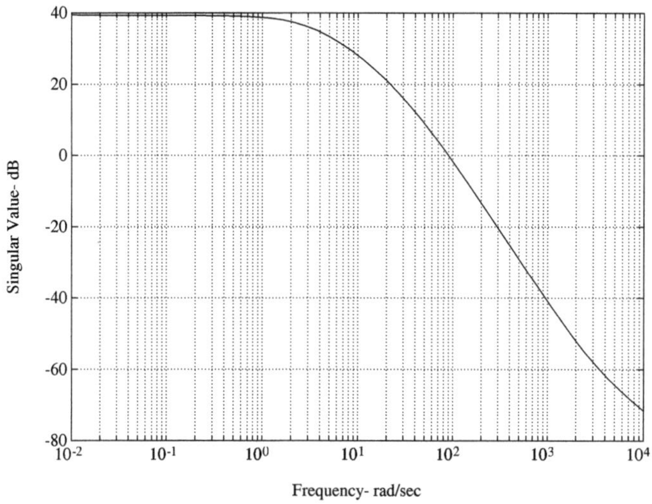

Figure 15.10: open loop magnitude Bode diagram of the controller

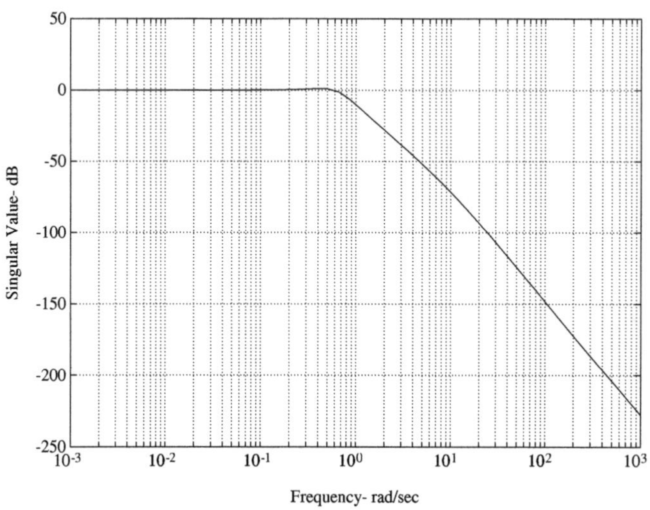

Figure 15.11: magnitude Bode diagram from d_c to d

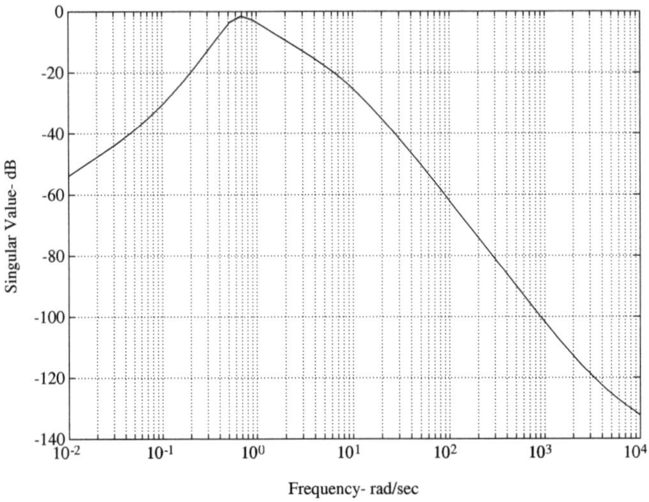

Figure 15.12: magnitude Bode diagram from d_c to u

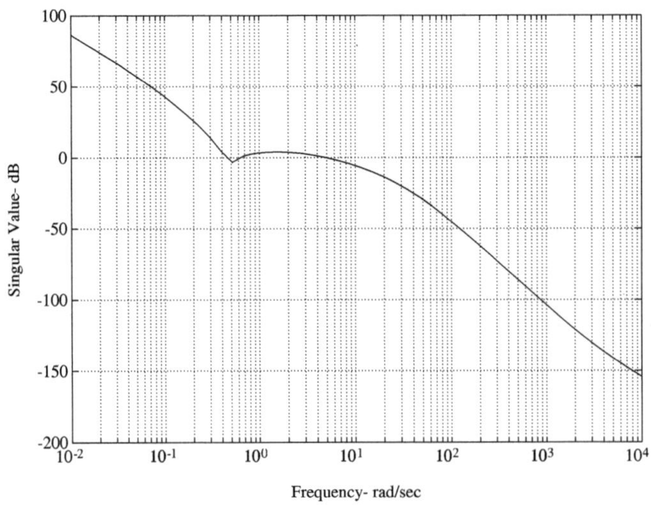

Figure 15.13: magnitude Bode diagram of the loop gain from u_c to u

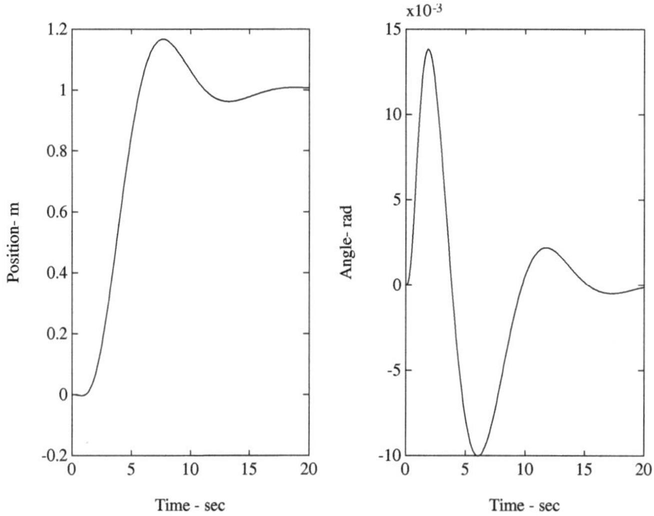

Figure 15.14: step response from d_c to d and φ

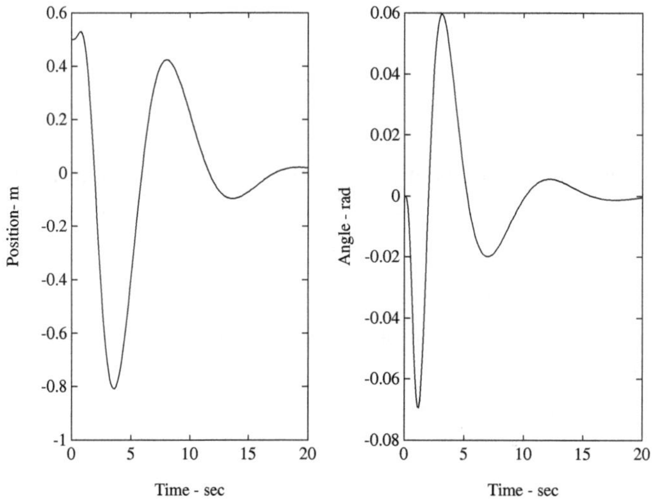

Figure 15.15: impulse response from d_c to d and φ

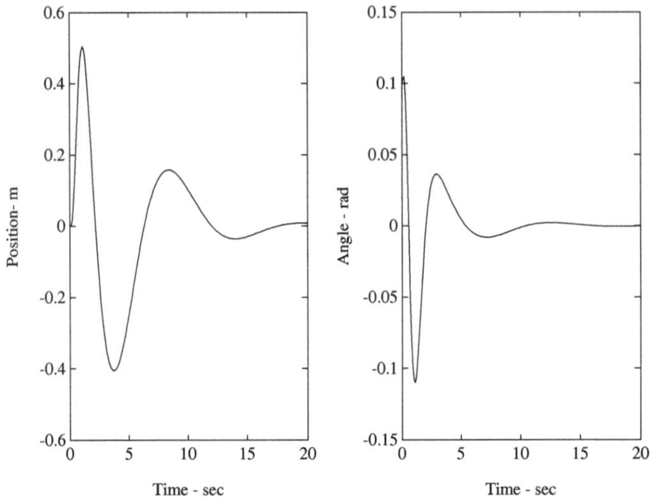

Figure 15.16: impulse response from φ to d and φ

To conclude, we would like to note that we can design controllers via H_∞ control similar as one used to do it via LQG control. However, it is our belief that the H_∞ norm makes it easier and more transparent to incorporate several performance requirements in our cost criterion, especially those performance requirements which are directly related to robustness and magnitude Bode diagrams. However a lot of work still needs to be done. In particular we have to gain experience similar to that we have gained for LQG to translate our criteria into a well-formulated H_∞ problem.

15.6 Exercises

15.1 In section 15.5 we have designed a controller for an inverted pendulum on a cart with parameter values $M = 1, m = 0.1, l = 1$ and $F = 0.1$. The specifications which we posed were:

- Tracking of a reference signal for the position up to 0.5 rad/sec.
- Open loop bandwidth of the controller less than 100 rad/sec with magnitude less than 50 dB.
- Robust stability with respect to fluctuations up to 200 % in the parameters F and m.
- Robust stability with respect to variations in the differential equation due to the nonlinearities of magnitude less than 0.02.

Design a controller using the techniques of H_∞ presented in the last chapters by choosing suitable weighting functions. The controller with measurements

$d, \dot{d}, \varphi$ should satisfy the above specifications as well as possible. Any additional freedom should be used to reduce the bandwidth of the controller. One is allowed all freedom and engineering tricks to do so.

15.2 Look at the setup of exercise 15.1. Design a controller without the robustness specifications. That is, try to achieve better tracking up to or above 1 rad/sec with a controller which has a similar bandwidth and magnitude compared to the controller designed in exercise 15.1. Check the robustness properties of the resulting closed loop system.

15.7 Notes and references

Additive model uncertainty has been discussed in Glover [61]. It shown there that our bound is equal to the smallest Hankel singular value for systems which have no stable poles. Hankel singular values are closely related to efficient techniques and finding approximate models of lower McMillan degree, see Glover [60]. This yields a nice connection since differences between models and the plant are often introduced by looking for a simple model, i.e. a model of low McMillan degree. Coprime factor model uncertainty is studied in detail in MacFarlane and Glover [123].

A more thorough understanding of singular H_∞ optimal control methods is important because the techniques presented in this chapter suffer from serious numerical difficulties if our choice of γ gets near the minimal achievable H_∞ norm. Especially, if we have infinite zeros or zeros on the imaginary axis of higher degree. Techniques presented in Gahinet, Apkarian [54] are very elegant but have the same numerical difficulties. Methods presented in Stoorvogel [184, 186] and Scherer [163] are much better suited to overcome these numerical difficulties. For recent results we refer to Chen [28] Xin, Anderson and Mita [225].

Minimum entropy as treated in this chapter was originate by Mustafa and Glover in [133]. Our approach follows Stoorvogel, [183]. The concepts of function theory used in some of the proofs can be found in Rudin [156] or Churchill and Brown [30].

In order to understand the concepts related to frequency-domain analysis as used in part of section we refer to the textbooks by Kuo [101], Franklin, Powell and Emami-Naeini [49], and van de Vegte [202]. Several recent textbooks discuss design of H_∞ controllers in much more detail as we have done in this book. Especially connections to classical control is presented in this book only in a very limited fashion. For more details, we refer to some of the recent textbooks such as Maciejowski [118], Zhou, Doyle and Glover [232], Green, Limebeer [66], Skogestad and Postlethwaite [180]

Appendix A

Distributions

In this appendix, we have collected the most important facts on distributions that we need in this book.

We will start by introducing some notation. If $f : \mathbb{R} \to \mathbb{R}$ is a function then its *support* is defined as the closure of the set of all t such that $f(t) \neq 0$:

$$\operatorname{supp}(f) := \overline{\{t \in \mathbb{R} \mid f(t) \neq 0\}}.$$

If $f \neq 0$ then we define

$$\lambda(f) := \inf \operatorname{supp}(f),$$
$$\rho(f) := \sup \operatorname{supp}(f).$$

It is understood that $\lambda(f)$ and $\rho(f)$ can take the values $-\infty$ and ∞, respectively. If $f = 0$ then we define $\lambda(f) := \infty$ and $\rho(f) := -\infty$.

For a given f we define $\check{f}$ by

$$\check{f}(t) := f(-t).$$

If, in addition, $\tau \in \mathbb{R}$ then $\sigma_\tau f$ is defined by

$$(\sigma_\tau f)(t) := f(t + \tau).$$

We denote by $\mathfrak{C}^\infty(\mathbb{R}, \mathbb{R})$ the space of all infinitely often differentiable functions $f : \mathbb{R} \to \mathbb{R}$. The following subspaces of $\mathfrak{C}^\infty(\mathbb{R}, \mathbb{R})$ are important to us:

$$\mathfrak{D}_- := \{\phi \in \mathfrak{C}^\infty(\mathbb{R}, \mathbb{R}) \mid \rho(\phi) < \infty\},$$

$$\mathfrak{D}_+ := \{\psi \in \mathfrak{C}^\infty(\mathbb{R}, \mathbb{R}) \mid \lambda(\psi) > -\infty\}.$$

Elements of $\mathfrak{D}_-$ are called *test functions*. There is a natural way to define a topology on the linear space $\mathfrak{D}_-$. The exact definition of this topology is not important to us and is therefore omitted. In any case, once we have put a topology on $\mathfrak{D}_-$, we can speak about continuity of functionals defined on $\mathfrak{D}_-$. A *distribution* is simply a continuous linear functional on the space of testfunctions $\mathfrak{D}_-$. The value of the distribution x at $\phi \in \mathfrak{D}_-$ is denoted by $\langle x, \phi \rangle$. The space of all distributions is denoted by $\mathfrak{D}'_+$. Although this is the dual space of $\mathfrak{D}_-$ the $+$ is used to indicate that these distributions have a support which is bounded on the right.

Example A.1 Every function $\psi \in \mathfrak{D}_+$ can be identified with a distribution by defining the value of ψ at $\phi \in \mathfrak{D}_-$ by

$$\langle \psi, \phi \rangle := \int_{-\infty}^{\infty} \psi(t)\phi(t)\,\mathrm{d}t.$$

The crucial point here is of course that the product function $\psi\phi$ is zero outside some compact interval. Since $\psi\phi$ is continuous on the interval, the integral must be finite. Thus, $\mathfrak{D}_+$ can be considered as a linear subspace of $\mathfrak{D}'_+$. Elements of $\mathfrak{D}_+$ are called *smooth* distributions.

Example A.2 An important example of a distribution that is not smooth is the *Dirac distribution*. The Dirac distribution, which is denoted by δ, is defined by

$$\langle \delta, \phi \rangle := \phi(0).$$

Intuitively, this distribution represents the (fictitious) real function $\delta(t)$ with the properties that $\delta(t) = 0$ for $t \neq 0$ while $\int_{-\infty}^{\infty} \delta(t)\phi(t)\,\mathrm{d}t = \phi(0)$ for all $\phi \in \mathfrak{D}_-$. The other common intuitive interpretation of the function $\delta(t)$ as a function with the properties that $\delta(0) = \infty$ while $\delta(t) = 0$ for $t \neq 0$ is easy to understand but not powerful enough to work with.

Example A.3 Another example of a distribution that is not smooth is the *Heaviside distribution*. It is denoted by h and its value at $\phi \in \mathfrak{D}_-$ is defined by

$$\langle h, \phi \rangle := \int_{0}^{\infty} \phi(t)\,\mathrm{d}t.$$

We now define what we mean by differentiation of distributions. If x is a distribution then its derivative $\dot{x}$ is defined as the distribution given by

$$\langle \dot{x}, \phi \rangle := -\langle x, \dot{\phi} \rangle.$$

Here, $\dot{\phi}$ denotes the ordinary derivative of the function ϕ. Note that, since $\phi \in \mathfrak{D}_-$, $\langle x, \dot{\phi} \rangle$ is well-defined. We claim that if x is the smooth distribution corresponding to the function $\psi \in \mathfrak{D}_-$, i.e. if

$$\langle x, \phi \rangle = \int_{-\infty}^{\infty} \psi(t)\phi(t)\,\mathrm{d}t,$$

then $\dot{x}$ is the smooth distribution corresponding to the function $\dot{\psi}$. Stated differently: for functions $\psi \in \mathfrak{D}_+$ differentiation in distributional sense coincides with ordinary differentiation. Indeed, if $\phi \in \mathfrak{D}_-$ then by partial integration we have

$$\begin{aligned}\langle \dot{x}, \phi \rangle &= -\langle x, \dot{\phi} \rangle \\ &= -\int_{-\infty}^{\infty} \psi(t)\dot{\phi}(t)\,\mathrm{d}t \\ &= \int_{-\infty}^{\infty} \dot{\psi}(t)\phi(t)\,\mathrm{d}t.\end{aligned}$$

Example A.4 The derivative of the Heaviside distribution is equal to the Dirac distribution. Indeed, for $\phi \in \mathfrak{D}_-$ we have

$$\langle \dot{h}, \phi \rangle = -\langle h, \dot{\phi} \rangle = -\int_0^{\infty} \dot{\phi}(t)\,\mathrm{d}t = \phi(0).$$

The derivative $\dot{\delta}$ of the Dirac distribution δ is given by $\langle \dot{\delta}, \phi \rangle = -\dot{\phi}(0)$. The derivative $\dot{\delta}$ is often denoted by $\delta^{(1)}$. The i-th derivative of the Dirac distribution is denoted by $\delta^{(i)}$ and is defined inductively by $\delta^{(i)} := (\delta^{(i-1)})^{\cdot}$. Keeping in line with this notation, the Dirac distribution δ is sometimes denoted by $\delta^{(0)}$. It is easily seen that $\langle \delta^{(i)}, \phi \rangle = (-1)^i \phi^{(i)}(0)$.

An important role is played by *convolution* of distributions. Recall that if f and g are functions, then their convolution is defined by

$$(f * g)(t) := \int_{-\infty}^{\infty} f(t-\tau)g(\tau)d\tau$$

(provided that, of course, the integral exists). The concept of convolution can be generalized to distributions. We first define convolution of a distribution with a smooth distribution. Let $x \in \mathfrak{D}'_+$ and $\psi \in \mathfrak{D}_+$. Then their convolution $x * \psi$ is defined as

$$(x * \psi)(t) := \langle x, \sigma^{-t}\check{\psi} \rangle \quad (t \in \mathbb{R}).$$

It can be shown that $x * \psi \in \mathfrak{D}_+$. Next, we define the convolution of arbitrary distributions. Let $x, y \in \mathfrak{D}'_+$. Then $x * y$ is defined as

$$\langle x * y, \phi \rangle := \langle x, (y * \check{\phi})\check{\ } \rangle \quad (\phi \in \mathfrak{D}_-).$$

Note that if $\phi \in \mathfrak{D}_-$ then $\check{\phi} \in \mathfrak{D}_+$. If x and y are smooth distributions, corresponding to the functions ψ and χ respectively, then $x * y$ is the smooth distribution corresponding to the ordinary convolution of ψ and χ, i.e., for all $\phi \in \mathfrak{D}_-$ we have

$$\langle x * y, \phi \rangle = \int_{-\infty}^{\infty} (\psi * \chi)(t)\phi(t)\,\mathrm{d}t.$$

It can be proven immediately from the definition that convolution of distributions is a commutative operation. In fact, with the pointwise addition and scalar multiplication $(x, y) \mapsto x + y$, $(\lambda, x) \mapsto \lambda x$, and with convolution as multiplication, $\mathfrak{D}'_+$ is a *commutative algebra* over $\mathbb{R}$. The *unit element* is the Dirac distribution δ. Indeed, if $x \in \mathfrak{D}'_+$ then we have

$$\begin{aligned}\langle \delta * x, \phi \rangle &= \langle \delta, (x * \check{\phi})\check{\ } \rangle \\ &= (x * \check{\phi})(0) = \langle x, \phi \rangle.\end{aligned}$$

A distribution $x \in \mathfrak{D}'_+$ is called *invertible* if there exists $y \in \mathfrak{D}'_+$ such that $x * y = y * x = \delta$. If x is invertible then there is exactly one such y. This distribution y is called *the inverse of* x and is denoted by x^{-1}.

It is straightforward to verify that the following rule for differentiation of a convolution product is valid:

$$(x * y)^{\cdot} = \dot{x} * y = x * \dot{y}.$$

As a consequence, differentiation of a distribution x is the same as taking the convolution of x with $\dot{\delta}$, the derivative of the Dirac distribution:

$$\dot{\delta} * x = \dot{x}$$

Example A.5 The distribution $\dot{\delta}$ is invertible. Its inverse is h, the Heaviside distribution. Indeed,

$$\dot{\delta} * h = \dot{h} = \delta$$

An important role in chapter 8 of this book is played by the subclass of $\mathfrak{D}'_+$ consisting of impulsive-smooth distributions. These are defined as follows:

Definition A.6 *A distribution $x \in \mathfrak{D}'_+$ is called* impulsive *if it has the form*

$$x = \alpha_0 \delta + \alpha_1 \delta^{(1)} + \alpha_2 \delta^{(2)} + \cdots + \alpha_k \delta^{(k)},$$

with $\alpha_i \in \mathbb{R}$. Keeping in line with the notation introduced above, we can write $x = \sum_{i=0}^{k} \alpha_i \delta^{(i)}$.

A distribution x is called smooth on $\mathbb{R}^+$ *if there exists a function $\psi \in \mathfrak{C}^\infty(\mathbb{R}, \mathbb{R})$ such that*

$$\langle x, \phi \rangle = \int_0^\infty \psi(t)\phi(t)\, dt, \quad \phi \in \mathfrak{D}_-$$

Thus, if x is smooth on $\mathbb{R}^+$, it can be identified with a function of the form $\psi(t) i_{\mathbb{R}^+}(t)$, where $\psi \in \mathfrak{C}^\infty(\mathbb{R}, \mathbb{R})$ and where $i_{\mathbb{R}^+}(t)$ denotes the indicator function of $\mathbb{R}^+$ (defined by $i_{\mathbb{R}^+}(t) = 0$ for $t < 0$ and $i_{\mathbb{R}^+}(t) = 1$ for $t \geqslant 0$).

Finally $x \in \mathfrak{D}'_+$ is called impulsive-smooth *if it has the form $x = x_1 + x_2$, where x_1 is impulsive and x_2 is smooth on $\mathbb{R}^+$. In this case, x_1 is called the* impulsive part of x *and x_2 is called the* smooth part of x. *The subclass of impulsive-smooth distributions is a sub-algebra of $\mathfrak{D}'_+$ and will be denoted by $\mathfrak{D}_0$*

Example A.7 The Heaviside distribution h is smooth on $\mathbb{R}^+$. It can be identified with the function $i_{\mathbb{R}^+}(t)$ which is often also called the Heaviside step function.

Example A.8 Let x be smooth on $\mathbb{R}^+$, so that x corresponds to the function

$$\psi(t)i_{\mathbb{R}^+}(t)$$

for some $\psi \in \mathfrak{C}^\infty(\mathbb{R}, \mathbb{R}))$. Then

$$\langle x, \phi\rangle = \int_0^\infty \psi(t)\phi(t)\,\mathrm{d}t.$$

Then we have:

$$\dot{x} = \psi(0)\delta + z,$$

where z is smooth on $\mathbb{R}^+$ and corresponds to the function $\dot{\psi}(t)i_{\mathbb{R}^+}(t)$, i.e.

$$\langle z, \phi\rangle = \int_0^\infty \dot{\psi}(t)\phi(t)\,\mathrm{d}t.$$

Stated loosely: the distributional derivative of the function $\psi(t)i_{\mathbb{R}^+}(t)$ is equal to the height of its jump at $t = 0$ times the Dirac distribution plus the function $\dot{\psi}(t)i_{\mathbb{R}^+}(t)$. Indeed, for $\phi \in \mathfrak{D}_-$ we have

$$\begin{aligned}
\langle \dot{x}, \phi\rangle &= -\langle x, \dot{\phi}\rangle \\
&= -\int_0^\infty \psi(t)\dot{\phi}(t)\,\mathrm{d}t \\
&= \psi(0)\phi(0) + \int_0^\infty \dot{\psi}(t)\phi(t)\,\mathrm{d}t \\
&= \langle \psi(0)\delta, \phi\rangle + \langle z, \phi\rangle \\
&= \langle \psi(0)\delta + z, \phi\rangle.
\end{aligned}$$

Example A.9 Let $a \in \mathbb{R}$. The impulsive distribution $\dot{\delta} - a\delta$ is invertible. Its inverse is smooth on $\mathbb{R}^+$: it is equal to the distribution x corresponding to the function $e^{at}i_{\mathbb{R}^+}(t)$. Indeed, the height of the jump of this function at $t = 0$ equals 1 so $\dot{x} = \delta + z$, where z corresponds to the function $ae^{at}i_{\mathbb{R}^+}(t)$. Obviously, $z = ax$. Thus

$$(\dot{\delta} - a\delta) * x = \dot{x} - ax = \delta,$$

from which we obtain: $(\dot{\delta} - a\delta)^{-1} = x$.

If $g(s)$ is a real rational function, i.e., if $g(s) = \frac{n(s)}{d(s)}$, with $n(s)$ and $d(s)$ polynomials with real coefficients, then we can associate with $g(s)$ a distribution $g(\dot{\delta})$, formally obtained by replacing the variable s by the distribution $\dot{\delta}$. If $g(s)$ is a polynomial, say $g(s) = g_0 + g_1 s + \cdots + g_l s^l$, then it is obvious how we should interpret $g(\dot{\delta})$: in this case we simple define $g(\dot{\delta})$ to be the impulsive distribution

$$g(\dot{\delta}) := g_0\delta + g_1\delta^{(1)} + g_2\delta^{(2)} + \cdots + g_l\delta^{(l)} = \sum_{i=0}^{l} g_i\delta^{(i)}.$$

In general, if $g(s) = n(s)/d(s)$ then we define

$$g(\dot{\delta}) := n(\dot{\delta}) * d(\dot{\delta})^{-1},$$

the convolution of the impulsive distribution $n(\dot{\delta})$ with the inverse of the impulsive distribution $d(\dot{\delta})$. Of course, we should verify that the distribution $d(\dot{\delta})$ is indeed invertible. To show this, note that we can always factor

$$d(s) = c(s - a_1)(s - a_2)\cdots(s - a_l),$$

with $c \in \mathbb{R}$, $a_i \in \mathbb{C}$. It is verified immediately that

$$d(\dot{\delta}) = c(\dot{\delta} - a_1\delta) * (\dot{\delta} - a_2\delta) * \cdots * (\dot{\delta} - a_l\delta).$$

In example A.9 we proved that $\dot{\delta} - a_i\delta$ is invertible (the extension to complex valued distributions being left to the reader). The proof is then completed by noting that if x and y are invertible, then $x * y$ is invertible, and $(x * y)^{-1} = x^{-1} * y^{-1}$. Note that, since $\mathfrak{D}_0$ is a sub-algebra of $\mathfrak{D}'_+$, $g(\dot{\delta})$ is an impulsive-smooth distribution.

Example A.10 Consider the rational function $g(s) = \frac{s}{s+1}$. Then $g(\dot{\delta}) = \dot{\delta}*(\dot{\delta}+\delta)^{-1}$. Now, $(\dot{\delta} + \delta)^{-1}$ is smooth on $\mathbb{R}^+$, and corresponds to the function $e^{-t}i_{\mathbb{R}^+}(t)$. Thus $g(\dot{\delta})$ is equal to $\delta + z$, where the distribution z is smooth on $\mathbb{R}^+$, and corresponds to the function $-e^{-t}i_{\mathbb{R}^+}(t)$.

All concepts that have been introduced in this appendix up to now can be extended to vectors and matrices with distributions as components. We denote by $\mathfrak{D}'^{p\times n}_+$ the space of all $p \times n$ matrices with components in $\mathfrak{D}'_+$. Likewise we define $\mathfrak{D}'^n_+$, $\mathfrak{D}'^{p\times n}_0$ and $\mathfrak{D}^n_0$.

If $K \in \mathfrak{D}'^{p\times n}_+$ then for $\phi \in \mathfrak{D}_-$ we define $\langle K, \phi\rangle \in \mathbb{R}^{p\times n}$ by $\langle K, \phi\rangle_{ij} := \langle K_{ij}, \phi\rangle$. Similarly, if $x \in \mathfrak{D}'^n_+$ then $\langle x, \phi\rangle \in \mathbb{R}^n$ is defined by $\langle x, \phi\rangle_i := \langle x_i, \phi\rangle$. Differentiation of elements in $\mathfrak{D}'^{p\times n}_+$ and $\mathfrak{D}'^n_+$ is defined componentwise. For $K \in \mathfrak{D}'^{p\times n}_+$ and $L \in \mathfrak{D}'^{n\times r}_+$ the convolution $K * L$ is the matrix in $\mathfrak{D}'^{p\times r}_+$ defined by

$$(K * L)_{ij} := \sum_{k=1}^{n} K_{ik} * L_{kj}.$$

In the same way, if $x \in \mathfrak{D}'^n_+$ then the convolution of K and x is the vector distribution in $\mathfrak{D}'^p_+$ defined by

$$(K * x)_i := \sum_{j=1}^{n} K_{ij} * x_j.$$

The space $\mathfrak{D}'^{n\times n}_+$ of all (square) $n \times n$ matrices with components in $\mathfrak{D}'_+$ is a (non-commutative) algebra over $\mathbb{R}$ (with pointwise addition and scalar multiplication). The unit element in this algebra is $I\delta$, where $I \in \mathbb{R}^{n\times n}$ denotes the identity matrix, and δ

is the Dirac distribution: if $A \in \mathbb{R}^{n\times n}$ and $x \in \mathfrak{D}'_+$ then $Ax \in \mathfrak{D}'^{\,n\times n}_+$ is defined by $(Ax)_{ij} := A_{ij}x$.

An element of $\mathfrak{D}'^{n\times n}_+$ is called *invertible* if there exists $L \in \mathfrak{D}'^{n\times n}_+$ such that $K * L = L * K = I\delta$. If such L exists then it is unique. We denote L by K^{-1}. Let $K \in \mathfrak{D}'^{p\times n}_+$. If $L \in \mathfrak{D}'^{n\times p}_+$ is such that $K * L = I$ then L is called a *right-inverse* of K. Likewise, K is called a *left-inverse* of L.

Example A.11 If $x \in \mathfrak{D}'^n_+$ then $\dot{x} = I\dot{\delta} * x$.

Example A.12 Let $A \in \mathbb{R}^{n\times n}$. The impulsive matrix distribution $I\dot{\delta} - A\delta$ is invertible. Its inverse is smooth on $\mathbb{R}^+$: $(I\dot{\delta} - A\delta)^{-1}$ is equal to the distribution corresponding to $e^{At}i_{\mathbb{R}^+}(t)$.

Example A.13 The derivative of $(I\dot{\delta} - A\delta)^{-1}$ is equal to $I\delta + (I\dot{\delta} - A\delta)^{-1}A$. Indeed, $(I\dot{\delta} - A\delta)^{-1}$ corresponds to the function $e^{At}i_{\mathbb{R}^+}(t)$. This function makes a jump I at $t = 0$, while the derivative of $e^{At}i_{\mathbb{R}^+}(t)$ on $\mathbb{R}^+$ is given by $e^{At}Ai_{\mathbb{R}^+}(t)$. The latter corresponds to the distribution $(I\dot{\delta} - A\delta)^{-1}A$. It is an easy exercise to verify that the ith derivative of $(I\dot{\delta} - A\delta)^{-1}$ equals

$$I\delta^{(i-1)} + A\delta^{(i-2)} + \cdots + A^{i-1}\delta + (I\dot{\delta} - A)^{-1}A^i.$$

If $G(s)$ is a real rational matrix then we can associate with $G(s)$ a matrix distribution $G(\dot{\delta})$. If $G_{ij}(s)$ is the (i, j)-component of $G(s)$ then we define $G(\dot{\delta})$ componentwise by

$$(G(\dot{\delta}))_{ij} := G_{ij}(\dot{\delta}).$$

The following observations are then important to us:

Theorem A.14

(i) Let $G_1(s)$ and $G_2(s)$ be real rational matrices of the same dimensions. Then $(G_1 + G_2)(\dot{\delta}) = G_1(\dot{\delta}) + G_2(\dot{\delta})$.

(ii) Let $G_1(s)$ and $G_2(s)$ be real rational matrices such that G_1G_2 exists. Then $(G_1G_2)(\dot{\delta}) = G_1(\dot{\delta}) * G_1(\dot{\delta})$.

(iii) Let $G(s)$ be a right-invertible real rational matrix and let $H(s)$ be a right-inverse of $G(s)$ (i.e. $GH = I$). Then $G(\dot{\delta})$ is right-invertible and has a right-inverse $H(\dot{\delta})$.

(iv) Let $G(s)$ be an invertible real rational matrix with inverse $G^{-1}(s)$. Then $G(\dot{\delta})$ is invertible and we have $G(\dot{\delta})^{-1} = G^{-1}(\dot{\delta})$.

Proof : Of course (iii) and (iv) are immediate consequences of (ii). For the proofs of (i) and (ii) we only need to consider the scalar case. The extension to the matrix case is then straightforward . First assume that $g_1(s)$ and $g_2(s)$ are real polynomials. It is easily verified that $(g_1 \cdot g_2)(\dot{\delta}) = g_1(\dot{\delta}) * g_2(\dot{\delta})$. Next, if $g_1 \neq 0$ and $g_2 \neq 0$ then we obviously have

$$(g_1 \cdot g_2)(\dot{\delta})^{-1} = g_1(\dot{\delta})^{-1} * g_2(\dot{\delta})^{-1}.$$

Now, let $g_1(s)$ and $g_2(s)$ be real rational functions. Let $g_i(s) = n_i(s)/d_i(s)$. We have

$$\begin{aligned}(g_1 \cdot g_2)(\dot{\delta}) &= (n_1 \cdot n_2)(\dot{\delta}) * (d_1 \cdot d_2)(\dot{\delta})^{-1} \\ &= n_1(\dot{\delta}) * n_2(\dot{\delta}) * d_1(\dot{\delta})^{-1} * d_2(\dot{\delta})^{-1} \\ &= g_1(\dot{\delta}) * g_2(\dot{\delta}).\end{aligned}$$

A proof of (i) can be given similarly. This details are left to the reader. ■

Example A.15 Let $T(s) := (Is - A)^{-1}$. We claim that $T(\dot{\delta}) = (I\dot{\delta} - A\delta)^{-1}$ (see A.12). Indeed, let $R(s) = Is - A$. Then $R(\dot{\delta}) = I\dot{\delta} - A\delta$. Also, $T(s) = R^{-1}(s)$. By theorem A.14 we therefore have

$$T(\dot{\delta}) = R^{-1}(\dot{\delta}) = R(\dot{\delta})^{-1} = (I\dot{\delta} - A\delta)^{-1}.$$

Example A.16 Let $G(s) := C(Is - A)^{-1}B + D$ be the transfer matrix of the system $\Sigma = (A, B, C, D)$. We claim that $G(\dot{\delta}) = C(I\dot{\delta} - A\delta)^{-1}B + D\delta$. The proof of this uses theorem A.14 and example A.15 and is left to the reader.

A.1 Notes and references

The delta distribution was introduced by Heaviside as an artifice for dealing with various mathematical and physical problems, in particular for systems with impulsive inputs. In the original treatment it was considered a function, and called the delta function. However, it was clear from the outset that no function can exist that satisfies the conditions imposed on the delta function.

The general theory of distributions due to Laurent Schwartz (see [175]) gives a mathematically satisfactory description of the impulse. The theory leans heavily on functional analytic methods, in particular locally convex spaces. Especially the definition of convolution was rather involved. A more accessible treatment is given in Schwartz [114]. In Hautus [70], the convolution was defined in an rather ad hoc, but quick way. The reader interested in more advanced properties of distributions is advised the get acquainted with the 'official' definition.

Bibliography

[1] J. Ackermann, "Der Entwurf linearer Regelungssysteme im Zustandsraum", Regelungstechnik, 20, 1972, pp. 297–300.

[2] H. Akashi and H. Imai, "Disturbance localization and output deadbeat through an observer in discrete-time linear multivariable systems", IEEE Trans. Aut. Contr., 24, 1979, pp. 621–627.

[3] H. Aling and J.M Schumacher, "A nine-fold canonical decomposition for linear systems", Int. J. Contr., 39(4), 1984, pp. 779–805.

[4] B.D.O. Anderson, "A note on transmission zeros of a transfer function matrix", IEEE Trans. Aut. Contr., 21(4), 1976, pp. 589–591.

[5] B.D.O. Anderson and J.B. Moore, *Linear optimal control*, Prentice-Hall, Englewood Cliffs, N.J., 1971.

[6] ——, *Optimal control: linear quadratic methods*, Prentice-Hall, Englewood Cliffs, 1989.

[7] B.D.O. Anderson and R.W. Scott, "Parametric solution of the stable exact model matching problem", IEEE Trans. Aut. Contr., 22(1), 1977, pp. 382–386.

[8] V.I. Arnol'd, *Ordinary differential equations*, Springer Verlag, 1992. Translated from the Russian.

[9] M. Athans, Guest Ed., *Special issue on the Linear-Quadratic-Gaussian problem*, IEEE Trans. Aut. Contr., 16, 1971.

[10] G. Basile and G. Marro, "Controlled and conditioned invariant subspaces in linear system theory", J. Optim. Th. & Appl., 3, 1969, pp. 306–315.

[11] ——, "On the observability of linear, time-invariant systems with unknown inputs", J. Optim. Th. & Appl., 3, 1969, pp. 410–415.

[12] ——, "Self-bounded controlled invariant subspaces: a straightforward approach to constrained controllability", J. Optim. Th. & Appl., 38(1), 1982, pp. 71–81.

[13] ——, "Stability without eigenspaces in the geometric approach: some new results", in Frequency domain and state space methods for linear systems, C.A. Byrnes and A. Lindquist, eds., North Holland, Amsterdam, 1986, pp. 441–450.

[14] ——, *Controlled and conditioned invariants in linear system theory*, Prentice Hall, Englewood Cliffs, 1992.

[15] V. Belevitch, *Classical Network Theory*, Holden-Day, San Francisco, 1968.

[16] S.P. Bhattacharyya, "Output regulation with bounded energy", IEEE Trans. Aut. Contr., 18(4), 1973, pp. 381–383.

[17] ——, "On calculating maximal (A, B) invariant subspaces", IEEE Trans. Aut. Contr., 20, 1975, pp. 264–265.

[18] ——, "Observer design for linear systems with unknown inputs", IEEE Trans. Aut. Contr., 23(3), 1978, pp. 483–484.

[19] ——, "Frequency domain conditions for disturbance rejection", IEEE Trans. Aut. Contr., 25(6), 1980, pp. 1211–1213.

[20] ——, "Parameter invariant observers", Int. J. Contr., 32(6), 1980, pp. 1127–1132.

[21] ——, "Generalized controllability, (A, B)-invariant subspaces and parameter invariant control", SIAM J. Algebraic & Discrete Methods, 4(4), 1983, pp. 529–533.

[22] S.P. BHATTACHARYYA, A.C. DEL NERO GOMES, AND J.W. HOWZE, "The structure of robust disturbance rejection control", IEEE Trans. Aut. Contr., 28(9), 1983, pp. 874–881.

[23] S.P. BHATTACHARYYA, J.B. PEARSON, AND W.M. WONHAM, "On zeroing the output of a linear system", Information and Control, 20(2), 1972, pp. 135–142.

[24] P. BOHL, "Über Differentialungleichungen", Journal für Reine Angew. Mathematik, 144, 1913, pp. 284–313.

[25] R.W. BROCKETT, *Finite dimensional linear systems*, J. Wiley & Sons, 1969.

[26] R. BYERS, "Solving the algebraic Riccati equation with the matrix sign function", Lin. Alg. Appl., 85, 1987, pp. 267–279.

[27] B.M. CHEN, "On properties of the special coordinate basis of linear systems", Int. J. Contr., 71(6), 1998, pp. 981–1003.

[28] ——, *Robust and H_∞ control*, Communication and Control Engineering Series, Springer Verlag, 2000.

[29] B.M. CHEN, A. SABERI, P. SANNUTI, AND Y. SHAMASH, "Construction and parameterization of all static and dynamic H_2 optimal state feedback solutions, optimal fixed modes, and fixed decoupling zeros", IEEE Trans. Aut. Contr., 38(2), 1993, pp. 248–261.

[30] R.V. CHURCHILL AND J.W. BROWN, *Complex variables and applications*, McGraw-Hill, 4th Ed., 1984.

[31] D.J. CLEMENTS AND B.D.O. ANDERSON, *Singular optimal control: the linear quadratic problem*, vol. 5 of Lecture Notes in Control and Information Sciences, Springer Verlag, 1978.

[32] D.J. CLEMENTS, B.D.O. ANDERSON, AND P.J. MOYLAN, "Matrix inequality solution to linear-quadratic singular control problems", IEEE Trans. Aut. Contr., 22(1), 1977, pp. 55–57.

[33] W.A. COPPEL, "Matrix quadratic equations", Bull. Austral. Math. Soc., 10, 1974, pp. 377–401.

[34] R.F. CURTAIN AND A.A. STOORVOGEL, Guest Eds., *Special Issue on infinite-dimensional systems*, Int. J. Robust & Nonlinear Control, 6(5), 1996.

[35] E.J. DAVISON, "The output control of linear time-invariant multi-variable systems with unmeasured arbitrary disturbances", IEEE Trans. Aut. Contr., 17, 1972, pp. 621–630. Correction: Vol. 20, No. 12, 1975, pp. 824.

[36] E.J. DAVISON AND A. GOLDENBERG, "The robust control of a general servomechanism problem: the servo compensator", Automatica, 11, 1975, pp. 461–471.

[37] C.A. DESOER AND J.D. SCHULMAN, "Zeros and poles of linear multivariable systems and their dynamical interpretation", IEEE Trans. Circ. & Syst., 21, 1974, pp. 3–8.

[38] C.A. DESOER AND Y.T. WANG, "On the minimum order of a robust servocompensator", IEEE Trans. Aut. Contr., 23(1), 1978, pp. 70–73.

[39] J.C. DOYLE, "Analysis of feedback systems with structured uncertainties", IEE Proc., Part D, 129, 1982, pp. 242–250.

[40] J.C. DOYLE, B.A. FRANCIS, AND A.R. TANNENBAUM, *Feedback control theory*, MacMillan, New York, 1992.

[41] J.C. DOYLE, K. GLOVER, P.P. KHARGONEKAR, AND B.A. FRANCIS, "State space solutions to standard H_2 and H_∞ control problems", IEEE Trans. Aut. Contr., 34(8), 1989, pp. 831–847.

[42] G.E. DULLERUD AND F. PAGANINI, *A course in robust control theory*, Springer Verlag, New York, 1999.

[43] E. EMRE AND M.L.J. HAUTUS, "A polynomial characterization of (A, B)-invariant and reachability subspaces", SIAM J. Contr. & Opt., 18(4), 1980, pp. 420–436.

[44] B.A. FRANCIS, "The linear multivariable regulator problem", SIAM J. Contr. & Opt., 15(3), 1977, pp. 486–505.

[45] ——, "The optimal linear-quadratic time invariant regulator with cheap control", IEEE Trans. Aut. Contr., 24, 1979, pp. 616–621.

[46] ——, *A course in H_∞ control theory*, vol. 88 of Lecture notes in control and information sciences, Springer-Verlag, 1987.

[47] B. FRANCIS, O.A. SEBAKHY, AND W.M. WONHAM, "Synthesis of multivariable regulators: the internal model principle", Appl. Math. & Optimization, 1(1), 1974, pp. 64–86.

[48] B.A. FRANCIS AND W.M. WONHAM, "The internal model principle for linear multivariable regulators", Appl. Math. & Optimization, 2(2), 1975, pp. 170–194.

[49] G.F. FRANKLIN, J.D. POWELL, AND A. EMAMI-NAEINI, *Feedback control of dynamic systems*, Addison-Wesley, 3rd Ed., 1995.

[50] J.S. FREUDENBERG AND D.P. LOOZE, *Frequency domain properties of scalar and multivariable feedback systems*, vol. 104 of Lecture notes in control and information sciences, Springer-Verlag, 1988.

[51] P.A. FUHRMANN AND J.C. WILLEMS, "A study of (A, B)-invariant subspaces via polynomial models", Int. J. Contr., 31(3), 1980, pp. 467–494.

[52] T. FUJII, "On the perfect regulation of optimal regulators", Syst. & Contr. Letters, 1(6), 1981/82, pp. 356–359.

[53] P. GAHINET, "Explicit controller formulas for LMI-based H_∞ synthesis", Automatica, 32(7), 1996, pp. 1007–1019.

[54] P. GAHINET AND P. APKARIAN, "A linear matrix inequality approach to H_∞ control", Int. J. Robust & Nonlinear Control, 4(4), 1994, pp. 421–448.

[55] F.R. GANTMACHER, *The theory of matrices*, Chelsea, New York, 1959.

[56] A.H.W. GEERTS AND M.L.J. HAUTUS, "The output-stabilizable subspace and linear optimal control", in Robust control of linear systems and nonlinear control, no. 4 in Progr. Systems Control Theory, Birkhäuser, Boston, MA, 1990, pp. 113–120.

[57] T. GEERTS, "All optimal controls for the singular linear-quadratic problem without stability; a new interpretation of the optimal cost", Lin. Alg. Appl., 122, 1989, pp. 65–104.

[58] ——, "A necessary and sufficient condition for the existence of a tightest local frame", Syst. & Contr. Letters, 22(2), 1994, pp. 131–136.

[59] E.G. GILBERT, "Controllability and observability in multivariable control systems", J. Soc. Indust. Appl. Math., Ser. A., 2, 1963, pp. 128–151.

[60] K. GLOVER, "All optimal Hankel-norm approximations of linear multivariable systems and their L^∞-error bounds", Int. J. Contr., 39, 1984, pp. 1115–1193.

[61] ——, "Robust stabilization of multivariable linear systems: relations to approximation", Int. J. Contr., 43, 1986, pp. 741–766.

[62] K. GLOVER AND J. DOYLE, "State-space formulaes for all stabilizing controllers that satisfy an H_∞ norm bound and relations to risk sensitivity", Syst. & Contr. Letters, 11, 1988, pp. 167–172.

[63] K. GLOVER, D.J.N. LIMEBEER, J.C. DOYLE, E.M. KASENALLY, AND M.G. SAFONOV, "A characterization of all solutions to the four block general distance problem", SIAM J. Contr. & Opt., 29, 1991, pp. 283–324.

[64] J. GOLDBERG AND M.C. POTTER, *Differential equations: a systems approach*, Prentice-Hall, Upper Saddle River, 1998.

[65] M. GREEN, K. GLOVER, D.J.N. LIMEBEER, AND J.C. DOYLE, "A J-spectral factorization approach to H_∞ control", SIAM J. Contr. & Opt., 28(6), 1990, pp. 1350–1371.

[66] M. GREEN AND D.J.N. LIMEBEER, *Linear robust control*, Information and System Sciences Series, Prentice Hall, 1994.

[67] PAUL R. HALMOS, *Finite-dimensional vector spaces*, Undergraduate texts in mathematics, Springer Verlag, Berlin, 1974.

[68] M.L.J. HAUTUS, "Controllability and observability conditions of linear autonomous systems", Proc. Nederl. Akad. Wetensch., Ser. A, 72(5), 1969, pp. 443–448.

[69] ——, "Stabilization, controllability and observability of linear autonomous systems", Proc. Nederl. Akad. Wetensch., Ser. A, 73(5), 1970, pp. 448–455.

[70] ——, "The formal Laplace transform for smooth linear systems", in Proc. International Symposium on Mathematical System theory, vol. 131, Udine, 1976, Springer Verlag. Lecture Notes in Economics and Mathematical Systems.

[71] ——, "A simple proof of Heymann's Lemma", IEEE Trans. Aut. Contr., 22, 1977, pp. 885–886.

[72] ——, "(A, B)-invariant and stabilizability subspaces, a frequency domain description", Automatica, 16(6), 1980, pp. 703–707.

[73] ——, "Linear matrix equations with applications to the regulator problem", in Outils et Modèles Mathématique pour l'Automatique, I.D. Landau, ed., C.N.R.S., Paris, 1983, pp. 399–412.

[74] ——, "Strong detectability and observers", Lin. Alg. Appl., 50, 1983, pp. 353–368.

[75] ——, "Operator substitution", Lin. Alg. Appl., 205-206, 1994, pp. 713–740. Third Special Issue on Systems and Control.

[76] M.L.J. HAUTUS AND L.M. SILVERMAN, "System structure and singular control", Lin. Alg. Appl., 50, 1983, pp. 369–402.

[77] M. HEYMANN, "Comments on 'Pole assignment in multi-input controllable linear systems"', IEEE Trans. Automat. Contr., 13, 1968, pp. 748–749.

[78] ——, "Controllability subspaces and feedback simulation", SIAM J. Contr. & Opt., 4, 1976, pp. 769–789.

[79] D. HINRICHSEN AND A.J. PRITCHARD, "Real and complex stability radii: a survey", in Control of Uncertain Systems, Proc. International Workshop Bremen, Boston, 1990, Birkhäuser, pp. 119–162.

[80] R. HORN AND C.R. JOHNSON, *Topics in matrix analysis*, Cambridge University Press, Cambridge, 1991.

[81] I.M. HOROWITZ, *Synthesis of feedback systems*, Academic Press, New York, 1963.

[82] S. HOSOE, "On a time-domain characterization of the numerator polynomials of the Smith McMillan form", IEEE Trans. Aut. Contr., 20(6), 1975, pp. 799–800.

[83] M. HOU, A. PUGH, AND P.C. MÜLLER, "Disturbance decoupled functional observers", IEEE Trans. Aut. Contr., 44(2), 1999, pp. 382–386.

[84] P.A. IGLESIAS AND K. GLOVER, "State space approach to discrete time H_∞ control", Int. J. Contr., 54(5), 1991, pp. 1031–1073.

[85] H. IMAI AND H. AKASHI, "Disturbance localization and pole shifting by dynamic compensation", IEEE Trans. Aut. Contr., 26, 1981, pp. 226–235.

[86] A. ISIDORI, *Nonlinear control systems*, Springer Verlag, 1989.

[87] A. ISIDORI AND C.I. BYRNES, "Output regulation of nonlinear systems", IEEE Trans. Aut. Contr., 35(2), 1990, pp. 131–140.

[88] T. IWASAKI AND R.E. SKELTON, "All controllers for the general H_∞ control problem: LMI existence conditions and state space formulas", Automatica, 30(8), 1994, pp. 1307–1317.

[89] A. JAMESON AND R.E. O'MALLEY JR., "Cheap control of the time-invariant regulator", Appl. Math. & Optimization, 1(4), 1974/1975, pp. 337–354.

[90] T. KAILATH, *Linear Systems*, Prentice-Hall, Englewood Cliffs, 1980.

[91] R.E. KALMAN, "Contributions to the theory of optimal control", Bol. Soc. Mat. Mexicana, 5, 1960, pp. 102–119.

[92] ——, "A new approach to linear filtering and prediction theory", ASME Transactions, Series D (Journal of Basic Eng.), 82, 1960, pp. 35–45.

[93] ——, "On the general theory of control systems", in Proc. First IFAC Congress, Butterworth, 1960, pp. 481–491.

[94] ——, "Mathematical description of linear dynamical systems", J. Soc. Indust. Appl. Math., Ser. A., 1, 1963, pp. 152–192.

[95] R.E. KALMAN AND R.S. BUCY, "New results in linear filtering and prediction theory", ASME Transactions, Series D (Journal of Basic Eng.), 83, 1960, pp. 95–108.

[96] R.E. KALMAN, Y.C. HO, AND K.S. NARENDRA, "Controllability of linear dynamical systems", Contrib. Diff. Eq., 1, 1963, pp. 189–213.

[97] E.W. KAMEN AND B.S. HECK, *Fundamentals of signals and systems*, Prentice-Hall, 1997.

[98] C. KENNEY, A.J. LAUB, AND M. WETTE, "Error bounds for Newton refinement of solutions to algebraic Riccati equations", Math. Contr. Sign. & Syst., 3(3), 1990, pp. 211–224.

[99] HASSAN K. KHALIL, *Nonlinear systems*, Prentice Hall, Upper Saddle River, 2nd Ed., 1996.

[100] P.P. KHARGONEKAR, "State-space H_∞ control theory", in Mathematical system theory: the influence of R.E. Kalman, A.C. Antoulas, ed., Springer Verlag, Berlin, 1991, pp. 159–176.

[101] BENJAMIN C. KUO, *Automatic control systems*, Prentice-Hall, 7th Ed., 1995.

[102] V. KUÇERA, "A contribution to matrix quadratic equations", IEEE Trans. Aut. Contr., 17, 1972, pp. 344–347.

[103] H. KWAKERNAAK, "A polynomial approach to minimax frequency domain optimization of multivariable feedback systems", Int. J. Contr., 41, 1986, pp. 117–156.

[104] ——, "Robust control and H_∞ optimization – tutorial paper", Automatica, 29(2), 1993, pp. 255–274.

[105] H. KWAKERNAAK AND R. SIVAN, *Linear optimal control theory*, Wiley, 1972.

[106] ——, "The maximally achievable accuracy of linear regulators and linear optimal filters", IEEE Trans. Aut. Contr., 17, 1972, pp. 79–86.

[107] ——, *Modern signals and systems*, Prentice-Hall, 1991.

[108] A.J. LAUB, "A Schur method for solving algebraic Riccati equations", IEEE Trans. Aut. Contr., 24(6), 1979, pp. 913–921.

[109] A.J. LAUB AND P. GAHINET, "Numerical improvements for solving Riccati equations", IEEE Trans. Aut. Contr., 42(9), 1997, pp. 1303–1308.

[110] D.J.N. LIMEBEER AND B.D.O. ANDERSON, "An interpolation theory approach to H_∞ controller degree bounds", Lin. Alg. Appl., 98, 1988, pp. 347–386.

[111] D.J.N. LIMEBEER, B.D.O. ANDERSON, P.P. KHARGONEKAR, AND M. GREEN, "A game theoretic approach to H^∞ control for time-varying systems", SIAM J. Contr. & Opt., 30(2), 1992, pp. 262–283.

[112] L. LJUNG, *System identification: theory for the user*, Prentice-Hall, Englewood Cliffs, 1987.

[113] LENNART LJUNG AND TORKEL GLAD, *Modeling of dynamic systems*, Prentice Hall, Englewood Cliffs, 1994.

[114] L.SCHWARTZ, *Mathematics for the physical sciences*, Addison-Wesley, 1966.

[115] D.G. LUENBERGER, "An introduction to observers", IEEE Trans. Aut. Contr., 16, 1971, pp. 596–602.

[116] D.G. LUENBURGER, *Optimization by vector space methods*, John Wiley & Sons, New York, 1968.

[117] A.G.J. MacFarlane and N. Karcanias, "Poles and zeros of linear multivariable systems: a survey of the algebraic, geometric and complex-variable theory", Int. J. Contr., 24(1), 1976, pp. 33–74.

[118] J.M. Maciejowski, *Multivariable feedback design*, Addison-Wesley, Reading, MA, 1989.

[119] M. Malabre, "Almost invariant subspaces, transmission and infinite zeros: a lattice interpretation", Syst. & Contr. Letters, 1(6), 1981/82, pp. 347–355.

[120] M. Malabre, J.C. Martinez-Garcia, and B. Del-Muro-Cuellar, "On the fixed poles for disturbance rejection", IEEE Trans. Aut. Contr., 33(6), 1997, pp. 1209–1211.

[121] K. Mårtenson, "On the matrix Riccati equation", Inform. Sci., 3, 1971, pp. 17–49.

[122] C.C. McDuffee, *The theory of matrices*, Chelsea, New York, 1960.

[123] D.C. McFarlane and K. Glover, *Robust controller design using normalized coprime factor descriptions*, vol. 138 of Lecture notes in control and information sciences, Springer Verlag, Berlin, 1990.

[124] B.P. Molinari, "Zeros of the system matrix", IEEE Trans. Aut. Contr., 21(5), 1976, pp. 795–797.

[125] ——, "The time-invariant linear-quadratic optimal control problem", Automatica, 13, 1977, pp. 347–357.

[126] B P. Molinari, "On strong controllability and observability in linear multivariable control", IEEE Trans. Aut. Contr., AC-26, 1976, pp. 761–764.

[127] B.C. Moore and A.J. Laub, "Computation of supremal (A, B)-invariant and controllability subspaces", IEEE Trans. Aut. Contr., 23(5), 1978, pp. 783–792.

[128] ——, "Correction to: "Computation of supremal (A, B)-invariant and controllability subspaces"", IEEE Trans. Aut. Contr., 24(4), 1979, p. 677.

[129] A.S. Morse, "Structural invariants of linear multivariable systems", SIAM J. Contr. & Opt., 11, 1973, pp. 446–465.

[130] ——, "Minimal solutions to transfer matrix equations", IEEE Trans. Aut. Contr., 21(1), 1976, pp. 131–133.

[131] A.S. Morse and W.M. Wonham, "Status of noninteracting control", IEEE Trans. Aut. Contr., 16, 1971, pp. 568–581.

[132] A. S. Morse and W. M. Wonham, "Decoupling and pole assignment by dynamic compensation", SIAM J. Contr. & Opt., 8, 1970, pp. 317–337.

[133] D. Mustafa and K. Glover, *Minimum entropy H_∞ control*, vol. 146 of Lecture notes in control and information sciences, Springer Verlag, Berlin, 1990.

[134] H. Nijmeijer and A.J. van der Schaft, *Nonlinear dynamical control systems*, Springer Verlag, 1990.

[135] D.Y. Ohm, S.P. Bhattacharyya, and J.W. Howze, "Transfer function conditions for (C, A, B) pairs", IEEE Trans. Aut. Contr., 29(2), 1984, pp. 172–174.

[136] R.E. O'Malley Jr., "A more direct solution of the nearly singular linear regulator problem", SIAM J. Contr. & Opt., 14(6), 1976, pp. 1063–1077.

[137] ——, *Singular perturbation methods for ordinary differential equations*, Springer Verlag, Berlin, 1991.

[138] R.E. O'Malley Jr. and A. Jameson, "Singular perturbations and singular arcs I", IEEE Trans. Aut. Contr., 20, 1975, pp. 218–226.

[139] ——, "Singular perturbations and singular arcs II", IEEE Trans. Aut. Contr., 22, 1977, pp. 328–337.

[140] N. Otsuka, "Generalized (C, A, B)-pairs and parameter insensitive disturbance-rejection problems with dynamic compensator", IEEE Trans. Aut. Contr., 44(11), 1999, pp. 2195–2200.

[141] H.K. Ozcetin, A. Saberi, and Y. Shamash, "An H_∞ almost disturbance decoupling problem for nonstrictly proper systems: a singular perturbation approach", Control – Th. & Adv. Techn., 9(1), 1993, pp. 203–246.

[142] J.B. Pearson, R.W. Shields, and P.W. Staats Jr., "Robust solutions to linear multivariable control problems", IEEE Trans. Aut. Contr., 19, 1974, pp. 508–517.

[143] L. Perko, *Differential equations and dynamical systems*, Springer Verlag, 1991.

[144] I.R. Petersen and C.V. Hollot, "A Riccati equation approach to the stabilization of uncertain linear systems", Automatica, 22, 1986, pp. 397–411.

[145] C.L. Phillips and R.C. Harbor, *Feedback control systems*, Prentice Hall, Englewood Cliffs, 3rd Ed., 1996.

[146] J.W. Polderman and J.C. Willems, *Introduction to mathematical system theory*, Springer Verlag, New York, 1998.

[147] V.M. Popov, "Hyperstability and optimality of automatic systems with several control functions", Rev. Roum. Tech. Ser. Electrotech. Energ., 9, 1964, pp. 913–949.

[148] ——, *Hiperstabilitatea Sistemelor Automate*, Editura Axcademiei Republicii Socialiste România, 1966. in Rumanian.

[149] ——, *Hyperstability of Control Systems*, Springer-Verlag, Berlin, 1973. translation of [148].

[150] J.E. Potter, "Matrix quadratic solutions", SIAM J. Applied Math., 14, 1966, pp. 496–501.

[151] A.C. Pugh, "Transmission and system zeros", Int. J. Contr., 26, 1976, pp. 315–324.

[152] R. Ravi, K.M. Nagpal, and P.P. Khargonekar, "H_∞ control of linear time-varying systems: a state space approach", SIAM J. Contr. & Opt., 29(6), 1991, pp. 1394–1413.

[153] J. Rissanen, "Control system synthesis by analogue computer based on 'the generalized linear feedback' concept", in Int. Sem. on Analogue Computation to the study of Chemical processes, Brussels, 1960.

[154] J.D. Roberts, "Linear model reduction and solution of the algebraic Riccati equation by use of the sign function", Int. J. Contr., 32, 1980, pp. 677–687.

[155] H. H. Rosenbrock, *State-space and multivariable theory*, John-Wiley, New York, 1970.

[156] W. Rudin, *Real and complex analysis*, McGraw-Hill, 3 Ed., 1987.

[157] A. Saberi, Z. Lin, and A.A. Stoorvogel, "H_2 and H_∞ almost disturbance decoupling problem with internal stability", Int. J. Robust & Nonlinear Control, 6(8), 1996, pp. 789–803.

[158] A. Saberi, P. Sannuti, and A.A. Stoorvogel, "H_2 optimal controllers with measurement feedback for continuous-time systems – flexibility in closed-loop pole placement", Automatica, 32(8), 1996, pp. 1201–1209.

[159] A. Saberi, A.A. Stoorvogel, and P. Sannuti, *Control of linear systems with regulation and input constraints*, Communication and Control Engineering Series, Springer Verlag, 2000.

[160] M.G. Safonov, D.J. Limebeer, and R.Y. Chiang, "Simplifying the H_∞ theory via loop-shifting, matrix-pencil and descriptor concepts", Int. J. Contr., 50, 1989, pp. 2467–2488.

[161] M.K. Sain and J.L. Massey, "Invertibility of linear time-invariant dynamical systems", IEEE Trans. Aut. Contr., 14(2), 1969, pp. 141–149.

[162] C. Scherer, "H_∞ control by state feedback and fast algorithms for the computation of optimal H_∞ norms", IEEE Trans. Aut. Contr., 35, 1990, pp. 1090–1099.

[163] ——, "H_∞ -optimization without assumptions on finite or infinite zeros", SIAM J. Contr. & Opt., 30(1), 1992, pp. 143–166.

[164] ——, "H_∞ control by state feedback for plants with zeros on the imaginary axis", SIAM J. Contr. & Opt., 30(1), 1992, pp. 123–142.

[165] ——, "The state feedback H_∞ -problem at optimality", Automatica, 30(2), 1994, pp. 293–306.

[166] J.M. Schumacher, "Compensator synthesis using (C, A, B) pairs", IEEE Trans. Aut. Contr., 25, 1980, pp. 1133–1138.

[167] ——, "A complement on pole placement", IEEE Trans. Aut. Contr., 25(2), 1980, pp. 281–282.

[168] ——, *Dynamic feedback in finite and infinite dimensional linear systems*, vol. 143, Math. Centre Tracts, Amsterdam, 1981.

[169] ——, "Regulator synthesis using (C, A, B)-pairs", IEEE Trans. Aut. Contr., 27(6), 1982, pp. 1211–1221.

[170] ——, "Algebraic characterizations of almost invariance", Int. J. Contr., 38(1), 1983, pp. 107–124.

[171] ——, "On the structure of strongly controllable systems", Int. J. Contr., 38, 1983, pp. 525–545.

[172] ——, "The role of the dissipation matrix in singular optimal control", Syst. & Contr. Letters, 2, 1983, pp. 262–266.

[173] ——, "Almost stabilizability subspaces and high gain feedback", IEEE Trans. Aut. Contr., 29(7), 1984, pp. 620–628.

[174] ——, "A geometric approach to the singular filtering problem", IEEE Trans. Aut. Contr., 30, 1985, pp. 1075–1082.

[175] L. Schwartz, *Théorie des Distributions I, II*, Hermann, Paris, 1951.

[176] L.M. Silverman, "Properties and application of inverse systems", IEEE Trans. Aut. Contr., 13, 1968, pp. 436–437.

[177] ——, "Inversion of multivariable linear systems", IEEE Trans. Aut. Contr., 14, 1969, pp. 270–276.

[178] L. Silverman, "Discrete Riccati equations: alternative algorithms, asymptotic properties, and system theory interpretations", Control and Dynamical Systems, 12, 1976, pp. 313–386.

[179] L.M. Silverman and H.J. Payne, "Input-output structure of linear systems with application to the decoupling problem", SIAM J. Contr., 9(2), 1971, pp. 199–233.

[180] S. Skogestad and I. Postlethwaite, *Multivariable feedback control - analysis and design*, John Wiley & Sons, 1996.

[181] E.D. Sontag, *Mathematical control theory : deterministic finite dimensional systems*, vol. 6 of Texts in applied mathematics, Springer Verlag, 1990.

[182] G.W. Stewart, "Algorithm 506-HQR3 and EXCHNG: Fortran subroutines for calculating and ordering the eigenvalues of a real upper Hessenberg matrix", ACM Transactions Math. Software, 2, 1976, pp. 275–280.

[183] A.A. Stoorvogel, "The singular minimum entropy H_∞ control problem", Syst. & Contr. Letters, 16(6), 1991, pp. 411–422.

[184] ——, *The H_∞ control problem: a state space approach*, Prentice-Hall, Englewood Cliffs, 1992.

[185] ——, "The singular H_2 control problem", Automatica, 28(3), 1992, pp. 627–631.

[186] ——, "The H_∞ control problem with zeros on the boundary of the stability domain", Int. J. Contr., 32(1), 1996, pp. 29–47.

[187] ——, "Stabilizing solution of the H_∞ algebraic Riccati equation", Lin. Alg. Appl., 240, 1996, pp. 153–172.

[188] A.A. Stoorvogel, A. Saberi, and B.M. Chen, "The discrete-time H_∞ control problem with measurement feedback", Int. J. Robust & Nonlinear Control, 4(4), 1994, pp. 457–480.

[189] A.A. Stoorvogel and H.L. Trentelman, "The quadratic matrix inequality in singular H_∞ control with state feedback", SIAM J. Contr. & Opt., 28(5), 1990, pp. 1190–1208.

[190] ——, "The finite horizon singular time-varying H_∞ control problem with dynamic measurement feedback", Lin. Alg. Appl., 187, 1993, pp. 113–161.

[191] A.A. Stoorvogel and J.W. van der Woude, "The disturbance decoupling problem with measurement feedback and stability including direct-feedthrough matrices", Syst. & Contr. Letters, 17(3), 1991, pp. 217–226.

[192] J.J. Sylvester, "Sur l'équation en matrices $px = xq$", Comptes Rendus XCIX, 115/116, 1884, pp. 67–71.

[193] G. Tadmor, "Worst-case design in the time domain: the maximum principle and the standard H_∞ problem", Math. Contr. Sign. & Syst., 3, 1990, pp. 301–324.

[194] H.L. Trentelman, "On the assignability of infinite root loci in almost disturbance decoupling", Int. J. Contr., 38(1), 1983, pp. 147–167.

[195] ——, "Almost disturbance decoupling with bounded peaking", SIAM J. Contr. & Opt., 24(6), 1986, pp. 1150–1176.

[196] ——, *Almost invariant subspaces and high gain feedback*, vol. 29 of CWI Tracts, Amsterdam, 1986.

[197] ——, "Families of linear-quadratic problems: continuity properties", IEEE Trans. Aut. Contr., 32, 1987, pp. 323–329.

[198] ——, "The regular free-endpoint linear quadratic problem with indefinite cost", SIAM J. Contr. & Opt., 27(1), 1989, pp. 27–42.

[199] H.L. TRENTELMAN AND J.C. WILLEMS, "Guaranteed roll-off in a class of high-gain feedback design problems", Syst. & Contr. Letters, 3(6), 1983, pp. 361–369.

[200] ——, "The dissipation inequality and the algebraic Riccati equation", in The Riccati Equation, S. Bittanti, A J. Laub, and J C. Willems, eds., Springer Verlag, sep 1991, pp. 197–242.

[201] C.C. TSUI, "A new design approach to unknown input observers", IEEE Trans. Aut. Contr., 41(3), 1996, pp. 464–468.

[202] J. VAN DE VEGTE, *Feedback control systems*, Prentice Hall, Englewood Cliffs, 3rd Ed., 1994.

[203] J.W. VAN DER WOUDE, "Disturbance decoupling and output stabilization by measurement feedback: a combined approach", Int. J. Contr., 47(1), 1988, pp. 393–412.

[204] ——, "Disturbance decoupling by measurement feedback for structured transfer matrix systems", Automatica, 32(3), 1996, pp. 357–363.

[205] B. VAN KEULEN, *H_∞-control for distributed parameter systems: a state space approach*, Birkhäuser, Boston, 1993.

[206] A.I.G. VARDULAKIS, *Linear multivariable control, algebraic analysis and synthesis methods*, Wiley, Chichester, 1991.

[207] M. VIDYASAGAR, *Control system synthesis: a factorization approach*, MIT Press, Cambridge, MA, 1985.

[208] ——, *Nonlinear systems analysis*, Prentice Hall, London, 2nd Ed., 1992.

[209] S.H. WANG AND E.J. DAVISON, "A minimization algorithm for the design of linear multivariable systems", IEEE Trans. Aut. Contr., 18(3), 1973, pp. 516–518.

[210] J.H.M. WEDDERBURN, *Lectures on Matrices*, American Mathematical Society, Providence R.I., 1934.

[211] S. WEILAND AND J.C. WILLEMS, "Almost disturbance decoupling with internal stability", IEEE Trans. Aut. Contr., 32, 1989, pp. 277–286.

[212] J.C. WILLEMS, *The analysis of feedback systems*, MIT Press, 1971.

[213] ——, "Least squares stationary optimal control and the algebraic Riccati equation", IEEE Trans. Aut. Contr., 16, 1971, pp. 621–634.

[214] ——, "Almost $A(\text{mod}\mathcal{B})$-invariant subspaces", Astérisque, 75-76, 1980, pp. 239–248.

[215] ——, "Almost invariant subspaces: an approach to high gain feedback design– Part I: almost controlled invariant subspaces", IEEE Trans. Aut. Contr., 26, 1981, pp. 235–252.

[216] ——, "Almost invariant subspaces: an approach to high gain feedback design– Part II: almost conditionally invariant subspaces", IEEE Trans. Aut. Contr., 27, 1982, pp. 1071–1084.

[217] ——, "Paradigms and puzzles in the theory of dynamical systems", IEEE Trans. Aut. Contr., 36(3), 1991, pp. 259–294.

[218] J.C. Willems and C. Commault, "Disturbance decoupling by measurement feedback with stability or pole-placement", SIAM J. Contr. & Opt., 19(4), 1981, pp. 490–504.

[219] J.C. Willems, A. Kitapçi, and L.M. Silverman, "Singular optimal control: a geometric approach", SIAM J. Contr. & Opt., 24, 1986, pp. 323–337.

[220] W.M. Wonham, "On pole assignment in multi-input controllable linear systems", IEEE Trans. Aut. Contr., 12, 1967, pp. 600–665.

[221] ——, "On a matrix Riccati equation of stochastic control", SIAM J. Contr., 6(4), 1968, pp. 681–697.

[222] ——, "On the separation theorem of stochastic control", SIAM J. Contr. & Opt., 6, 1968, pp. 312–326.

[223] ——, *Linear multivariable control: a geometric approach* , Springer Verlag, New York, Third Ed., 1985.

[224] W.M. Wonham and A.S. Morse, "Decoupling and pole assignment in linear multivariable systems: A geometric approach", SIAM J. Contr. & Opt., 8, 1970, pp. 1–18.

[225] X. Xin, B.D.O. Anderson, and T. Mita, "Complete solution of the 4-block H^∞ control problem with infinite and finite $j\omega$-axis zeros", Int. J. Robust & Nonlinear Control, 10(2), 2000, pp. 59–81.

[226] D.C. Youla, J.J. Bongiorno, and H.A. Jabr, "Modern Wiener-Hopf design of optimal controllers, Part I", IEEE Trans. Aut. Contr., 21, 1976, pp. 3–14.

[227] ——, "Modern Wiener-Hopf design of optimal controllers, Part II", IEEE Trans. Aut. Contr., 21, 1976, pp. 319–330.

[228] N. Young, *An introduction to Hilbert space*, Cambridge University Press, Cambridge, 1988.

[229] L.A. Zadeh and C.A. Desoer, *Linear Systems Theory – The State-Space Approach*, McGraw-Hill, New York, 1963.

[230] G. Zames, "Feedback and optimal sensitivity: model reference transformations, multiplicative seminorms, and approximate inverses", IEEE Trans. Aut. Contr., 26, 1981, pp. 301–320.

[231] G. Zames and B.A. Francis, "Feedback, minimax sensitivity and optimal robustness", IEEE Trans. Aut. Contr., 28, 1983, pp. 585–600.

[232] K. Zhou, J.C. Doyle, and K. Glover, *Robust and optimal control*, Prentice Hall, Upper Saddle River, NJ, 1996.

Index

MIX
Papier aus verantwortungsvollen Quellen
Paper from responsible sources
FSC® C105338

If you have any concerns about our products,
you can contact us on
ProductSafety@springernature.com

In case Publisher is established outside the EU,
the EU authorized representative is:
Springer Nature Customer Service Center GmbH
Europaplatz 3, 69115 Heidelberg, Germany

Printed by Libri Plureos GmbH
in Hamburg, Germany